SPEZIALFAHRZEUGE 1

Time table

	1929
End of year	The Army Weapons Office, Ballistics Department, is commissioned by the Reichswehr Minister to develop solid-fuel rockets and test liquid-propellant jet engines.
	1930
Spring	Walter R. Dornberger joins the Army Weapons Office Ballistics Department.
	1932
January	Wernher von Braun receives an employment contract at the Army Weapons Office.
	1933
Spring	»Machtergreifung« (»seizure of power«) of the national socialist party. Adolf Hitler is appointed German Reich Chancellor.
Autumn	Experiments with Aggregate 1 (A1)
	1934
December	The Aggregate 2 (A2) is successfully tested at Borkum.
	1936
Spring	Definition of the main dimensions of the A4
August	Start of construction work at the Army resesarch center *Peenemünde* / Usedom; construction period 1936 -1940 (1943)
	1937
December	Aggregate 3 (A3) successfully launched at Greifswalder Oie.
	1938
Summer	Aggregate 5 (A5) successfully launched at Greifswalder Oie.
	1940
August	Study and draft drawings for an A4 transport and launch trailer
	1941
August	Hitler authorizes testing of the A4 until operational readiness.
Oct.-Dec.	FXM Company builds the Meillerwagen MA4 prototype. Delivery to the OKH at the turn of the year 41/42.
	1942
February	Determination of scope and purpose of the special vehicles for motorized units
Spring	The ground organisation receives its first special vehicles.
October	October 3, 1942 - First successful launch of the A4
December	Order given for serial production of the A4
	1943
January	FXM delivers the reworked Meillerwagen MA4-2, 8 more deliveries will follow until May.
June	June 11, 1943 – First field launch of an A4 (V29) from P X with special vehicles of the ground organization of the test department
June	June 18, 1943 – Delivery of the first *Feuerleitpanzer* Fz.101
August	Forming of the first Art. Abteilungs. Establishment of a FR school in Köslin
August	August 18, 1943 - Heavy bombing raid of the RAF (Operation Hydra) on *Peenemünde*
September	FXM starts delivery of Meillerwagens MA4-3. 317 orders in total, including license production by Lindner and Schenk.
Oct./Nov.	Start of final outfitting of various special vehicles at Lager Rebstock
November	First field test firing at training base »Heidelager«, Blizna
	1944
April	Status of individual vehicle groups
Jan.-June	Further test firing of Art.Abt.836 and 444, 485 + 500 at »Heidelager« (Blizna).
June	Determination of the leading firms
August	Mission preparation carrying out necessary modifications on the vehicles.
September	Assumption of command by the SS, SS-Gruppenführer Dr. Hans Kammler is appointed »Sonderbevollmächtigter 2«. Frontline operation of the Division z.V. organized into the groups North and South with the Art.Abt.485, 836, 444, 500. The first V2 was fired on 8.9.1944.
Sep.-Jan.	Further test firings are carried out on the training base »Tucheler Heide« until the beginning of January 1945.
Oct.-Dec.	Further development mot.II to mot.III. Delivery of test vehicles to mot.III.
November	Railroad firing test with 2/836 at *Peenemünde*
	1945
January	Reorganization of the Division z.V. First successful launch of A4b by test department on P X
Feb./March	Evacuation of *Peenemünde.* Transfer of the EW departments to the »Mittelraum«.
March	Launching of the last A4 by Gruppe Nord
April	Dissolution of Division z.V. and reorganization into Art. Abt. z.V.
May	May 8, 1945 – End of war

Zeittafel

	1929
Jahresende	Das Heereswaffenamt, Abteilung Ballistik, erhält vom Reichswehrminister den Auftrag zur Entwicklung von Feststoffraketen und Erprobung von Flüssigkeits-Strahltriebwerken.
	1930
Frühjahr	Walter R. Dornberger tritt in das Heereswaffenamt, Abteilung Ballistik, ein.
	1932
Januar	Wernher von Braun erhält einen Anstellungsvertrag im Heereswaffenamt.
	1933
Frühjahr	»Machtergreifung« der nationalsozialistischen Partei – Adolf Hitler wird zum Deutschen Reichskanzler ernannt.
Herbst	Versuche mit Aggregat 1 (A1).
	1934
Dezember	Das Aggregat 2 (A2) wird erfolgreich auf Borkum erprobt.
	1936
Frühjahr	Definition der Hauptabmessungen der A4
August	Beginn der Bauarbeiten der Heeresversuchsanstalt Peenemünde / Usedom; Bauzeit 1936 - 1940 (1943)
	1937
Dezember	Das Aggregat 3 (A3) startet erfolgreich auf der Greifswalder Oie.
	1938
Sommer	Das Aggregat 5 (A5) startet erfolgreich auf der Greifswalder Oie.
	1940
August	Studie und Entwurfszeichnungen für einen A4-Transport- und Abschussanhänger
	1941
August	Hitler genehmigt die Erprobung der A4 bis zur Einsatzreife.
Okt.-Dez.	Fa. FXM baut den Prototyp Meillerwagen MA4. Auslieferung zum Jahreswechsel 1941/42 an das OKH.
	1942
Februar	Festlegung von Umfang und Einsatzzweck der Spezialfahrzeuge der mot. Einheiten
Frühjahr	Die Bodenorganisation erhält ihre ersten Spezialfahrzeuge.
Oktober	3. Oktober 1942 – Erster erfolgreicher Start der A4
Dezember	Der Befehl zur Serienproduktion der A4 wird erteilt.
	1943
Januar	FXM liefert den überarbeiteten Meillerwagen MA4-2 aus, es folgen acht weitere bis Mai.
Juni	11. Juni 1943 – Erster feldmäßiger Start einer A4 (V29) vom P X mit Spezialfahrzeugen der Bodenorganisation der Versuchsabteilung
Juni	18. Juni 1943 – Erste Lieferung des Feuerleitpanzers Fz.101
August	Aufstellung der ersten Art. Abteilung und Errichtung einer FF-Schule in Köslin
August	18. August 1943 – Schwerer Bombenangriff der RAF (Operation Hydra) auf Peenemünde.
September	FXM beginnt mit der Auslieferung des Meillerwagens MA4-3. Insgesamt 317 Bestellungen, einschließlich Lizenzfertigung durch Lindner und Schenk.
Okt./Nov.	Beginn der Endausrüstung verschiedener Spezialfahrzeuge im Lager Rebstock
November	Erstes Versuchsschießen auf dem Truppenübungsplatz »Heidelager« bei Blizna
	1944
April	Sachstand-Statusbericht zu einzelnen Fahrzeuggruppen
Jan.-Juni	Weiteres Versuchsschießen der Art.Abt.836, 444, 485 und 500 in »Heidelager«
Juni	Festlegung der Leitfirmen
August	Einsatzvorbereitung. Durchführung notwendiger Änderungen an den Fahrzeugen.
September	Übernahme der Befehlsgewalt durch die SS, SS-Gruppenführer Dr. Hans Kammler wird zum »Sonderbevollmächtigter 2« ernannt. Fronteinsatz der Division z.V. aufgeteilt in die Gruppen Nord und Süd mit den Art.Abt.485, 836, 444, 500. Die erste V2 wurde am 8.9.1944 verschossen.
Sept- Jan	Weitere Versuchsschüsse werden bis Anfang Januar 1945 auf dem Truppenübungsplatz »Tucheler Heide« durchgeführt.
Okt.-Dez.	Weiterentwicklung mot.II nach mot.III. Auslieferung der Versuchsfahrzeuge nach mot.III
November	Eisenbahnversuchsschiessen mit der 2./836 in Peenemünde
	1945
Januar	Umgliederung der Division z.V. Erster Start der A4b durch Versuchsabteilung auf P X
Feb./März	Räumung von Peenemünde. Verlegung der Entwicklungsabteilungen in den »Mittelraum«.
März	Verschießen der letzten A4 durch die Gruppe Nord
April	Auflösen der Division z.V. und Umstrukturierung in eine Art.Abt. z.V.
Mai	8. Mai 1945 – Kriegsende

SPECIAL VEHICLES OF DIVISION z.V. A4 ROCKET UNITS • DESIGN AND VEHICLE CONSTRUCTION

VOLKER RUFF

SPEZIALFAHRZEUGE 1

PEENEMÜNDE 1942-45

Publisher/Herausgeber

Volker Ruff, strabokran.de, Am Erlach 20, 86675 Buchdorf, Germany

Credit/Bildnachweis

Cover/Titel:
BArch, RH8II Bild-B0517-44, o.Ang.

Backside/Rückseite:
L: BArch, RH8II Bild-D12-44, o.Ang
M: BArch, RH8II Bild-B1145-44, o.Ang
R: BAMA RH8-1262 Bild-B186-44 TD5224

1. Auflage 2025

ISBN 978-3-9816908-3-5

Design/Gestaltung

Günther Nord, 69434 Heddesbach
info@formwaerts.de

Distribution/Vertrieb

strabokran.de
Volker Ruff
Am Erlach 20
86675 Buchdorf
Germany
www.strabokran.de

RZM Imports, Inc.
PO Box 522
Riverside, CT 06878
US
www.rzm.com

Panzerwrecks Limited
Great Priors, Church Street
Old Heathfield
Sussex TN21 9AJ
United Kingdom
www.panzerwrecks.com

Printed by/Druck

Druckerei & Verlag Steinmeier GmbH & Co. KG
86738 Deiningen

Printed in Germany

The A4 *Gerät* is a long-range ballistic missile self-propelled by liquid propellants. Propulsion during the firing period is at an almost constant rate, regardless of altitude, because the rocket-propelled projectile carries its own oxygen supply with it.

The firing range is determined by cutting off propulsion.

The direction is given to the rocket by a previously set steering device, which in special cases is assisted by remote control.

Trajectory and flight attitude are not maintained by spin, but by fins and rudders in conjunction with the steering apparatus.

The A4 *Gerät* is used to engage area targets at long distances.

Das Gerät A4 ist ein Fernraketen-Geschoß, das sich durch Eigenantrieb mit flüssigen Treibstoffen fortbewegt. Der Antrieb während der Brennzeit erfolgt mit annähernd gleichbleibender Stärke, unabhängig von der Flughöhe, da das Raketen-Geschoß seinen Sauerstoffträger mit sich führt.

Die Schußweite wird durch Abschalten des Antriebes bestimmt.

Die Richtung wird dem Geschoss durch eine vorher eingestellte Steuerapparatur verliehen, die in Sonderfällen durch Fernlenkung unterstützt wird.

Flugbahn und Fluglage werden nicht durch Drall, sondern durch Flossen und Ruder in Verbindung mit der Steuerapparatur eingehalten.

Das Gerät A4 dient zur Bekämpfung von Flächenzielen auf große Entfernungen.

OKH/We A/Wa Prüf
Anlage zu Bd.Nr19/45 gK

A wide variety of special vehicles were developed, tested and prepared for these operations. The purposes and use of these vehicles are presented in this book.

Zur Vorbereitung dieser Einsätze wurde eine umfangreiche Palette von Spezialfahrzeugen entwickelt, getestet und eingesetzt. Der Zweck und die Verwendung der Fahrzeuge wird in diesem Buch vorgestellt.

Deutsches Museum, München, Archiv- FA014/B683/43 BSM (CD82908)

1

1 Launch preparations of Prototype 29 (V29) on the 11th June 1943 at Test Stand X.

1 Startvorbereitungen des Versuchsmusters 29 (V29) am 11. Juni 1943 auf dem Startplatz X.

Content

Inhalt

ACKNOWLEDGEMENT
DANKSAGUNG

The idea of writing a book about the *Peenemünde* special vehicles, the ground vehicles of the A4 (V2) rocket, was born in 2012.

A first look at the documents of the »German Documents« (GD) in the archives of the Deutsches Museum Munich, BAMA Freiburg and BArch Koblenz showed how extensive the project would be. Further archive visits, including to the company archive of F.X. Meiller, Munich and the Saxony-Anhalt State Archives were to follow. Thus thousands of pages of copied documents and hundreds of illustrations were to form the basis for a manuscript.

After initially working through classifying and merging the documents in 2013/2014, it became clear that a multi-volume book series should be created. The project name chosen for the book series was »The *Peenemünde* Triple«. It was to consist of the volumes »Special vehicles«, »Meillerwagen« and »Vidalwagen«.

In the years of 2015 to 2018, detailed work began on the content, layout concept and design. The time frame was completely underestimated and had to be reorganized. The original scope expanded so quickly that at the end of 2019 it was decided to split the »Special vehicles« book series into

Special vehicles 1
Design and vehicle construction
Special vehicles 2
Railroad vehicles, training, operational preparation.

Now, in autumn 2025, the first volume »Special vehicles 1« has been completed. In the next few months, the next volume »Special vehicles 2« and the books on the Meillerwagen and Vidalwagen will be published. The series will be supplemented with a volume dealing with the tubular swing axle trailers produced by the firm of F.X. Meiller in Munich.

For my book project, I was supported from the very beginning by Mike Imhoff, Bert Koopman, Ed Straten, all members of the IV2RG.

Die Idee, ein Buch über die Spezialfahrzeuge »Peenemünde«, den Bodenfahrzeugen der A4 (V2) Rakete zu schreiben, entstand in 2012. Ein erstes Sichten der Unterlagen der »German Documents« (GD) in den Archiven des Deutschen Museum München, BAMA Freiburg und BArch Koblenz zeigte, wie umfangreich sich das Vorhaben gestalten sollte. Weitere Archivbesuche u.a. im Firmen-Archiv von F.X. Meiller, München und dem Landesarchiv Sachsen-Anhalt sollten folgen. So sammelten sich sehr schnell einige Tausend Seiten kopierter Dokumente und Hunderte von Abbildungen an, die die Grundlage für ein Manuskript bilden sollten.

Nach einem ersten Durcharbeiten, Einordnen und Zusammenführen der Dokumente in 2013/2014 wurde klar, dass hier eine mehrbändige Buchreihe entstehen kann. Für die Buchreihe wurde als Projektname »Das Peenemünde Triple« gewählt. Es sollte aus den Bänden »Spezialfahrzeugen«, »Meillerwagen« und »Vidalwagen« bestehen. In 2015 bis 2018 begann dann die Detailarbeit an Inhalten, Layoutkonzept und Ausgestaltung. Dabei wurde der zeitliche Rahmen vollkommen unterschätzt und musste neu aufgestellt werden. Der ursprüngliche Umfang erweiterte sich so rasch, dass Ende 2019 entschieden wurde, die Buchreihe »Spezialfahrzeuge« aufzuteilen in

Spezialfahrzeuge 1
Entwicklung und Fahrzeugbau
Spezialfahrzeuge 2
Eisenbahnfahrzeuge, Training, Einsatzvorbereitung.

Jetzt, im Herbst 2025, ist der erste Band »Spezialfahrzeuge 1« fertiggestellt. In den nächsten Monaten wird der Band »Spezialfahrzeug 2« und die Bücher zum Meillerwagen und Vidalwagen erscheinen. Ergänzt wird die Serie mit einem Band über die Rohrschwingachs-Anhänger der Firma F.X. Meiller, München.

Für mein Buchvorhaben wurde ich von Mike Imhoff, Bert Koopman, Ed Straten, alle Mitglieder der IV2RG, unterstützt.

In particular, I have to thank Mike Imhoff who provide me with documents of his meticulous preparatory work on the illustrated book *Peenemünde* at Fort Eustis and the documentation of the Meillerwagen Restoration Project of the USAF Museum at Wright Patterson Airfield in Dayton (OH).

During my visits to the archive of the F.X. Meiller company, I was able to view a large number of original drawings of the Meillerwagen. From the holdings of the BAMA, BArch and DMA archives. Images were selected from the stock of the original negatives of the »Image Archive Pee«, newly digitized and used in my books. Surviving drawings of the ground vehicles in the BAMA and Deutsches Museum, data from measured original vehicles such as Meiller- and Vidalwagen from the collections in Cosford and Melbourne, Oorlogsmuseum Arnhem and the Wehrtechnische Museum Röthenbach were used for the creation of the scale drawing in 1/35th or reconstructed from photos. Layout drawings have been added for a better understanding.

This was followed by an intensive exchange of information on the ground facilities and the Peenemünde Army Research Center with members of the International V2 Research Group (IV2RG), namely Mike Imhoff, Bert Koopman, Ed Straten, Thomas Köhler, Stan Druzynski, Tracy Dungan, Detlev Paul, Wolfgang Gückelkorn and Reiner Sigmund. Further documents were provided by Frank Schein, Klaus Schrader, Christian Suhr, Heiner Duske, Darius Stucinskas. Thanks to Holger Erdmann for providing drawings of the closed box bodies (305) and Falk Springer for the service regulations on the base vehicles from his archive.

With the help of Hilary Doyle, the production, equipment and delivery figures of the Fz.101 Feuerleitpanzer were compiled. Many thanks also to all unnamed persons for their contributions to the project.

Documents from German Sources and image documents were used from the following archives: Bundesarchiv Militär Archiv, Freiburg (BAMA), Federal Archives (BArch) Koblenz with the support of Ms. Walter, Deutsches Museum Munich, Archive (DMA) with the support of Ms. Schwarzenau and Ms. Püttner, Archive F.X. Meiller (FXM) with the support of Ms. Korenjak and the Meiller family, Landesarchiv Sachsen-Anhalt (LASA), Imperial War Museum (IWM), US Army Heritage and Education Center at Carlisle Barracks, PA (USAHEC) Aberdeen Proving Ground (APG), Archive of Modern Conflict, London (AMC), PA Media Group Limited (PA), Ullstein Bild, Master Sergeant Paul Cernak via Grandson Christopher R. Van Valkenburgh, King's College London (KCL), Bombardier Transportation GmbH, Bautzen (BTB).

For the English text editing I would like to thank John Prigent, Jon Feenstra, Tom Cockle and Claus Reuter. Thanks to Günther Nord for the professional layout and to the team at Druckerei & Verlag Steinmeier for their printing work.

Im Besonderen habe ich Mike Imhoff zu danken, der mir Unterlagen seiner akribischen Vorarbeit über den Bildband »Peenemünde« in Fort Eustis und der Dokumentation zur Meillerwagen Restauration – Museum der USAF Wright Patterson Airfield Dayton (OH), zur Verfügung gestellt hat.

Bei meinen Besuchen im Firmenarchiv von F.X. Meiller konnte ich ein Großteil original erhaltener Zeichnungen zu den Meillerwagen einsehen. Aus den Beständen des BAMA, BArch und DMA wurden ausgewählte Bilder des »Bildbestandes Pee« von den Originalnegativen neu digitalisiert und in meinen Büchern verwendet. Überlieferte Zeichnungen, soweit diese zu den Bodenfahrzeugen im BAMA und Deutschen Museum vorhanden sind, Daten vermessener Originalfahrzeuge wie Meiller- und Vidalwagen aus den Sammlungen in Cosford und Melbourne, Oorlogsmuseum Arnhem und ebenfalls im Wehrtechnisches Museum Röthenbach stehende Basisfahrzeuge wurden für die Erstellung der Maßstabszeichnung in 1/35 verwendet bzw. über Bilder rekonstruiert. Zum besseren Verständnis sind ggf. Beladepläne hinzugefügt.

Es folgte ein intensiver Austausch von Informationen zu den Bodenanlagen und der Heeresversuchsanstalt Peenemünde mit Mitgliedern der International V2 Research Group (IV2RG), namentlich Mike Imhoff, Bert Koopman, Ed Straten, Thomas Köhler, Stan Druzynski, Tracy Dungan, Detlev Paul, Wolfgang Gückelkorn und Reiner Sigmund. Weitere Unterlagen wurden bereitgestellt von Frank Schein, Klaus Schrader, Christian Suhr, Heiner Duske, Darius Stucinskas. Mein Dank geht an Holger Erdmann für die Bereitstellung der Zeichnungen der geschlossenen Kofferaufbauten (305) und an Falk Springer für die Dienstvorschriften von Basisfahrzeugen aus seinem Archiv. Mit Hilfe von Hilary Doyle konnten die Fertigungs-, Ausrüstungs- und Auslieferungszahlen des Fz.101 Feuerleitpanzer zusammengeführt werden. Vielen Dank auch an alle ungenannten Personen für ihre Beiträge zum Projekt.

Unterlagen aus den German Documents und Bilddokumente wurden aus folgende Archiven verwendet: Bundesarchiv Militär Archiv, Freiburg (BAMA), Bundesarchiv (BArch) Koblenz mit Unterstützung von Frau Walter, Deutsches Museum München, Archiv (DMA) mit Unterstützung von Frau Schwarzenau und Frau Püttner, Archiv F.X. Meiller (FXM) mit Unterstützung von Frau Korenjak und der Familie Meiller, Landesarchiv Sachsen-Anhalt (LASA), Imperial War Museum (IWM), US Army Heritage and Education Center at Carlisle Barracks, PA (USAHEC) Aberdeen Proving Ground (APG), Archive of Modern Conflict, London (AMC), PA Media Group Limited (PA), Ullstein Bild, Master Sergeant Paul Cernak via Grandson Christopher R. Van Valkenburgh, King's College London (KCL), Bombardier Transportation GmbH, Bautzen (BTB).

Ich bedanke mich bei John Prigent, Jon Feenstra, Tom Cockle und Claus Reuter für die englische Textbearbeitung. Danke an Günther Nord für das professionell erstellte Layout, sowie an das Team der Druckerei & Verlag Steinmeier für deren drucktechnische Umsetzung.

INTRODUCTION
EINLEITUNG

The use of solid rockets (black powder) has been known for several centuries. Militarily, solid rockets were used successfully by some European armies in the 18th and 19th centuries. First experiments with liquid propellants were done in laboratories before the First World War and resumed after the end of the war. Various dissertations, papers and books dealing with different rocket models, propulsion techniques with solid or liquid propulsion were published [1].

The HWA of the Reichswehr recognized the potential for a new weapons technology that was not subject to the restrictions of the Treaty of Versailles. A development order for a long range rocket was issued and the leadership was assigned to Captain Walter Dornberger. For its purposes, the HWA sought contact with the VfR [2] community, where Wernher von Braun came to their attention of the HWA. Von Braun was hired by the HWA in October 1932 as an employee/assistant and assigned to the Rocket Weapons Department headed by Colonel Dornberger. The first liquid-propellant rockets were projected and tested by Wernher von Braun, Klaus Riedel and Arthur Rudolph at the Kummersdorf Heeresversuchsanstalt near Berlin. However, the Kummersdorf Heeresversuchsanstalt was too small for testing a long range rocket. A new development center was built at Peenemünde starting in 1936, where new types of weapons were to be developed for the army and air force. New propulsion systems, projectiles, materials and electrical components were conceived, manufactured and brought to operational readiness by the technicians in the laboratories, supported by the scientific elite at the universities and the research departments of industry. The first rocket airplanes, guided missiles, controlled and uncontrolled rockets and antiaircraft missiles, including their new propulsion and control technology, were developed from design to feasibility and technical implementation.

This basic research became the pioneer for manned spaceflight and for many of today's missiles and cruise missiles.

Die Verwendung von Feststoffraketen mit Schwarzpulverantrieb ist seit mehreren Jahrhunderten bekannt. Militärisch wurden diese von einigen europäischen Armeen im 18. und 19. Jahrhundert mit Erfolg eingesetzt. Erste Versuche mit Flüssigkeitsantrieben wurden noch vor dem ersten Weltkrieg in Laboren unternommen und nach Kriegsende wieder aufgenommen. Diverse Dissertationen, Aufsätze und Bücher, die sich mit verschiedenen Raketenmodellen, Vortriebstechniken mit Feststoff- oder Flüssigkeitsantrieben befassten, wurden veröffentlicht [1].

Im HWA der Reichswehr erkannte man das Potenzial für eine neue Waffentechnik, die nicht den Beschränkungen des Versailler Vertrages unterworfen war. Ein Entwicklungsauftrag für eine Fernrakete wurde erteilt und die Leitung Hauptmann Walter Dornberger übertragen. Für seine Zwecke suchte das HWA den Kontakt zum Umfeld des VfR [2], wo man auf Wernher von Braun aufmerksam wurde. Von Braun wurde im Oktober 1932 vom HWA als Angestellter/Assistent eingestellt und der Abteilung für Raketenwaffen unter Leitung von Oberst Dornberger zugeteilt. In der Heeresversuchsanstalt Kummersdorf bei Berlin wurden erste Flüssigkeitsraketen von Wernher von Braun, Klaus Riedel und Arthur Rudolph projektiert und erprobt. Für die Erprobung einer Fernrakete war die Heeresversuchsanstalt Kummersdorf jedoch zu klein. Ein neues Entwicklungszentrum wurde ab 1936 in Peenemünde errichtet, wo jetzt neuartige Waffen für Heer und Luftwaffe entwickelt werden sollten. Neue Antriebe, Geschosse, Werkstoffe und elektrische Bauteile wurden von den Technikern in den Laboren erdacht, gefertigt und zur Einsatzreife gebracht. Unterstützt wurden sie von der wissenschaftlichen Elite an den Universitäten und Forschungsabteilungen der Industrie. Erste Raketenflugzeuge, Lenkflugkörper, gesteuerte und ungesteuerte Raketen sowie Flakraketen wurden einschließlich deren neuen Antriebs- und Steuertechnik von der Auslegung auf Machbarkeit und technische Umsetzung entwickelt.

Diese Grundlagenforschung war der Wegbereiter für die bemannte Raumfahrt und für viele der heutigen Raketen und Marschflugkörper.

Development Plant Peenemünde Ost

Entwicklungswerk (EW) Peenemünde Ost

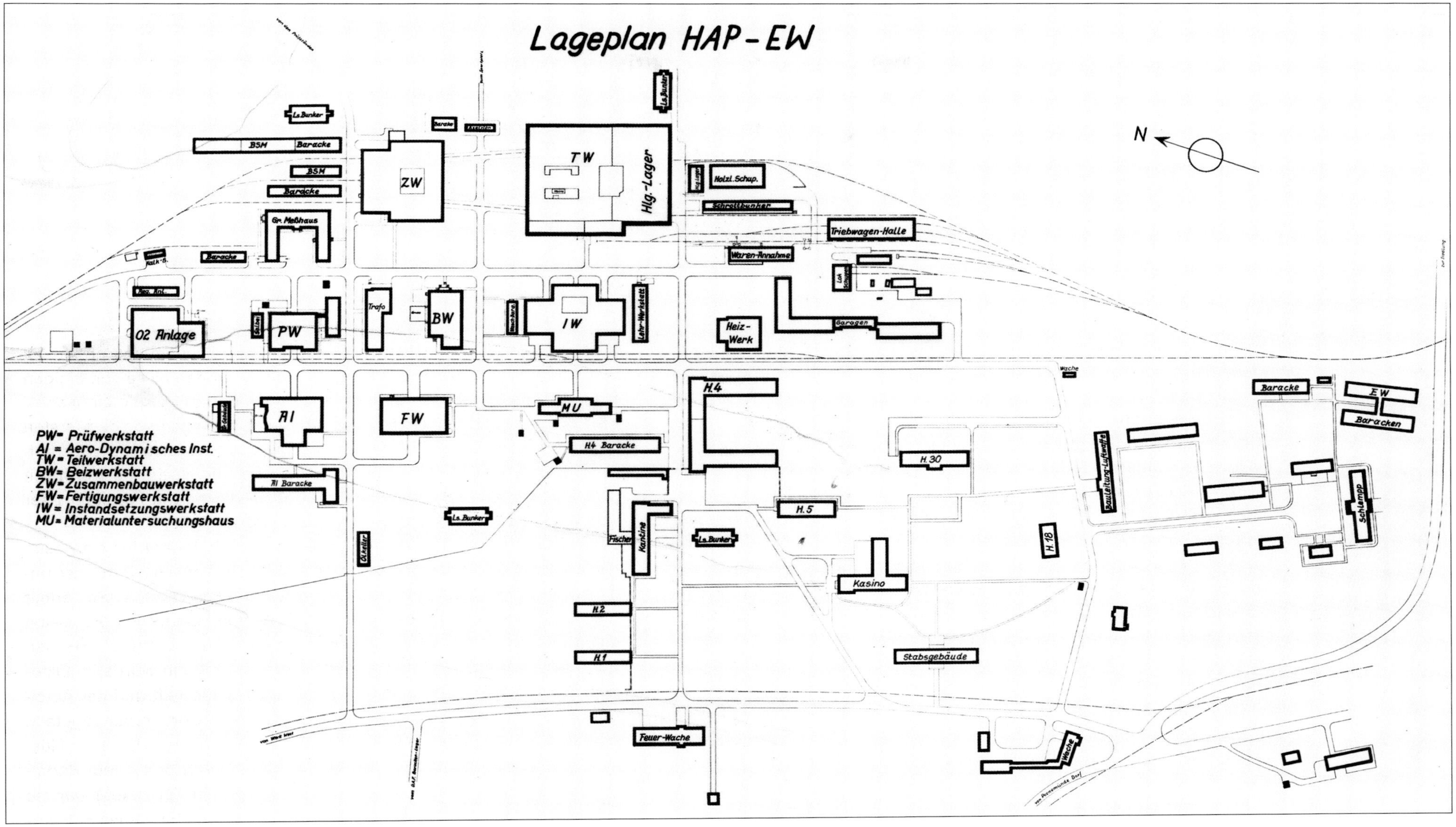

BArch, RH8II Bild-B0633-42 BSM, o.Ang.

2

Site plan HAP – EW
Lageplan HAP – EW

2
(PW) Test workshop
(AI) Aero-Dynamic Institute
(TW) Parts workshop
(BW) Aluminum treatment (pickling) workshop
(ZW) Assembly workshop
(FW) Production workshop
(IW) Repair workshop
(MU) Materials testing house
(O2) O_2-facility

2
(PW) Prüfwerkstatt
(AI) Aero-Dynamisches Institut
(TW) Teilewerkstatt
(BW) Beizwerkstatt
(ZW) Zusammenbauwerkstatt
(FW) Fertigungswerkstatt
(IW) Instandsetzungswerkstatt
(MU) Materialuntersuchungshaus
(O2) O_2-Anlage

DESIGN AND SPECIFICATIONS
AUSLEGUNG UND VORGABE

In the spring of 1936 Dr. Walter Dornberger, Wernher von Braun and Klaus Riedel met in Kummersdorf [1] to plan the *Peenemünde* Project. The building activities, test stands, engine design and basics for the large rocket were discussed. The development target for the large rocket was a range of 275 km, with accuracy of 2-3 % of the range and impact width at that distance and a payload of 1 ton. This target was a clear improvement in performance and precision compared to the »Paris cannon« [2] in the First World War. *»External dimensions were limited so that the unassembled Gerät suitable for transportation on roads and not exceed the maximum width prescribed for road vehicles. When transported by rail, the rocket should be able to be driven through all tunnels. The main dimensions of the rocket were determined by this requirement«.** The rocket had to have a diameter of more than 1.60 m and a length of more than 14 m. The diameter measured over the tail fins was not allowed to exceed 3.5 m.

The Reichsamt für Fahrzeugwesen, section for trailers, was involved in the planning of road vehicles at an early stage. The first proposals, which took special account of the requirements for the transportability and positioning of the rocket, were worked out and evaluated and followed up until the beginning of 1940. As the development of the FR rocket progressed, the focus also shifted to the various possible applications, such as shooting from bunkers, field positions and from rail.

In spring 1942 a conference was convened to decide on the initial demand for vehicles for the shooting batteries from field positions and on rail. The following excerpts show the initial demand.

* *Peenemünde* history of V-weapons, Walter Dornberger, Ullsteinverlag, p.60 and p.61

Im Frühjahr 1936 trafen sich die Herren Dr. Walter Dornberger, Wernher von Braun und Klaus Riedel in Kummersdorf [1] zu Planungen zum Projekt Peenemünde. Es wurden die baulichen Maßnahmen, die Triebwerksauslegung und die Grundlagen für die Großrakete besprochen. Als Entwicklungsziel wurde eine Reichweite von 275 km bei nur zwei bis drei Prozent Längen- und Breitenstreuung auf diese Entfernung und eine Nutzlast von einer Tonne angestrebt. Diese Vorgabe war gegenüber der »Pariser Kanone« [2] im ersten Weltkrieg eine deutliche Verbesserung in der Leistung und Präzision. *»Die äußeren Maße beschränkte ich dahin, dass das Gerät unzerlegt zum Transport auf Straßen geeignet sei und die für Straßenfahrzeuge vorgeschriebene maximale lichte Weite nicht überschreiten dürfe. Beim Transport auf Bahn sollte die Rakete durch alle Tunnel gefahren werden können. Durch diese Forderung waren die Hauptabmessungen der Rakete festgelegt«.** Die Rakete musste bei einem Durchmesser von mehr als 1,60 m eine Länge über 14 m haben. Der über das Leitwerk gemessene Durchmesser durfte 3,5 m nicht überschreiten.

In die Planungen für die Straßenfahrzeuge wurde bereits frühzeitig das Reichsamt für Fahrzeugwesen, Abteilung Anhänger, eingebunden. Erste Vorschläge, die speziell die Vorgaben an die Transportfähigkeit und das Aufstellen der Rakete berücksichtigten, wurden bis Anfang 1940 erarbeitet und beurteilt. Mit fortschreitender Weiterentwicklung der FR-Rakete rückten auch die verschiedenen Einsatzmöglichkeiten in den Fokus. So sollte ein Schießen aus Bunkern, Feldstellungen und von der Schiene möglich sein.

Im Frühjahr 1942 wurde eine Tagung einberufen, um den Erstbedarf an Fahrzeugen für die aus Feldstellungen und auf Schiene schießenden Batterien festzulegen. In Auszügen ist dieser Erstbedarf nachfolgend wiedergegeben.

* Peenemünde Geschichte der V-Waffen, Walter Dornberger, Ullsteinverlag, S. 60, 61

The first Meillerwagen
Der erste Meillerwagen

3 The first Meillerwagen is completed. After several designs in 1940, (drawings 8015 and 8197) the F.X. Meiller Maschinenfabrik und Wagenbauanstalt from Munich was commissioned by the OKH to build a »special six-wheel trailer with swing axles«. The Meiller »three-axle special trailer« type MA4 was manufactured as factory number 5253 and officially handed over to the OKH, Berlin, on 16.1.1942.

The photograph from the winter of 1941/42 was taken on the company's premises in Munich. The tilting frame over 9 meters long gives an idea of the dimensions of the special trailer. The launch table on this special construction (SK) model type MA4 was still part of the trailer. Very early on, after the first loading and setting up exercises, it was decided to separate the launch table from the trailer and to continue developing it further independently as the Fz.103 launch table.

3 Der erste Meillerwagen ist fertiggestellt. Nach mehreren Entwürfen von 1940 (Zeichnung 8015 und 8197) wurde die Firma F.X. Meiller Maschinenfabrik und Wagenbauanstalt aus München mit dem Bau von einem Exemplar »Sechsrad-Sonderanhänger mit Schwingachsen« durch das OKH beauftragt. Der Meiller »Dreiachs-Sonder-Anhänger«-Typ MA4 wurde als Fabriknr. 5253 hergestellt und am 16.1.1942 offiziell an den Auftraggeber, das OKH Berlin, überstellt.

Die Aufnahme aus dem Winter 1941/42 ist auf dem Betriebsgelände der Firma in München entstanden. Der aufgestellte Kipprahmen von über neun Meter Länge lässt die Dimensionen des Sonderanhängers erahnen. Der Starttisch war an dieser Sonderkonstruktion (SK) Typ MA4 noch Bestandteil des Anhängers. Schon sehr früh wurde nach den ersten Lade- und Aufstellübungen entschieden, den Starttisch vom Anhänger zu trennen und eigenständig als Starttisch Fz.103 weiterzuentwickelt.

FXM

3

PROJECTED PROCUREMENT
VORAUSSICHTLICHE BESCHAFFUNG

14

BAMA, RH 8/1217
Wa Prüf 11/III
Berlin, den 2.3.1942
Abt.: Wa Prüf11/III
Az.: 76 a/p 1070
Bd.Nr. 243/42 g.Kdos.

Minutes

of the meeting between Wa Prüf 11/III Ing. Pflugk and HAP EW/T Ing. Riedel II at 27.02.42 in Berlin (excerpt)

6) Projected procurement for the mission [1]:

Total requirement:

a) For supplies of fuels and aggregates:

1)	162	SS-railroad carriages
2)	18	5,000 l O_2-trailers
3)	18	heavy tractors
4)	42	27,000 l railway carriages for O_2
5)	36	2,900 spirit truck
6)	42	18,000 l railway carriages for methylated spirits
7)	18	2,100 l T-Stoff trucks
8)	324	10 l cans for Z-Stoff
9)	324	4 ½ l cans for Z-Stoff

b) Equipment of 3 Batteries:

1) 18 Meillerwagen
2) 18 tractors
3) 18 MAN railroad vehicles. The MAN vehicles are only needed for missions when fired from railroad. It must be decided whether all 3 batteries are to be equipped for this or whether, for example 1 Battery is sufficient. For 1 Battery the requirement reduces to 6 MAN railway vehicles.
4) 3 power supply vehicles
5) 3 cable trailers for power supply vehicles
6) 3 heavy tractors (belongs to points 4 and 5)

Niederschrift

über die Besprechung zwischen Wa Prüf 11/III Ing. Pflugk und HAP-EW/T Ing. Riedel II am 27.02.1942 in Berlin (Auszug der Niederschrift)

6.) Voraussichtliche Beschaffung für den Einsatz [1]:

Insgesamt werden benötigt:

a) Für den Nachschub an Treibstoffen und Aggregaten:

1)	162 Stück	SS-Schienentransportwagen
2)	18 Stück	5.000 l O_2-Anhänger
3)	18 Stück	schwere Radschlepper
4)	42 Stück	27.000 l Eisenbahnwagen für O_2
5)	36 Stück	2.900 l Spiritus-Lkw
6)	42 Stück	18.000 l Eisenbahnwagen für Spiritus
7)	18 Stück	2.100 l T-Stoff-Lkw
8)	324 Stück	10 l Kanister für Z-Stoff
9)	324 Stück	4 ½ l Kanister für Z-Stoff

b) Ausrüstung von 3 Batterien:

1) 18 Stück Meillerfahrzeuge
2) 18 Stück Zugmaschinen
3) 18 Stück MAN-Eisenbahnfahrzeuge. Die MAN-Fahrzeuge werden beim Einsatz nur für das Schießen vom Gleis benötigt. Es muss entschieden werden, ob alle 3 Batterien hierfür ausgerüstet werden sollen oder ob beispielsweise 1 Batterie genügt. In diesem Falle vermindert sich die Stückzahl auf 6 MAN-Wagen.
4) 3 Stück Stromversorgungswagen
5) 3 Stück Kabelanhänger für Stromversorgungswagen
6) 3 Stück schwere Radschlepper (gehört zu Punkt 4 und 5)

7) 3 armored cars. As 2 of these vehicles are already being procured, it is recommended that for the third required vehicle again an armored car will be selected, so as not to have to reschedule the already defined installations, which would be necessary if another vehicle were to be used.
A decision on this must be made as soon as possible. Ing. Pflugk gives a presentation to the head of the department.
8) 9 motorized ladders (mobile French Army observation tower, french model)
9) 3 maintenance units each consisting of 1 maintenance vehicle, 1 equipment truck, 1 heavy generator set A (standard trailer) and 1 four wheel trailer for spares. For this Wa Prüf 11/III needs information from HAP EW/T as soon as possible about overall dimensions and weights.
10) 3 transmitter vehicles for combustion cut off (Opel Blitz) with single axle trailers (light power generator)
11) 3 cable trucks with single axle trailers (combustion cut-off secondary site)
12) 6 guide beam transmitter vehicles
13) 3 single axle trailers (for the guide beam secondary site)
14) 3 trucks with closed bodies and heaters to maintain 25° room temperature in the winter. The heaters must not need an external power supply. It is necessary to investigate whether exhaust heating is sufficient or whether additional heating with bottled gas is required.
15) 3 cable vehicles for the guide beam service (probably Kfz. 77)
16) ? Personnel carriers. The number of personnel carriers depends on the selection of the tractors for the Meillerwagens, since there are already a large number of people to be carried. The number of vehicles indicated in this listing will carry approximately 300 men. The rest of the men, who cannot be transported in those, must be carried in personnel carrier vehicles. The numbers for this must be specified. Presumably 3 vehicles will be sufficient.
17) 9 Volkswagens.

The vehicle requirements for b) are for 3 Batteries, but since the establishment of a training Battery is intended it must be decided whether it is to be equipped from these numbers, for example as an advance requirement, or whether the training Battery requirements are to be separately procured.
For these the listed quantities divided by 3 are valid.
Wa Prüf 11/III is to take all necessary steps to prepare and clarify this provisional listing, so that the purchase order can be placed as soon as possible after it is ordered by the Department head.

Sign.: Pflugk
draft seen by: Dornberger

7) 3 Stück Bedienungspanzerwagen. Da sich bereits 2 dieser Fahrzeuge in Beschaffung befinden, wird empfohlen, auch für das 3. benötigte Fahrzeug wiederum einen Panzerwagen zu nehmen, um die bereits festgelegten Einbauten nicht noch einmal umplanen zu müssen, was bei der Verwendung eines anderen Wagen erforderlich wäre. Entscheidung hierfür muss baldigst erfolgen. Über die Beschaffungslage der Bedienungspanzerwagen hält Ing. Pflugk dem Abteilungschef Vortrag.
8) 9 Stück Motorleitern (fahrbarer franz.Beob.Turm)
9) 3 Stück Werkstattzüge bestehend aus je 1 Werkstattkraftwagen, 1 Geräte-Lkw, 1 schwerer Maschinensatz A (Einheitsanhänger) und 1 zweiachsiger Anhänger für Ersatzteile. Für diesen benötigt Wa Prüf 11/III baldigst Angaben über Platzbedarf und Gewicht durch HAP EW/T.
10) 3 Stück Sendewagen für Brennschluss (Opel-Blitz) mit Einachsanhänger (leichter Maschinensatz)
11) 3 Stück Kabel-Lkw mit Einachsanhänger (Brennschlussnebenstelle)
12) 6 Stück Leitstrahlsendewagen
13) 3 Stück Einachsanhänger (Leitstrahlnebenstelle) 15
14) 3 Stück Lkw mit geschlossenen Aufbau und Heizeinrichtung für 25° C Raumtemperatur im Winter. Die Heizeinrichtung darf keine fremde Stromquelle benötigen. Es ist zu untersuchen, ob Auspuffheizung ausreicht oder ob zusätzlich Heizung mit Flaschengas erfolgen muss.
15) 3 Stück Kabelfahrzeuge für Leitstrahl (wahrscheinlich Kfz. 77)
16) ? Stück Mannschafts-Lkw. Die Anzahl der Mannschafts-Lkw. hängt von der Auswahl der Zugmaschinen für die Meillerwagen ab, da sich dort schon eine große Anzahl von Leuten unterbringen lassen. Mit der in dieser Aufstellung angegebenen Anzahl von Fahrzeugen sind rund 300 Mann mitzubefördern. Der nicht mehr unterzubringende Rest an Mannschaften muss auf den Mannschafts-Lkw befördert werden. Hiernach ist die Stückzahl festzulegen. Voraussichtlich werden 3 Stück genügen.
17) 9 Stück Volkswagen.

Die Fahrzeugbeschaffung zu b) erfolgt für 3 Batterien, da die Aufstellung einer Lehrbatterie beabsichtigt ist, muss entschieden werden, ob diese aus den aufgeführten Stückzahlen entnommen werden soll und beispielsweise als Vorablieferung dringlich betrieben, oder ob die Lehrbatterie besonders beschafft werden soll.
Für diese gelten die aufgeführten Stückzahlen geteilt durch 3.
Wa Prüf 11/III trifft nach dieser vorläufigen Aufstellung sämtliche zur Vorbereitung und Klärung erforderlichen Maßnahmen, damit die Bestellung nach erfolgtem Befehl des Abteilungschefs baldigst erfolgen kann.

Gez.: Pflugk
im Entwurf gesehen: Dornberger

DIVISION OF DUTIES
ARBEITSAUFTEILUNG

BAMA, RH 8/1954
HAP
Peenemünde, 14.10.1942
Az.: 72p 1095
Bd.-Nr. 02022/42g

Minutes
of the meeting on October 14, 1942 in Peenemünde (Excerpt)

Subject: division of duties within the ground support organization

Subj.: Wa Prüf 11 76 a p1050 Wa Prüf 11/III 114/10/42 from 6.10.42 und HAP/EW Bd.nr.: 01470/42g

I) proposal for division of duties

1) Wa Prüf 11/III Reg. Baurat Schwab and staff:
Area of responsibility: Procurement of the vehicles of ground support organization A4. Design and development of the vehicles in scope of responsibility for working area vehicles. Further development of the vehicles based on experience and documents from the tests by HAP/EW/Vers/D. Contact with HAP/EW: Ing. Heinisch Contact at HAP/EW: Ing. Riedel II, T/Ass.

2) T/Ass. Ing. Riedel II:
Area of responsibility: Organizing the tasks of the ground support organization from the point of view of the units. Processing all questions related to the ground support organization by BSM, BN and Test departments. All written communication to Wa Prüf 11/III to be sent via T/Ass. Signature by EW/L and/or HAP/L.

3) Department Vers./D: Head of department Ing. Teßmann, Supervising Office Ing. Hüter as permanent representative of Vers./L.
Area of responsibility: Design, development and testing of the vehicles and installation of ground support accessories A4 except for work directly affecting the vehicles.

Niederschrift
der Besprechung am 14.10.1942 in Peenemünde (Auszug)

Betreff: Arbeitsaufteilung auf dem Gebiet der Bodenorganisation

Vorg.: Wa Prüf 11 76 a p1050 Wa Prüf 11/III 114/10/42 vom 6.10.42 und HAP/EW Bd.nr.: 01470/42g

I) Vorschlag für die Aufteilung der Arbeitsgebiete

1) Wa Prüf 11/III Reg. Baurat Schwab und Mitarbeiter:
Aufgabengebiete: Beschaffung der Fahrzeuge der Bodenorganisation A4 Konstruktion und Entwicklung der Fahrzeuge im Rahmen des Arbeitsbereiches Fahrzeuge. Weiterentwicklung der Fahrzeuge an Hand der von HAP/EW/Vers/D anfallenden Erfahrung und Unterlagen aus der Erprobung. Verbindungsmann zu HAP/EW: Ing. Heinisch Verbindungsmann bei HAP/EW: Ing. Riedel II, T/Ass.

2) T/Ass. Ing. Riedel II:
Aufgabengebiete: Bearbeitung der Aufgaben der Bodenorganisation vom Standpunkt des Truppeneinsatzes. Zusammenfassende Bearbeitung aller Bodenorganisationfragen der Abteilung BSM. BN und Vers. Sämtliche Schreiben an Wa Prüf 11/III sind über T/Ass. zu leiten. Unterschrift durch EW/L bzw. HAP/L.

3) Abteilung Vers./D: Abteilungsleiter Ing. Teßmann, vorgesetzte Stelle Ing. Hüter als ständiger Vertreter von Vers./L.
Aufgabengebiet: Konstruktion, Entwicklung und Erprobung der Fahrzeuge und Einrichtung des Bodenzubehörs A4 mit Ausnahme der Arbeitsgebiete, die direkt die Fahrzeuge betreffen.

The results of testing are documented:
1) For Wa Prüf 11/III vehicle development and procurement.
2) For T/Ass. Ing. Riedel II technical and military questions for operating missions.
3) Consequences to design of the whole *A4 Gerät* (chef of design).

4) HAP/EW/V K is under the direction of Inspector Grützner, a special district with the above-mentioned office designation has been established.
Area of responsibility:
Administration and maintenance of the vehicles of the ground support organization.
Comprising:
a) Vehicles for the test operation EW, the BSM and BN departments.
b) Development vehicles for military use.
c) Vehicles of the training battery.

The administrative tasks include both motor vehicles and as well as rail vehicles of the ground organization. Administrative registration of vehicles.. Procuring fuel and spare parts as well as accessories apply directly by Wa Prüf 11/III or HKP (requirement ticket for Wa Pr.11/III ü. VA/EK).
Assignment of the vehicles with drivers for testing required by departments.
To distingish the vehicles in groups 4 a) b)and c) they are to be marked separately by color.
The use of vehicles in group b in a is only allowed with permission from the head of department Ing. Hüter, verified by Inspector Grützner.

Markings:
- Vehicles of group a (test vehicles) receive a suitable spot, a clearly visible white crossbar.
- Vehicles of group b (development vehicles of the ground support organization) no markings.
- Vehicles of group c (training Battery) red* crossbar.

* As an alternative, the marking was also executed as a black crossbar. An »X« was painted on in white to mark the training battery and a number for the vehicle.

Die Ergebnisse der Erprobung sind Unterlagen:
1) für Wa Prüf 11/III Fahrzeug-Entwicklung und Beschaffung.
2) für T/Ass. Ing. Riedel II technische und militärische Einsatzfragen.
3) Auswirkungen auf Konstruktion des Gesamtgerätes A4 (Chefkonstrukteur)

4) HAP/EW/V K: Dort ist unter Leitung von Inspektor Grützner ein besondere Bezirk mit der o.a. Dienststellenbezeichnung eingerichtet worden.
Aufgabengebiete:
Verwaltung und Pflege der Fahrzeuge der Bodenorganisation.
Umfassen:
a) Fahrzeuge für Versuchsbetrieb EW, der Abteilungen BSM und BN.
b) Entwicklungsfahrzeuge für den Truppeneinsatz.
c) Fahrzeuge der Lehrbatterie.

Die Verwaltungsaufgaben umfassen sowohl Kraftfahrzeuge als auch Schienenfahrzeuge der Bodenorganisation. Verwaltungsmäßige Registrierung der Fahrzeuge. Beschaffung von Betriebsstoffen und Ersatzteilen sowie Zubehörteilen direkt über Wa Prüf 11/III oder HKP (Forderungszettel für Wa Pr.11/III ü. VA/EK)
Ausgabe der Fahrzeuge mit Fahrpersonal an die zur Erprobung anfordernden Abteilungen.
Zur Unterscheidung sind die zu Ziffer 4 a), b) und c) genannten Fahrzeuggruppen durch Farben getrennt zu kennzeichnen.
Die Verwendung von Fahrzeugen der Gruppe b in a ist nur mit besonderer Genehmigung von Abteilungsleiter Ing. Hüter gestattet. Überprüf. durch Insp. Grützner.

Kennzeichnung:
- Fahrzeuge der Gruppe a (Versuchs-Fahrzeuge) erhalten an geeigneter Stelle gut sichtbarer weißer Querbalken.
- Fahrzeuge der Gruppe b (Entwicklungsfahrzeuge für Bodenorganisation) keine Kennzeichnung.
- Fahrzeuge der Gruppe c (Lehrbatterie) roter* Querbalken.

* Alternativ wurde die Kennzeichnung auch als schwarzer Querbalken ausgeführt. Ein »X« als Markierung für die Lehrbatterie und eine Zählnummer für das Fahrzeug wurden in weißer Farbe aufgemalt.

II.) On the basis of the list at annex 8 of HAP/EW, Bd.Nr. 01417/42g (Bd.Nr.1541/42 Wa Prüf/II, Bd.Nr.1125/42g HAP) the actual status and whereabouts of each vehicles is given. The numbers shown are those in the list:

II.) Anhand der Aufstellung, Anlage 8 zu HAP/EW,Bd.Nr. 01417/42g (Bd. Nr.1541/42 Wa Prüf/II, Bd.Nr.1125/42g HAP) wird der derzeitige Stand und Verbleib der einzelnen Fahrzeuge durchgesprochen. Die laufenden Nummern betreffen die Nummern der zugehörigen Aufstellung:

Special vehicles
Spezialfahrzeuge

item	object	comment	location
1	Transport- and dump vehicles	delivered	Pr.VII
2 u. 3	Alcohol tank vehicles	not yet delivered	
4 u. 5	5,000 l liquid oxygen transport-trailer	delivered	O_2-facility
6, 7, 8	5,000 l liquid oxygen transport-trailer	not yet delivered	1 trailer advised by Wa Pr.11/III
9	Transport vehicle for T-Stoff	2 delivered	Garage Service station
10	Test superstructure for operating vehicle	delivered, used for conversations	
11, 12	Operating vehicle	1 delivered	behind EW
13	Current supply vehicle		assembly hall Pr.VII
14	Cable reel trailer	1 delivered	EW
15	Heavy wheeled tractor	15 delivered	10 Pee-Dorf, 5 EW gas station
16	Opel-Blitz 3 t chassis	3 delivered	BSM
17	ditto	ditto	ditto
18	French mobile observation towers	not yet delivered	whereabouts unknown
19	T-Stoff transport vehicle	not yet delivered	
20	Superstructure for operating vehicle	not yet delivered	
21	FR vehicle	not yet delivered	
22	Twin axle trailer	not yet delivered	
23	Communication vehicle	1 delivered	EW gas station
24	4 ½ t Lkw Typ A	not yet delivered	
25	ditto	ditto	
26	ditto	ditto	
27	4 ½ t Lkw Typ A with flatbed (maintenance unit)	1 delivered	EW gas station
28	4 ½ t Lkw Typ A FF-cable vehicles	3 delivered	EW gas station
29	Single axle trailer for BSM	1 delivered	BSM
30	Opel-Blitz 3 t chassis	3 delivered	BSM

Gegenstand	Bemerkung	Standort
Transport- und Kippfahrzeug	ausgeliefert	Pr.VII
Tankwagen f. Spi.	noch nicht ausgeliefert	
Flüss. O_2-Transport-Anhänger für 5.000 l	ausgeliefert	O_2-Anlage
Flüss. O_2-Transport-Anhänger für 5.000 l	noch nicht ausgeliefert	1 Anh. durch Wa Pr.11/III avisiert
Transportfahrzeug für T-Stoff	2 ausgeliefert	1 Garage Tankstelle
Probe Aufbau für Bedienungsfahrzeug	ausgeliefert, zu Umbauten verwendet	
Bedienfahrzeug	1 ausgeliefert	Hinter EW
Stromversorgungswagen	1 ausgeliefert	Montagehalle Pr.VII
Kabeltrommelanhänger	1 ausgeliefert	EW
Schwerer Radschlepper	15 ausgeliefert	10 Pee Dorf 5 Tankstelle EW
Opel-Blitz 3 t Fahrgestelle	3 ausgeliefert	BSM
dgl.	dgl.	dgl.
Fahrbarer französischer Beobachtungsturm	noch nicht ausgeliefert	Verbleib unbekannt
Transportfahrzeug für T-Stoff	noch nicht ausgeliefert	
Aufbau für Bedienungsfahrzeug	noch nicht ausgeliefert	
FR-Wagen	noch nicht ausgeliefert	
Zwei-Achsanhänger	noch nicht ausgeliefert	
Nachrichten-Kraftwagen	1 ausgeliefert	Tankstelle EW
4 ½ t Lkw Typ A	noch nicht ausgeliefert	
dgl.	dgl.	
dgl.	dgl.	
4 ½ t Lkw Typ A mit Pritschen Aufbau (Werkstattzug)	1 ausgeliefert	Tankstelle EW
4 ½ t Lkw Typ A FF-Kabelkraftwagen	3 ausgeliefert	Tankstelle EW
Ein-Achsanhänger für BSM	1 ausgeliefert	BSM
Opel-Blitz 3 t Fahrgestelle	3 ausgeliefert	BSM

Railroad vehicles
Eisenbahnfahrzeuge

item	object	comment	location
1	MAN transport- and dump truck	1 delivered	Pr. VII
2	ditto for EW	not yet delivered	
3	Alcohol transport wagon	1 delivered	Pr. VII
4, 5	27,000 l O_2-Wagon	not yet delivered	
6, 7, 8	T-Stoff-Wagon	8 delivered	factory south
9	Transport and dump truck for Proj. »F«	not yet delivered	
10	Transport tank wagon for liquid oxygen	not yet delivered	

Gegenstand	Bemerkung	Standort
Transport- u. Kippfahrzeug MAN	1 ausgeliefert	Pr.VII
dgl. für EW	noch nicht ausgeliefert	
Transportwagen für Spiritus	1 ausgeliefert	Pr.VII
27.000 l O_2-Wagen	noch nicht ausgeliefert	
T-Stoff-Transportwagen	8 ausgeliefert	Werk Süd
Transport- u. Kippfahrzeug Proj. »F«	noch nicht ausgeliefert	
Transporttankwagen für flüss. O_2	noch nicht ausgeliefert	

First Vehicles
Erste Fahrzeuge

Deutsches Museum, München, Archiv- FA014/B695/42 BSM (CD49013)

Operating vehicle
Bedienfahrzeug

4 The operating vehicle is parked on the grounds of test stand X for a test. The large assembly hall of test stand VII can be seen in the background. Numerous cables lead from the junction boxes on the operating vehicle to the machine set 11, from there to the B.A. test vehicle and the *Gerät* at the Meillerwagen. Since there were too few junction boxes, other openings such as the lateral viewing slit, the hatch, and the rear openings were used to lead cables into the vehicle interior.

4 Auf dem Gelände des Prüfstand X ist das Bedienfahrzeug für einen Test abgestellt. Im Hintergrund ist die große Montagehalle am Prüfstand VII zu sehen. Unzählige Kabel führen von den Anschlussdosen am Bedienfahrzeug zum Maschinensatz 11, von dort zum B.A. Prüffahrzeug und zum Gerät am Meillerwagen. Da zu wenige Anschlussdosen vorhanden waren nutzte man für die Verlegung der Kabel in den Fahrzeuginnenraum die Öffnungen der seitlichen Sehschlitze und der Luke, oder verlegte sie einfach über Öffnungen am Fahrzeugheck.

4

BArch, RHBII Bild-B0767-42 BSM, o.Ang

5

5 The operating vehicle with an armored superstructure for the operating crew was mounted on the chassis of the m. ZgKw. 5 t (HKp 902). The prototype of the later *Feuerleitpanzer* Fz.101 was used until spring 1944 for the launch tests at *Peenemünde.* The body is divided into two parts. The lower part is bolted to the chassis as an inverted truncated pyramid on which the upper part (truncated pyramid) was placed and bolted. The armoured superstructure was made of 8 to 10 mm thick sheet plate. In the divided front panel vision blocks are cut and convered with bullet-proof glass. They serve in the lower area for the driver/co-driver to see ahead and in the upper area for the operating crew to observe. There is a hatch for the commander and an access door for the crew on the left side. Cable connection feedthroughs were provided in the right front of the superstructure.

5 Der gepanzerte Aufbau des Fahrzeugs für die Bedienmannschaft wurde auf das Fahrgestell des m. ZgKw. 5 t (HKp 902) montiert. Der Vorläufer des späteren Feuerleitpanzers Fz.101 wurde bis zum Frühjahr 1944 für die Startversuche auf Peenemünde verwendet. Der Aufbau ist zweigeteilt. Der untere Teil ist als umgedrehter Pyramidenstumpf mit dem Fahrgestell verbolzt. Darauf wurde das Oberteil (Pyramidenstumpf) aufgesetzt und verschraubt. Der gepanzerte Aufbau wurde aus 8 bis 10 mm dicken Blechen gefertigt. In die geteilte Frontplatte sind mit Glasblöcken ausgerüstete Sichtfenster geschnitten. Sie dienen im unteren Bereich dem Fahrer/Beifahrer zur Sicht nach vorne und im oberen Bereich der Bedienmannschaft zur Beobachtung. Für den Kommandanten ist eine Luke vorhanden. Eine Zugangstür für die Mannschaft befindet sich auf der linken Seite. Für Kabelanschlüsse sind Durchführungen im Aufbau an der rechten Vorderseite vorhanden.

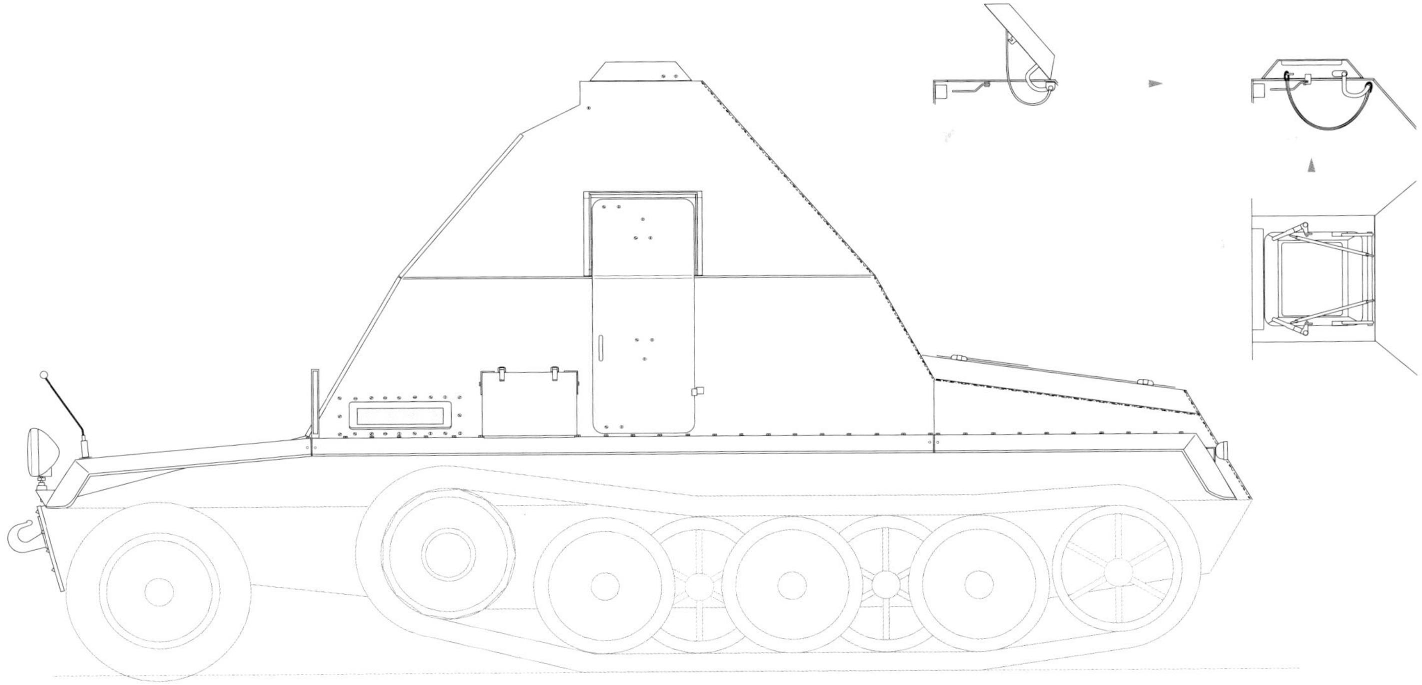

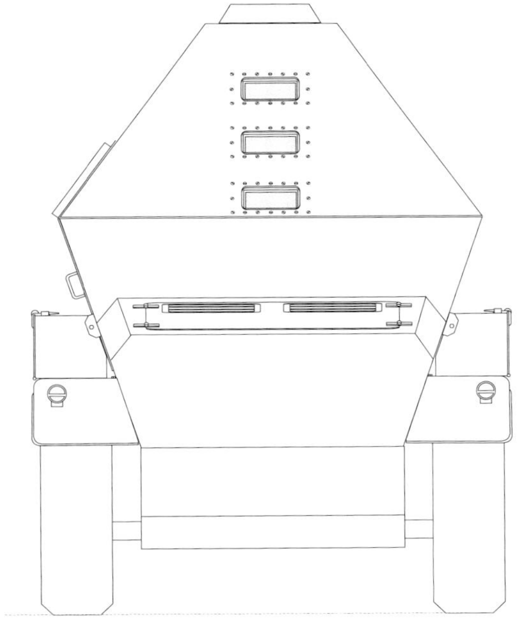

1 m

Scale / Maßstab 1:35 © Volker Ruff

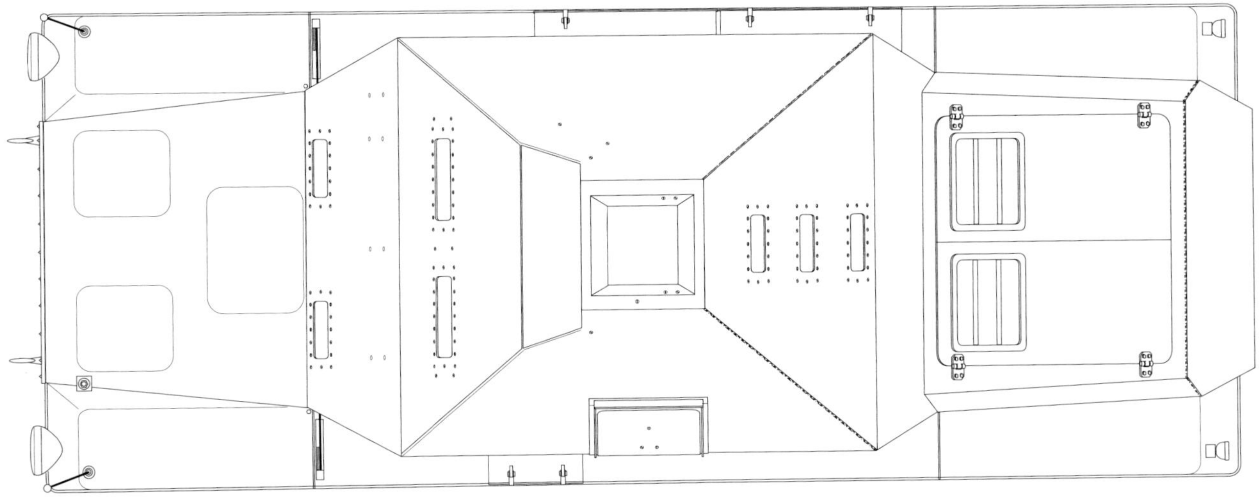

Historiathek

6

Bedienfahrzeug mit gepanzerten Aufbau

Drawing set used n.a.
Verwendeter Zeichnungssatz k.A.

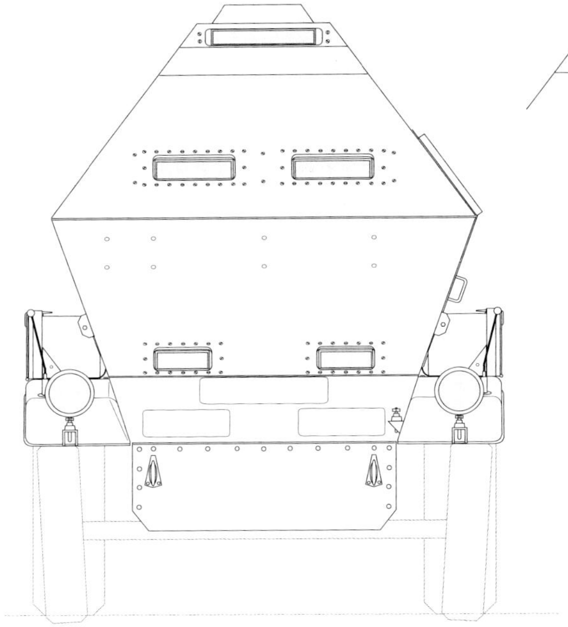

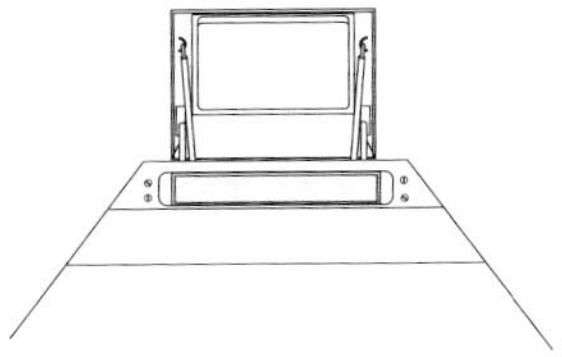

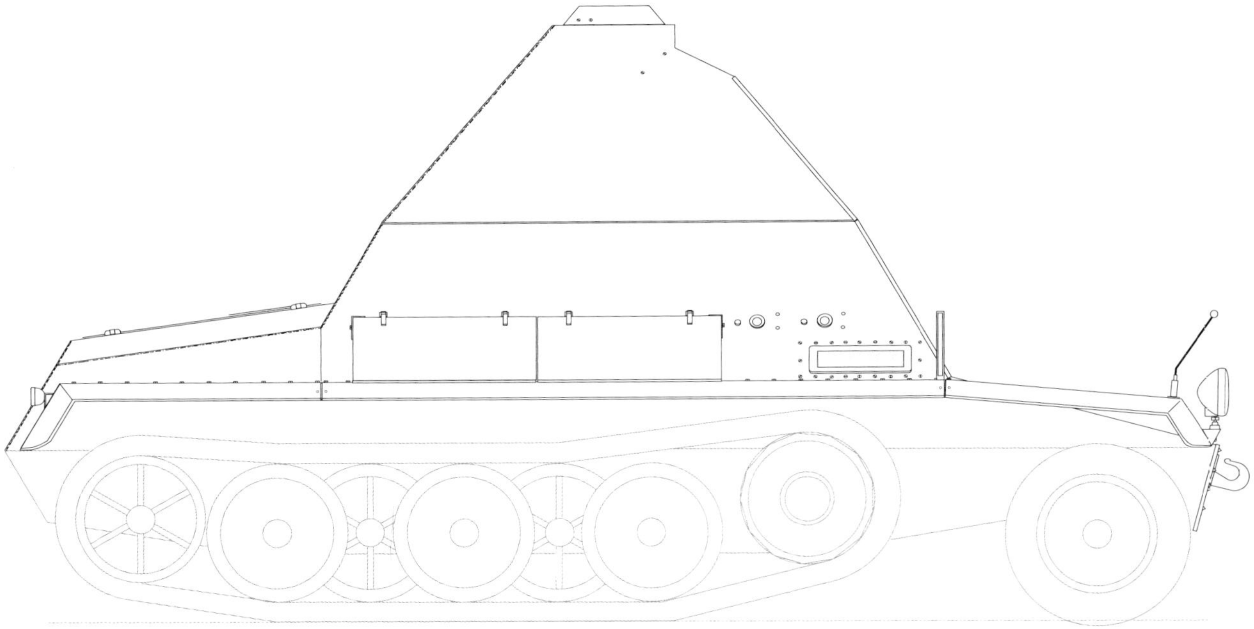

1 m

Scale / Maßstab 1:35 © Volker Ruff

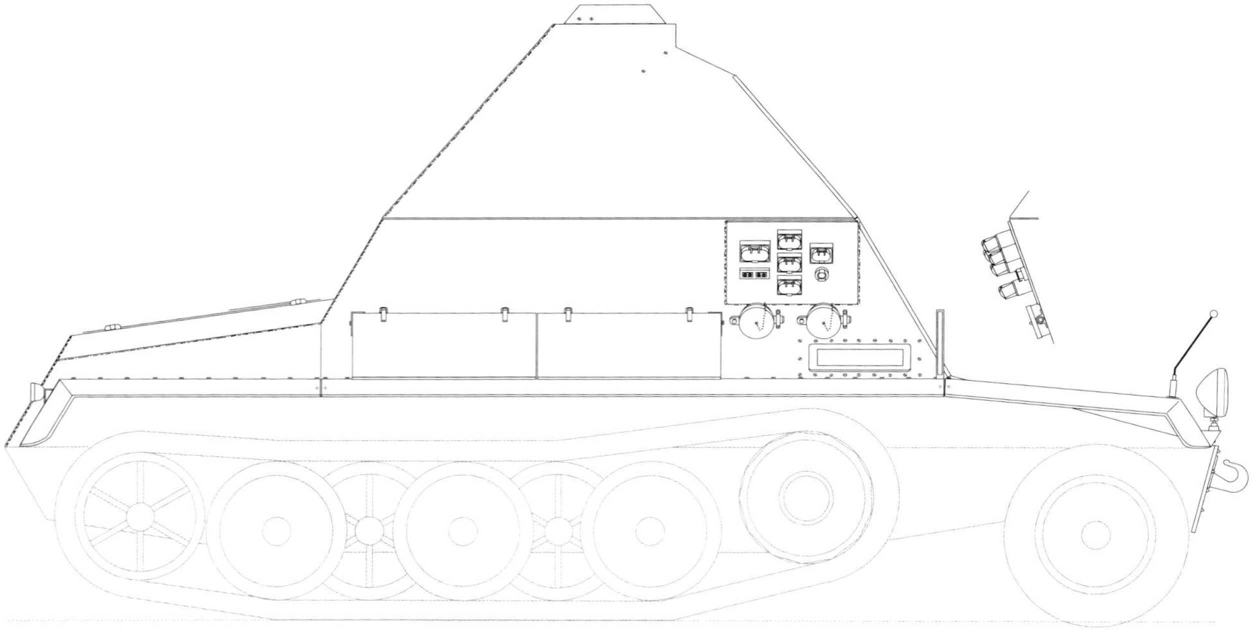

Development variant
Entwicklungsvariante

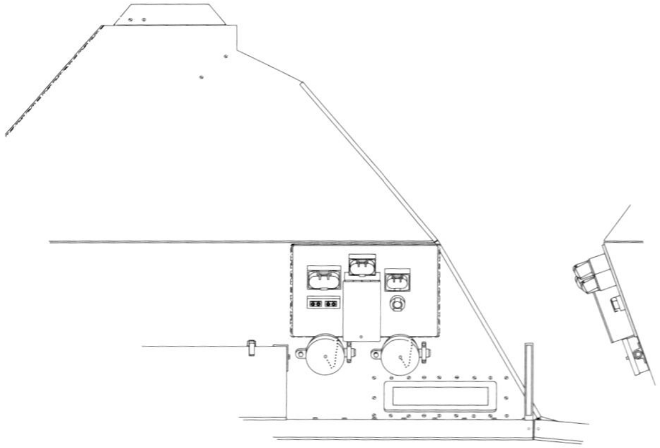

Bedienfahrzeug mit gepanzerten Aufbau

Drawing set used n.a.
Verwendeter Zeichnungssatz k.A.

Deutsches Museum, München, Archiv- FA014/B536/43 BSM [CD49013]

7

Deutsches Museum, München, Archiv- FA014/B535/43 BSM [CD49013]

8

7 The connections were continuously increased to during the first tests. Two 104-pin Flak connectors (*Flaksteckdose*) were installed. Additional various connections for power, a telephone connection and a power strip, were mounted to a welded on sheet metal panel.

8 Through the access door you have a view of the installations. The radio panel is located at the top, located under it are the installation frames for the control panels for starting and control commands. However, the panels are missing. The provisional cabling leads to the radio panel and to the installation boxes for starting and control. It corresponds to the current state of development.

7 Die Anschlüsse wurden während der ersten Erprobungen laufend erweitert. An den vorhandenen Durchbrüchen wurden zwei 104-polige Flaksteckdosen montiert. Auf eine zusätzlich angeschweißte Blechtafel befestigte man verschiedene Stecker, Kraftsteckdosen, eine Telefonanschlussdose und eine Steckerleiste.

8 Durch die Zugangstür hat man einen Blick auf die Einbauten – oben befindet sich das Funkpanel, darunter die Einbaurahmen für die Bedientafeln für Startvorgang und Steuerbefehle. Die Tafeln fehlen jedoch. Die provisorische Verkabelung führt zur Funktafel und zu den Einbaukästen für Start und Steuerung. Sie entspricht dem jeweiligen Entwicklungsstand.

BArch, RH8II Bild-B0771-42 BSM, o.Ang.

9

Power plant vehicle
Stromversorgungswagen

9 A machine set 11 was used for the power supply requirements of the vehicles – *Feuerleitpanzer*, test vehicles I / III and the A4 *Gerät*. A body was mounted on a two-axle trailer to contain a »Strüver« diesel direct injection unit of 48 kW, a four-field switch cabinet, a direct injection converter and a Citomat system.

The machine set 11 (Fz.104) was used for firing mot I, which was replaced in 1944 by the power supply vehicle Fz.105 for firing mot.II.

9 Für den Strombedarf der Fahrzeuge – Feuerleitpanzer, Prüfwagen I / III und dem Gerät A4 – wurde ein Maschinensatz 11 verwendet. Auf einen zweiachsigen Anhänger wurde ein Aufbau montiert, der ein »Strüver« Diesel-Gleichstrom-Aggregat 48 kW, einen vierfeldrigen Schaltschrank, einen Gleichstromumformer und eine Citomat-Anlage umgab.

Der Maschinensatz 11 (Fz.104) wurde für das Schießen nach mot.I verwendet. Dieser wurde 1944 für das Schießen nach mot. II durch den Stromversorgungswagen Fz.105 abgelöst.

BArch, RH8II Bild-B0766-42 BSM, o.Ang.

10

Test vehicle
Prüfwagen

10 The picture shows an early test/evaluation vehicle on the chassis of the Opel Blitz 3 t 4x4 truck with a closed superstructure and attachment walls. The attachment walls were here firmly connected to the side walls and additionally covered with sheet metal. An identical vehicle is in the background, showing the white night driving marking on the mudguard. Both vehicles belong to group »b« (development vehicles for ground organization) and have no other markings.

10 Das Bild zeigt einen frühen Prüf- /Auswertewagen auf dem Fahrgestell des Opel Blitz 3 t-Lkw 4x4 mit einem geschlossenen Aufbau und Aufsatzwänden. Die Aufsatzwände waren hier mit den Seitenwänden fest verbunden, die zusätzlich mit Blech verkleidet wurden. Ein baugleiches Fahrzeug steht im Hintergrund, die weiße Markierung am Kotflügel ist eine Nachtfahrmarkierung. Beide Fahrzeuge gehören zur Gruppe »b« (Entwicklungsfahrzeuge für Bodenorganisation) und haben keine Kennzeichnung.

Cable reel trailer
Kabeltrommelanhänger

11 The early cable trailer on the chassis of the standard trailer E3 had a superstructure with only three cable reel compartments; the following model Fz.108 on the standard trailer E5 had four compartments to accommodate the cable reels for connection and connecting cables in the firing position.

11 Der frühe Kabelwagen auf dem Fahrgestell des Einheitsanhänger E3 hatte einen Aufbau mit nur drei Kabeltrommelabteilen. Das Nachfolgemodell Fz.108 auf dem Einheitsanhänger E5 hatte vier Abteile zur Aufnahme der Kabeltrommeln für Anschluß- und Verbindungsleitungen in der Feuerstellung.

BArch, RH8II Bild-B0764-42 BSM, o.Ang.

11

Single axle trailer for BSM
Einachsanhänger für BSM

12 A simple sheet metal body was placed on a single-axle trailer for the B-stoff pump. The housing is not yet bolted to the chassis frame and was provisionally fixed to the frame with two wires. An A-Stoff pump for the A-Stoff trailer 5 was also carried along.

12 Für die B-Stoff-Pumpe wurde ein einfacher Blechaufbau auf einen Einachsanhänger gesetzt. Das Gehäuse ist noch nicht mit dem Fahrgestellrahmen verschraubt und wurde mit zwei Drähten auf dem Rahmen provisorisch fixiert. Im Gehäuse wurde auch die A-Stoff-Pumpe für den A-Stoff-Anhänger 5 mitgeführt.

BArch, RH8II Bild-B0765-42 BSM, o.Ang.

12

Deutsches Museum, München, Archiv- FA014/B291/44 TD5224 [BN10847]

Deutsches Museum, München, Archiv- FA014/B292/44 TD5224 [CD75022]

14

Mobile french observation tower
Fahrbarer französischer Beobachtungsturm

13-15 A mobile French observation tower on the chassis of the Citroen T45 was to provide access to the hatches of the of the deployed A4 *Gerät*. The vehicle was listed in the overview of motor vehicles in Group I as vehicle Fz.109 Sonder-Kfz (Turmwagen), but was never used and was removed from the list of Mot.-vehicles. As its replacement a Magirus ladder, and later an assembly platform hanging from the rocket top, came into use. The tower lift was placed directly behind the driver's cab. Before raising, the support outriggers mounted on the chassis were extended and braced to the ground. Pendulums on the tower frame show the position of the danger zone for safe extension. The tower control operators were housed in the rear-mounted cabin.

13-15 Ein fahrbarer französischer Beobachtungsturm auf dem Fahrgestell des Citroen T45 sollte den Zugang zu den Klappen des aufgestellten A4-Geräts ermöglichen. Das Fahrzeug war in der Übersicht der Mot.-Fahrzeuge in der Gruppe I als Fahrzeug Fz.109 Sonder-Kfz (Turmwagen) gelistet, aber nie verwendet worden. Es wurde wieder aus der Übersicht der Mot.-Fahrzeuge gestrichen. Als Ersatz kam eine Magirusleiter, später eine Montagebühne zum Einhängen an der Raketenspitze in Gebrauch. Der Turmaufzug befindet sich direkt hinter dem Führerhaus. Vor dem Anheben wurden die am Fahrgestell montierten Stützausleger ausgefahren und zum Boden hin abgestützt. Pendel am Turmgerüst zeigten den Gefahrenbereich für ein sicheres Ausfahren. Die Turmbedienung war in der am Heck montierten Kabine untergebracht.

Deutsches Museum, München, Archiv- FA014/B293/44 TD5224 [CD75023]

15

Deutsches Museum, München, Archiv- FA014/B494/44 TD5224 [CD74755]

16

Liquid oxygen transport trailer
Transport-Anhänger für flüssigen Sauerstoff

16 For the transport of liquid oxygen (LOX temperature -183 °C) to the firing position, tankers with specially insulated containers were developed. On a special trailer manufactured from the Waggon- und Maschinenfabrik AG, formerly Busch in the city of Bautzen, a frame with an outer container and a tank with a volume of 5,000 liters was mounted suspended on chains inside the container. For insulation purposes the space between the tank and the outer container was filled with Iporka*.

* *Iporka* (trade name) is a plastic foam made from formaldehyde resin and urea.

16 Für den Transport von flüssigem Sauerstoff (Temperatur -183 °C) zur Feuerstellung wurden Tankfahrzeuge mit besonders isolierten Behältern entwickelt. Auf einen Spezial-Anhänger der Waggon- und Maschinenfabrik AG, vormals Busch, Bautzen wurde ein Gerüst mit einem Außenbehälter aufgesetzt und ein Tank mit 5.000 l Volumen an Ketten im Behälter-Innenraum eingehängt. Zur Isolierung wurde der Raum zwischen Tank und Außenbehälter mit Iporka* befüllt.

* *Iporka* (Handelsname) ist ein Schaumstoff, der aus Formaldehydharz und Harnstoff hergestellt wurde.

BArch, RH8II Bild-B0745-42BSM, o.Ang.

17

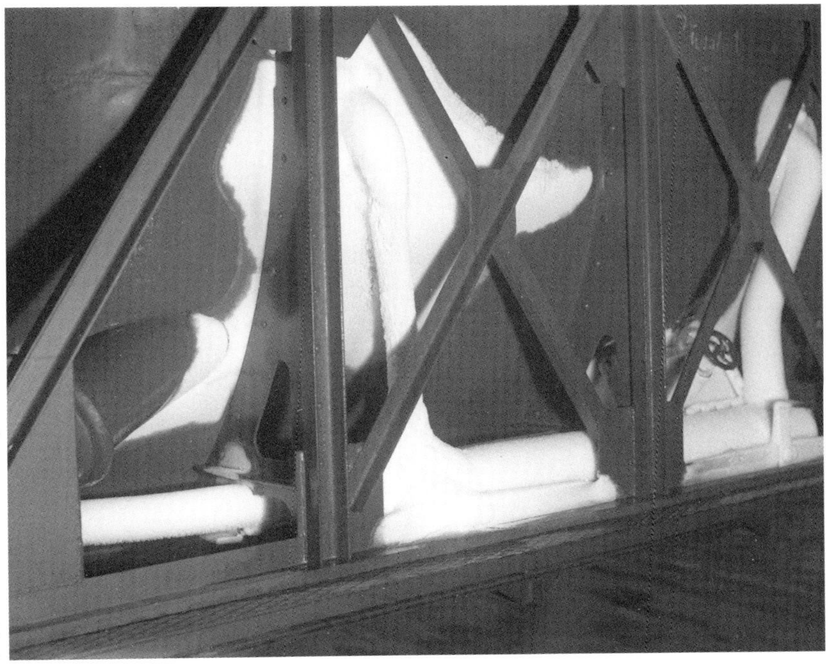

BArch, RH8II Bild-B0744-42BSM, o.Ang.

18

BArch, RH8II Bild-B0747-42BSM, o.Ang.

19

17-19 On one of the first A-Stoff trailers, the pipes were still routed along the outer container. Inadequate insulation led to considerable ice build-up. On later built Fz.401 models and its successor the model Fz.402, the pipes were located in the space between the outer container and inner tank. The outlet overflow valve (picture on the right) shows a clear ice build-up.

17-19 An einem der ersten A-Stoff-Anhänger wurden die Mantelrohre noch am Außenbehälter entlang geführt. Mangelhafte Isolierung führte zu erheblichem Eisansatz. Bei später gebauten Fz.401 und dessen Nachfolgemodell Fz.402 wurden die Mantelrohre in den Zwischenraum von Außenbehälter und Innentank verlegt. Das Auslass-Überlaufventil (Bild rechts) zeigt einen deutlichen Eisaufbau.

Deutsches Museum, München, Archiv- FA014/B1057/43 TD5224 [CD74959]

20

T-Stoff transport vehicle
Transportfahrzeug für T-Stoff

20 The round tank (approx. 2.000 l) of the early T-Stoff tanker was mounted on the frame of an Opel Blitz 3 t truck chassis with tension straps. Directly behind the cab is a water tank with hand pump for flushing and neutralizing the T-Stoff. Behind the windshield at left and right are the markings of training battery »c«, a white »X« and a white serial number on a black background.

20 Der runde Tankbehälter (ca. 2.000 l) des frühen T-Stoff-Tankers wurde mit Spanngurten auf den Rahmen eines Opel Blitz 3 t Lkw-Fahrgestells montiert. Direkt hinter dem Führerhaus war ein Wassertank mit Handpumpe zum Spülen und zur Neutralisierung von T-Stoff befestigt. Hinter der Windschutzscheibe sind links und rechts die Kennzeichen der Lehrbatterie »c«, ein weißes »X« und eine weiße Zählnummer auf schwarzem Grund aufgemalt.

21 A housing for the fittings is mounted on the tank. Here, on this point, hose connections for refueling and a hand pump for pumping T-substance are installed. In storage boxes, hoses and tools were stored on the vehicle. Mounted on the tank on the driver side is a hose container and driving ramps. The later T-Stoff vehicle Fz.405 had an oval tank diameter.

21 Am Tankkessel ist ein Armaturengehäuse angeschraubt. An dieser Stelle sind Schlauchanschlüsse zum Betanken und eine Handpumpe zur Förderung von T-Stoff eingebaut. Schläuche und Werkzeug sind in den Staukästen am Fahrzeug untergebracht. Auf der Fahrerseite sind ein Schlauchbehälter und Auffahrrampen am Tankkessel befestigt. Der spätere T-Stoff Wagen Fz.405 hatte einen ovalen Tankdurchmesser.

PRODUCTION
FERTIGUNG

BAMA, RH 8/1954
HAP11
Karlshagen, 5.12.1944
Bd.Nr.E1981/43 gK
Lagebericht A4

With progress in its development and the successful launch of an A4 rocket on 3 October 1942, the production of the rocket was again given the highest priority by the high command. The few ground vehicles produced to date, as defined for a FR-Abteilung (mot) in December 1942, were far from being available in the planned quantities by the end of 1943. Only the Meiller wagons and launch tables were available in small quantities. With the establishment of the special purpose artillery Abteilungs (mot), the production of the ground vehicles became a top priority for Dornberger and Wernher von Braun.

After the RAF's heavy air raid on *Peenemünde* in mid-August 1943 the search for alternative production facilities for series production of the rocket and ground vehicles was intensified. The relocation to the Harz (Kohnstein*) was pushed forward. Under the leadership of the SS**, Kohnstein was quickly expanded into a underground production facility*** for A4 production without regard to resources. A few railway tunnels (Lager Rebstock) in the Ahr valley were used for ground vehicle production. The first vehicles were completed there in autumn 1943 and handed over to the newly established Abteilungs.

Constant changes to the *Gerät* also had a considerable impact on the ground vehicles. The needed circuit changes, changes to the installations, loading plans, and changes to the scope of equipment all led to delays in deliveries. Changes from mot.I to mot.II and the future mot.III operation of their planning and modifications to the vehicles and placing into service at the beginning of 1944 were already being dealt with.

* The plant was owned by the Wirtschaftliche Forschungsgesellschaft (Wifo) (Economic Research Society).
** Accountable SS-Gruppenführer and Generalleutnant Dr. Hans Kammler
*** Mittelwerk GmbH

Mit Fortschritt in der Entwicklung und dem erfolgreichen Start einer A4-Rakete am 3. Oktober 1942 wurde von der obersten Führung der Fertigung der Rakete wieder die höchste Priorität gewährt. Die wenigen bisher gefertigten Bodenfahrzeuge, gemäß Festlegung für eine FR-Abteilung (mot) vom Dezember 1942, waren bei weitem nicht in den geplanten Stückzahlen bis Ende 1943 verfügbar. Lediglich der Meillerwagen und die Starttische standen in geringen Stückzahlen zur Verfügung. Mit der Aufstellung der Artillerie Abteilungen (mot) wurde die Fertigung der Bodenfahrzeuge zur Chefsache von Dornberger (B.z.b.V) und Wernher von Braun erklärt.

Nach dem schweren Angriff der Royal Air Force (RAF) auf Peenemünde Mitte August 1943 verstärkte man die Suche nach alternativen Fertigungsstätten für die Serienfertigung der Rakete und der Bodenfahrzeuge. Die Verlagerung in den Harz (Kohnstein*) wurde vorangetrieben. Ohne Rücksicht auf Ressourcen wurde in kürzester Zeit unter Führung der SS** der Kohnstein zur Untertage-Fertigungsstätte*** für das A4-Programm ausgebaut. Für die Bodenfahrzeuge nutzte man zunächst einige Eisenbahntunnel (Lager Rebstock) im Ahrtal. Dort wurden bereits im Herbst 1943 die ersten Fahrzeuge endausgerüstet und an die neu aufgestellten Art. Abteilungen übergeben.

Ständige Änderungen am »Gerät« hatten auch erhebliche Auswirkungen auf die Bodenfahrzeuge. Die Folge waren Änderungen an der Schaltung, den Geräte-Gestellen, Beladeplänen und am Ausrüstungsumfang. Sie führten zu Verzögerungen in der Auslieferung. Auch befasste man sich bereits mit Änderungen an der Schaltung mot. I nach mot. II und dem zukünftigen mot. III Betrieb sowie an der damit verbundenen weiteren Planung und Änderung der Fahrzeuge bis zur Einführung Ende 1944.

* Eigentümer der Anlage war die Wirtschaftliche Forschungsgesellschaft (Wifo).
** Verantwortlich SS-Gruppenführers und Generalleutnant Dr. Hans Kammler
*** Mittelwerk GmbH

Overview of the special vehicles delivered to Abt.485, Abt.836, Spare Vehicles, Bat.444, SS-Battery 500, Abt.900, Abt.271

– without vehicles for EW and Kdo.-Stelle S

Übersicht der ausgelieferten Sonderfahrzeuge an die Abt.485, Abt.836, Ersatzfahrzeuge, Bat.444, SS-Batterie 500, Abt.900, Abt.271

– ohne EW und Kdo.-Stelle S

Fahrzeug	Bezeichnung	delivered until end of year 1943 bis Ende 1943 ausgeliefert	delivered by end of March 1944 bis Ende März 1944 ausgeliefert	delivered by end of July 1944 bis Ende Juli 1944 ausgeliefert	delivered by end of August 1944 or after modification bis Ende August 1944 ausgeliefert bzw. nach Modifikation	by end of 1944 planned delivery geplante Auslieferung bis Ende 1944	Order total requirements Wa Prüf 11/III Bestellung Gesamtbedarf Wa Prüf 11/III	
Fz.101	Sonder-Kfz.7 m. gep. Spezialaufbau	3	5	18	20	56	56	
Fz.102	FR-Anhänger (S) (3-achs.)	3	27	75*	k.A.	259	317	* 55 pieces cannot be used due to hydraulic problems, send to Kleinbodungen for rework 55 Stück wegen Hydraulikproblemen nicht einsetzbar, kommen zur Nacharbeit nach Kleinbodungen
Fz.103	Abschußplattform Anhänger (1-achs.)	3	15	51	k.A.	201	317	
Fz.104	Masch.Satz 11(Strüver)	3	3	4*	4*	4*	4	* no new production keine Neufertigung
Fz.105	Lkw-2 t (Steyr) als Stromversorgungswagen		6	17	19	90	100	
Fz.106	Lkw 3 t geschl. (305) Prüfwagen I		4	9	9	53		
Fz.107	Lkw 3 t geschl. (305) Prüfwagen III		5	11	11	40	115	
Fz.108	Kabeltrommel Anhänger (2-achs.)	4	4	11	15	39	k.A.	
Fz.109	Turmwagen						*	* cancelled entfallen
Fz.110	Kabeltrommel Anhänger (1-achs.) zu Fz.105		6	17	19	90	100	
Fz.201	Lkw 3 t gl. geschl. (305) f. Bs-Anlage 1		2	6	7	30	k.A.	
Fz.202	Lkw 3 t gl. geschl. (305) f. Bs-Anlage		2	4	7	30	k.A.	
Fz.203	BS-Anhänger, Typ C (1-achs.)						*	* cancelled entfallen
Fz.204	BS-Antennen-Anhänger (1-achs.)		2	7	7	30	k.A.	
Fz.205	Lkw 3 t gl geschl. (Maultier) Ls-Sende-Lkw			2	3*	25	k.A.	* unusable, 4 each in new design to be delivered from 15.8.44 each 1 Fz. per week until mid September unbrauchbar, 4 Stück in neuer Ausführung ab 15.8.44 je 1 Fz. pro Woche bis Mitte September
Fz.206	Lkw 3 t gl geschl. (Maultier) Ls-Geräte-Lkw			2	3*	25	k.A.	* unusable, 4 each in new design to be delivered from 15.8.44 each 1 Fz. per week until mid September unbrauchbar, 4 Stück in neuer Ausführung ab 15.8.44 je 1 Fz. pro Woche bis Mitte September
Fz.207	LS-Ant.-Anhänger (1-achs.)			4	6*	50	k.A.	* unusable, 8 each in new design to be delivered from 15.8.44 each 2 Fz. per week until mid September unbrauchbar, 8 Stück in neuer Ausführung ab 15.8.44 je 2 Fz. pro Woche bis Mitte September
Fz.208	LS-Kontroll-Ant.-Anhänger (1-achs.)			2	3*	25	k.A.	* unusable, 4 each in new design to be delivered from 15.8.44 each 1 Fz. per week until mid September unbrauchbar, 4 Stück in neuer Ausführung ab 15.8.44 je 1 Fz. pro Woche bis Mitte September
Fz.301	Werkstatttrupp f. Geräte u.Kfz. Lkw 4,5 t geschl. 1	7	7	7	7	9	k.A.	
Fz.302	Lkw 4,5 t geschl. 2	7	7	7	7	9	k.A.	
Fz.303	Lkw 4,5 t offen	7	7	7	7	9	k.A.	
Fz.304	Lkw 3 t gl. geschl. (305) f. Ersatzteile			4	4	49	k.A.	
Fz.401	A-Stoff-Anhänger 5 (2-achs.)	28	28	28*	28*	28*	28	* no new production keine Neufertigung
Fz.402	A-Stoff-Anhänger 6 (2-achs.)		25	41	48	243	344	
Fz.403	Kessel-Kw 2.900 l B-Stoff		40	40	42*	42*	40	* production overrun, no new production Überlieferung, keine Neufertigung
Fz.404	Kessel-Kw 3.500 l B-Stoff		72*	130*	68	284	284	* incl. 4 (March 44) 62 (July 44) under acceptance + further 154 each waiting for acceptance inkl. 4 (März 44) 62 (Juli 44) in der Abnahme + weitere 154 Stück warten auf Abnahme
Fz.405	Kessel-Kw 2.100 l T-Stoff		40	40	40	117	117	
Fz.406	Pumpenanhänger Typ A (1-achs.)	14	14	14	15	138	138	
Fz.407	T-Stoff-Vorwärm-Anh. (1-achs.)		13	16	18	72	150	
Fz.408	Lkw 4,5 t geschl. als Sammler-Kw, heizbar		27	16	k.A.	40	k.A.	
Fz.409	Feuerlösch-Anhänger (1-achs.)			30*	k.A.	30*	30*	* Luftwaffen standard extinguishing trailer is used Luftwaffen Standard Löschanhänger wird verwendet
Fz.501	Lkw 3 t gl. geschl. (305)		4	4	k.A.	28	k.A.	
Fz.502	Lkw 3 t gl. geschl. (305) Kabelwagen II		4	4	k.A.	28	k.A.	
Fz.503	Lkw 3 t gl. geschl. (305) el. Ersatzteile		4	4	k.A.	28	k.A.	
Fz.504	schw. Masch.Satz I		7	*	*	*	*	* ersetzt – heeresüblicher Sd.Ah.24 verwendet replaced – Army standard Sd.Ah.24 used
Fz.505	Umformer-Anhänger 220/380 auf 27 V		7	8	k.A.	28	k.A.	
Fz.506	Luftverdichter-Anhänger (2-achs.)		7	9	k.A.	80	80	
Fz.507	Lkw 3 t gl offen Werkzeug und Ersatzg. (Akku)		7	*	*	*	*	* replaced – Army standard Lkw 3 t gl offen used ersetzt – heeresüblicher Lkw 3 t gl offen verwendet
Fz.508	Lkw 4,5 t geschl. als Sammler-Kw, heizbar		4	4	k.A.	66	k.A.	
Fz.509	Ladestation-Anhänger Typ B (2-achs.)	3	3	3	k.A.	34	k.A.	
Fz.510	Lkw 3 t geschl. (305) FR-Ers.Teile (Werkstatt)		4	5	k.A.	29	k.A.	
Fz.511	Zwei-Achs-Anhänger Werkstattwagen		3	3	k.A.	21	k.A.	
Fz.601	FR-transportanhänger (Vidal)		11	200	200*	*	200*	* retrofitting chassis, drawbar and saddle Umrüstung Fahrgestell, Zugdeichsel und Auflagen
Fz.601b	FR-transportanhänger (Vidal)					250	250	
Fz.701	FR-Wagen (E)					12	k.A.	
Fz.702	A-Stoff-Kesselwagen 27			101	k.A.	158	202	
Fz.703	B-Stoff-Kesselwagen			76	k.A.	k.A.	k.A.	
Fz.704	T-Stoff-Kesselwagen			30	k.A.	k.A.	k.A.	

Special Committee A4

To ensure the mass production of the A4 *Gerät* and its ground systems, a Special Committee A4 was formed in mid-January 1943 by the Reichsminister for Armament and Ammunition.

Sonderausschuss A4

Zur Sicherstellung der Massenproduktion des A4-Gerätes und der Bodenanlage wurde Mitte Januar 1943 der Sonderausschuss A4 durch den Reichsminister für Bewaffnung und Munition eingesetzt.

Geheim

Sonderausschuß A4
Leiter: Dir. G. Degenkolb
Berlin-Charlottenburg, Bismarckstr. 112
Telefon: 318121 Fernschreiber: 011435
Vertreter: Dir. H. Kunze
Berlin-Charlottenburg, Bismarckstr. 112
Telefon: 318121 Fernschreiber: 011435

Arbeitseinsatz	Kontingente	Bauten	Einrichtungen u. Betriebsmittel	Energie	Prüf~~stands~~-einrichtungen	Fertigungs-Planung	Rohmaterial	Fahrzeuge u. Zubehör	Treibstoff	Elektr. Geräte	Zelle
Leiter Jaeger	Leiter Beyer	Leiter Scheiblauer	Leiter Reinhardt	Leiter Lehmann	Leiter Dr. Simon	Leiter Stahlknecht	Leiter Roth	Leiter Dr. Schwab	Leiter Dr. Ehmann	Leiter Storch	Leiter Schlosser
Bln.-Charlottenburg 2 Bismarckstraße 112 Telefon: 318121 Drahtw.: Lokwaggon Bln. Fernschrb.: 011435 Bln.-Charlottenburg 9 Bolivar-Allee 22 Telefon: 993651	Bln.-Charlottenburg 2 Bismarckstraße 112 Telefon: 318121 Drahtw.: Lokwaggon Bln. Fernschrb.: 011435 Potsdam-Babelsberg Göringstraße 59 Telefon:	Reichsministerium Speer Bln-Charlottenburg 1 verl. Jebenstr. Baracken am Zoo Telefon: 110052/2290 Drahtw.: – Fernschrb.: – Bln.-Charlottenburg 2 Herderstr. 2 Telefon: 315941	Bln.-Charlottenburg 2 Bismarckstraße 112 Telefon: 318121 Drahtw.: Lokwaggon Bln. Fernschrb.: 011435 Bln.-Charlottenburg Pension Bünder Giesebrechtstr. 1 Telefon: 322074	Deutsche Reichsbahn Berlin SW11 Hallesches Ufer 76 Telefon: Ortsv. 195271/31493 Fernv. 196901/31493 Drahtw.: [illegible] Fernschrb.: K1 1124 od. 1562 [illegible] Bln.-Lichterfelde Ost Bismarckstraße 10a Telefon: 732384	Deutsche Telefonwerke u. Kabelindustrie A.G. Berlin SO 36 Zeughofstr. 6-10 Telefon: 680011 Drahtw.: Fernschrb.: Bln.-Charlottenburg 9 Insterburg-Allee 20 Telefon: 992192	HAP TDS Telefon: HAP 264-266/141 Drahtw.: HAP a. U. Fernschrb.: über OKH an HBPE HAP a. U. Haus Hubertus 5 Telefon: HAP 264/383	Henschel und Sohn Kassel Henschelstraße 2 Telefon 24151 Drahtw.: Fernschrb.: 03224 Kassel Weinbergstraße 14 Telefon: 33066	Wa Prüf 11 Bln.-Schmargendorf Kranzerstraße 3 Telefon: 897631/182 Drahtw.: OKH-Wa Prüf 11 Fernschrb.: 1433 Bln.-Charlottenburg 9 Bayernallee 19a Telefon: 990820	Wa 7 Rü Mun 3 Berlin W35 Tirpitzufer 38/40 Telefon: Ortsv. 218191/62311 Fernv. 218091/62311 Drahtw.: OKH Mun 3 Fernschrb.: 1322 (OKH) Berlin W50 Eislebenerstraße 12 Telefon: 241289	Siemens u. Halske Berlin-Siemensstadt Wernerwerk T Telefon: 340010/1028 Drahtw.: Wernerwerk T Bln. Fernschrb.: 011201 Berlin Zehlendorf-West Wolzogenstraße 6 Telefon: 851933	Daimler Benz Sindelfingen/Wttbg. Telefon: Amt Stuttgart 91156/57 Drahtw.: Daimlerbenz Fernschrb.: 06941 Stuttgart/Vaihingen Wörthstraße 9 Telefon: Stuttgart 789324
Vertreter Schulte	Vertreter Dobler	Vertreter Neubauer	Vertreter Poenicke	Vertreter Becker	Vertreter Dr. Kühle	Vertreter Dellmaier	Vertreter Voy	Vertreter Pflugk	Vertreter Dr. Holl	Vertreter Dr. Kühle	Vertreter Maurer
Mitarbeiter	Mitarbeiter	Mitarbeiter	Mitarbeiter	Mitarbeiter	Mitarbeiter	Mitarbeiter	Mitarbeiter	Mitarbeiter	Mitarbeiter	Mitarbeiter	Mitarbeiter
Schulte Wa Wessling S.u.H.	Matschky Wa J Rü Mun 3	Matschky Wa J Rü Mun 3 Maurer Wa Chef Ing 4 Neubauer Wa Prüf 11 Thom Gottron Min. Sp. Jebenstr.	Martin HAP VW Dr. Engel Demag Maurer Wa Chef Ing 4	Elektrizitätsfragen. Planungen und Neubau: Dr. Bock Generalinspektor f. Wasser u. Energie Elektrizitätsfragen für lfd. Versorgung: Goerke Reichsstelle f. Elektrizitätswirtschaft Gas- und Wasserfragen: Knappe Generalinspektor f. Wasser u. Energie Kohlenfragen: Heil Reichsstelle f. Kohle Abt. Wehrmacht Sauerstoffbelieferung f. d. Fertigung des Gerätes Kalbfleisch Vereinigte Sauerstoffwerke Matschky Wa J Rü Mun 3	Dr. Thiel HAP Dr. Schilling HAP Gegelbach II HAP Arbeitsunterausschuß Maschinen, Schalttafeln usw. Leiter Still Fa. Still Arbeitsunterausschuß Messgeräte, Tintenschreiber, Druckmesser usw. Leiter Dr. Georgi AEG FKU	Gruppe: Geräte Obmann: Krüger HAP NB Mitarb.: Kruck HAP VW de Vries HAP EW Heim II Daiml. Benz Sindelfingen Poenicke Wa Malauschek Wa Gruppe: Betriebsmittel Obmann: ~~Maasing~~ HAP NB Mitarb.: ~~Martin HAP VW~~ Kowall HAP EW ~~Petzold Daiml. Benz~~ Sindelfingen Poenicke Wa Malauschek Wa	Hartmann R.Z.A. Berlin Becker Duisburg	Nobel Munimin (Normal-Fahrzeuge) Rahm Lindner Meiller (Spezial-Fahrzeuge) Still Fa. Still (Sonderaufbauten f. elektr. Teile) Dr. Poetter Mako-Werke Dr. Schlaich Ziemann (Sonderaufbauten f. Spezial-Behälteraufbauten) Riedel II HAP EW (Allgemeine Fragen)	Dr. Thiel HAP EW	Betz Wa Chef Ing 4 Arbeitsunterausschuß Hochfrequenzgeräte Leiter: Storch Siemens Vertr.: Dr. Bocker Siemens Arbeitsunterausschuß Elektro-Mechan.-Geräte Leiter: ~~Wilde~~ Askania Vertr.: Tränkner LGW Arbeitsunterausschuß Elektrisches Zubehör Leiter: Still Fa. Still Vertr.: Dr. Möller Stotz-Kontakt Arbeitsunterausschuß Elekr. Prüf-Geräte	Maurer Wa Chef Ing 4 (Maschinen u. Geräte) Kruck HAP VW Miller Friedrichshafen (Mittelteile) Roth Roth Elektrik München (Fragen d. Schweißtechnik) EMTER (VOSS) (Gerätespitze) Deppenbrock HAP EW (Materialprüfungsfragen) ~~Rehs~~ HAP

22

Auszug aus Schreiben
Sonderausschuß A4
Beim Reichsminister für
Beschaffung und Munition
Direktor Degenkolb
Berlin-Charlottenburg 2
29. Januar 1943
Schreiben an Herrn Dr. v. Braun

Arbeitsausschüsse

	Turbag	Tube	Zulieferungen	Rationalisierung	Verlagerung	Zusammenbau
	Leiter Radtke	Leiter Föckeler	Leiter Wiedenmann	Leiter Rees	Leiter Figge	Leiter Stahlknecht
	Mabag Nordhausen/Harz Telefon: 1570 Drahtw.: Radtke Mabag Nordhausen Nordhausen/Harz Kneiffstr. 6 Telefon: 1533	M A N Mainz / Gustavsburg Telefon: 41241 Drahtw.: Manwerk Mainz/Gustavsburg Fernschrb.: 03482 Mainz/Gustavsburg Jahnstr. 4 Telefon: 41241	HAP Z EK Telefon: HAP 262/153 Drahtw.: HAP a. U. Fernschrb.: über OKHanHBPE Zinnowitz a. U. Haus Meeresblick Telefon: Zinnow. 385	HAP Telefon: HAP 264/171 Drahtw.: HAP a. U. Fernschrb.: über OKHanHBPE HAP a. U. Haus Godenwind Telefon: HAP 264/385	Bln.-Charlottenburg Bismarckstraße 112 Telefon: 318121 Drahtw.: Lokwaggon Bln. Fernschrb.: 011435	H A P TDS Telefon: HAP 264-266/141 Drahtw.: HAP a. U. Fernschrb.: über OKHanHBPE HAP a. U. Haus Hubertus 5 Telefon: HAP 264/383
	Vertreter ~~Malauschek~~ [illegible]	Vertreter Malauschek	Vertreter Stahlknecht	Vertreter Ohrnberger	Vertreter Dr. Müller I	Vertreter Rudolph
	Mitarbeiter	Mitarbeiter	Mitarbeiter	Mitarbeiter	Mitarbeiter	Mitarbeiter
	~~Malauschek~~ Wa Chef Ing 4 Maurer Wa Chef Ing 4 [illegible]	Tietz Weigelwerke Kohls Krupp Berndorf Wirth Ziemann Haensch Schlegel Leipzig Hallesche Masch. F. A. u. C. Kosik Ardeltwerke Junker u. Ruh Graud. Wende HAP EW Dr. Thiel HAP EW Grabowsky HAP EW Deppenbrock HAP EW [illegible]	Herterich Daimler Benz Sindelf. (Heck) Rehm Friedrichshafen (Mittelteile) Stuhlfauth HAP VW	Riedel III HAP EW Kruck HAP VSW Reinhardt S.A. A IV Miller Friedrichshaf. [illegible]	Stahlknecht HAP TDS Stuhlfauth HAP VW Dr. Müller I HAP	Eckener Friedrichshafen (Mittelteile) Heim II Daimler Benz Sindel. (Heck) ~~Martin HAP VW~~ Poenicke Wa (Pumpen) (Spitze) [illegible] (Triebwerke) [illegible]

	Endabnahme	Transport	Statistik	Sauerstoff	Armaturen	Funkgeräte
	Leiter Dr. v. Braun	Leiter Koesters	Leiter Norpoth	Leiter Matschky	Leiter Lindenmayr	Leiter
	HAP Telefon: HAP 262-264 Drahtw.: HAP a. U. Fernschrb.: über OKHanHBPE HAP a. U. Haus Godenwind Telefon: HAP 262-264	Bln.-Charlottenburg Bismarckstraße 112 Telefon: 318121 Drahtw.: Lokwaggon Bln. Fernschrb.: 011435 Berlin W 15 Düsseldorferstraße 41 Telefon: 911841	Bln.-Charlottenburg Bismarckstraße 112 Telefon: 318121 Drahtw.: Lokwaggon Bln. Fernschrb.: 011435 Berlin N.W 87 Elberfelderstraße 29 Telefon: 397314	Wa J Rü Mun 3 Berlin W 35 Tirpitzufer 38/40 Telefon: Ortsv. 218191/62738, Fernv. 218091/62738 Berlin-Wilmersdorf Rudolstädterstr. 127 Telefon: 87 2085	HAP TBA 3 Telefon: HAP 264/231 Drahtw.: HAP a. U. Fernschrb.: über OKHanHBPE HAP a. U. Telefon: HAP 264/7034	
	Vertreter Dr. Thiel	Vertreter Hartmann	Vertreter Pliquet	Vertreter	Vertreter Ehrlicher	Vertreter
	Mitarbeiter	Mitarbeiter	Mitarbeiter	Mitarbeiter	Mitarbeiter	
	Reinhardt HAP Maganja Friedrichsh. Strohmeyer Friedrichsh. Wackernagel HAP VSW Dr. Schilling HAP EW Maetzke HAP EW Ludewig HAP EW Steinecke HAP EW	Riedel II HAP EW		Stolle Heylandt Sedlmayer Heylandt Hüter HAP Utpatel HAP Funk Wa J Rü Mun 3		

Reichsminister Speer has appointed and instructed to lead this special committee and to find a solution in the continuation of the war. An extremely important task of personal responsibility handed over into the trustworthy hands of the industry involved. In fulfillment of this I ask you* for your message and at the same time call on you to the head of the final acceptance working committee.

Sign: Degenkolb

* Remark by author – means Dr. v. Braun

Herr Reichsminister Speer hat mich mit der Führung dieses Sonderausschusses beauftragt und damit die Lösung einer für die weitere Kriegsführung außerordentlich wichtigen Aufgabe der Selbstverantwortung der beteiligten Industrie zu treuen Händen übergeben. In Erfüllung dieser Aufgabe bitte ich Sie* um ihre Nachricht und berufe Sie zugleich zum Leiter des Arbeitsausschusses Endabnahme.

Gez.: Degenkolb

* Anmerkung des Autors – gemeint ist hier Dr. v. Braun

LAGER REBSTOCK
LAGER REBSTOCK

Schriftverkehr mit
Wolfgang Gückelhorn

BAMA, RH8/1267
Chef H Rüst u. BdE
o.O, den 4.3.1944

Chef H Rüst u BdEAz:
79 Wa Prüf 10 TD5224
Bd.Nr.. E756/44 gk
Photographische
Aufnahmen
von Lager Rebstock

Under the code name »Lager Rebstock«, the Gollnow & Sohn company used from October 1943 to mid-December 1944 , all five tunnels of the unfinished railroad of the Rheinbach-Ringen-Rech [1] cross connection into the Ahr valley to the Remagen-Duempelfeld railroad line (the Silberberg, Kuxberg, Trotzenberg, Sonderberg and Herrenberg tunnels) for the final outfitting of A4 ground vehicles. The required empty vehicles were provided by the manufacturers Opel, Meiller, Lindner and other companies via the army ordnance depots, and delivered to Ahrweiler-Marienthal. Under difficult working conditions, the final equipment, the installation of switchboards, equipment cabinets, electrical wiring, generators and final mechanical modifications and fixtures, was carried out under the guidance of EW specialists and lead hands from Gollnow & Sohn and by requisitioned workers from the Ahrweiler district. Between August 26 and December 13, 1944, forced laborers, including up to 213 prisoners from the Buchenwald concentration camp on duty as production assistants, took over the skilled work in the final finishing. [2] Front-line use of the A4 (V2) was delayed until early September 1944 because technical problems could not be solved until the summer during firing tests in Blizna. As a result, production of A4 ground equipment at Lager Rebstock ran at low capacity. After the successful first mission on September 8, 1944, it was ordered to equip all 8 batteries with three firing platoons each. Now Gollnow & Sohn had to ramp up production. Missing parts (bottleneck parts) and constant modifications led to delays in the delivery of the vehicles. The acceptance took place at the EW plant and the army acceptance station on site. Until the transfer of the final equipment in mid-December 1944 to the new location at Artern (code name Adorf) in the Mittelwerk area, many of the vehicles delivered so far had to be returned to Rebstock in order to implement final modification requests. It was not until the beginning of December 1944 that all 24 firing platoons were completely equipped.

Unter dem Decknamen »Lager Rebstock« nutzte die Firma Gollnow & Sohn für die Endausrüstung der A4-Bodenfahrzeuge von Oktober 1943 bis Mitte Dezember 1944 alle fünf Tunnels der nicht fertiggestellten Trasse der Querverbindung Rheinbach-Ringen-Rech [1] in das Ahrtal zur Bahnstrecke Remagen-Dümpelfeld (Silberberg-, Kuxberg-, Trotzenberg-, Sonderberg- und Herrenbergtunnel). Benötigte Leerfahrzeuge wurden von den Herstellerfirmen Opel, Meiller, Lindner und weiterer Firmen über die Heereszeugämter bereitgestellt und nach Ahrweiler-Marienthal angeliefert. Unter schwierigen Arbeitsbedingungen erfolgte unter der Anleitung von EW-Spezialisten, Vorarbeitern der Firma Gollnow & Sohn und von dienstverpflichteten Arbeitskräften aus dem Kreis Ahrweiler die Endausrüstung, der Einbau von Schalttafeln, Geräteschränken, elektrischer Verkabelung, Generatoren und letzter mechanischer Änderungen und Einbauten. Zwischen dem 26.08. und dem 13.12.1944 übernahmen Zwangsarbeiter, darunter bis zu 213 Häftlinge aus dem KZ-Buchenwald als Produktionshelfer im Einsatz, die qualifizierten Arbeiten in der Endausrüstung. [2] Der Fronteinsatz der A4 (V2) verzögerte sich bis Anfang September 1944, weil technische Probleme erst im Sommer bei Schießversuchen in Blizna gelöst werden konnten. Deshalb lief die Produktion der A4-Bodenanlagen in Lager Rebstock nur mit geringer Auslastung. Nach dem erfolgreichen Ersteinsatz am 08.09.1944 wurde befohlen, alle acht Batterien mit je drei Schießzügen auszustatten. Nun musste Gollnow & Sohn die Produktion hochfahren. Fehlteile (Engpassteile) und ständige Änderungen führten zu Verzögerungen bei der Fahrzeugauslieferung. Die Abnahme erfolgte durch das EW-Werk und der Heeres-Abnahmestelle vor Ort. Bis zur Verlagerung der Endausrüstung Mitte Dezember 1944 an den neuen Standort Artern (Deckname Adorf) im Bereich des Mittelwerkes mussten viele der bisher ausgelieferten Fahrzeuge wieder nach Rebstock zurückgeführt werden um letzte Änderungswünsche zu realisieren. Erst Anfang Dezember 1944 waren alle 24 Schießzüge komplett ausgerüstet.

23

Fz.101 Armored fire control vehicle
Fz.101 Feuerleitpanzer

23 On three armored fire control vehicles Fz.101 the attachments and fittings in the interior of the armored body, mounted on the chassis of the medium half-track, 8 t KM 11m (Sd.Kfz.7) were completed. Furthermore installations, control and control panels from the production of Gollnow & Sohn (G+S) as well as a telephone connection were mounted.

23 An drei Feuerleitpanzern Fz.101 werden die An- und Einbauten vervollständigt. Im Innenraum des gepanzerten Aufbaus, der auf das Fahrgestell des mittleren geländegängigen Zugkraftwagen m. Zgkw. 8 t KM 11m (Sd.Kfz.7) montiert war, rüstete man die Einbauten, Schalt- und Bedienpulte aus der Fertigung von Gollnow und Sohn (G+S) sowie eine Fernsprechvermittlung ein.

24

Fz.102 FR-trailer (S) (3-axle)
Fz.102 FR-Anhänger (S) (3-achs.)

24 The »Meillerwagen« or FR-trailer were manufactured as empty vehicles by the companies of Meiller, Lindner and Schenk. Here a »Meillerwagen« type MA4-3 from the production of the Lindner Company is worked on. After completing and modifying the mechanical special equipment, the wiring and socket for 104 pin (Flaksteckdose104-polig) anti aircraft cable 108 x 0.5 mm^2 were installed.

24 Die »Meillerwagen« wurden als Leerfahrzeuge von den Firmen Meiller, Lindner und Schenk hergestellt. Auf dem Bild wird ein »Meillerwagen« Typ MA4-3 aus der Fertigung der Firma Lindner ausgerüstet. Nach Ergänzung und Änderungen an der mechanischen Sonderausrüstung wurde die Verkabelung 108 x 0,5 mm^2 und die 104-poligen Flaksteckdosen montiert.

Deutsches Museum, München, Archiv- FA014/B1246/43 TD5224 [CD74767]

25

Fz.104 Machine set 11 (Strüver)
Fz.104 Maschinensatz 11 (Strüver)

25 G+S were responsible for all installations including the »Citomat« switchboard. Here three trailers are equipped at the same time. These vehicles are at the Lager Rebstock for refitting and are being re-equipped there. They are part of the inventory of the training and test battery. The »Strüverwagen« was built only in small numbers and later replaced by the Fz.105. The special circuit on the Fz.104 had to be tuned with the switchboard in the Fz.101 *Feuerleitpanzer*, in a so-called mot.I circuit.

25 Für die Fahrzeug-Einbauten mit Telefonzentrale »Citomat« war ebenfalls G+S verantwortlich. Auf dem Bild zu sehen ist die gleichzeitige Ausrüstung von drei Anhängern. Die Fahrzeuge befinden sich zur Umrüstung im Lager Restock und werden dort neu ausgerüstet. Sie gehören zum Bestand der Lehr- und Versuchsbatterie. Der »Strüverwagen« wurde nur in geringer Stückzahl gebaut und später durch das Fz.105 ersetzt. Die spezielle Schaltung am Fz.104 musste mit dem Schaltpult im Fz.101 Feuerleitpanzer abgestimmt werden, der sog. mot.I-Schaltung.

Deutsches Museum, München, Archiv- FA014/B1245/43 TD5224 [R0159_7]

26

Deutsches Museum, München, Archiv- FA014/B1248/43 TD5224 [CD74768]

27

Fz. 108 Cable reel trailer
Fz. 108 Kabeltrommelanhänger

26 Mobile generators were mounted on a workbench and tested. A lead hand (with light coat) monitors the activities of the workers.

27 G+S modified a trailer (5 t) with bodywork (manufactured by: Buhne) with slots for the cable reels and equipped them with 56 cable reels used for current-, FF- and FL-cables (total ca. 2.850 m). In addition to new equipment, older cable trailers were also delivered for refitting and re-equipment. Here, two cable trailers of an older design from the inventory of the training and test battery are being processed; in front of them, two new vehicles are lined up. The laying cart is part of the delivery, two of these laying carts were carried disassembled on the rear of vehicle 108.

26 Mobile Generatoren werden auf einer Werkbank montiert und getestet. Ein Vorarbeiter (mit hellem Mantel) überwacht die Aktivitäten der Arbeiter.

27 Gollnow und Sohn (G+S) baut auf einen Anhänger (5 t) mit Aufbau (Hersteller Fa. Buhne) Einschübe für die Kabel- und Telefonrollen ein. Das Fahrzeug war mit 56 Kabeltrommeln für Strom-, FF- und FL-Kabel (insgesamt ca. 2.850 m) ausgerüstet. Neben Neuausrüstungen wurden auch Kabeltrommelanhänger zur Umrüstung angeliefert und neu ausgerüstet. Hier sind zwei Kabelwagen einer älteren Ausführung aus dem Bestand der Lehr- und Versuchsbatterie in Bearbeitung, davor stehen in der Reihe zwei Neufahrzeuge. Der Verlegewagen gehört zum Lieferumfang, zwei dieser Verlegewagen wurden zerlegt am Heck von Fahrzeug 108 mitgeführt.

Fz.509 Charging Station Trailer Type B
Fz.509 Ladestation-Anhänger Typ B

28 A Luftwaffe special trailer type B2 with a closed body is converted with parts from G+S into a loading station. The structure is divided in two rooms. One for the lead collectors and one for the NC collectors. Racks with various test devices, power switches and ammeters are part of G+S's final equipment.

28 Ein Sonderanhänger B2 der Luftwaffe mit geschlossenem Aufbau wird mit Einbauten von G+S zum Ladestation-Anhänger umgerüstet. Der Aufbau ist in zwei Räume getrennt, jeweils für die Blei-Sammler und NC-Sammler unterteilt. Gestelle mit diversen Prüfgeräten, Netzschalter und Amperemeter gehören zum Umfang der Endausrüstung von G+S.

GROUND PRODUCTION SERVICE
FERTIGUNGSDIENST BODEN

DMA, GD639.5.21
TD1
E1745/44g
Karlshagen,
29.2.1944
Auszug

Directive – TD1
Ref.: Ground production service
sign: v. Braun

In order to ensure proper and on-schedule completion of the special vehicles for Batt. mot. and techn. Batt., the »ground production service« is introduced as a new institution. In order to relieve the developed units of production tasks as far as possible, the tasks associated with production essentially pass into the hands of those departments which carry out the testing of the ground equipment.

The main tasks of the ground production service are the following:

a) Preparation and supervision of the proper and punctual production process of the vehicles with FR ground equipment.
b) Inspection of the finished vehicle. Prepare the examination of the inspection by the army approval office.
c) Processing and implementation of proposed modifications.
d) Processing of all other questions in connection with vehicles occur with FR ground equipment.

Dr. Mahlo, TD507, is in charge of all questions relating to the ground production service. At the same time, the TD507 is the liaison office to the HAP/Ia.

The following gentlemen will be appointed as the administrator for the individual vehicle groups or vehicles.

Direktive – TD1
Betreff: Fertigungsdienst Boden
gez: v. Braun

Um eine ordnungsgemäße und termingerechte Fertigstellung der Sonder-Kfz. für Batt. mot. und techn. Batt. sicherzustellen, wird als Neueinrichtung der »Fertigungsdienst Boden« eingeführt. Um die entwickelten Stellen nach Möglichkeit von Fertigungsaufgaben zu entlasten, gehen die mit der Fertigung verbundenen Aufgaben im Wesentlichen in die Hände derjenigen Stellen über, die die Erprobung der Bodenanlagen durchführen.

Die wesentlichen Aufgaben des Fertigungsdienstes Boden sind die folgenden:

a) Vorbereitung und Überwachung des ordnungsgemäßen und termingerechten Herstellungsganges der Kfz. mit Bodengerät FR.
b) Prüfung der fertigen Kfz., Vorbereitung der Übernahme der Prüfung durch die Heeresabnahme.
c) Bearbeitung und Verwirklichung von Änderungsvorschlägen.
d) Bearbeitung aller sonstigen Fragen, die im Zusammenhang mit Kfz. mit Bodengerät FR auftreten.

Die Federführung in allen Fragen des Fertigungsdienstes Boden liegt bei Dr. Mahlo, TD507, gleichzeitig ist die TD507 Verbindungsstelle zum HAP/Ia.

Als Sachbearbeiter für die einzelnen Fahrzeuggruppen bzw. Fahrzeugen werden die folgenden Herren bestimmt.

List of administrator per vehicles groups
Liste der Sachbearbeiter je Fahrzeuggruppe

Fahrzeuggruppe 1	**Feuerstellung**	**Firing position**	**Administrator Sachbearbeiter**	Ing. Vogt	TD531
	101	Sonder-Kfz.7 m.gep. Spezialaufbau		Ing. Staadt	TD531
	102	FR-Anhänger (S) (3-achs.)		Ing. Heinisch	z. Zt. WaPrüf 11/III
	103	Abschußplattform Anhänger (1-achs.)		Ing. Heinisch	z. Zt. WaPrüf 11/III
	104	Masch.Satz 11 (Strüver)		Dipl.-Ing. Pauls	TD531
	105	Lkw 2 t (Steyr) als Stromversorgungswagen		Dipl.-Ing. Pauls	TD531
	106	Lkw 3 t geschl. (305) Prüfwagen I		Ing. Decker / Ing. Edler	TD5313 / TD41
	107	Lkw 3 t geschl. (305) Prüfwagen III		Ing. Decker / Ing. Edler	TD5313 / TD41
	108	Kabeltrommel Anhänger (2-achs.)		Dipl.-Ing. Pauls	TD531
Fahrzeuggruppe 2	**Fernlenkstellung**	**Remote controll position**	**Administrator Sachbearbeiter**	Dipl.-Ing. Hoberg	TD531
	201	Lkw 3 t gl. geschl. (305) f. BS-Anlage 1		Dipl.-Ing. Scheibe	TD5313
	202	Lkw 3 t gl. geschl. (305) f. BS-Anlage		Dipl.-Ing. Scheibe	TD5313
	203	BS-Anhänger, Typ C (1-achs.)		Dipl.-Ing. Scheibe	TD5313
	204	BS-Antennen-Anhänger (1-achs.)		Dipl.-Ing. Scheibe	TD5313
	205	Lkw 3 t gl geschl. (Maultier) LS-Sende-Lkw		Dipl.-Ing. Handl	TD5313
	206	Lkw 3 t gl geschl. (Maultier) LS-Geräte-Lkw		Dipl.-Ing. Handl	TD5313
	207	LS-Ant.-Anhänger (1-achs.)		Dipl.-Ing. Handl	TD5313
	208	LS-Kontroll-Ant.-Anh. (1-achs.)		Dipl.-Ing. Handl	TD5313
Fahrzeuggruppe 3	**Protzenstellung**	**Limber position**	**Administrator Sachbearbeiter**	Ing. Edler	TD41
	301	Werkstatttrupp f. Geräte u.Kfz. Lkw 4,5 t geschl. 1		Uffz. Streit	z. Zt. WaPrüf 11/III
	302	Lkw 4,5 t geschl. 2		Uffz. Streit	z. Zt. WaPrüf 11/III
	303	Lkw 4,5 t offen		Uffz. Streit	z. Zt. WaPrüf 11/III
	304	Lkw 3 t gl. geschl. (305) f. Ersatzteile		Ing. Schwedes	TD 5313
Fahrzeuggruppe 4	**T-Kolonne**	**Fuel column**	**Administrator Sachbearbeiter**	Ing. Edler	TD41
	401	A-Stoff-Anhänger 5 (2-achs.)		Ing. Nettesheim	TD41
	402	A-Stoff-Anhänger 6 (2-achs.)		Ing. Nettesheim	TD41
	403	Kessel-Kw 2.900 l B-Stoff		Ing. Busekros	TD41
	404	Kessel-Kw 3.500 l B-Stoff		Ing. Busekros	TD41
	405	Kessel-Kw 2.100 l T-Stoff		Ing. Busekros	TD41
	406	Pumpenanhänger Typ A (1-achs.)		Ing. Busekros	TD41
	407	T-Stoff-Vorwärm-Anh. (1-achs.)		Ing. Nettesheim	TD41
	408	Lkw 4,5 t geschl. als Sammler-Kw, heizbar		wird bei Fahrzeuggruppe 5 bearbeitet	TD41
Fahrzeuggruppe 5	**Technische Batterie**	**Technical battery**	**Administrator Sachbearbeiter**	Ing. Fichtner	TD531
	501	Lkw 3 t gl. geschl. (305)		Dipl.-Ing. Pauls	TD531
	502	Lkw 3 t gl. geschl. (305) Kabelwagen II		Ing. Schwedes	TD531
	503	Lkw 3 t gl. geschl. (305) el. Ersatzteile		Ing. Schwedes	TD531
	504	schw. Masch.Satz I		Dipl.-Ing. Pauls	TD531
	505	Umformer-Anhänger 220/380 auf 27 V		Dipl.-Ing. Pauls	TD531
	506	Luftverdichter-Anhänger (2-achs.)		Ing. Papenhausen	TD531
	507	Lkw 3 t gl offen Werkzeug und Ersatzg. (Akku)		Ing. Schwedes	TD531
	508	Lkw 4,5 t geschl. als Sammler-Kw, heizbar		Ing. Schwedes / Ing. Heller	TD531 / TD41
	509	Ladestation-Anhänger Typ B (2-achs.)		Ing. Schwedes	TD531
	510	Lkw 3 t geschl. (305) FR-Ers.Teile (Werkstatt)		Dipl.-Ing. Pauls	TD531
	511	Zwei-Achs-Anh. Werkstattwagen		Ing. Schwedes	TD531

WORKING CONFERENCE GROUND VEHICLES
ARBEITSTAGUNG BODENFAHRZEUGE

At the invitation of Wernher von Braun, all vehicle group administrators met in Karlshagen and Lubmin from 25.3.44 to 1.4.44 for workshops »to clean up the production services of ground vehicles«.

DMA, GD639.5.21
TD1
Bd.Nr. E928/44 g.Kdos.
Karlshagen, 17.3.1944
Auszug
Arbeitstagung
siehe Geheimschreiben
E1745/44g vom 29.2.44

The following were discussed

Vehicle group I – Firing position	
representative Ing. Vogt	TD551
Vehicle group II – Remote control position	
representative Dipl.Ing. Hoberg	TD531
vehicle group III – Limber position	
representative Ing. Edler	TD41
Vehicle group IV – Fuel column	
representative Ing. Edler	TD41
Vehicle group V – Technical battery	
representative Ing. Fichtner	TD531

Remark:
Vehicles which are not mentioned in the subject, i.e. railway vehicles, Vidal trailers, new combustion cut-off and guide beam developments, are not to be discussed.
Signed: v. Braun

Note:
The individual administrators in the vehicle groups each provided a status report on the production situation, defects and remedies, as well as the outlook for the development status of the vehicles concerned. Open questions, requests and suggestions were discussed; any decisions made were recorded in a final meeting and production aligned accordingly.

Auf Einladung von Wernher von Braun trafen sich in der Zeit vom 25.3.44 bis 1.4.44 alle Sachbearbeiter der Fahrzeuggruppen zu Arbeitstagungen »zur Bereinigung der Fertigungsdienste der Bodenfahrzeuge« in Karlshagen und Lubmin.

Besprochen wurden

Fahrzeuggruppe I – Feuerstellung	
Sachbearbeiter Ing. Vogt	TD551
Fahrzeuggruppe II – Fernlenkstellung	
Sachbearbeiter Dipl.Ing. Hoberg	TD531
Fahrzeuggruppe III – Protzenstellung	
Sachbearbeiter Ing. Edler	TD41
Fahrzeuggruppe IV – T-Kolonne	
Sachbearbeiter Ing. Edler	TD41
Fahrzeuggruppe V – Technische Batterie	
Sachbearbeiter Ing. Fichtner	TD531

Anmerkung:
Fahrzeuge, die im Vorgang nicht genannt sind, also Eisenbahnfahrzeuge, Vidalanhänger, Brennschluss- und LS-Neuentwicklungen, wurden nicht besprochen.
Gezeichnet: v. Braun

Bemerkung:
Die einzelnen Sachbearbeiter in den Fahrzeuggruppen gaben jeweils einen Sachstand zur Fertigungslage, Mängel und Behebung und Ausblick zum Entwicklungsstand der betreuten Fahrzeuge ab. Offene Fragen, Wünsche und Anregungen wurden diskutiert; getroffene Festlegungen in einer Schlussbesprechung zu Protokoll genommen und die Fertigung hiernach ausgerichtet.

First Meeting

Tag:	24.3.44 and 25.3.44		
Zeit:	24.3.	at 7.30	Vehicles 101 und 102
	24.3.	at 14.00	Vehicles 104 und 105
	25.3.	at 7.00	Vehicles 106, 107 und 108
	25.3.	at 14.30	Final discussion and minutes of the meeting
Location:	Lager Rebstock		
Betreff:	Vehicle group I (Firing position)		

Second Meeting

Tag:	29.3.44	
Zeit:	at 8.00	Vehicles 201-204 (BS)
	at 14.00	Vehicles 205-08 (LS)
Location:	Lubmin	
Betreff:	Vehicle group II (Remote control position)	

Third Meeting

Tag:	31.3.44	
Zeit:	at 7.30	Vehicles 301-304
Location:	Karlshagen I, Community welfare area Pr. VII.	
Betreff:	Vehicle group III (Limber position)	

Third Meeting (continued)

Tag:	31.3.44
Zeit:	at 14.30
Location:	Karlshagen I, Community welfare area Pr. VII.
Betreff:	Vehicle group IV (Fuel column) vehicles 401-408

Third Meeting (continued)

Tag:	1.4..44	
Zeit:	at 7.30	Vehicles 501-511
	at 14.00	Final discussion and minutes of the meeting
Location:	Karlshagen I, Community welfare area Pr. VII. Karlshagen I, meeting room building 4	
Betreff:	Vehicle group V (technical battery)	

Erste Besprechung

Tag:	24.3.44 und 25.3.44		
Zeit:	24.3.	7.30 Uhr	Fahrzeug 101 und 102
	24.3.	14.00 Uhr	Fahrzeug 104 und 105
	25.3.	7.00 Uhr	Fahrzeug 106, 107 und 108
	25.3.	14.30 Uhr	Schlussbesprechung und Aufnahme des Protokolls
Ort:	Lager Rebstock		
Betreff:	Fahrzeuggruppe I (Feuerstellung)		

Zweite Besprechung

Tag:	29.3.44	
Zeit:	8.00 Uhr	Fahrzeug 201-204 (BS)
	14.00 Uhr	Fahrzeug 205-208 (LS)
Ort:	Lubmin	
Betreff:	Fahrzeuggruppe II (Fernlenkstellung)	

Dritte Besprechung

Tag:	31.3.44	
Zeit:	7.30 Uhr	Fahrzeug 301-304
Ort:	Karlshagen I, Wohlfahrtsraum Pr. VII.	
Betreff:	Fahrzeuggruppe III (Protzenstellung)	

Dritte Besprechung (Fortsetzung)

Tag:	31.3.44
Zeit:	14.30 Uhr
Ort:	Karlshagen I, Wohlfahrtsraum Pr. VII.
Betreff:	Fahrzeuggruppe IV (T-Kolonne) Fahrzeug 401-408

Dritte Besprechung (Fortsetzung)

Tag:	1.4..44	
Zeit:	7.30 Uhr	Fahrzeug 501-511
	14.00 Uhr	Schlussbesprechung und Aufnahme des Protokolls
Ort:	Karlshagen I, Wohlfahrtsraum Pr. VII. Karlshagen I, Sitzungssaal Haus 4	
Betreff:	Fahrzeuggruppe V (technische Batterie)	

Note by author: Minutes of the meetings of each vehicles group I, III, IV and V were assigned to the individual vehicles descriptions with remark. A MoM for vehicle group II was not found.

Anmerkung des Autors: Die einzelnen Protokollpunkte zu den Fahrzeugen der Gruppen I, III, IV und V sind den Fahrzeugbeschreibungen zugeordnet und dort vermerkt. Ein Protokoll zur Fahrzeuggruppe II wurde nicht gefunden.

FINAL MEETING
ABSCHLUSSBESPRECHUNG

46 DMA, GD639.5.18
OKH/Wa Prüf 10
O.U. 1. April 1944
Bd.Nr. 802/44 g.Kdos.

Minutes
Final meeting of the progress of the ground support group

1. General:
The meeting on ground support was very important because some questions were unclear and could not be settled by extensive correspondence but only by discussion of the subject between the tactical command, development division, production and the units.

2. Changes to A, B, C:
Those attending were immediately made responsible for meeting the definite due dates agreed in the meeting, without waiting for the receiving the minutes.

From the specific compilation, the necessary (case A) and desirable (case B) changes with all specifications were planned resulting in a similar systematic end to design as was done in the *Geräte* malfunction meetings.

The case »A« changes can mostly be done quickly without delaying production, and many case »B« changes can be done by the units themselves by modifying devices already delivered.

Change case »C« collects all final development planning, whether for a new design, because of significant simplification in handling, or production enhancements. Not to forget savings in material and vehicles.

The aim is to ensure that the equipment of the units is completely standardized. The vehicles and devices supplied are also brought up to this standard.

Niederschrift
Schlussbesprechung der Arbeitstagung Bodenanlagen

1. Allgemeines:
Die Besprechung war auf dem Sektor der Bodenanlagen besonders wichtig, da manche Fragen unklar waren, die nicht durch einen umfangreiche Schriftverkehr, sondern nur durch eine Aussprache am Gerät zwischen taktischer Führung, Entwicklungsstelle, Fertigung und Truppe stattfinden konnte.

2. Änderungen nach A, B, C:
Für die in der Besprechung festgelegten Termine sind die entsprechenden Besprechungsteilnehmer ab sofort verantwortlich gemacht worden, nicht erst nach Erhalt der Niederschriften.

Aus der gesonderten Zusammenstellung, der notwendigen (Fall A) und erwünschten Änderungen (Fall B) mit den dazugehörigen Daten, ergibt sich ähnlich wie bei der Pannenbesprechung für das Gerät selbst eine Planung für systematischen Entwicklungsabschluss.

Die Änderungsfälle »A« lassen sich erfreulicherweise fast alle in sehr kurzer Zeit berücksichtigen ohne Fertigungsverzögerung, vielfach auch ebenso rasch für ausgelieferte Geräte durch Formänderungen (Fall B) bei der Truppe durchführen.

Als Änderungsfall »C« werden letzten Endes alle Entwicklungsplanungen zusammengefasst, für die sich eine gänzliche Neukonstruktion ergibt, sei es um wesentliche Bedienungsvereinfachung bzw. Fertigungsverbesserungen zu erreichen. Hinzu kommt noch Einsparung von Werkstoffen und Fahrzeugen.

Es wird durchweg angestrebt und erreicht, dass die Ausrüstung der Truppe völlig einheitlich wird. Auch die ausgelieferten Fahrzeuge und Geräte werden auf diesen Zustand gebracht.

3. Development outlook (change case C):

a) The following proposals were made:
Case I) Combine the Feuerleitpanzer and the electrical parts of the test vehicle. Combine the power supply vehicle and the mechanical parts of the test vehicle. For both vehicles (now 2 instead of 3) a 3 t Lkw with closed body will sufficient. Advantage: simpler ordering, easy production, simpler transport of the equipment. Introduction of a reliable army power supply system, saving of cables needed by a Battery. Personnel reduction of one driver and co-driver. Case II) like Case I) but retaining the Feuerleitpanzer with integration of the B-Gestell of the current test vehicle (if necessary in a smaller version).

b) Simplification of the firing process, eliminating all petrol engines in the units and replacing with electric motors.

c) Instead of the current refueling with 2100L T-Stoff vehicles, T-Stoff and also Z-Stoff will be carried in closed containers for filling. These containers should then be hung in a horizontal position on the girder of the FR trailer and are erected with the *Gerät*.

d) Test vehicles for 3 *Geräte*: Development of a test vehicle to test 3 *Geräte* at the same time.

4. The following are to be continuously involved during development:

a) for motor vehicle equipment: OKW/Chef WK)W/mot-Entw.

b) the Special Committee for Vehicle Superstructures Ing. Horn.

5. Staff:

It is urgently necessary that the processing of special vehicles be transferred with personnel from Wa Prüf 11 to HAP11. These personnel can take over additional development work and production supervision at HAP 11 in cooperation with WuG 10 for other vehicles not yet processed by Wa Prüf 11.

6. Procurement:

The meeting minutes provide a clear picture of the production situation and documents. It goes without saying that production and acceptance can only be based on the production documents currently available. The only exception to this rule is special radio equipment, such as the guidance and firing systems. The vehicles must come to HAP 11 for testing at random after acceptance. For the guide beam and combustion cut-off vehicles, this testing takes place at HAP 11 as the acceptance test.

7. Procurement is transferred to WuG 10.

For this transfer, Wa Prüf 11/III, HAP 11 and Wa Prüf 10 provide a status quo up to 1.5.44.

Draft signed: Roßmann
For correctness: Heinisch

3. Entwicklungsplanung auf längere Sicht (Änderungsfall C):

a) Hierzu werden folgende Vorschläge gemacht:
Fall I) Feuerleitpanzer und elektrische Teile des Prüfwagens vereinen. Stromversorgungswagen mit mechanischem Teil des Prüfwagens vereinigen. Für beide Fahrzeuge (es sind also statt 3 nunmehr 2) kommen 3 t Lkw mit geschl. Aufbau in Frage. Vorteil: Viel einfachere Beschaffung, leichtere Fertigung, einfacher Aufbau der transportablen Einzelgeräte. Einführung einer zuverlässigen heeresüblichen Stromversorgungsanlage, Einsparung an Leitungen innerhalb der Batterie. Personelle Einsparung zumindest eines Fahrers und Begleiters. Fall II) wie Fall I) jedoch unter Beibehaltung des Feuerleitpanzers, in dem aber das B-Gestell des bisherigen Prüfwagens (notfalls in kleinerer Ausführung) untergebracht wird.

b) Vereinfachungen des Schießbetriebes durch Wegfall aller Otto-Motoren in den Fahrzeugen und Ersatz durch elektrische Motore.

c) Anstelle des bisherigen Tankbetriebes mit dem 2100L T-Stoff-Lkw wird der T-Stoff ebenso wie der Z-Stoff in geschlossenen Gefäßen mit Normalfüllung eingeschüttet. Diese Gefäße sollen dann in liegendem Zustand am Kipprahmen des FR-Anhängers eingehängt und mit dem Gerät aufgerichtet werden. 47

d) Entwicklung eines Prüfwagens zum Prüfen von 3 Geräten.

4. Bei der Entwicklung sind laufend einzuschalten:

a) auf kraftfahrtechnischem Gerät OKW/Chef WK)W/mot-Entw.

b) Sonderausschuß für Fahrzeugaufbauten Ing. Horn.

5. Personal:

Es ist dringend notwendig, daß die Bearbeitung der Spezialfahrzeuge von Wa Prüf 11 mit Personal zum HAP11 überführt wird. Von diesem Personal kann beim HAP 11 zusätzliche Entwicklungsarbeit und Fertigungsaufsicht in Zusammenarbeit auch mit WuG 10 noch für andere, bisher nicht vom Wa Prüf 11 bearbeitete Fahrzeuge übernommen werden.

6. Beschaffung:

Durch die Besprechungsniederschriften ist über die Fertigungslage und Unterlagen ein klares Bild gegeben. Es ist selbstverständlich, daß nur nach den z.Zt. vorhandenen Fertigungsunterlagen gefertigt und abgenommen werden kann. Eine Ausnahme ist nur bei den Spezial-Funkeinrichtungen wie Leitstrahl- und Brennschlußanlage. Die Fahrzeuge müssen stichprobenweise nach der Abnahme zum HAP 11 zur Erprobung kommen. Bei den Fahrzeugen LS und BS erfolgt diese Erprobung beim HAP 11 als Abnahme.

7. Die Beschaffung wird an WuG 10 übergeleitet

Für diese Übergabe stellen Wa Prüf 11/III, HAP 11 und Wa Prüf 10 einen Status quo bis zum 1.5.44 auf.

im Entwurf gezeichnet: Roßmann
F.d.R. : Heinisch

TRANSFER OF THE GROUND PROGRAM
ÜBERLEITUNG DER BESCHAFFUNG DES BODEN-PROGRAMMS

BAMA, RH 8/1266
Wa J Ru (WuG 10)

Gr.IIId- Az.: 72 p 5801
o.U. den 22.7.1944
Stä/Re.
Nr.1317a/44 g.Kdos

Minutes of Meeting

about the meeting held at WuG 10 on 21.7.44
regarding the transfer of the ground program from Wa Prüf 4/V (formerly Prüf. II/III) to WuG-Abtlg.
Meeting starts at 8.40
Meeting ends at 17.30
Major Starkloff welcomed the participants. He stresses the secrecy obligation and the significance of the ground program's difficulties for the overall program.

A. General information

1. The MW has been instructed to supervise the manufacturing of a number of vehicles which are directly connected to the *Gerät*, namely: Nos. 101, 102, 103, 105, 106, 107, 110, 201, 202, and 204.

   B.z.b.V. Heer (Major Eberlein) pointed out that the activation of the MW will not affect the tasks of the testing department for development and the WuG departments for procurement. The objective, the quick manufacturing of the vehicles, will be accomplished through cooperation of the parties involved.

   MW is not present at the meeting because Dr. Simon is prevented from attending due to an urgent official trip.

2. Commercial and military standard vehicles will be requested through WuG 10 directly from AHA 1b/5: the deadline for 1945 requirements is the end of July 1944.

Besprechungs-Niederschrift

über die am 21.7.44 bei WuG 10 stattgefunden Sitzung
betr. Überleitung der Beschaffung des Boden-Programms von Wa Prüf 4/V (vorm. Prüf.II/III) an die WuG-Abtlg.
Beginn der Sitzung um 8.40 Uhr
Ende der Sitzung um 17.30 Uhr
Major Starkloff begrüßt die Teilnehmer. Er weist auf die Geheimhaltungsverpflichtung hin und weist auf die Bedeutung der Schwierigkeiten des Bodenprogramms für die Gesamtprogramm hin.

A. Allgemeines

1. Das MW ist vom B.z.b.V. beauftragt worden die Fertigung einer Anzahl von Fahrzeugen zu betreuen, die mit dem Gerät in unmittelbarer Verbindung stehen und zwar: Nr.101, 102, 103,105 106,107,110, 201, 202 und 204.

   B.z.b.V. Heer (Major Eberlein) führt aus, dass durch die Einschaltung des MW sich an den Aufgaben der Prüf Abteilung für Entwicklung und der WuG-Abteilungen für die Beschaffung nichts ändert. Das Ziel, die schnelle Erstellung der Fahrzeuge, wird sich durch Zusammenarbeit der beteiligten Stellen erreichen lassen.

   MW ist bei der Besprechung nicht vertreten, da Dr. Simon wegen dringender Dienstreise verhindert ist.

2. Handels- und heeresübliche Kfz werden durch das WuG 10 unmittelbar bei AHA 1b/5 angefordert: Termin für Anforderung 1945 bis Ende Juli 1944.

Major Eberlein criticized the concentration of empty trucks in Koblenz. He demands their immediate release and future adjustment of assignments and manufacturing requirements because, apart from the danger of air attacks, the hoarding of vehicles is irresponsible.

H. Abn. LR, Hptm. Spang is responsible for the accommodation of empty vehicles and completed vehicles until their loading. He has already dispersed empty vehicles to 4 different places in the vicinity and wider surroundings of Koblenz.

3. Under an order of Wa A dated June 13, 1944 the H. Abn. LR department has been assigned to insure the accelerated activation of other H. Abn. dispersal locations.

4. A vehicle is considered »ready« after its successful acceptance.

5. Major Eberlein plans for 1945 to increase the target figures for vehicles 101, 102, and 103. It is expected that the numbers of those will be doubled for 1945.

6. For army standard vehicles new tires will be delivered along with the vehicles, and replacements will be acquired by the units them-selves.

   For the special vehicles, both the new and replacement tires are ordered directly by the WuG department responsible for the vehicle directly from WuG 12, Rittm. Kossan, who is responsible for the vehicle. (Deadline by the 5th of each month on two copies of the form).

7. Spare parts and accessories are requested by B.z.b.V. Heer from WuG 10 and ordered by the latter for procurement from the relevant WuG departments. The B.z.b.V. is striving to procure subgroup components instead of the large number of individual spare parts in order to ease the workload for the units in the event of defects and ordering more spare parts.

8. Baurat Schwab of Wa Prüf 4 advises that Gen. Schneider is proposing to carry out the development of vehicles Nos. 102, 402, and 403 and the 27 t A-Stoff railroad carriage for the time being. The final decision will be announced.

9. Wa Prüf 4/V will make the final transfer of the contract documents dependent on the payment of the invoices for the contracts still in progress for 1944. All involved parties agree that the financial settlement of the current contracts for 1944 has to be executed by the respective responsible WuG department.

Major Eberlein bemängelt die Anhäufung von Leer Kfz in Koblenz. Er fordert sofortige Abstellung und künftigen Abgleich von Zuweisungen und Fertigungsbedarf, da abgesehen von Luftgefährdung eine Hortung von Kfz gegenwärtig nicht zu vertreten ist.

H. Abn. LR, Hptm. Spang ist für die Unterbringung der Leer Kfz und fertiggestellten Kfz bis zur Verladung verantwortlich. Er hat bereits eine Aufteilung der Leer Kfz in der näheren und weiteren Umgebung von Koblenz auf 4 verschiedene Stellen durchgeführt.

3. Lt. Verfügung der Wa A vom 13.6.44 ist die Dienststelle H. Abn. LR beauftragt, beschleunigte Einschaltung der anderen für die einzelnen Firmen in Betracht kommenden H. Abn. Stellen für Bodenanlagen sicherzustellen.

4. Als »fertig« gilt ein Fahrzeug nach erfolgter Abnahme.

5. Major Eberlein stellt für 1945 Erhöhung der Planzahlen der Fz 101, 102 und 103 in Aussicht. Es ist zu erwarten, dass die Stückzahlen für 1945 für die genannten Fahrzeuge verdoppelt werden.

6. Für heeresübliche Kfz wird die Neubereifung mit den Fahrzeugen geliefert und Ersatz von der Truppe selbst beschafft.

   Für die Spezialfahrzeuge wird sowohl die Neu- als auch die Ersatzbereifung von der für das Fahrzeug zuständigen WuG-Abt. unmittelbar bei WuG 12, Rittm. Kossan, angefordert. (Termin bis zum 5. jedes Monats auf Formblatt zweifach).

7. Ersatzteile und Zubehör werden vom B.z.b.V. Heer bei WuG 10 angefordert und von dieser zur Beschaffung bei den in Frage kommenden WuG-Abteilungen bestellt. Vom B.z.b.V. wird die Beschaffung von Untergruppenbauteilen anstelle der großen Zahl von Einzelersatzteilen angestrebt, um die Arbeit der Truppe bei Defekten zu erleichtern und die Ersatzteilbestellung im Ganzen übersichtlicher zu gestalten.

8. Baurat Schwab, Wa Prüf 4, teilt mit, dass Gen. Schneider vorgeschlagen wird, die Durchführung der Entwicklung für die Fahrzeuge Nr. 102, 402, 403 und den 27 t A-Stoff-Eisenbahnwagen zunächst bei Wa Prüf 4 zu belassen. Endgültige Entscheidung wird bekanntgegeben.

9. Wa Prüf 4/V macht die endgültige Überleitung der Auftragsunterlagen von der Regelung der Bezahlung der Rechnungen der noch für 1944 laufenden Aufträge abhängig. Alle Beteiligten kommen überein, dass die geldliche Abwicklung auch der laufenden Aufträge von 1944 von der jeweils zuständigen WuG Abt. zu erledigen ist.

10. Hptm. v. d. Tann of WuG 10 explains the WuG-Stab order dated June 27, 1944 whereby the other WuG departments have to notify WuG 10 of the production reported by the 3rd of each month with the Pos. number, WH number, and chassis number of each completed vehicle plus the target and actual numbers completed in the past month and the estimated actual numbers for the current month.

11. Hptm. Schulte of WuG 10 asks the responsible WuG departments of the companies to collate the raw material requirements as quickly as possible and send them (on Forms A and C) to WuG 10 after checking and approval.

12. Major Starkloff points out that changes to local conditions at the company Gollnow & Sohn are feared. A relocation of the company would cause great delay to the vehicles' delivery. B.z.b.V. (Major Eberlein) will exert its influence in order to avoid a disruption to production.

10. Hptm. v. d. Tann, WuG 10, erläutert den Befehl WuG-Stab vom 27.6.1944 wonach der federführenden WuG 10 von den anderen WuG-Abteilungen über die Fertigung Meldungen zu machen sind und zwar: mit FS bis zum 3. jeden Monats Pos. Nr. WH Nr. und Fahrgestell Nr. der fertigen Fahrzeuge zuzüglich Soll und Ist des abgelaufenen Monats und voraussichtliches Ist des laufenden Monats.

11. Hptm. Schulte, WuG 10, bittet die zuständigen WuG-Abteilungen von den Firmen die Rohstoffforderungen schnellstens einzuziehen und nach Prüfung mit Anerkennungsvermerk (auf Formblatt A und C) der WuG 10 zuzustellen.

12. Major Starkloff weist darauf hin, dass Änderungen in den örtlichen Verhältnissen der Fa. Gollnow & Sohn, LR, befürchtet werden. Eine Umlegung der Firma würde einen weiteren großen Verzug in der Fahrzeuglieferung hervorrufen, B.z.b.V. (Major Eberlein) wird seinen Einfluss geltend machen, damit eine Störung der Fertigung vermieden wird.

B. Individual vehicles

The information provided by B.z.b.V. Heer is for vehicles already with the units.

Kfz.-Nr. 101: (Feuerleitpanzer = armored fire control vehicle)

WuG 6 gives its consent for the acquisition of this vehicle conversion for 1945, but mentions that it will probably be impossible for the supplying company to produce even the planned one vehicle per month. Moreover, WuG 6 has taken over the continuing supervision of the contracts for 1944 commissioned by Wa Prüf II. After the conversion, carried out by WuG 6, WuG 10 will take over the acquisition of the final equipment.

B. Einzelfahrzeuge

Als »geliefert« gelten die Angaben B.z.b.V. Heer für bereits bei der Truppe befindliche Fahrzeuge.

Kfz.-Nr. 101: (Feuerleitpanzer)

WuG 6 gibt Einverständnis mit der Beschaffung der Kfz-Umrüstung für 1945, erwähnt jedoch, dass es für die Lieferfirma nicht gut möglich ist, monatlich nur ein einziges Kfz. gem. Planung herauszubringen. WuG 6 hat auch darüber hinaus die Weiterbetreuung der für 1944 seinerzeit von Wa Prüf II veranlassten Aufträge übernommen. Nach Umrüstung durch WuG 6 übernimmt WuG 10 die Beschaffung der Endausrüstung.

Delivery quantities / Lieferzahlen units / Stück	to end of July 44 / bis Ende Juli 44	to end of 1944 / bis Ende 1944
requirement / Forderung	44	56
delivered / geliefert	18	
shortage / Fehl	26	38

Empty vehicles beyond the B.z.b.V. requirement are present in Koblenz. Manufacturing problems: Provision of measuring instruments (microammeter from the company Sochor, Blanz, and carrier frequency device TFb1). Delays will be made up in agreement with B.z.b.V. Heer and WuG 10 by distributing the work over several months. Production is currently running smoothly apart from the provision of measuring instruments.

One TFb1 army standard carrier frequency device and one special transmission system are required for each vehicle among other things.

Leer Kfz. sind über B.z.b.V. Forderung hinaus in Koblenz vorhanden. Fertigungsschwierigkeiten: Beistellung der Messinstrumente (Mikroamperemeter d. Firma Sochor, Blanz und Trägerfrequenzgerät TFb1). Rückstände werden im Einvernehmen mit B.z.b.V. Heer und WuG 10 unter Aufteilung auf die einzelnen Monate aufgeholt. Fertigung läuft z.Zt., abgesehen von der Beistellung der Messinstrumente, reibungslos.

Zum Einbau erforderlich sind pro Kfz. u.a. ein Trägerfrequenzgerät TFb1 (heeresüblich) und eine Spezialübertragungsanlage.

TFb1 currently in short supply. Provision has been by Wa Prüf 10 until now. Provision has so far been made by Wa Prüf 10. The requirement of 70 pieces of TFb1 equipment registered by Wa Prüf 10 with B.z.b.V. Heer will be maintained. Control of the the equipment by WuG 10/Ib. Finding that vehicle no. 101 has so far been delivered by LR without carrier frequency device TFb1. Transmission device: 19 systems delivered to ASL (Prüf 10) to date. VTL not available for this. VTL not available for this.
B.z.b.V. Heer requires the elimination of this problem and the installation of the carrier frequency device as well as the transmission device, in situ at the LR in order to avoid further time-consuming transport of the affected vehicles. H. Abn. at the LR resumes pressure for the mounting of a VTL. Designs in very close cooperation with Wa Prüf 10. In the future the Gollnow & Sohn company will take over the Wa Prüf 10 contracts for the transmission equipment which has been provided by Wa Prüf 10 until now. Agreement that army standard communication devices of the Gollnow & Sohn company will be provided by WuG 10 (except the special transmission device).

TFb1 z.Zt. Engpass. Beistellung bisher durch Wa Prüf 10 erfolgt. Der von Wa Prüf 10 bei B.z.b.V. Heer angemeldete Bedarf von 70 Stück TFb1 Geräten wird weiterhin aufrechterhalten. Steuerung dieser Geräte durch WuG 10/Ib. Feststellung, dass das Kfz-Nr. 101 bislang ohne Trägerfrequenzgerät TFb1 von LR abgeliefert wurde. Übertragungsgerät: bisher 19 Anlagen an ASL (Prüf 10) geliefert. VTL dafür nicht vorhanden.
B.z.b.V. Heer fordert Abstellung dieses Übelstandes und Einbau des Trägerfrequenzgerätes wie auch der Übertragungsanlage an Ort und Stelle in LR, um einen nochmaligen zeitraubenden Transport der betr. Fahrzeuge zu vermeiden. H. Abn. in LR übernimmt die Veranlassung der Aufstellung einer VTL. Entwürfe in engster Zusammenarbeit mit Wa Prüf 10. Für das bisher von Wa Prüf 10 beigestellte Übertragungsgerät übernimmt in Zukunft Firma Gollnow & Sohn die Wa Prüf 10 Aufträge.
Übereinkunft, dass heeresübliches Nachrichtengerät der Firma Gollnow & Sohn von WuG 10 beigestellt wird (ausgenommen das spez. Übertragungsgerät).

Kfz.-Nr. 102 and Nr. 103: (FR-trailers and platform trailers)

Total takeover of the vehicles for 1945 as well for 1944 carried out by WuG 2 (according to Ing. Kötzsche). Acceptance takes place at the involved manufacturing sites with technical supervision by H. Abn. F.P.Nr. 68890.

Kfz.-Nr. 102 und Nr. 103: (FR Anhänger u. Plattformanhänger)

Restlose Übernahme der Fahrzeuge für 1945 als auch für 1944 von WuG 2 erfolgt (lt. Ing. Kötzsche). Abnahme erfolgt bei den betr. Fertigungsstätten, techn. Betreuung abnahmemäßig durch H. Abn. F.P.Nr. 68890.

Delivery quantities / Lieferzahlen	to end of July 44 / bis Ende Juli 44		to end of 1944 / bis Ende 1944	
units / Stück	Nr. 102	Nr. 103	Nr. 102	Nr. 103
requirement / Forderung	159	101	259	201
delivered / geliefert	75	51		
shortage / Fehl	84	56	184	150

Presumably the target figures will double. It is certain that there are currently ca. 22 Meiller vehicles completed except for the hydraulics, at the Meiller company in Munich.
No manufacturing difficulties but exclusively lack of raw materials (description by Wa Prüf 4/V). Immediate and most precise examination of the status of the raw material allocation will be carried out by WuG 10.

Vermutlich tritt Verdoppelung der Planzahlen ein. Feststellung, dass bei Firma Meiller in München z.Zt. ca. 22 Meiller-Fahrzeuge, bis auf die Hydraulik fertiggestellt vorhanden sind.
Keine Fertigungsschwierigkeiten, sondern ausschließlich Rohstoffmangel (Darstellung Wa Prüf 4/V). Umgehende genaueste Überprüfung der Rohstoffzuteilungslage wird durch WuG 10 vorgenommen.

Kfz.-Nr. 104: (Masch.Satz 11)
Requirement 4 units, delivered 4 units.
Cancelled for acquisition through WuG.

Kfz.-Nr. 104: (Masch. Satz 11)
Forderung 4 Stück, geliefert 4 Stück.
Entfällt für Beschaffung durch WuG.

Kfz.-Nr. 105: (Steyr power supply vehicle)
Conversion WuG 12, final outfitting WuG 10.

Kfz.-Nr. 105: (Steyr-Stromversorgungswagen)
Umrüstung WuG 12, Endausrüstung WuG 10.

Delivery quantities / Lieferzahlen units / Stück	to end of July 44 / bis Ende Juli 44	to end of 1944 / bis Ende 1944
requirement / Forderung	58	90
delivered / geliefert	17	
shortage / Fehl	41	73

The vehicles' delivery depends on the timely provision of electrical components to the Gollnow & Sohn company. The target was not reached due to a shortage of empty vehicles at that time; currently enough empty vehicles are available.
Reason: suspension of delivery (enemy action at Steyr company).
Wa Prüf 10 anticipates handing over the required component lists very soon.
The delivery of the 73 missing vehicles for 1944 is uncertain. The Gollnow & Sohn company is checking the status of deliveries in the context of the manufacturing of Kfz.-Nr. 101 and will inform WuG 10.
H. Abn. at the Karosserie Buhne company, Berlin and at the LR.

Kfz.-Nr. 106, 107, and 110: (testing vehicles I, II, and cable reel trailer)

Fitting out WuG. Transfer from Wa Prüf 11 to the Arbeitsgemeinschaft (working group) Schlosser München. Awarding contracts for valve boxes, test panels and A-Stoff refueling hoses to company Gollnow & Sohn is required.

Die Auslieferung der Fahrzeuge ist von der rechtzeitigen Beistellung el. Bauteile an Firma Gollnow & Sohn abhängig. Soll seinerzeit nicht erfüllt durch Mangel an Leerfahrzeugen, z.Zt. Leerfahrzeuge wieder genügend vorhanden.
Begründung: Lieferstop (Feindeinwirkung bei Firma Steyr).
Wa Prüf 10 stellt Übergabe erforderlicher Stücklisten für die allernächste Zeit in Aussicht.
Auslieferung von 73 fehlenden Kfz. für 1944 ist fraglich. Firma Gollnow & Sohn untersucht Lieferlage im Zusammenhang mit Fertigung Kfz.-Nr. 101 und unterrichtet WuG 10.
H. Abn. bei Firma Karosserie Buhne, Berlin, und in LR.

Kfz.-Nr. 106, 107 und 110: (Prüfwagen I, II und Kabeltrommelanhänger)

Ausrüstung WuG. Überleitung der Wa Prüf 11 an Arbeitsgemeinschaft Schlosser München vergebene Aufträge auf Ventilkästen, Prüftafel und A-Stoff-Betankungsschlauch an Firma Gollnow und Sohn erforderlich.

Delivery quantities / Lieferzahlen units / Stück	to end of July 44 / bis Ende Juli 44 Nr. 106	Nr. 107	to end of 1944 / bis Ende 1944 Nr. 106	Nr. 107
requirement / Forderung	35	26	53	40
delivered / geliefert	9	11		
shortage / Fehl	26	15	44	29

Delivery quantities for Kfz.-Nr. 110 exactly as Kfz.-Nr. 105

Shortages: 1.) B-racks to be provided by Wa Prüf 10
2.) Engine related equipment

Inspection of the B-racks by H. Abn. is planned to take place in the LR; but so far still undertaken by Prüf 10. Reference is made to the five disc clutch as a special consumable item. After eliminating the shortages it is possible to make up the 1944 delivery shortfall at Gollnow & Sohn.
Acceptance of vehicle Nr. 106 and 107 currently at Prüf 10. Acceptance of vehicle Nr. 110 already at H. Abn. at LR. Acceptance of vehicle Nr. 106/107 in approx. 6 weeks by H. Abn. in LR.

Kfz.-Nr. 108: (two-axle cable reel trailer)

Conversion WuG 12, final outfitting WuG 10. The vehicle stock is exhausted and it is no longer manufactured. B.z.b.V. Heer is asked to notify spare cable stocks in writing to WuG 10.

Lieferzahlen für Kfz.-Nr. 110 genau wie Kfz.-Nr. 105

Engpässe: 1.) Von Wa Prüf 10 beizustellende B-Gestelle
2.) Triebwerkseitige Ausrüstung

Abnahme der B-Gestelle durch H Abn. in LR vorgesehen, bisher jedoch noch von Prüf 10. Auf fünffach-Kupplung als besonderen Verbrauchsgegenstand wird hingewiesen. Aufholung der Lieferrückstände 1944 bei Firma Gollnow & Sohn nach Behebung der Engpässe möglich.
Abnahme von Fahrzeug Nr. 106 und 107 z.Zt. bei Prüf 10. Abnahme von Fahrzeug Nr. 110 bereits bei H. Abn. in LR. Abnahme von Fahrzeug Nr. 106/107 in ca. 6 Wochen durch H. Abn. in LR.

Kfz.-Nr. 108: (Kabeltrommelanhänger 2-achsig)

Umrüstung WuG 12, Endausrüstung Wug 10. Fahrzeug läuft aus und wird nicht mehr gefertigt. B.z.b.V. Heer wird gebeten, die Kabelreserve schriftlich im WuG 10 bekannt zu geben.

Delivery quantities / Lieferzahlen units / Stück	to end of July 44 / bis Ende Juli 44	to end of 1944 / bis Ende 1944
requirement / Forderung	23	29
delivered / geliefert	11	
shortage / Fehl	12	18

Additional target figures from October are not available. The delivery shortfall was caused by missing construction documents and sluggish delivery of cable reels. The delivery shortfall will be made up. The Carosserie Buhne, Berlin, does not have the drawing for the cable bridges which are now attached to the vehicle. Wa Prüf 10 is asked to provide drawings for them. To be printed at WuG 10. Drawings for cable bridges already in existence at the firm of Gollnow & Sohn at LR have to be taken into consideration by Prüf 10. Wa Prüf 10 agrees to hand over the parts list to WuG 10 as quickly as possible. Inspection by H. Abn. at the company Buhne and H. Abn. in LR.

Weitere Planzahlen ab Oktober 1944 liegen nicht vor. Lieferrückstand verursacht durch fehlende Bauunterlagen und schleppende Zulieferung von Kabeltrommeln. Lieferrückstand wird aufgeholt. Für neuerdings an das Fahrzeug anzubringende Kabelbrücken fehlen bei der Firma Carosserie Buhne, Berlin, die Zeichnungsunterlagen, Wa Prüf 10 wird gebeten, Zeichnungen für die Kabelbrücken zur Verfügung zu stellen. Nebenabdruck an WuG10. Bereits bei Firma Gollnow & Sohn in LR vorhandenen Zeichnungsunterlagen über Kabelbrücken sind von Prüf 10 zu berücksichtigen. Kabelbrücken sollen in größerer Anzahl bei Firma Gollnow & Sohn vorhanden sein. Wa Prüf 10 sagt zu, Teileliste an WuG 10 schnellstens zu übergeben. Abnahme durch H. Abn. bei Firma Buhne und H. Abn. in LR.

Kfz.-Nr. 201, 202, and 208: (trucks for combustion cut-off system, guide beam-receiver trailer)

Fitting out WuG. Inspection of Kfz.-Nr. 201 and Nr. 202 currently by Wa Prüf 10; vehicle Nr. 208 already by H. Abn. at LR; however, testing is by Wa Prüf 10.

Kfz.-Nr. 201, 202 und 208: (Lkw für BS-Anlage, LS-Kontrollempfänger-Anh.)

Ausrüstung WuG 10. Abnahme von Kfz.-Nr. 201 und Nr. 202 z.Zt. durch Wa Prü 10, Fahrzeuge Nr. 208 bereits durch H.Abn. in LR, Erprobung jedoch durch Wa Prüf 10.

Delivery quantities / Lieferzahlen units / Stück	to end of July 44 / bis Ende Juli 44 Nr. 201	Nr. 202	to end of 1944 / bis Ende 1944 Nr. 201	Nr. 202
requirement / Forderung	21	21	30	30
delivered / geliefert	6	4		
shortage / Fehl	15	17	24	26

Delivery quantities / Lieferzahlen units / Stück	to end of July 44 / bis Ende Juli 44 Nr. 208	to end of 1944 / bis Ende 1944 Nr. 208
requirement / Forderung	16	25
delivered / geliefert	2	
shortage / Fehl	14	23

6 vehicles have been completed, of which 4 units are still undergoing testing.

Fertiggestellt sind insgesamt 6 Fahrzeuge wovon sich 4 Stück noch in der Prüfung befinden.

Kfz.-Nr. 201: So far 11 BS systems have been delivered, 12th and 13th system are being tested. The 12th and 13th are being tested. The 14th and 15th systems are being equipped. 4 of these systems are equipped with a home frequency; 1 device is designated for transformer station C. These systems are not designated as unit equipment by B.z.b.V. Heer. The contract has to specify the number of home frequencies. WuG 10 has to order 5 additional combustion cut-off systems from the company Gollnow & Sohn consisting of vehicles no. 201, 202, and 204. For vehicles no. 201, 202, and 208, WuG10 has so far obtained from Prüf 10 parts lists only for the electrical equipment. Prüf 10 is responsible for delivery of the parts lists for the mechanical sector.
Delivery shortfalls were caused by a two-month suspension of deliveries by Prüf 10 due to fundamental design changes. Kfz.-Nrs. 201/202 are currently in production. However, a new prototype vehicle of Kfz.-Nr. 208 is currently under construction. If bottlenecks are eliminated the Gollnow company will be able to make up the existing backlog.

Kfz.-Nr. 201: Bis jetzt sind 11 BS-Anlagen ausgeliefert, 12. und 13. Anlage zur Erprobung. Die 14. und 15. Anlage ist in Ausrüstung begriffen. Davon sind 4 Anlagen mit Heimatfrequenz ausgerüstet, 1 Anlage für Umspannwerk C vorgesehen. Diese Anlagen werden vom B.z.b.V. Heer nicht als Truppenanlagen berechnet.
Auftrag muss um die Anzahl der Heimatfrequenzen aufgestockt werden. Es sind bei der Firma Gollnow von WuG 10 zusätzlich zu bestellen 5 BS-Anlagen bestehend aus Fahrzeug Nr. 201, 202 und 204. Fahrzeug Nr. 201, 202 und 208 hat WuG 10 bisher von Prüf 10 Teilelisten nur für die el. Ausrüstung erhalten. Prüf 10 veranlasst Zustellung der Teilelisten für den mech. Sektor.
Lieferverzögerungen begründet durch 2-monatlichen Lieferstopp von Prüf 10 wegen grundlegender Konstruktionsänderungen. Kfz.-Nr. 201/202 befinden sich z.Zt. in laufender Fertigung. Von Kfz.-Nr. 208 ist dagegen z.Zt. ein neues Musterfahrzeug im Bau. Bei Beseitigung der Engpässe kann Firma Gollnow bestehende Rückstände aufholen.

Kfz.-Nrs. 204, 205, 206, and 207: Combustion cut-off antenna trailer, guide beam transmitter vehicle, guide beam equipment vehicle and guide beam antenna trailer

Kfz.-Nr. 204, 205, 206 und 207: (BS-Ant.Anh., LS-Sende-KW, LS-Geräte-KW u. LS-Anh.)

Outfitting at WuG 10.

Ausrüstung WuG 10.

Delivery quantities / Lieferzahlen units / Stück	to end of July 44 / bis Ende Juli 44 Nr. 204	to end of 1944 / bis Ende 1944 Nr. 204
requirement / Forderung	21	30
delivered / geliefert	7	
shortage / Fehl	14	23

Delivery quantities / Lieferzahlen units / Stück	to end of July 44 / bis Ende Juli 44 Nr. 205	Nr. 206	to end of 1944 / bis Ende 1944 Nr. 205	Nr. 206
requirement / Forderung	16	16	25	25
delivered / geliefert	2	2		
shortage / Fehl	14	14	23	23

Delivery quantities / Lieferzahlen units / Stück	to end of July 44 / bis Ende Juli 44 Nr. 207	to end of 1944 / bis Ende 1944 Nr. 207
requirement / Forderung	32	50
delivered / geliefert	4	
shortage / Fehl	28	46

Backlog caused by current technical changes and insufficient delivery of construction components.

Shortages: Measuring instruments, resistors, photo cell compensator, moving coil meter, voltmeter (already been fixed) at Kfz.-Nr. 205. Clarification of existing difficulties directly between WuG 10 and the company Schulze-Wechsungen to be transferred to SAERG. Preliminary inspection of the vehicles in Görkau, final inspection at Prüf 10.

Rückstand bedingt durch laufende techn. Änderungen und mangelnde Zulieferung von Bauteilen.

Engpässe: Bei Kfz.-Nr. 205 Messinstrumente, Widerstände; Photozellen Kompensator, Produktenmesser, Prozentenmesser (bereits behoben). Klarstellung bestehender Schwierigkeiten unmittelbar zwischen WuG 10 und Firma Schulze-Wechsungen zur Weitergabe an SAERG. Vor Abnahme der Fahrzeuge in Görkau, Endabnahme bei Prüf 10.

Kfz.-Nr. 301: (for Vehicle Workshop Platoon)

Kfz.-Nr. 301: (m. Kfz. Inst. Zug)

Delivery quantities / Lieferzahlen units / Stück	to end of July 44 / bis Ende Juli 44 Nr. 301	to end of 1944 / bis Ende 1944 Nr. 301
requirement / Forderung	9 Medium Platoons / 9 mittlere Züge	in total / insgesamt
delivered / geliefert	7 Medium Platoons / 7 mittlere Züge	

Eight system stationary in Rebstock
Ninth system being delivered

Achte Anlage stationär in Rebstock
Neunte Anlage in Auslieferung

Kfz.-Nr. 303: (Light Vehicle Workshop Platoon)

Requirement for 9 by the end of 1944. Output according to the announcement of Fz In presumably in July 1944; the requirement still remains in effect. Requested directly through B.z.b.V. Heer by WuG 10. Notices of missing tools as well as additional tools to be given to WuG 5 for their acquisition.

Kfz.-Nr. 303: (le. Kfz.- Inst. Zug)

Forderung bis Ende 1944 9 Stück. Ausstoß lt. Mitteilung Fz In voraussichtlich im Juli 44; Forderung bleibt nach wie vor bestehen. Abruf unmittelbar durch B.z.b.V. Heer von WuG 10. Festgestellte fehlende Werkzeuge wie auch zusätzliche Werkzeuge werden an WuG 5 zwecks Beschaffung bekanntgegeben.

B.z.b.V. Heer recommends calling to account those responsible for loss or damage.

B.z.b.V. Heer regt an, die für den Verlust bzw. die Verrottung der Werkzeuge Verantwortlichen zur Rechenschaft zu ziehen.

Kfz.-Nr. 304: (trucks for spare parts)

Kfz.-Nr. 304: (Lkw für Ersatzteile)

Delivery quantities / Lieferzahlen units / Stück	to end of July 44 / bis Ende Juli 44 Nr. 304	to end of 1944 / bis Ende 1944 Nr. 304
requirement / Forderung	32	49
delivered / geliefert	3	
shortage / Fehl	29	46

For the purpose of simplification B.z.b.V. Heer intends to remove the vehicles Nr. 304 and Nr. 510 from the present framework, and just to order complete interior sets. Complete sets consisting of a shelf and boxes are to be put into storage at Fz In. While 4000 boxes have already been ordered at (from) HAP, the shelves are being manufactured by the Gollnow & Sohn company at LR.
WuG 10 will take over Prüf 10's current order, and will arrange the placing of an order for shelves with Gollnow & Sohn.

Larger shortage caused by delays in development. Vehicles without internals are available in sufficient quantities.

B.z.b.V. Heer beabsichtigt zwecks Vereinfachung die Fahrzeuge Nr. 304 und 510 aus dem bisherigen Rahmen herauszunehmen und lediglich die Einbauten satzweise komplett in Auftrag zu geben. Komplette Einbausätze bestehend aus Regal und Kisten sind bei Fz In einzulagern. Von HAP sind bereits 4000 Stück Kisten bestellt, während die Regale von der Firma Gollnow & Sohn in LR gefertigt werden.
Den laufenden Auftrag von Prüf 10 übernimmt WuG 10 und veranlasst Auftragserteilung für Regale bei Firma Gollnow & Sohn.

Größerer Fehl durch Entwicklungsverzögerung begründet. Fahrzeuge ohne Einbauten in genügendem Umfang vorhanden.

Kfz.-Nr. 401: (A-Stoff trailer)

Kfz.-Nr. 401: (A-Stoff-Anhänger)

Delivery quantities / Lieferzahlen units / Stück	Nr. 401
requirement / Forderung	28
delivered / geliefert	28

Not applicable for procurement through WuG.

Entfällt für Beschaffung durch WuG.

Kfz.-Nr. 402: (A-Stoff trailer)

Chassis is commercially available insofar all standardized parts can be used. parts can be used. Not to be requested from AHA Ib/5, but to be ordered from the Ackermann company, Wuppertal. Order processing for the year 1944 at Prüf 4/V for the next of 3 weeks due to developmental reasons. WuG 5 in charge in regard to the summary of all construction components.

Kfz.-Nr. 402: (A-Stoff-Anhänger)

Fahrgestell handelsüblich insoweit als sämtliche genormten Teile Verwendung finden können. Nicht bei AHA Ib/5 abzurufen, sondern bei Firma Ackermann, Wuppertal, in Auftrag zu geben. Auftragsabwicklung für das Jahr 1944 aus entwicklungsmäßigen Gründen bei Prüf 4/V auf die Dauer von weiteren 3 Wochen. Federführend WuG 5 hinsichtlich der Zusammenfassung aller erforderlichen Bauelemente.

Delivery quantities / Lieferzahlen units / Stück	to end of July 44 / bis Ende Juli 44 Nr. 304	to end of 1944 / bis Ende 1944 Nr. 304
requirement / Forderung	123	243
delivered / geliefert	41	
shortage / Fehl	82	202

Shortfall caused by insufficient delivery of raw materials according to statement of Wa Prüf 4.

Fehl durch mangelnde Zulieferung an Rohstoffen bedingt nach Angabe von Wa Prüf 4.

The Mako company, Erfurt has undertaken to deliver 35 units for each month of the July-August time period. Delivery of the rest is questionable; clarification is to be sought by WuG 5.
Developmental documents now clarified by Prüf 4. Vehicle inspection at the manufacturing companies by H. Abn. Final inspection of completed refueling vehicles is tied to locations producing the A-Stoff. Concerning this matter, final inspection of refueling vehicles also at the Mako company, Erfurt by the end of the year is in preparation (previously only Karlshagen and Friedrichshafen). The requested quotas for the second quarter for manufacturing at Ackermann could not be allocated before the third quarter for lack of available quantities.

Firma Mako, Erfurt, hat sich verpflichtet für die monatliche Juli-August Zeitspanne je 35 Stück zu liefern. Restauslieferung fraglich, Klarstellung durch WuG 5 zu veranlassen.
Entwicklungsmäßige Unterlagen nach Abgabe von Prüf 4 nunmehr klargestellt. Abnahme fahrzeugmäßig bei den Herstellerfirmen durch H. Abn. Endabnahme kompletter betankter Fahrzeuge ist an A-Stoff erzeugende Stellen gebunden. Diesbezüglich Endabnahme betankter Fahrzeuge auch bei Firma Mako, Erfurt, bis Ende des Jahres in Vorbereitung (bisher Karlshagen bzw. Friedrichshafen). Für Fertigung Ackermann für zweites Quartal angeforderte Kontingente konnten mangels Verfügungsmengen erst im dritten Quartal zugewiesen werden.

Kfz.-Nr. 403: (B-Stoff-tank KW 2900 l).

Kfz.-Nr. 403: (B-Stoff-Kessel KW 2900 l)

Delivery quantities / Lieferzahlen units / Stück	Nr. 403
requirement / Forderung	40
delivered / geliefert	40

Not applicable for procurement through WuG.

Entfällt für Beschaffung durch WuG.

Kfz.-Nr. 404: (B-Stoff-tank KW 3500 l).

Kfz.-Nr. 404: (B-Stoff-Kessel KW 3500 l)

Delivery quantities / Lieferzahlen units / Stück	to end of July 44 / bis Ende Juli 44 Nr. 404	to end of 1944 / bis Ende 1944 Nr. 404
requirement / Forderung	164	284
delivered / geliefert	130	
shortage / Fehl	34	154

Difficulties with vehicle Nr. 404 are traceable back to the stop by B.z.b.V. Heer. According to Wa Prüf 4 the present shortage can be made up by the end of the year. Inspection by H. Abn. at the manufacturing companies. WuG 10 requests the chassis from AHA 1b/5; WuG 5 undertakes the complete outfitting of the vehicle. Inspection by Prüf 4 follows the Luftwaffe version. Conversion unnecessary.

Schwierigkeiten bei Fahrzeug Nr. 404 sind auf Stopp B.z.b.V. Heer zurückzuführen. Nach Ansicht von Wa Prüf 4 lässt sich bestehendes Fehl bis Ende des Jahres aufholen. Abnahme durch H Abn. bei den Herstellerfirmen. WuG 10 fordert Fahrgestelle bei AHA 1b/5 an, WuG 5 übernimmt komplette Ausstattung des Fahrzeuges. Nach Abnahme von Prüf 4 Luftwaffenausführung. Umrüstung entfällt.

Kfz.-Nr. 405: (T-Stoff-tank KW 2100 l)

Regular chassis; request through WuG 10 at AHA 1b/5. WuG undertakes complete outfitting at the firm Volkmar Hänig, Dresden.

Kfz.-Nr. 405: (Kessel KW 2100 l T-Stoff)

Normales Fahrgestell, Anforderung durch WuG 10 bei AHA 1b/5 WuG übernimmt volle Ausrüstung bei Firma Volkmar Hänig, Dresden.

Delivery quantities / Lieferzahlen units / Stück	to end of July 44 / bis Ende Juli 44 Nr. 405	to end of 1944 / bis Ende 1944 Nr. 405
requirement / Forderung	77	117
delivered / geliefert	40	
shortage / Fehl	37	77

Restocking not necessary because this vehicle can probably be dropped. Inspection by manufacturing company.

Aufstockung nicht erforderlich, da dieses Fahrzeug vermutlich gestoppt wird. Abnahme Herstellerfirma.

Kfz.-Nr. 406: (trailer with pump type A)

Army standard chassis, requested through WuG 10 by AHA. WuG 5 undertakes the outfitting. Chassis conversion not necessary. Clarification of the immediate assignment by Prüf 4 to the processing WuG department by WuG 10.

Kfz.-Nr. 406: (Pumpen-Anh. Type A)

Heeresübliches Fahrgestell, durch WuG 10 bei AHA angefordert. WuG 5 übernimmt Ausrüstung. Umrüstung des Fahrgestells nicht erforderlich. Klarstellung der sofortigen Auftragserteilung von Prüf 4 an die bearbeitende WuG-Abteilung nach WuG 10 zu veranlassen.

Delivery quantities / Lieferzahlen units / Stück	to end of July 44 / bis Ende Juli 44 Nr. 406	to end of 1944 / bis Ende 1944 Nr. 406
requirement / Forderung	108	183
delivered / geliefert	14	
shortage / Fehl	94	169

14 vehicles already delivered have to be converted. The reason for the delivery delay is bomb damage at the Siemens company. Shortage: Siemens dial indicator and A-Stoff pump. Stated by Prüf 4 that the dial indicator shortage is partially dealt with. Possibility of delivering the remainder still to be clarified.
Correspondence about this handed over from Prüf 4 to the investigating WuG 5.
Note from B.z.b.V. Heer that this vehicle is second only to the testing vehicle in terms of priority. Inspection at the manufacturing company.

14 bereits gelieferte Fahrzeuge müssen umgebaut werden. Grund der Lieferverzögerung Bombenschaden bei Firma Siemens. Engpass: Siemens-Messuhr und A-Stoff-Pumpe. Nach Angabe von Prüf 4 teilweise Behebung des Engpasses Messuhr. Liefermöglichkeit des Rückstandes ist noch zu klären.
Diesbezügl. Schriftwechsel von Prüf 4 an bearbeitende WuG 5 übergeben.
Hinweis von B.z.b.V. Heer, dass es sich bei diesem Fahrzeug nächst dem Prüfwagen um das vordringlichste Kfz. handelt. Abnahme bei Herstellerfirma.

Kfz.-Nr. 407: (T-Stoff preheater trailer)

No army standard chassis to date; Wa Prüf 10 to clarify if the use of a Luftwaffe standard trailer is possible.

Kfz.-Nr. 407: (T-Stoff-Vorwärmer-Anh.)

Bisher kein normales heeresübliches Fahrgestell, Wa Prüf 10 klärt, ob Verwendung eines heeresüblichen Luftwaffenanhängers möglich ist.

Delivery quantities / Lieferzahlen units / Stück	to end of July 44 / bis Ende Juli 44 Nr. 407	to end of 1944 / bis Ende 1944 Nr. 407
requirement / Forderung	50	72
delivered / geliefert	16	
shortage / Fehl	34	56

Conclusive clarification is to take place at Wa Prüf 10 at a date yet to be determined (Major Kleidörfer). According to WaChef Ing 4 (Ing. Jahnke), the Pintsch company is capable of delivering 30 vehicles per month without difficulties. Required involvement of WuG 5 and check-over at the Pintsch company necessary. No manufacturing difficulties. Approval unclear.

Endgültige Klarstellung erfolgt in einer bei Wa Prüf 10 noch festzusetzenden Besprechung. (Maj. Kleidörfer). Firma Pintsch ist nach Angabe von WaChef Ing 4 (Ing. Jahnke) ohne weiteres in der Lage, monatlich 30 Fahrzeuge zu fertigen. Geforderte Einschaltung von WuG 5 und Nachprüfung bei Firma Pintsch erforderlich. Keine Fertigungsschwierigkeiten. Abnahme unklar.

Kfz.-Nr. 408 and Nr. 508: (accumulator truck)

Conversion by WuG 12; final outfitting by WuG 10. According to WaChef Ing 4, the shelves which have been installed by the Gollnow & Sohn company in LR until now can also be installed without difficulties by the Buhne company, Berlin, in the course of the general conversion. The Gollnow company in LR only manufactures fittings according to the agreement to the current extent. Vehicles are located there. This work will be carried out in future by the Buhne, Berlin, company. According to B.z.b.V. Heer is expected to shorten the program

Kfz.-Nr. 408 und Nr. 508: (Sammler-Lkw)

Umrüstung WuG 12, Endausrüstung WuG 10. Die bisher von Firma Gollnow & Sohn in LR eingebauten Regale können nach Angabe von WaChef Ing 4 ohne weiteres von Firma Buhne, Berlin, im Zuge der allgemeinen Umrüstung miteingebaut werden. Firma Gollnow in LR fertigt Einbauten vereinbarungsgemäß nur noch im Umfange z.Zt. dort befindlicher Fahrzeuge. Ausführung dieser Arbeiten in Zukunft von Firma Buhne, Berlin. Lt. B.z.b.V. Heer ist mit Kürzung des Programmes zu rechnen.

Delivery quantities / Lieferzahlen units / Stück	to end of July 44 / bis Ende Juli 44 Nr. 408	 Nr. 508	to end of 1944 / bis Ende 1944 Nr. 408	 Nr. 508
requirement / Forderung	35	54	40	66
delivered / geliefert	16	4		
shortage / Fehl	19	50	24	62

Delivery delays were caused by technical changes. No additional manufacturing difficulties at present. B.z.b.V. Heer agrees to the output of 40 vehicles until the end of the year. Acceptance by army (Heer) by the army (Heer). In the course of simplification and the reduction in the number of specialized vehicles whether standard army vehicles can be used.

Lieferverzögerung durch techn. Änderungen begründet. Weitere Fertigungsschwierigkeit bestehen z.Zt. nicht. B.z.b.V. Heer erklärt sich mit Ausbringung von 40 Fahrzeugen bis Ende des Jahres einverstanden. Abnahme heeresmäßig. Im Zuge der Vereinfachung und gewünschten Herabminderung der Spezialfahrzeuge ist anzustreben, ob heeresübliches Kfz. Verwendung finden kann.

Kfz.-Nr. 409: (single-axle fire-extinguishing trailer)

Single purchase of 30 units. Further procurement not necessary. According to B.z.b.V. Heer the spare parts issue is not acute. If necessary spare parts will be provided by B.z.b.V. Heer.

Kfz.-Nr. 409: (Lösch-Anhänger einachsig)

Einmalige Anschaffung von 30 Stück. Neubeschaffung nicht erforderlich. Lt. B.z.b.V. Heer ist die Ersatzteilfrage nicht akut. Erforderliche Ersatzteile werden im Bedarfsfalle durch B.z.b.V. Heer gedeckt.

Kfz.-Nr. 501: (testing vehicle II)

Obtained by WuG 10 at AHA; fitting out by the Gollnow & Sohn company in LR.

Kfz.-Nr. 501: (Prüfwagen II)

Abruf durch WuG 10 bei AHA, Ausrüstung durch Firma Gollnow & Sohn in LR.

Delivery quantities / Lieferzahlen units / Stück	to end of July 44 / bis Ende Juli 44 Nr. 501	to end of 1944 / bis Ende 1944 Nr. 501
requirement / Forderung	19	28
delivered / geliefert	4 built by Kdo. Basse / v. Kdo. Basse gebaut	
shortage / Fehl	15	24

Additional deliveries currently impossible due to the lack of technical documents. Kommando Basse is responsible for these. The missing parts list will be compiled by Prüf 10. Usual H. Abn. After completion of development, proposal of Wa Prüf 10 (Major Kleidörfer) to deal with all vehicles of the 500-group separately in a meeting with Kommando Basse. The issue of the time relay's installation is not yet clarified.

Weitere Auslieferung mangels techn. Unterlagen z.Zt. nicht möglich. Zuständig hierfür Kommando Basse. Fehlende Teileliste wird durch Prüf 10 erstellt. Übliche H. Abn. Nach vollendeter Entwicklung, Vorschlag von Wa Prüf 10 (Maj. Kleidörfer) sämtliche Fahrzeuge der fünfhunderter Gruppe getrennt in einer Besprechung mit Kommando Basse zu verhandeln. Frage des Einbaus der Zeitrelais noch ungeklärt.

Kfz.-Nr. 502, Nr. 503: (cable and spare device parts vehicles)

Figures and other information exactly as Kfz.-Nr. 501.

Kfz.-Nr. 502, Nr. 503: (Kabelwagen und Ersatzteilgeräte)

Zahlen und sonstige Angaben genau wie bei Kfz.-Nr. 501.

Kfz.-Nr. 504: (heavy machinery set I)

Cancelled for acquisition by WuG. Army standard, available direct from B.z.b.V. Heer Fz In an. Requisition at the time of deploying the units.

Kfz.-Nr. 504: (schwerer Maschinensatz I)

Entfällt für Beschaffung durch WuG. Heeresüblich, fordert B.z.b.V. Heer unmittelbar bei Fz In an. Anforderung im Augenblick der Truppenaufstellung.

Kfz.-Nr. 505: (transformer trailer)
WuG 7 acquires the transformer units. Acquisition of the required specialized vehicle (a Luftwaffe trailer) will be arranged directly between WuG 7 and Prüf 4/V. Prüf 10 is to clarify if a standard army trailer can be used.

Kfz.-Nr. 505: (Umformer-Anhänger)
Umformer-Aggregat beschafft WuG 7. Beschaffung des dazu erforderlichen Sonderfahrzeuges (Luftwaffenanhänger) wird in unmittelbarem Einvernehmen zwischen WuG 7 und Prüf 4/V veranlasst. Prüf 10 klärt, ob heeresüblicher Anhänger Verwendung finden kann.

Delivery quantities / Lieferzahlen units / Stück	to end of July 44 / bis Ende Juli 44 Nr. 505	to end of 1944 / bis Ende 1944 Nr. 505
requirement / Forderung	19	28
delivered / geliefert	8	
shortage / Fehl	11	20

According to the planning of B.z.b.V (Heer) an increase of orders from WuG 10 to WuG 7 is necessary.
WuG 10 places a purchase order for 1945 with WuG 7. From WuG 7's point of view there are no delivery problems. Maintenance is the responsibility of processing WuG department. Kfz.-Nr. 505 fully with WuG 7.

Auftragserhöhung von WuG 10 an WuG 7 lt. B.z.b.V. Planung erforderlich. WuG 10 veranlasst Bestellverfügung für 1945 an WuG 7. Lieferschwierigkeiten nach Ansicht von WuG 7 nicht vorhanden. Instandsetzung obliegt der bearbeitenden WuG-Abteilung. Kfz.-Nr. 505 bei WuG 7 vollkommen kontengiert.

Kfz.-Nr. 506: (trailer for air compressor)

Conversion by WuG 12; final fitting out by WuG 5. The E3-trailer is considered for future manufacture. From B.z.b.V. Heer's point of view a requisition or request at AHA is not currently necessary. Wa Prüf 10's decision awaited.

Kfz.-Nr. 506: (Anh. für Luftverdichter)

Umrüstung WuG 12, Endausrüstung WuG 5. Für laufende Fertigung kommt der E3-Anhänger in Betracht. Anforderung bzw. Abruf nach Ansicht von B.z.b.V. Heer bei AHA im Augenblick nicht erforderlich. Entscheidung von Wa Prüf 10 abwarten.

Delivery quantities / Lieferzahlen units / Stück	to end of July 44 / bis Ende Juli 44 Nr. 506	to end of 1944 / bis Ende 1944 Nr. 506
requirement / Forderung	50	80
delivered / geliefert	9	
shortage / Fehl	41	71

Bottleneck and reason for delayed delivery: the air compressor. Manufacturer is Junkers, Munich-Allach. Action by WuG 10 at RLM, Berlin regarding the allocation of additional air compressors immediately necessary, because the prospective complete output has been confiscated by the RLM. Production is dependent on the release of the air compressors by the RLM. Because the air compressor requires an overhaul after about 200 operating hours, an appropriate reserve has to be included.

Engpass und Grund der Lieferverzögerung: Luftverdichter. Herstellfirma Junkers, München-Allach. Einschaltung von WuG 10 beim RLM, Berlin, hinsichtlich Zuteilung weiterer Luftverdichter umgehend erforderlich, da gesamte in Aussicht stehende Ausbringung vom RLM beschlagnahmt. Ausbringung von der Freigabe der Luftverdichter im RLM abhängig. Der Luftverdichter macht nach etwa 200 Betriebsstunden eine Überholung notwendig, sodass entsprechende Reserve einzuplanen ist.

Kfz.-Nr. 507: (truck for tools and spare parts)

Vehicle cancelled. Open truck immediately requested by B.z.b.V. Heer and made available to the units.

Kfz.-Nr. 507: (Lkw für Werkzeug u. Ersatzteile)

Fahrzeug entfällt. Offener Lkw von B.z.b.V. Heer unmittelbar angefordert und der Truppe zur Verfügung gestellt.

Kfz.-Nr. 509: (trailer for charging station)

Conversion by WuG 12, final outfitting by WuG 10.

Kfz.-Nr. 509: (Anh. für Ladestation)

Umrüstung WuG 12, Endausrüstung WuG 10.

Delivery quantities / Lieferzahlen units / Stück	to end of July 44 / bis Ende Juli 44 Nr. 509	to end of 1944 / bis Ende 1944 Nr. 509
requirement / Forderung	19	34
delivered / geliefert	3	
shortage / Fehl	16	31

E-3 trailers in the future. Manufacturing currently stopped due to lack of technical data. Furnishings and fittings are completed, documents for

In Zukunft E-3-Anhänger. Fertigungsstopp z.Zt. mangels techn. Angaben. Einrichtungsgegenstände sämtlich fertiggestellt, es fehlt jedoch die Uninstallation is missing. Wa Prüf 10 to provide the drawings. The shortfall in devices required for the installation will be provided by the Gollnow company to the Fageb company. H. Abn. not available at Fageb company. In the future this will be handled by Abnahme via LR.

terlage für die Einteilung derselben. Wa Prüf 10 veranlasst Beistellung der Zeichnungsunterlagen. Für Einbau erforderliche Engpassgeräte werden zukünftig von Firma Gollnow an Firma Fageb beigestellt. H. Abn. bei Firma Fageb nicht vorhanden. Erfolgt in Zukunft durch Abnahme über LR.

Kfz.-Nr. 510: (truck for FR spare parts)

Fitting out by WuG 10. Vehicle to be delivered to the units by B.z.b.V. Heer. Installation sets (shelves and boxes) have to be provided separately.

Kfz.-Nr. 510: (Lkw für FR-Ersatzteile)

Ausrüstung WuG 10. Fahrzeug durch B.z.b.V. Heer der Truppe zugestellt. Einbausätze (Regale und Kisten) müssen getrennt z. Verfügung gestellt werden.

Delivery quantities / Lieferzahlen units / Stück	to end of July 44 / bis Ende Juli 44 Nr. 510	to end of 1944 / bis Ende 1944 Nr. 510
requirement / Forderung	20	29
delivered / geliefert	5	
shortage / Fehl	15	24

Reason for delay: no manufacturing documents yet. Usual H. Abn.

Grund der Verzögerung : Noch keine Bauunterlagen. Übliche H. Abn.

Kfz.-Nr. 511: (trailer for workshop vehicle)

B.z.b.V. Heer attempting to simplify of the vehicle.

Kfz.-Nr. 511: (Anh. Werkstattwagen)

B.z.b.V. Heer strebt Vereinfachung des Fahrzeuges an.

Delivery quantities / Lieferzahlen units / Stück	to end of July 44 / bis Ende Juli 44 Nr. 511	to end of 1944 / bis Ende 1944 Nr. 511
requirement / Forderung	16	21
delivered / geliefert	3	
shortage / Fehl	13	18

E3 trailer has to be requested by WuG 10 from AHA. Closed superstructure required. Drawings and documents are missing and have to be provided by Wa Prüf 10 in order to place the order through WuG 12 as quickly as possible. H. Abn.

E3-Anhänger ist von WuG 10 bei AHA anzufordern. Geschlossener Aufbau erforderlich. Zeichnungsunterlagen fehlen und sind von Wa Prüf 10 zwecks Erteilung des Auftrages durch WuG 12 schnellstens zu erbringen. H. Abn.

Kfz.-Nr. 601: (FR transport trailer Vidal)

Kfz.-Nr. 601: (FR-Transportanhänger) Vidal

Delivery quantities / Lieferzahlen units / Stück	to end of July 44 / bis Ende Juli 44 Nr. 601
requirement / Forderung	200
delivered / geliefert	200

Still acquired by WuG 11. Vehicle currently being converted (strengthening of chassis).

Noch durch Prüf 11 beschafft. Kfz. z. Zt. im Umbau (Fahrgestellverstärkung).

Kfz.-Nr. 601 b: (FR transport trailer Vidal)

WuG 12 has already placed a new order for 250 units at the Vidal company, Hamburg-Harburg.

Kfz.-Nr. 601 b: (FR-Transportanhänger Vidal)

Neuer Auftrag durch WuG 12 an Firma Vidal, Hamburg-Harburg, auf 250 Stück bereits erteilt.

Delivery quantities / Lieferzahlen units / Stück	Nr. 601 b
requirement for August, Sept., Oct. 1944 / Forderung für August, Sept., Okt. 1944	250
delivery in 4 weeks, beginning Sept. / Lieferung 4 wöch. Anlauf Sept.	100
remainder to be delivered in October / im Oktober Lieferung restl.	150

Follow-up order for Kfz.-Nr. 601 b by B.z.b.V. Heer not yet finalized.

Anschlussauftrag für Kfz.-Nr. 601 b lt. B.z.b.V. Heer noch nicht entschieden.

C Railway vehicles

A-Stoff tanker waggon 27

Processing by WuG 5 in agreement with Prüf 4/V.

C Eisenbahnfahrzeuge

A-Stoff-Kesselwagen 27

Bearbeitung von WuG 5 im Einvernehmen mit Prüf 4/V.

Delivery quantities / Lieferzahlen units / Stück	
requirement until end of July 1944 / Forderung bis Juli 1944	100
delivered armored units / ausgeliefert gepanzert	101
completed wagons not yet armored / fertige Wagen vor der Panzerung	57

Inspection by Reichsbahn and steam boiler surveillance authority as well as H. Abn. – allocation to be carried out during the third quarter of 1944.

Abn. durch Reichsbahn und Dampfkesselüberwachungsverein sowie H. Abn. – Kontingentierung erfolgt per III. Quartal 1944

B-Stoff tanker waggons

24,000 l volume. Exclusively controlled and distributed by the OKW. So far 76 units available.

B-Stoff-Kesselwagen

24.000 l Inhalt ausschließlich von OKW zentral gesteuert und verteilt. Bisher 76 Stück vorhanden.

T-Stoff waggon 15000 l

30 units ordered from Vollmar Haenig company, Dresden and all delivered. Sufficient for 1944 and 1945.

T-Stoff-Wagen 15000 L

30 Stück bei Firma Vollmar Hänig, Dresden, bestellt und sämtlich ausgeliefert. Für 1944 und 1945 ausreichend.

8 t gantry crane from Ardelt company, Eberswalde

Transfer to WuG 5 from Prüf 4 already carried out.

Bockkran 8t der Firma Ardelt, Eberswalde

Übernahme durch WuG 5 von Prüf 4 bereits erfolgt.

Delivery quantities / Lieferzahlen units / Stück	
ordered / bestellt	44
delivered / geliefert	16
monthly delivery quota / monatliches Liefersoll	4

Order in progress.

Backorder caused by shortage of ball bearings. 10 cranes were most urgently needed, so the requirement for the time being is covered.

signed Goerke Hptm.

Bestellung wird aufrechterhalten.

Rückstand durch Kugellagerengpass begründet. 10 Stück werden vordringlich benötigt, womit Bedarf vorerst gedeckt sein dürfte.

gez. Goerke Hptm.

THE VEHICLE GROUPS
DIE FAHRZEUGGRUPPEN

The use of the A4 *Gerät** was planned for various applications:

Stationary operation	- firing from bunker positions.
Motorized operation	- firing from other prepared positions, field positions, or from rail vehicles

The special vehicles of the ground facility (BA) required for a mobile launch of the A4 *Gerät* were divided into six vehicle groups:

They are

Vehicle group I	Firing position
Vehicle group II	Remote control position
Vehicle group III	Limber position
Vehicle group IV	Fuel column
Vehicle group V	Technical battery
Vehicle group VI	Supply unit

The vehicles for railroad use were included at a later date as vehicle group VII rail vehicles.

Für das Gerät A4* wurden folgende Einsatzmöglichkeiten geplant:

Ortsfester Einsatz	- Verschießen aus Bunkerstellungen
Motorisierter Einsatz	- Verschießen aus vorbefestigter Stellung, Feldstellung oder von der Schiene

Die für einen Abschuss des A4-Gerätes dafür notwendigen Spezialfahrzeuge der Bodenanlage (BA) der mot. Abteilungen wurden in sechs Fahrzeuggruppen unterteilt:

Es sind

Fahrzeuggruppe I	Feuerstellung
Fahrzeuggruppe II	Fernlenkstellung
Fahrzeuggruppe III	Protzenstellung
Fahrzeuggruppe IV	Tankkolonne
Fahrzeuggruppe V	Technische Batterie
Fahrzeuggruppe VI	Nachschubeinrichtung

Die Fahrzeuge für den Schieneneinsatz (Eisenbahn) wurden zu einem späteren Zeitpunkt als Fahrzeuggruppe VII Schienenfahrzeuge aufgenommen.

* The code name for the long-range rocket was *Rauchspurgerät II* (Smoke Tracking Device, type 2). In technical documents, device descriptions and correspondence classified as Geheime Kommandosache (g or gK or gKom) (Command level secrets) between Wa Prüf and EW (HAP), the term rocket was not used and it was only referred to as A4 (Aggregat 4), *Gerät* A4, *Gerät* or *Sondergerät* (A4 (aggregate 4), device A4, device or special device). Wartime propaganda promoted the A4 *Gerät* to V2 (Vergeltungs waffe 2) (Retaliation or Vengeance Weapon 2) and it was widely known by that name.

The term *Gerät* (device) is retained in this book when excerpts from documents are used.

* Rauchspurgerät II ist der Tarnname der Fernrakete. In technischen Unterlagen, Gerätebeschreibungen und Schriftverkehr mit Einstufung als Geheime Kommandosache (g oder gK oder gKom) zwischen dem Wa Prüf und EW (HAP) wurde der Begriff Rakete nicht verwendet und nur mit A4 (Aggregat 4), Gerät A4, Gerät oder Sondergerät bezeichnet. Von der Kriegspropaganda wurde das A4-Gerät zur V2 (Vergeltungswaffe 2) erhoben und unter diesem Namen weitgehend bekannt.

Der Begriff Gerät wird im Buch bei Verwendung von Dokumenten in Auszügen beibehalten.

BArch, RH8II Bild-B1433-44, o.Ang.

29

29 On the square in front of the test workshop and the yard in front of the the large measuring house (BSM) a number of vehicles are parked). These vehicles belonging to the inventory of the test branch. A boarding ladder for access was positioned on the Fz.101 fire control vehicle.
An italian Fiat S.P.A. 38R (light truck) was selected for a loading test with two HB50 fan heaters, three A-Stoff pumps, various tombak hoses and accessory boxes. Similar cargo was also reserved for the Fz.507 which was replaced by standard 3 t trucks.

29 Auf dem Platz vor der Prüfwerkstatt und dem Hof vor dem großem Messhaus (BSM) sind eine Reihe von Fahrzeuge abgestellt. Sie gehören zum Inventar der Versuchsabteilung.
Am Feuerleitpanzer Fz.101 wurde eine Einstiegsleiter angestellt.
Ein italienischer Fiat S.P.A. 38R (leichter Lkw) wurde für einen Beladeversuch ausgewählt und mit zwei Heizlüftern HB50, drei A-Stoffpumpen, diversen Tombak-Schläuchen und Zubehörkisten beladen. Ähnliches Ladegut war auch dem Fz.507 vorbehalten, der aber durch einen Standard 3 t Lkw's ersetzt wurde.

General information

For the vehicles of groups I to VI, truck chassis with different payloads of 2 t, 3 t, 4.5 t manufactured by Steyr, Opel, Buessing-NAG and Mercedes were chosen, some of which were equipped with four-wheel drive. Single-axle chassis for 900 kg and 1.500 kg payloads and 2-axle chassis with 3 t and 5 t payloads were available as trailers. Trailers and vehicles already in use by the Heer and Luftwaffe were supplemented or converted for their tasks (e.g. the vehicles 106, 107, 201, 202, 205, 206, 301, 302, 303, 304, 408, 501, 502, 503, 504, 505, 508, 509, 510 and 511). The installations, transmitters, receivers, equipment racks, transport racks etc. were integrated into existing box bodies (Kfz.305) and placed on the chassis. Tanker vehicles of the Luftwaffe were adapted for their use of B- and T-Stoff. Only a few bodies were newly developed and mounted on selected chassis and trailers (e.g. vehicle 101, 108, 110, 208, 406, 407 and 506). New developments were limited to vehicles 102, 103, 204, 207, 401 and 402.

The individual vehicles are presented in detail for each vehicle group. In addition to the official designation Fz.-Pos.-Nr., the texts also include the internal designation of the production service, such as *Feuerleitpanzer, Meillerwagen, Meillertisch, Strüver, Campina I*, etc., as well as in reports on the development status, production figures and other special features. In the documents, various terms were used for naming the vehicles, such as Pos., Pos.Nr. or Fz.-Pos.-Nr. which were rewritten as Fahrzeug xxx or Fz.xxx.

Allgemeines

Für die Fahrzeuge der Fahrzeuggruppe I bis VI wurden Lkw-Fahrgestelle mit verschiedenen Nutzlasten von 2 t, 3 t, 4,5 t der Hersteller Steyr, Opel, Büssing-NAG und Mercedes gewählt. Diese waren teilweise mit Allradantrieb ausgerüstet. Als Anhänger standen Ein-Achs-Fahrgestelle für 900 kg und 1.500 kg Nutzlast, für Zwei-Achs-Fahrgestelle solche mit 3 t und 5 t Nutzlast zur Verfügung. Bereits bei Heer und Luftwaffe in Verwendung befindliche Anhänger und Fahrzeuge wurden für ihre Aufgabe entsprechend ergänzt oder umgebaut (z.B. die Fahrzeuge 106, 107, 201, 202, 205, 206, 301, 302, 303, 304, 408, 501, 502, 503, 504, 505, 508, 509, 510 und 511). Die Einbauten, Sender, Empfänger, Gerätegestelle und Transportgestelle wurden in vorhandene Kofferaufbauten (Kfz.305) integriert und auf die Fahrgestelle gesetzt. Die Tankfahrzeuge der Luftwaffe passte man zur Verwendung von B- und T-Stoffen an. Nur wenige Aufbauten wurden neu entwickelt und auf ausgewählte Fahrgestelle und Anhänger montiert (z.B. Fahrzeug 101, 108, 110, 208, 406, 407 und 506). Neuentwicklungen beschränkten sich auf die Fahrzeuge 102, 103, 204, 207, 401 und 402.

Die einzelnen Fahrzeuge werden innerhalb der jeweiligen Fahrzeuggruppe im Detail vorgestellt. In den Texten werden neben der offiziellen Benennung Fz.-Pos.-Nr. auch die interne Bezeichnung des Fertigungsdienstes wie z.B. *Feuerleitpanzer, Meillerwagen, Meillertisch, Strüver, Campina I* usw. genannt und überlieferte Berichte über den Entwicklungstand, Fertigungszahlen und andere Besonderheiten angeführt. In den Dokumenten finden für die Bezeichnung der Fahrzeuge verschiedene Begriffe wie Pos., Pos.-Nr und Fz.-Pos.-Nr. Verwendung. Diese Bezeichnungen wurden auf Fahrzeug xxx oder Fz.xxx umgeschrieben.

VEHICLE GROUP I
FAHRZEUGGRUPPE I

Firing position

Intended use
The firing position (launching site) contains the vehicles necessary for firing in order to perform the launch operation. An essential part of the firing position is the vehicle 101, which was used for the necessary tests, telephone connection to the BA vehicles and the launch command.

In training and deployment preparations, the crews were drilled on the vehicles and in the starting procedures with starting positions I to III.

Start procedures for the operation according to mot.I with Strüver circuitry, the Strüver machine set 11 as central power supply for up to three starting positions, and operations according to mot.II with a Steyr power supply car for each of the starting positions I to III were developed, tested and practiced.

BAMA, RH8-1310
Spezialfahrzeuge
Baugruppen und
wesentliche Bauteile
der Bodenanlage FR

Special vehicles in the fire position

Fz.-Pos.-Nr.101*	Special vehicle with armored special body
Fz.-Pos.-Nr.102	FR trailer (S) (3-axle)
Fz.-Pos.-Nr.103	Launching platform trailer (1-axle)
Fz.-Pos.-Nr.104	Machine set 11
Fz.-Pos.-Nr.105	Power supply car on 2 t truck (Steyr)
Fz.-Pos.-Nr.106*	Test vehicle I on 3 t truck closed (305)
Fz.-Pos.-Nr.107*	Test vehicle I/III on 3 t truck closed (305)
Fz.-Pos.-Nr.108	Cable reel trailer (2-axle)
Fz.-Pos.-Nr.109**	Special-Kw Tower car
Fz.-Pos.-Nr.110	Small cable reel trailer (1-axle)

* Operation mot.I add. »A« with Strüver circuitry in connection with Fz.104
** Vehicle not applicable

Feuerstellung

Verwendungszweck
Die Feuerstellung (Startplatz) besteht aus den für den Abschuss notwendigen Fahrzeugen um den Startbetrieb durchzuführen. Wesentlicher Bestandteil der Feuerstellung war das Fahrzeug 101, über das die notwendigen Prüfungen, Fernsprechverbindung zu den BA-Fahrzeugen und das Startkommando ausgeführt wurden.

In den Trainings- und Einsatzvorbereitungen wurden die Abteilungen an den Fahrzeugen und in den Startabläufen mit Startplätzen I bis III gedrillt.

Es wurden Startverfahren für den Betrieb nach mot.I mit Strüverschaltung, dem Strüver Maschinensatz 11 als zentrale Stromversorgung für bis zu drei Startplätzen, und der Betrieb nach mot.II mit einem Steyr Stromversorgungswagen für jeden der Startplätze I bis III entwickelt, erprobt und eingeübt.

Spezialfahrzeuge in der Feuerstellung

Fz.-Pos.-Nr.101*	Sonder-Kfz. mit gep. Spezialaufbau	Feuerleitpanzer
Fz.-Pos.-Nr.102	FR-Anhänger (S) (3-achs.)	Meillerwagen
Fz.-Pos.-Nr.103	Abschuß-Plattform-Anh. (1-achs.)	Meillertisch
Fz.-Pos.-Nr.104	Maschinen-Satz 11 Strüver	Strüver
Fz.-Pos.-Nr.105	Lkw 2 t Stromversorg. Wg.	Steyr
Fz.-Pos.-Nr.106*	Lkw geschl. (305) Prüfwg. I	Prüfwagen I
Fz.-Pos.-Nr.107*	Lkw geschl. (305) Prüfwg. III	Prüfwagen III
Fz.-Pos.-Nr.108	Kabeltrommel Anhänger (2-achs.)	
Fz.-Pos.-Nr.109**	Sonder-K Turmwagen	
Fz.-Pos.-Nr.110	kl. Kabeltrommel Anhänger (1-achs.)	

* Betrieb mot.I Zusatz »A« mit Strüverschaltung in Verbindung mit Fz.104
** Fahrzeug entfallen

BArch, RH8II Bild-B0517-44, o.An

30

30 March 16, 1944 – with a signal rocket in a descending smoke trail, the launch of an A4 rocket (17018) was announced on test stand VII. The launch took place at 11:04. Briefly the exercise on test stand X was halted for a firing to mot.II, with several launching sites. One A4 *Gerät* is already in place, other *Gerät* are being moved in or are still waiting to be set up. The *Feuerleitpanzer* is in position, the »Steyr« power supply vehicle is parked on the right. For the operation in September all Abteilungen were trained and equipped on mot.II. The A4 *Gerät* was mostly fired from only one launch site during the operation.

30 16. März 1944 – mit einer Signalrakete in niedergehender Rauchspur wurde der Start einer A4-Rakete (17018) auf dem Prüfstand VII angekündigt. Der Start erfolgte um 11:04. Kurzzeitig wurde die Übung auf dem Prüfstand X für ein Schießen nach mot.II, der Verwendung mehrerer Startstellen unterbrochen. Ein A4-Gerät ist bereits aufgestellt, weitere Geräte werden herangefahren bzw. warten noch auf ihre Aufstellung. Der Feuerleitpanzer ist in Stellung gebracht, rechts ist der Steyr-Stromversorgungswagen geparkt. Für den Einsatz im September wurden alle Abteilungen auf mot.II trainiert und ausgerüstet. Das A4-Gerät wurde im Einsatz meistens nur von einem Startplatz verschossen.

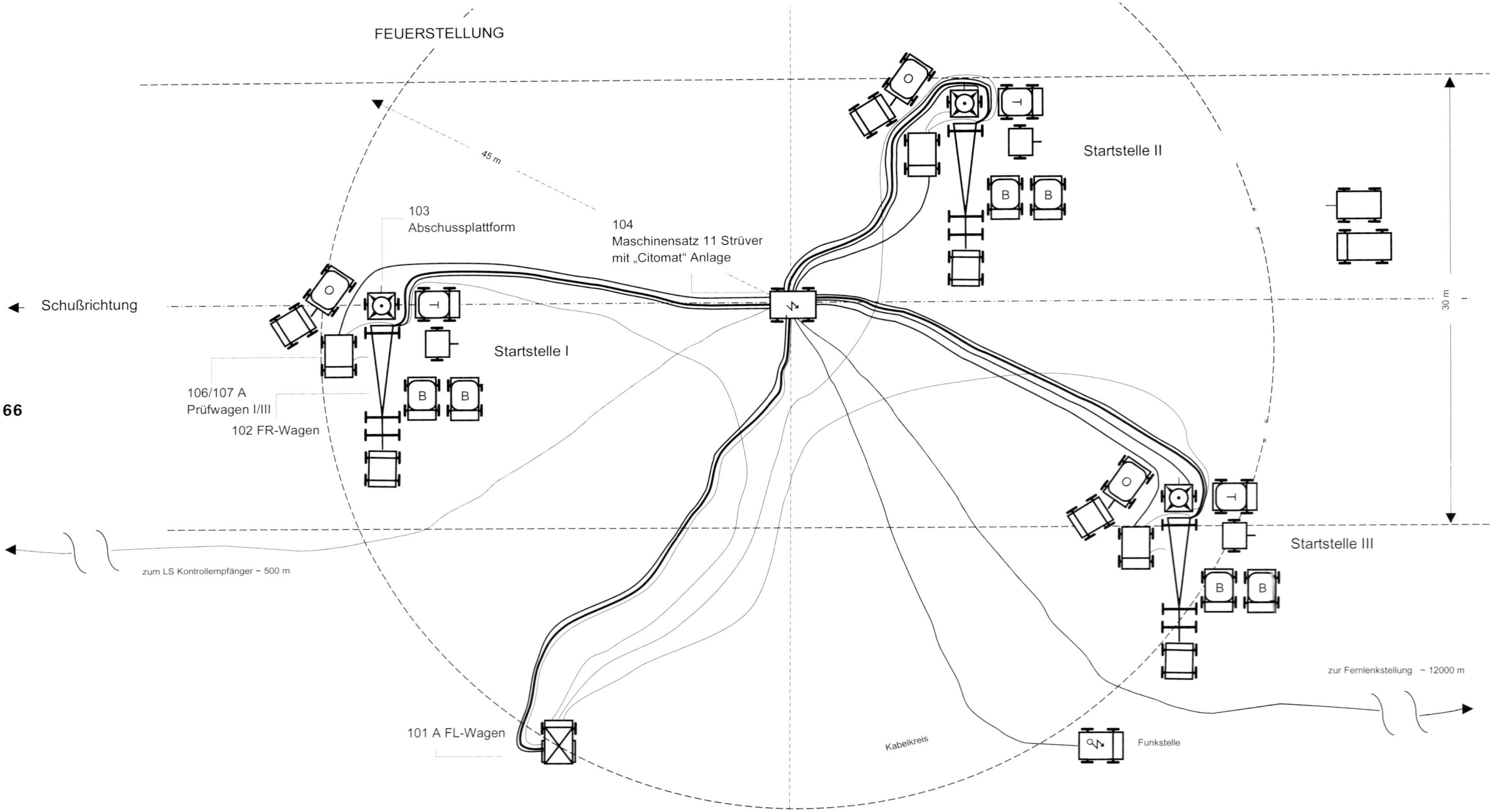

31

Diagram of a FR battery mot.I in the firing position

Schema einer FR-Batterie mot.I in Feuerstellung

31 The necessary vehicles are set up in the firing position. It is marked by an approx. 30 m wide firing corridor aligned in a radius of 45 m around the power generation vehicle Fz.104 »Strüver«. From the Fz.104 FF cables lead from the Citomat unit to the vehicles of the firing position and connect the telephone lines. In the mot.I circuit the cable connections lead for the test and start commands between the vehicle Fz.101 via Fz.104., that ensures the commands via heavy flak cables to the Fz.102 and the necessary power supply and communication to the vehicles of the firing position. Vehicles Fz.101 and Fz.106/107 carry the suffix A.

31 Die erforderlichen Fahrzeuge sind innerhalb der Feuerstellung aufgestellt. Sie wird durch einen ca. 30 m breiten Schusskorridor bestimmt, der in einem Radius von 45 m um den Stromerzeugungswagen Fz.104 »Strüver« ausgerichtet ist. Aus dem Fz.104 führen FF-Kabel von der Citomat-Anlage zu den Fahrzeugen der Feuerstellung und stellen die Fernsprechverbindungen her. In der mot.I-Schaltung führen die Kabelverbindungen für die Prüf- und Startbefehle zwischen dem Fahrzeug Fz.101 über das Fz.104, das die Befehle über schwere Flakkabel zu den Fz.102 und die notwendige Stromversorgung und Kommunikation über Kabel zu den Fahrzeugen der Feuerstellung sicherstellt. Die Fahrzeuge Fz.101 und Fz.106/107 führen den Zusatz A.

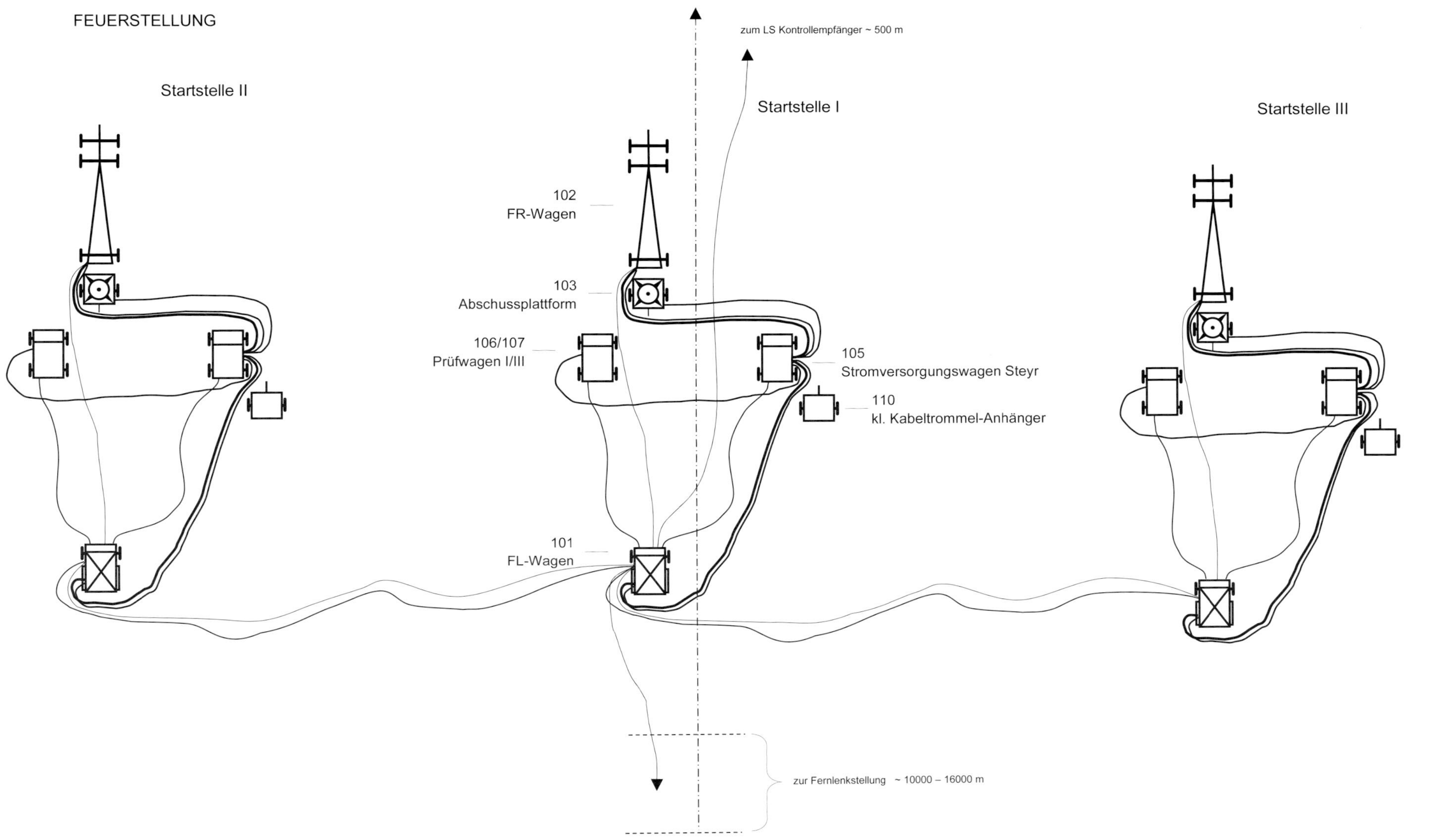

32

Diagram of a FR battery mot.II with up to three starting positions

Schema einer FR-Batterie mot.II mit bis zu drei Startpositionen

32 For each launch site, power is supplied by a Steyr power generator vehicle Fz.105 to the Fz.102, Fz.106/107 and *Feuerleitpanzer* Fz.101. When a second or third launch site was set up, the FF cable connections were routed via the Fz.101 located in launch site one in the fire control center. In the mot.II circuit, the cable connections for the test and firing commands between the vehicles Fz.101 are routed via Fz.105, which forwards the commands via flak cables to Fz.102 and there to the Stotz I + Stotz II connections of the A4. An exchange of the vehicle mot.I to mot.II is not possible.

32 Je Startplatz erfolgt die Stromversorgung über ein Steyr Stromerzeugerfahrzeug Fz.105 an die Fahrzeuge Fz.102, Fz.106/107 und das Feuerleitpanzer Fz.101. Bei Errichtung eines zweiten oder dritten Startplatzes werden die Verbindungen der FF-Kabel über das Fz.101 geführt, das sich am Startplatz eins in der Feuerstellung befindet. In der mot.II-Schaltung werden pro Startplatz die Kabelverbindungen für die Prüf- und Startbefehle zwischen den Fahrzeugen Fz.101 über das Fz.105 geführt, das die Befehle über Flakkabel an das Fz.102 und dort an die Stotz I + Stotz II Anschlüsse der A4 weiterleitet. Ein Tauschen der Fahrzeug mot.I nach mot.II ist nicht möglich.

DMA, GD639.5.18
HAP11/T 507
Karlshagen, 1.6.1944
Bd.Nr. E 3195/44 g

DMA, GD639.5.1
EW234
Fz.-Pos.-Nr.101

DMA, GD639.51.6
Auszug
Bedienungsanweisung
Feuerleitpanzer

DMA, GD630.02
A4 Lehrbuch
der FR-Schule
Stand Mai 1944
Auszug aus
Technische Fahrzeuge

Vehicle 101

Internal or former designation: Fz.-Pos.-Nr.101
Special vehicle with armored special body
***Feuerleitpanzer* (armored fire control vehicle)**

Intended use

The purpose of the *Feuerleitpanzer* (armored fire control vehicle) is to operate the *Gerät* in firing position and at the same time to protect the operator during firing.

It is its task to check the aggregate before firing and to perform the actual firing. In order to maintain mutual communication between the individual vehicles in the firing position, vehicle 101 is equipped with a flap cabinet for 20 lines with which the telephone connections are made. The X-times are transmitted via telephone lines.

The *Feuerleitpanzer* has the following dimensions and weights:

Total length:	6,850 mm
Width:	2,400 mm
Height:	3,000 mm
Track front wheels:	2,000 mm
Track chaine:	1,800 mm
Tare weight:	12,000 kg
Weight:	~ 12,920 kg
Wheels:	270-20 or 9.75-20

The *Feuerleitpanzer* consists of

- Chassis
- Armored body
- Installations

Chassis

Scope of supply

The *Feuerleitpanzer* consists of the chassis of the 8 t medium tractor vehicle (SdKfz.7) with special superstructure.

Mechanical construction

Motor vehicle Fz.101 has the motor vehicle designation:
»Special vehicle 7« with armored body and consists of the chassis of the 8 t medium tractor KM 11m. It is driven by caterpillar tracks, so that sufficient off-road mobility is guaranteed.
The total weight is 12 t.
The cable winch normally installed on the SdKfz.7 has been omitted.

Fahrzeug 101

Interne oder frühere Bezeichnung: Fz.-Pos.-Nr.101
Sonder-Kfz.7 mit gepanzertem Aufbau
Feuerleitpanzer

Verwendungszweck

Der Feuerleitpanzer hat den Zweck, das Gerät in Feuerstellung zu bedienen und bietet gleichzeitig der Bedienmannschaft Schutz beim Abschuss.

Es ist dessen Aufgabe die Aggregate vor dem Abschuss zu prüfen und den eigentlichen Abschuss durchzuführen. Um eine gegenseitige Verständigung zwischen den einzelnen Fahrzeugen in der Feuerstellung zu erhalten, ist in Fahrzeug 101 ein Klappenschrank für 20 Leitungen vorgesehen, mit dem die Fernsprechverbindungen durchgeführt werden. Die X-Zeiten werden über Fernsprechleitungen weitergegeben.

Der Feuerleitpanzer hat folgende Maße und Gewichte:

Gesamtlänge:	6.850 mm
Breite:	2.400 mm
Höhe:	3.000 mm
Spur Vorderrad:	2.000 mm
Spur Laufkette:	1.800 mm
Leergewicht:	12.000 kg
Gefechtsgewicht:	~ 12.920 kg
Räder:	270-20 oder 9,75-20

Der Feuerleitpanzer besteht aus

- Fahrgestell
- Gepanzerter Aufbau
- Einbauten

Fahrgestell

Lieferumfang

Der Feuerleitpanzer besteht aus dem Fahrgestell des mittleren Zugkraftwagens 8 t (SdKfz.7) mit Spezialaufbau.

Mechanischer Aufbau

Das Fahrzeug 101 hat die kraftfahrzeugmäßige Benennung:
»Sonder-Kfz.7« mit gepanzertem Aufbau und besteht aus dem Fahrgestell des mittleren Zugkraftwagens 8 t KM 11m. Der Antrieb erfolgt über Gleisketten, sodass eine hinreichende Geländegängigkeit gewährleistet ist.
Das Gesamtgewicht beträgt 12 t.
Die normalerweise am SdKfz.7 angebrachte Seilwinde wurde nicht montiert.

33 Armored body
On the right side of the special armored body (10 mm thickness) is located a door for the operating crew. Several armored glass vision ports are provided to observe the processes within the firing position. A hinged commander's hatch is installed on the roof of the special superstructure.

33 Gepanzerter Aufbau
Auf der rechten Seite des gepanzerten Spezialaufbaus (Stärke 10 mm) befindet sich eine Eingangstür für die Bedienungsmannschaften. Zur Beobachtung der Vorgänge innerhalb der Feuerstellung sind mehrere verglaste Sehschlitze vorhanden. Für den Kommandanten ist an der Decke des Spezialaufbaus eine aufklappbare Kommandantenluke eingebaut.

The first *Feuerleitpanzer* (armored fire control vehicle) is completed.
The handover of the vehicle took place on 18.6.1943 on the factory premises of Waggonbau Busch at the Bautzen site. The vehicle was then transferred to Peenemünde, where the installation of the control panels and telephone system was arranged.

Der erste Feuerleitpanzer ist fertiggestellt.
Die Übergabe des Fahrzeuges fand am 18.6.1943 auf dem Werksgelände der Firma Waggonbau Busch am Standort Bautzen statt. Anschließend wurde das Fahrzeug nach Peenemünde überführt und dort die Einrüstung der Steuerpulte und Telefonanlage veranlasst.

BTB

34

34 On the first completed body the armored glass vision port on the right side is missing. However, film footage shows the vision port on the left side.

34 Am ersten Aufbau fehlt noch der verglaste Sehschlitz auf der rechten Seite. Filmaufnahmen zeigen den Sehschlitz jedoch auf der linken Seite eingerüstet.

DMA, GD639.5.11
T404
o.O, den 15.6.1944
Bd.Nr. E 2197/44 gK
Auszug
Teileliste Fz.101

Vehicle order

Fahrzeugbestellung

101 **Parts list** / Teileliste	**Item** Gegenstand Stückzahl	**Quantity ordered** Bestellung je Fahrzeug	**Quantity per vehicle** Stück	**Subject** **Drawing number** Gegenstand Zeichnungs-Nummer	**ordered** **by order number** Besteller Bestell-Nummer	**Supplier** Lieferfirma
Fahrzeugbestellung	1	56	1	FL-Panzer-Aufbau	OKH Wa Prüf 11/III	Busch, Bautzen
Ausrüstungsfirmen	2	56	1	Triebwerkstaffel		Gollnow & Sohn, Koblenz

Production overview

Fertigungsübersicht

Remark / Anmerkung	9/43	10/43	11/43	12/43	1/44	2/44	3/44	4/44	5/44	6/44	7/44	8/44	9/44	10/44	11/44	12/44	1/45	2/45	3/45
Fgst delivered by Borgward to Busch Fgst Lieferung von Borgward an Busch	3				5	7		15					12	*)					
Superstructure assembly by Busch Aufbau montiert bei Busch		3				5	7		15					12					
Final outfitting by Gollnow & Son Endausrüstung durch Gollnow & Sohn			3			3	2	4	3	3	4	5	3	3	3	3	3		
Cumulative total delivered to Abt of Division z.V. Kumulierte Menge ausgeliefert an Abt Division z.V.				3*	3	3	6*	8	12	15	18*	22*	27*	30**	33**	36**	39**	42**	

*) Production stop by bombing raid
* Proved by records
** No records found found – Qty calculated

*) Fertigungsstopp durch Bombenschäden
* Belegt durch Aufzeichnungen
** Keine Aufzeichnungen vorhanden – Stückzahl geschätzt

The overview was compiled with the help of production documents, acceptance and delivery figures of the Borgward company for the Zgkw 8 t chassis for *Feuerleitpanzer* in 1943 and 1944*. These are prepared by the documented numbers of finished vehicles delivered to the mot Abteilungs. Vehicles in use by EW and the command center S are not included.

* Provided by Hilary Doyle.

Die Übersicht wurde mit Hilfe von Fertigungsunterlagen, Abnahme- und Lieferzahlen der Firma Borgward zum Zgkw 8 t Fahrgestell für Feuerleitpanzer im Jahr 1943 und 1944 zusammengestellt.* Diese sind um die belegten Stückzahlen an ausgelieferten Fertigfahrzeugen an die Abteilungen der Division z.V. erstellt. Fahrzeuge in Verwendung durch EW und der Kommandostelle S sind nicht enthalten.

* Zur Verfügung gestellt von Hilary Doyle.

BArch, RH8II Bild-D12-44, o.Ang.

35

DMA, GD639.5.21
OKW/Wa Prüf10
Karlshagen, 25.03.1944
Bd.Nr. 747/44 gK
Auszug – Besprechung der Fahrzeuggruppe 1 vom 24./25.3.44 im Lager Rebstock

Production status: Vehicle 101

3 FL-Panzer have been delivered to Abteilung 836 to date.
2 FL-Panzer for Abteilung 485 are currently in K. (Karlshagen) for final testing. The third FL-Panzer for this Abteilung has been checked in the LR (Lager Rebstock) and is ready for transport.

Subsequent completions will take place according to the following plan: 1 FL-Panzer with the same layout as the one for Abteilung 485 is currently located at the test group; an additional one will arrive at the test group on March 28. Subsequently, a FL tank in the version for department 836 will be delivered, which will be sent to the test group by 5.4. Ready for shipment (according to immediate program) is this FL tank on 7.4. A further 3 FL tanks in the version for the Abteilung 485 were to join the test group on 4, 8 and 10.4. Due to the insertion of the FL-Panzer for Abteilung 836 these dates will be delayed by about 6 days.

Current defects and their remedy

In terms of circuitry, no complaints were made by the units or at the meeting. The previous danger of an unintended program sequence caused by an incautious touch of the switch is now eliminated by the switch position »Fire«. The current drawback of the FL-Panzer is the penetration of moisture through the observation slits and the occurrence of water condensation. In the view of those attending, the two lower observation slits are not essential for operation and can therefore subject to the decision of B.z.b.V. Heer be omitted, whilst in the event of an accident or for those in the FL-Panzer already delivered their glass blocks can be removed. Then, an insulating intermediate wall can be fitted to the sloping interior wall, thus providing a drainage channel for the condensation water underneath it. Change Case A. Execution during manufacturing is possible without major delays. Change by the units according to instructions through HAP 11.

The units still complain about the clutch being too weak. Pursuit of the matter by Wa Prüf 6. d. Hauptmann Huck of Abteilung 836 suggests that an attachment device for the luffing gear is provided close to the protection box of the FF cable connections. This will make the allocation of telephone lines of the Battery network easier.

The decision in this matter is to be made by B.z.b.V. Heer. The change can be executed by the units.

Fertigungslage: Fahrzeug 101

Bisher sind 3 FL-Panzer für die Abteilung 836 ausgeliefert.
2 FL-Panzer für die Abteilung 485 sind in K. (Karlshagen) zur Schlusserprobung. Der dritte FL-Panzer dieser Abteilung steht geprüft im LR (Lager Rebstock) zum Abtransport bereit.

Die weitere Fertigstellung erfolgt nach folgendem Plan: In gleicher Ausführung wie für die Abteilung 485 befindet sich 1 FL-Panzer z.Zt. in der Prüfgruppe, ein weiterer kommt am 28.3 zur Prüfgruppe. Anschließend wird ein FL-Panzer in der Schaltung für Abteilung 836 nachgeliefert, der bis zum 5.4 zur Prüfgruppe kommt. Versandbereit (gem. Sofortprogramm) ist dieser FL-Panzer am 7.4. Weitere 3 FL-Panzer in der Schaltung der Abteilung 485 sollten am 4., 8 und 10.4 zur Prüfgruppe kommen. Durch die Zwischenlegung des FL-Panzers für Abt.836 verschieben sich diese Termine etwa um 6 Tage.

Derzeitige Mängel und ihre Behebung

Schaltungsmäßig wurden seitens der Truppe und bei der Besprechung keine Beanstandungen vorgebracht. Die frühere Gefahr des ungewollten Programmablaufes, bei unvorsichtiger Berührung des Schalters, ist bei Schalterstellung »Schießen« nun behoben. Nachteilig ist z.Zt. beim FL-Panzer das Eindringen von Feuchtigkeit durch die Sehschlitze und das Auftreten von Schwitzwasser. Da nach Ansicht der anwesenden Truppe die unteren beiden Sehschlitze für den Betrieb nicht erforderlich sind, können sie die Entscheidung durch den B.z.b.V. Heer vorausgesetzt – in Fortfall kommen bzw. bei bisher ausgelieferten FL-Panzern abgedeckt werden unter Entfernung der eingesetzten Glasblöcke. An der schrägen Innenwand lässt sich dann eine isolierende Zwischenwand einbauen und unter dieser einen Schwitzwasserablauf schaffen. Änderungsfall A. Durchführung in Fertigung ohne wesentliche Verzögerung möglich. Formänderung durch Truppe nach Anleitung durch HAP 11.

Die Truppe beanstandet noch die zu schwache Kupplung. Verfolgung der Angelegenheit durch Wa Prüf 6. Hauptmann Huck von Abt.836 schlägt vor, dass in der Nähe des Schutzkastens für die FF-Kabelanschlüsse eine Befestigungsvorrichtung für den Abspannbock vorgesehen wird. Damit die Zuführung der Fernsprechleitungen des Batterienetzes erleichtert wird.

Entscheidung auch in diesem Punkt durch B.z.b.V. Heer. Die Formänderung kann durch die Truppe durchgeführt werden.

BArch, RH8II Bild-B1782-44, o.Ang.

36

36 The armored superstructure was mounted with metal angles directly on the chassis frame and a mudguard was screwed on. On the left mudguard plate a storage box, a 5 t jack and a crowbar with length 180 cm were mounted. A telephone connection box is mounted on the superstructure, above it is a small cover which closes a lead-through. A power socket 4-polig and an Flaksteckdose 104-polig are mounted towards the rear. A power line leads from the illumination battery in the storage box to the rear, in addition two Wehrmacht canisters for water and petrol are inserted in holders at the body. On the greyish green primed superstructure, the mounting holes and welding points sprayed or painted over in dark yellow stand out.

36 Der gepanzerte Aufbau wurde mit Hilfe von Metallwinkeln direkt auf den Fahrgestellrahmen montiert und eine Kotflügelabdeckung angeschraubt. Auf dem linken Kotflügelblech wurde ein Staukasten, eine 5 t-Stockwinde, eine 180 cm lange Brechstange und am Aufbau ein Telefonanschlusskasten montiert. Darüber befindet sich ein kleiner Deckel, der eine Durchführung verschließt. Zum Heck hin sind eine 4-polige Kraftsteckdose und eine 104-polige Flaksteckdose angebracht. Eine Kabelleitung führt von der Beleuchtungsbatterie im Staukasten zum Heck. Zusätzlich sind zwei Wehrmachtskanister für Wasser und Benzin in Halterungen am Aufbau eingesetzt. Auf dem graugrün grundierten Aufbau heben sich die mit dunkelgelber Farbe übersprühten bzw. überstrichenen Befestigungslöcher und Schweißpunkte ab.

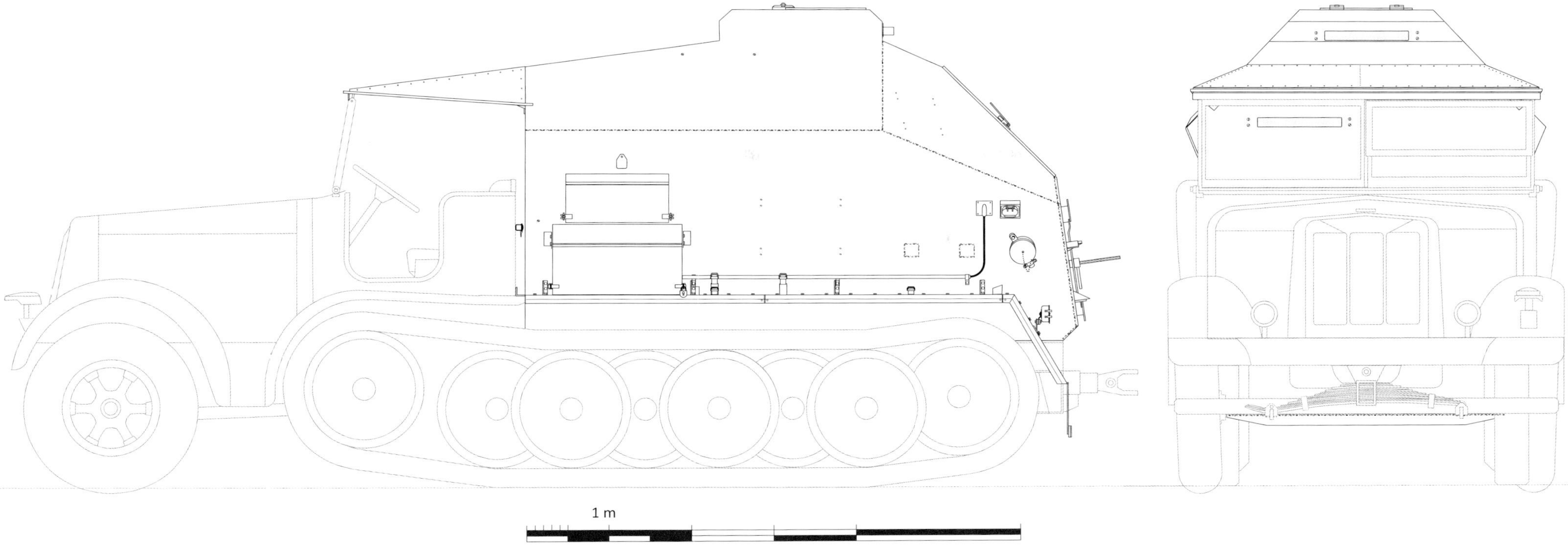

Scale / Maßstab 1:35 © Volker Ruff

BArch, RH8II Bild-B1386-44, o.Ang.

37

BArch, RH8II Bild-B1784-44, o.Ang.

38

BArch, RH8II Bild-B1785-44, o.Ang.

39

37, 38 On the left side of the armored structure there was the junction box for the telephone connections, here in the final version with the connections to Fz.102 for the launch sites I, II and III as well as to Fz.201 and Fz.208. Beside the junction box three telephone sockets were still attached to earlier *Feuerleitpanzern*, these were later omitted.

39 Below on the left mudguard was a storage box mounted. The illumination battery for supplying the interior lighting, and five to six oil bottles were located there.

37, 38 Auf der linken Seite des gepanzerten Aufbaus befand sich der Telefonanschlusskasten für die Fernsprechverbindungen, hier in der Abschlussausführung mit den Anschlüssen zum Fz.102 für die Startplätze I, II und III, sowie zum Fz.201 und Fz.208. An früheren Feuerleitpanzern waren neben dem Telefonanschluß-kasten noch drei Telefonsteckdosen angebracht. Diese entfielen später.

39 Darunter befand sich auf dem linken Kotflügelblech ein Staukasten. Dort wurde die Beleuchtungsbatterie für die Versorgung der Innenbeleuchtung und fünf bis sechs Ölflaschen untergebracht.

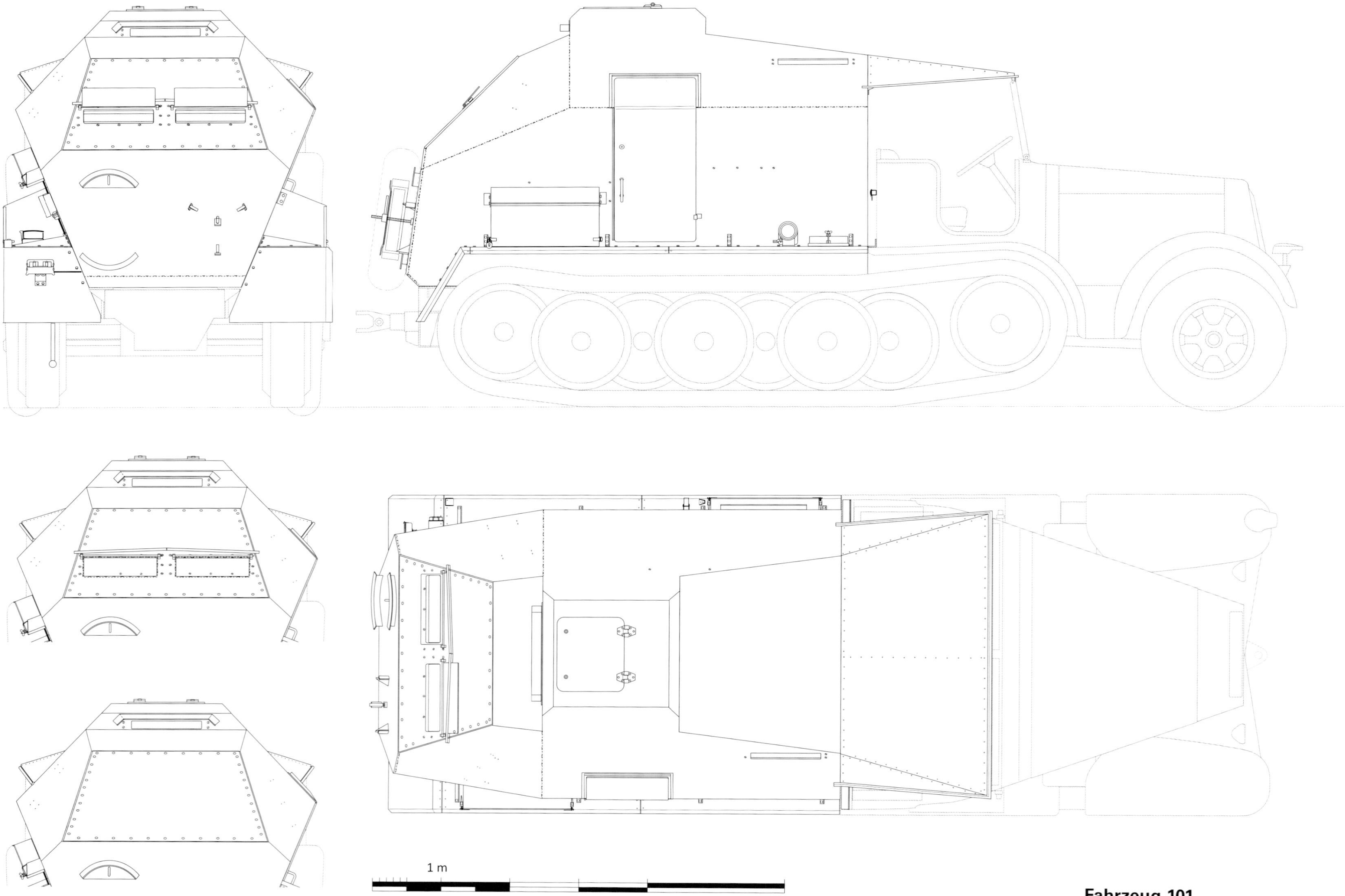

Scale / Maßstab 1:35 © Volker Ruff

Fahrzeug 101
Sonder-Kfz.7
mit gepanzertem Aufbau

Partially used drawings acc. to group listing 21Gr213

Teilweise verwendete Zeichnungen nach Gruppenliste 21Gr213

BArch, RH8II Bild-B1783-44, o.Ang.

40

40 On the right side of the Fz.101 is the access to the inner rooms. On the left of the entrance door a tool box with three drawers is mounted on the mudguard. The support block and a 4 t jack were mounted on the mudguard to the right of the entrance door and a hoe and spade were held against the back wall of the crew bench. The tractor has a camouflage paint and was obviously supplied from the repair shop for the change into a Fz.101. On both fenders are camouflage net made of Astralon. The inscription »35 km« for a speed limit of 35 km/h can be seen on the side wall of the cabin as well as a marking »1B« which indicates a test wiring with the »Wierer circuit« (mot.II). For differentiation of mot.I and mot.II the Fz.101 was called 1A and 1B.

40 Auf der rechten Seite des Fz.101 befindet sich der Zugang zu den inneren Räumen. Links neben der Eingangstür ist auf dem Kotflügel ein Werkzeugkasten mit drei Schubfächern, auf der rechten Seite ein Auflageklotz und ein 4 t-Wagenheber auf das Kotflügelblech montiert. An der Rückwand der Mannschaftssitzbank wurde eine Kreuzhacke und ein Spaten gehaltert.
Der Tarnanstrich der Zugmaschine weist darauf hin, dass das Fahrzeug aus der Instandsetzung zugeführt wurde. Auf beiden Kotflügeln liegen Tarnmatten aus Astralon. An der Seitenwand ist die Aufschrift »35 km« zur Geschwindigkeitsbeschränkung auf 35 km/h zu erkennen. Ebenso eine Markierung »1B«, was auf eine Testverkabelung mit der »Wierer-Schaltung« (mot.II) schließen lässt. Zur Unterscheidung von mot.I und mot.II wurde das Fz.101 als 1A und 1B bezeichnet.

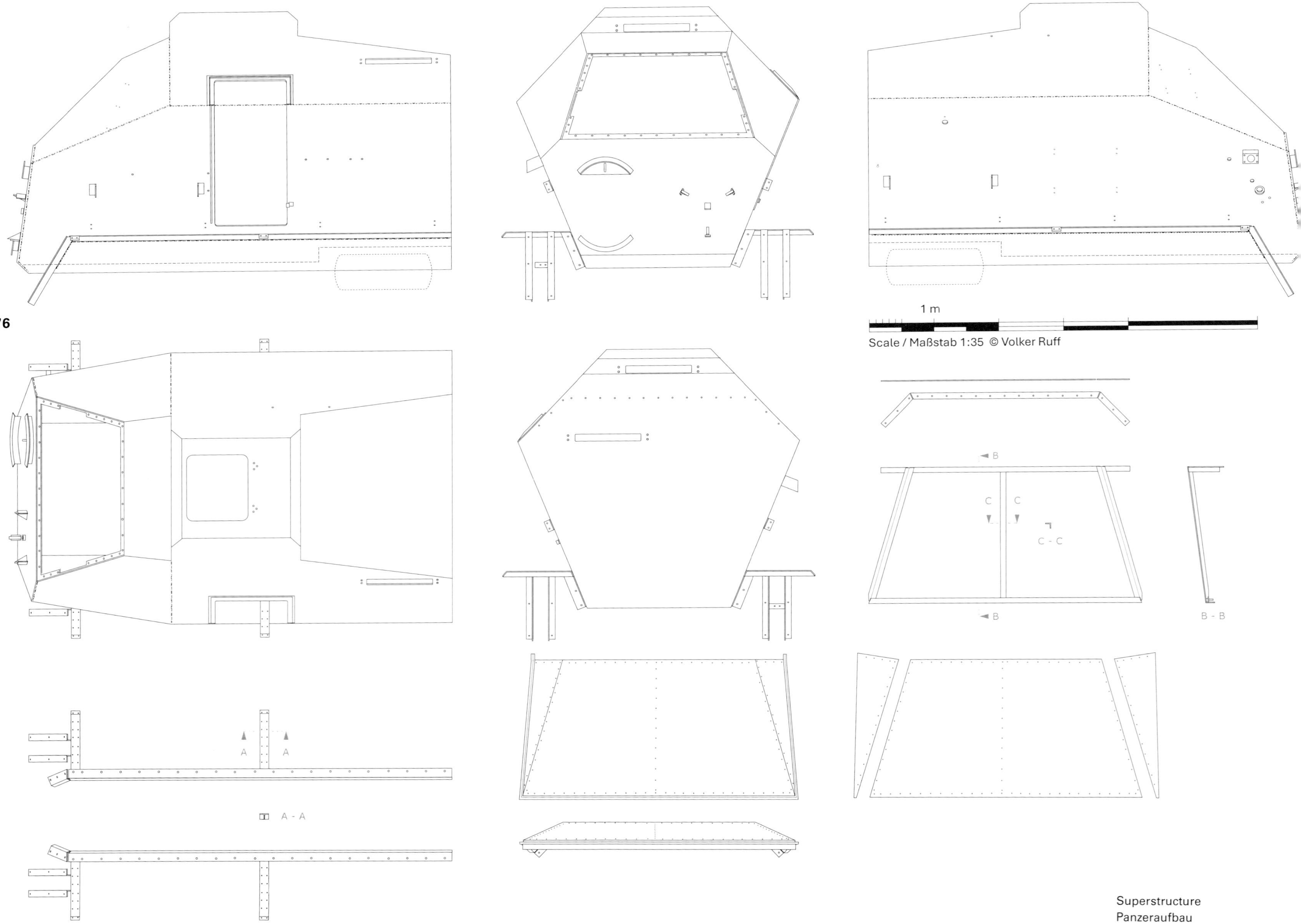

Scale / Maßstab 1:35 © Volker Ruff

Superstructure
Panzeraufbau

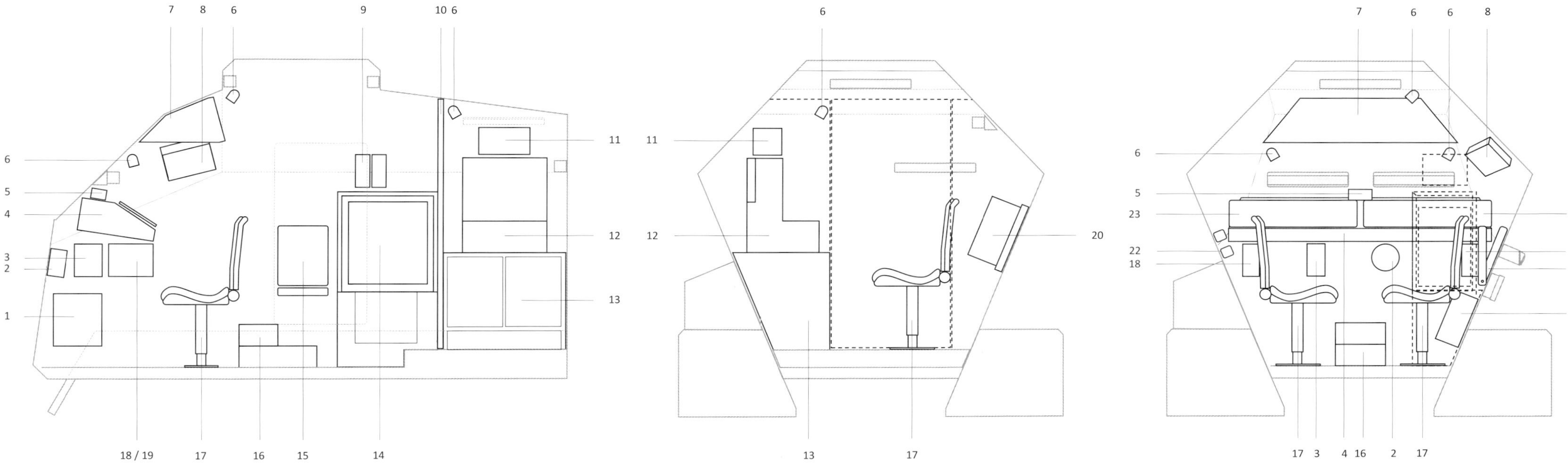

Fahrzeug 101
Sonder-Kfz.7 mit
gepanzertem Aufbau

Loading plan
Beladeplan

1 Power supply unit
2 Horn for clear signal
3 6V measuring battery
4 Control console
5 Lighting switch
6 Work lamp
7 F-radio panel (control unit)
8 Test box for I-Geräte
9 FF33 field telephone
10 Partition wall with door
11 telephone exchange »Amtszusatz« box
12 20-way field switch board
13 Console with doors
14 carrier frequency unit TFb1 with rack and console
15 Folding seat
16 Step
17 Chair column with seat for operating crew
18 Field telephone for control
19 Field telephone for engine
20 Storage for field flap cabinet cover
21 S-control panel (steering control)
22 Set of armored glass (replacement)
23 T-control panell (engine control)
X Accessory cabinet – assumed place

1 Netzgerät
2 Hupe für Klarmeldung
3 Messbatterie 6V
4 Bedienkonsole
5 Beleuchtungsschalter
6 Arbeitsleuchte
7 F-Steuergerät (Funk)
8 Prüfkasten für I-Geräte
9 Feldtelefon FF33
10 Trennwand mit Türe
11 Amtszusatz zum großen Feldklappenschrank
12 Feldklappenschrank zu 20 Leitungen
13 Konsole mit Türen
14 Trägerfrequenzgerät TFb1 mit Gestell und Konsole
15 Klappsitz
16 Trittstufe
17 Stuhlsäule mit Sitz für Bedienmannschaft
18 Feldtelefon für Steuerung
19 Feldtelefon für Triebwerk
20 Lagerung für Deckel Feldklappenschrank
21 S-Steuergerät (Steuerung)
22 Ersatzpanzerglas 2 Stück
23 T-Steuergerät (Triebwerk)
X Zubehörschrank - mögliche Einbauposition

Installations

The interior of the special superstructure is divided into two rooms: the larger room serves as a command room for the commander and work room for the engine and control man, while the smaller room separated by a door serves as a telephone exchange room.

Einbauten

Das Innere des Spezialaufbaues ist in zwei Räume unterteilt: Der größere Raum dient als Befehlsraum für den Kommandanten und Arbeitsraum für den Triebwerks- und Steuerungsmann, während der durch eine Tür getrennte kleinere Raum als Fernsprechvermittlungsraum vorgesehen ist.

BArch, RH8II Bild-B1386-44, o.Ang.

41

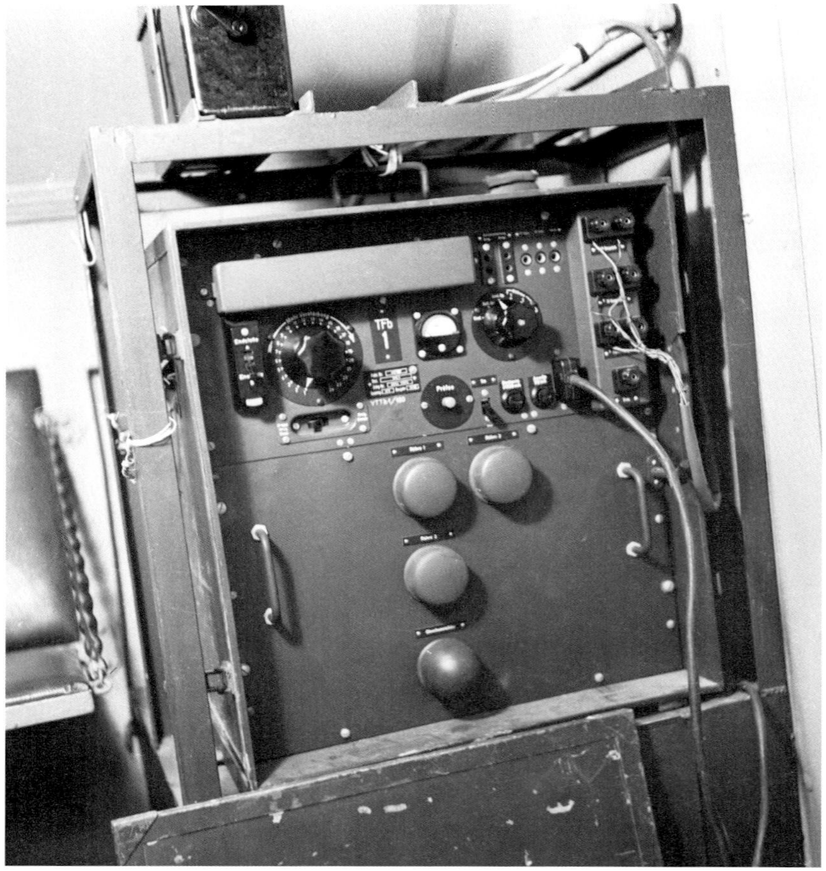

Deutsches Museum, München, Archiv- FA014/B980/44T5224 [CD74703]

42

Command room Befehlsraum

41, 42 View into the command room with view to the rear. Above armored glass vision ports there is the radio panel control unit F, to the right of it on the inner wall the test box for I-Geräte. Below the vision ports is the control panel, on the left the engine control panel T with the start key and switches for the start commands, on the right the control panel S for the control commands of the steering gear. Small work lamps were used to create a cone of light on the control panels and telephone exchanges.

Through the entrance door one looks directly at the carrier frequency unit TFb1, which is mounted in a frame on a console in front of the side wall/partition wall. Up to 2 field telephones of type FF33 can be placed on the frame. On the left in the picture the folding seat for the commander.

41, 42 Blick in den Befehlsraum mit Blickrichtung zum Heck. Oberhalb der Glasblöcke (Sehschlitz) befindet sich die Funktafel mit dem Steuergerät F, rechts davon an der Innenwand der Prüfksten für die I-Geräte. Unterhalb der Glasblöcke ist das Bedienpult zu sehen, links die Triebwerkstafel, das Steuergerät T mit dem Startschlüssel und Schaltern für die Startkommandos, rechts die Steuertafel, das Steuergerät S für die Steuerungsbefehle der Ruderanlage. Kleine Arbeitslampen sorgten für einen Lichtkegel auf die Bedientafeln und Fernsprechvermittlung.

Durch die Eingangstür blickte man direkt auf das Trägerfrequenz-Gerät TFb1, das in einem Rahmen auf einer Konsole vor der Seitenwand/Trennwand montiert war. Bis zu zwei Feldfernsprecher FF33 konnten auf dem Rahmen platziert werden. Links im Bild der Klappsitz für den Kommandanten.

BArch, RH8II Bild-B1786-44, o.Ang.

43

Deutsches Museum, München, Archiv- FA014/B970/44 T5224 [BN49010]

44

Telephone switch board
Fernsprechvermittlung

43, 44 View into the telephone switch board with the 20-way field switch board for OB*-switching. On the flap cabinet there is an additional telephone exchange »Amtszusatz« (box with three dialing discs), which could be mounted if required. The hexagon screw heads on the rear wall are the fastenings of the support profile for the steel roof above the driver's cab.

* OB (local battery operation) FF (field telephone)

43, 44 Blick auf die Fernsprechvermittlung mit dem 20-teiligen Feldklappenschrank zur OB*-Vermittlung. Auf dem Klappenschrank ist ein Amtszusatz (Kasten mit drei Wählscheiben) eingesetzt. Dieser konnte bei Bedarf montiert werden. Die Sechskantschraubenköpfe an der Rückwand sind die Befestigungen des Auflageprofiles für das Blechdach über dem Führerhaus.

* OB (Ortsbatteriebetrieb) FF (Feldfernsprecher)

Deutsches Museum, München, Archiv- FA014/B554/44 TD5224 [CD74813] Ausschnitt

45

45 At the end of 1943 Abteilung 836 received her first three Feuerleitpanzers. The photo was taken in March 1944 on the training area of Art.Abt.836 in Gross-Born. The Fz.101 received a large pattern of camouflage colours in olive green (RAL 6010) and reddish brown (RAL 8017) on a dark yellow base coat (RAL 7028). The front axle on the tractor has no reinforcement yet which has been retrofitted to all Fz.101 delivered to date since August 44.

45 Ende 1943 erhielt die Abteilung 836 ihre ersten drei Feuerleitpanzer. Die Aufnahme aus dem März 1944 ist auf dem Übungsgelände der Art.Abt.836 in Gross-Born entstanden. Das Fz.101 hat auf einem dunkelgelben Grundanstrich (RAL 7028) ein großflächiges Muster aus Tarnfarben in olivgrün (RAL 6010) und rotbraun (RAL 8017) erhalten. Die Vorderachse an der Zugmaschine weist noch keine Verstärkung auf. Sie wurde ab August 1944 bei allen bis dato gelieferten Fz.101 nachgerüstet.

BArch, RH8II Bild-B1945-44, o.Ang.

46

46 A Fz.101 Feuerleitpanzer in the final version. The vehicle belongs to the 2./836 which participated in the firing test »Feuerstellung Schiene« at Peenemünde in November 1944. The Fz.101 received a stain pattern of camouflage colours in olive green (RAL 6010) and reddish brown (RAL 8017) on a dark yellow ground coat.

46 Ein Fz.101 Feuerleitpanzer in der Abschlussausführung. Das Fahrzeug gehört zur 2./836, die im November 1944 an der Schusserprobung »Feuerstellung Schiene« auf Peenemünde teilnahm. Das Fz.101 hat auf einem dunkelgelben Grundanstrich ein Fleckenmuster aus Tarnfarben in olivgrün (RAL 6010) und rotbraun (RAL 8017) erhalten.

OTIT Depot O-644 Paris (Vincennes), France

In September 1944 the US Army established the O-644 depot at Paris, France. This was the largest supply depot for US units ever operated in Europe. The Ordnance Technical Intelligence Team (OTIT) stationed there had the task of collecting Wehrmacht vehicles. From October onwards captured vehicles, tanks, guns, V2 rockets, V1 flying bombs and other Wehrmacht weapons gradually arrived there. The collected items were technically examined and temporarily stored for further transport to Aberdeen Proving Ground in the USA.

OTIT Depot O-644 Paris (Vincennes), Frankreich

Im September 1944 wurde von der US Army das Depot O-644, Paris, France eingerichtet. Es war das größte in Europa jemals betriebene Versorgungsdepot der US-Truppen. Das dort stationierte Ordnance Technical Intelligence Team (OTIT) hatte die Aufgabe Fahrzeuge der Wehrmacht einzusammeln. Ab Oktober trafen dort nach und nach erbeutete Fahrzeuge, Panzer, Geschütze, V2-Raketen, V1-Flugbomben und andere Waffen der Wehrmacht ein. Die Sammelstücke wurden technisch untersucht und für den Weitertransport in die USA nach Aberdeen Proving Ground zwischengelagert.

Photo by Master Sergeant Paul Cernak via grandson Christopher R. Van Valkenburgh

47

Photo by Master Sergeant Paul Cernak via grandson Christopher R. Van Valkenburgh

48

47, 48 Summer 1945. – The OTIT-C collection point was located in a separate area on the depot site of Vincennes*. Among the collectibles were other rare vehicles such as a prototype of the light 10.5 cm Pz FH18/40/2 on self-propelled carriage III/IV. The self-propelled carriage was parked in front of a Fz.101 *Feuerleitpanzer* (armored fire control vehicle).

* Today the area with its buildings is used by a riding school.

47, 48 Sommer 1945 – In einem abgetrennten Bereich auf dem Depot Gelände im Wald von Vincenne* befand sich die Sammelstelle der OTIT-C. Unter den Sammelstücken waren auch seltene Fahrzeuge wie ein Prototyp der leichten 10,5 cm Pz FH18/40/2 auf Geschützwagen III/IV Selbstfahrlafette dabei. Die Selbstfahrlafette wurde vor einem Feuerleitpanzer Fz.101 abgestellt.

* Heute wird der Bereich mit seinen Gebäuden von einer Reitschule genutzt.

AMC

49

49 The collection of captured vehicles has grown considerably. New verhicles were constantly being delivered, others prepared for collection and transported away, so that the parking positions were being changed all the time. Among the various battle tanks, guns and other vehicles, parked here, we see a Jg.Pz.Kpfw. V Jagdpanther, two Pz.Kpfw. V Ausf. G Panther and an Fz.101 (*Feuerleitpanzer*).

49 Die Sammlung an Beutefahrzeugen ist beachtlich gewachsen. Ständig wurden neue Fahrzeuge angeliefert, andere zur Abholung vorbereitet und abtransportiert, so dass die Abstellpositionen ständig wechselten. Unter den verschiedensten Kampfpanzern, Geschützen und anderen Fahrzeugen ist hier neben einem abgestellten Jg.Pz.Kpfw. V Jagdpanther und zwei Pz.Kpfw. V Ausf. G Panther auch ein Fahrzeug 101 (Feuerleitpanzer) zu sehen.

Photo by Master Sergeant Paul Cernak via grandson Christopher R. Van Valkenburgh

50

50 The Fz.101 *Feuerleitpanzer* was undoubtedly one of the most exotic vehicles at the assembly point. The vehicle belonged to Abteilung 836 and was left behind on the retreat from the operational area of Army Group South in the Marburg / Kassel area. There the vehicle was found by advancing US units and taken by OTIT to the O-644 depot collection point for looted vehicles in Paris.

50 Zu den Exoten auf dem Sammelplatz gehörte zweifellos auch der Feuerleitpanzer Fz.101. Das Fahrzeug aus dem Bestand der Abt.836 wurde auf dem Rückzug aus dem Einsatzgebiet der Gruppe Süd im Raum Marburg / Kassel zurückgelassen. Dort wurde das Fahrzeug von vorrückenden US-Truppen aufgefunden und von der OTIT zur Sammelstelle für Beutefahrzeuge in das Depot O-644, Paris gebracht.

USAHEC

51

51 The parked vehicle was only painted in a dark yellow (RAL 7028) paint scheme. This vehicle was handed over to the British Army in June and repaired, defective parts such as the radiator grille were replaced. This Feuerleitpanzer Fz.101 was used in October 45 in Operation Backfire. As reinforcement a beveled flat bar was welded onto the front axle tube.

51 Das abgestellte Fahrzeug war nur mit einem dunkelgelben Anstrich (RAL 7028) versehen. Es wurde im Juni 1945 an die britische Armee übergeben und instandgesetzt. Defekte Teile wie der Kühlergrill wurden ausgetauscht. Danach kam es im Oktober 1945 bei der Operation Backfire zum Einsatz. Am Rohr der Vorderachse wurde ein abgeschrägtes Flacheisen zur Verstärkung angeschweißt.

KCL

52

KCL

53

VRU

54

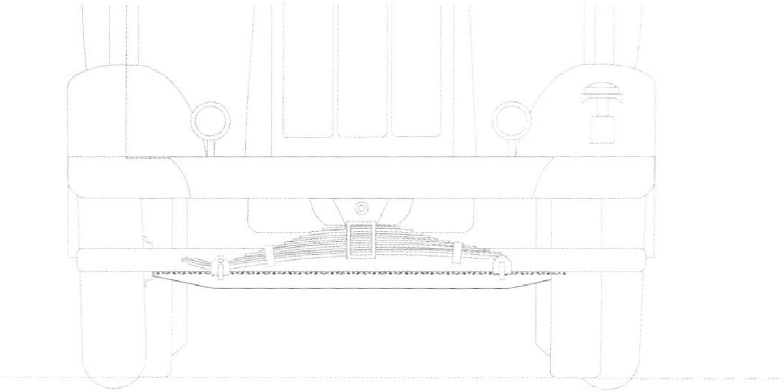

52-54 At the Krupp shooting range Altenwalde near Cuxhaven vehicles of the Division z.V. were collected by the British army after the end of the war. This Feuerleitpanzer was painted in a striking camouflage. The armoured body is mounted here on a late 8to Zugmaschine. For the post-testing of the A4 rockets in October 1945 (Operation Backfire), spare parts were taken from the remains of the collected, partly damaged vehicles and, as far as possible, operational vehicles were assembled. In front of the vehicle lies the reinforced front axle – this was also installed on other 8 t Zugmaschinen at the end of 1944, proven on the late Sd.Kfz.7 (with flatbed) and Sfl. Sd.Kfz.7/2 with 3.7 cm Flak 43.

52-54 Auf dem Krupp-Schießplatz Altenwalde bei Cuxhaven wurden nach Kriegsende von der britischen Armee aufgefundene Fahrzeuge der Division z.V gesammelt. Der abgebildete Feuerleitpanzer wurde in einem auffälligen Tarnanstrich bemalt. Der gepanzerte Aufbau ist hier auf einer 8 t-Zugmaschine der späten Ausführung montiert. Für die Nacherprobung der A4-Raketen im Oktober 1945 (Operation Backfire) wurden aus Resten der gesammelten, teilweise beschädigten Fahrzeuge Ersatzteile entnommen und soweit es möglich war, einsatzfähige Fahrzeug zusammengebaut. Vor dem Fahrzeug liegt die verstärkte Vorderachse, welche auch an anderen 8 t-Zugmaschinen Ende 1944 verbaut wurde. Beispiele hierfür sind das späte Sd.Kfz.7 (mit Pritsche) und die Sfl. Sd.Kfz.7/2 mit 3,7 cm Flak 43.

AMC

55

55 A *Feuerleitpanzer* of the final design was found by US Army units near Hofgeismar in the Kassel area. On the rear armor plate of the final version the openings for the armored glass vision ports have been omitted or closed over with a plate. See defects and their remedies on page 71.

55 Ein Feuerleitpanzer der Abschlussausführung wurde von Einheiten der US Army bei Hofgeismar, Landkreis Kassel, vorgefunden. An der Heckpanzerplatte der Abschlussausführung sind die Ausschnitte für die Sehschlitze mit Glasblöcken entfallen, bzw. durch eine Platte verschlossen. Siehe Mängel und ihre Behebung, Seite 71.

Vehicle 102

Internal or former designation: Fz.-Pos.-Nr.102
FR trailer (S) (3-axle)
Meillerwagen

DMA, GD639.5.18
HAP11/T 507
Karlshagen, 1.6.1944
Bd.Nr. E 3195/44 g

DM GD639.5.1
EW234
Fz.-Pos.-Nr. 102

DMA, GD639.51.7
Auszug
Bedienungsanweisung
Meillerwagen

DMA, GD639.5.6
Auszug
Vorläufige Vorschrift
Der FR-Anhänger (S)
(3achs)

DMA, GD630.02
A4 Lehrbuch
der FR-Schule
Stand Mai 1944
Auszug aus
Technische Fahrzeuge

Intended use

The FR trailer (S) is used to transport and errect the A4 *Gerät*. Furthermore for the last installation, filling and testing as well as for the start. The FR trailer (S) is a pneumatic 3-axle motor vehicle trailer with a self-tracking chassis and equipped with a hydraulic tilting device driven by a VW engine. A wheeled tractor (s) is provided as the towing device for the FR trailer (S). On roads of good condition, the loaded FR trailer (S) may travel at a speed of up to 41 km per hour. For all other road conditions and off-road the travel speed must be reduced accordingly, if necessary to walking speed.

The Meillerwagen MA4-3 has the following dimensions and weights:

Total length:	14,800 mm	
Width:	2,800 mm	
Height:	3,340 mm	
Wheelbase:	7,200/2,600 mm	boogie frame
Track width:	2,000/1,250 mm	center distance between dual tires
Tare weight:	11,300 kg	
Payload:	4,500 kg	
Service weight:	15,470 kg	
Wheels:	270-20	

The Meillerwagen consists of the following main components:

- Chassis consisting of a central tube with ball bearing slewing ring, steering column, fork, steering frame, bogie frame, turntable, right and left tube side parts, front endcarriage, journal bogie bearing, journal bogie and drawbar
- Tilting frame, with platforms, tongs and pendulum frame
- Hydraulic drive unit StaMo Kdf type 120/15
- Attachments: A-, B-, T-Stoff tubing, T-Stoff filling tank, cable, cable mast

Archiv F.X.M
Fabrikationsbuch
Anhängerbuch 10 + 11

Development variants

The company F.X. Meiller developed the FR wagon (S) on behalf of the OKH and manufactured it in three versions, which were designated MA4, MA4-2 and MA4-3. The Meillerwagen is based on the principle of the tubular swing axle trailer type developed by Meiller. In the MA4 and MA4-2 versions, the FR trailers were exclusively developed by F.X. Meiller. The companies Lindner and Schenk were also commissioned with the construction of the final version MA4-3 by Wa Prüf 11/III as a reconstruction company, whereby Lindner was designated as the leading firm (page 383) for the Meillerwagen.

Fahrzeug 102

Interne oder frühere Bezeichnung: Fz.-Pos.-Nr.102
FR-Anhänger (S) (3-achs.)
Meillerwagen

Verwendungszweck

Der FR-Anhänger (S) dient zum Transport und Aufstellen des Gerätes A4. Ferner zur letzten Installation, Füllung und Prüfung sowie zum Start. Der FR-Anhänger (S) ist ein luftbereifter 3-achsiger Kraftfahrzeuganhänger mit selbstspurendem Laufgestell und mit einer mittels VW-Motor angetriebenen hydraulischen Kippvorrichtung ausgerüstet. Als Zugmittel für den FR-Anhänger (S) ist ein (s) Radschlepper vorgesehen. Auf Straßen von guter Beschaffenheit ist für den belasteten FR-Anhänger (S) eine Fahrgeschwindigkeit bis 41 km pro Stunde zulässig. Bei allen anderen Wegverhältnissen und im Gelände muss die Fahrgeschwindigkeit entsprechend herabgemindert werden, erforderlichenfalls bis auf Schrittgeschwindigkeit.

Der Meillerwagen MA4-3 hat folgende Maße und Gewichte:

Gesamtlänge:	14.800 mm	
Breite:	2.800 mm	
Höhe:	3.340 mm	
Radstand:	7.200/2.600 mm	Laufgestell
Spurweite:	2.000/1.250 mm	Mittelabstand zwischen Doppelreifen
Leergewicht:	11.300 kg	
Nutzlast:	4.500 kg	
Gefechtsgewicht:	15.470 kg	
Räder:	270-20	

Der Meillerwagen besteht aus folgenden Hauptbauteilen:

- Fahrgestell aus einem Mittelrohr mit Kugeldrehkranz, Lenkstock, Gabelstück, Lenkgestell, Laufgestellrahmen, Drehschemel, rechtes und linkes Rohrseitenteil, vorderer Kopfträger, Zapfendrehgestelllager, Zapfendrehgestell und Zuggabel
- Kipprahmen, mit Podesten, Zange und Pendelrahmen
- Hydraulische Antriebseinheit StaMo Kdf Typ 120/15
- Anbauten: A-, B-, T-Stoff, T-Stoff Fülltank, Kabel, Kabelmast

Entwicklungsvarianten

Die Firma F.X. Meiller hatte im Auftrag des OKH den FR-Wagen (S) entwickelt und in drei Ausführungen gefertigt, die als MA4, MA4-2 und MA4-3 bezeichnet wurden. Der Meillerwagen basierte auf dem Prinzip des Rohrschwingachsers, einem von Meiller entwickelten Anhängertyp. In der Ausführung MA4 und MA4-2 wurden die FR-Anhänger exklusiv von F.X. Meiller gefertigt. Mit dem Bau der Abschlussausführung MA4-3 wurden auch die Firmen Lindner und Schenk als Nachbaufirmen durch Wa Prüf 11/III beauftragt, wobei die Firma Lindner als Leitfirma (Seite 383) für den Meillerwagen bestimmt wurde.

Meillerwagen MA4

The first Meillerwagen version MA4, built by F.X. Meiller in 1941, was delivered to the HAP in *Peenemünde* in January 1942 on behalf of the OKH. The main features of the Meillerwagen MA4 were the launch table hinged into the chassis and the short tipping frame equipped only with support shells. How the A4 *Gerät* was to be transported on the Meillerwagen was not yet determined at the time of this recording in winter 1941/42.

Meillerwagen MA4

Der erste Meillerwagen Ausführung MA4 gebaut von F.X. Meiller in 1941 wurde im Januar 1942 im Auftrag des OKH an das HAP in Peenemünde ausgeliefert. Hauptmerkmal am Meillerwagen MA4 war der am Fahrgestell angeschlagene Abschusstisch und der nur mit Auflageschalen ausgerüstete kurze Kipprahmen. Wie das A4-Gerät auf dem Meillerwagen transportiert werden sollte war zum Zeitpunkt dieser Aufnahme im Winter 1941/42 noch nicht festgelegt.

89

Deutsches Museum, München, Archiv- FA014/B83/42 BSM [BN09653]

Deutsches Museum, München, Archiv- FA014/B139/42 BSM [CD74644]

57

Meillerwagen MA4 Modification 1 EW Vers/D.
Meillerwagen MA4 Modifikation 1 EW Vers/D.

57 For the upcoming ground test with a dummy A4 the tipping frame was modified. The modification* to the frame concerned the rear and front support, which was replaced by a frame with a half shell and second saddle. The fixed support, in front at the middle tube of the chassis, was removed and replaced by an adjustable support with a contour piece. The dummy A4 was fixed to the added half-shells with straps.

57 Für den anstehenden Bodentest mit einem A4-Testkörper wurde der Kipprahmen umgebaut. Der Umbau* am Kipprahmen betraf die hintere und vordere Auflage, die durch ein Gerüst mit einer Halbschale und einem zweiten Sattel ersetzt wurde. Die feste Auflage, vorne am Mittelrohr des Fahrgestelles, wurde entfernt und durch eine verstellbare Auflage mit Konturstück ersetzt. Die A4-Testkörper waren mit Spannbändern an den hinzugefügten Halbschalen fixiert.

* The modifications were carried out in February 1942 according to drawings E1541 to E1544.

* Der Umbau erfolgte im Februar 1942 nach Zeichnungen E1541 bis E1544.

BArch, RH8 Bild-B0673-42BSM, o.Ang.

58

Meillerwagen MA4 Modification 2 EW Vers/D.
Meillerwagen MA4 Modifikation 2 EW Vers/D.

58 The start table* integrated on the Meillerwagen MA4 had proved to be unsuitable and was removed. The half-shells on the tilting frame proved to be an hindrance setting the rocket down on the launch table. A new solution** was developed. For this purpose the tilting frame was extended to the front. There a movable tongs (so-called Vowe tongs) was attached and a pendulum frame was mounted to hold the middle section. The tongs and the pendulum frame were connected with a rod and crank drive which allowed opening and closing.

* As a result, the launching table was further developed as an independent vehicle Fz.103.
** The modifications were carried out in July 1942 according to drawings O-4829A and O-4829B.

58 Der am Meillerwagen MA4 integrierte Starttisch* hatte sich als nicht zweckmäßig herausgestellt und wurde entfernt. Die Halbschalen am Kipprahmen erwiesen sich für das Absetzen der Rakete auf den Starttisch als hinderlich. Eine neue Lösung** wurde entwickelt. Dazu verlängerte man den Kipprahmen nach vorne. Dort wurde eine bewegliche Zange (sog. Vowe-Zange) befestigt und ein Pendelrahmen zum Halten des Mittelteiles montiert. Zange und Pendelrahmen waren mit einem Stangen- und Kurbelantrieb verbunden, die ein Öffnen und Schließen erlaubten.

* Daraufhin wurde der Starttisch als eigenständiges Fahrzeug Fz.103 weiterentwickelt.
** Der Umbau erfolgte im Juli 1942 nach Zeichnungen O-4829A und O-4829B.

Meillerwagen MA4-2

The first experiences with the FR trailer (S) version MA4 required a complete redesign of the tipping frame, with adjustments also on the chassis. For the new tipping frame, the frame girders were elaborately welded together from sheet metal and flat iron to form the torsion box. The Vowe tongs designed by EW were adapted accordingly by the maker and fixed to the tilting frame. By mid 1943 a total of nine FR trailers in the MA4-2 version were manufactured and delivered by F.X. Meiller.

Meillerwagen MA4-2

Die ersten Erfahrungen mit dem FR-Anhänger (S) Ausf. MA4 erforderten eine komplette Neuauslegung des Kipprahmens mit Anpassungen. Für den neuen Kipprahmen wurden die Rahmenträger aus Blechen und Flacheisen aufwendig zum Torsionskasten zusammengeschweißt. Die vom EW konstruierte Vowe-Zange wurde durch die Industrie entsprechend angepasst und auf dem Kipprahmen befestigt. Bis Mitte 1943 wurden insgesamt neun Stück der FR-Anhänger in der Ausführung MA4-2 von F.X. Meiller gefertigt und ausgeliefert.

BArch, RH8II Bild-B0266-43-BSM, o.Ang.

59

BArch, RH8II Bild-B0265-43-BSM, o.Ang.

59, 60 The FR trailer MA4-2 was delivered by the makers without attachments, such as working platforms, connection pipes for A-, B-Stoff, T-Stoff tank, and P-Stoff battery. Only with the experience gained during the countless ground tests could EW finally define the attachments have them integrated piece by piece on the vehicles by the contracted companies. For example, the hydraulic tank on the first MA4-2 was still attached transversely to the central tube of the chassis, so it had to be moved forwards parallel to the direction of travel once the central working platform had been defined. This FR trailer is painted in a two-tone paint scheme of RAL 8020 with spots in RAL 7008.

59, 60 Die FR-Anhänger MA4-2 wurden von der Industrie ohne Anbauten wie Arbeitsbühnen, Anschlussrohre für A-, B-Stoff, T-Stofffülltank und P-Stoff Batterie ausgeliefert. Erst mit den Erfahrungen während den unzähligen Bodenversuchen konnten auch die Anbauten durch das EW endgültig definiert und sukzessive durch die beauftragten Firmen ausgeführt werden. Der Hydrauliktank war beispielsweise an den ersten MA4-2 noch quer auf dem Fahrgestell-Mittelrohr befestigt. So musste dieser nach Festlegung der mittleren Arbeitsbühne nach vorne parallel zur Fahrtrichtung verlegt werden. Der FR-Anhänger ist in einem zweifarbigen Anstrich in RAL 8020 mit Flecken in RAL 7008 lackiert.

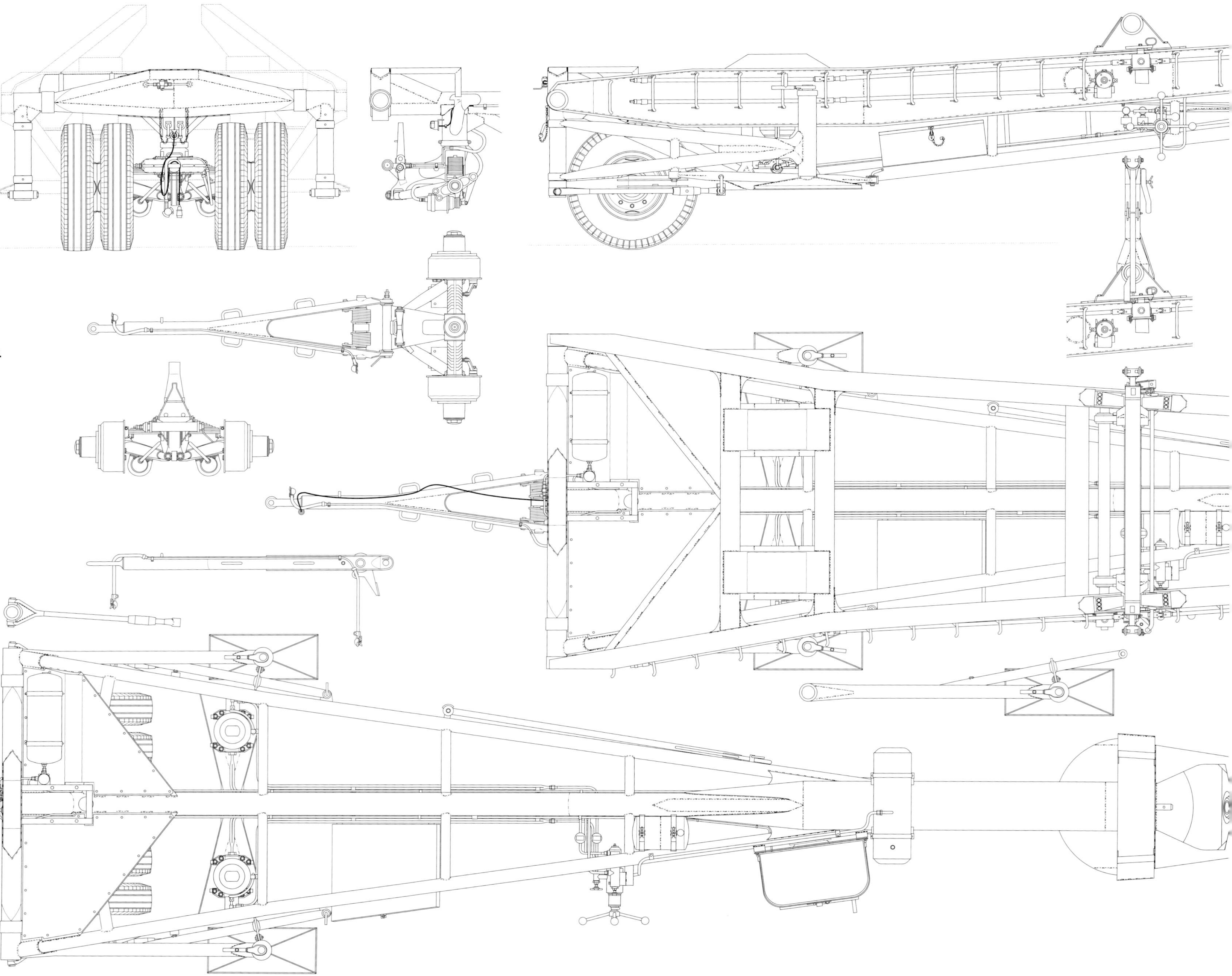

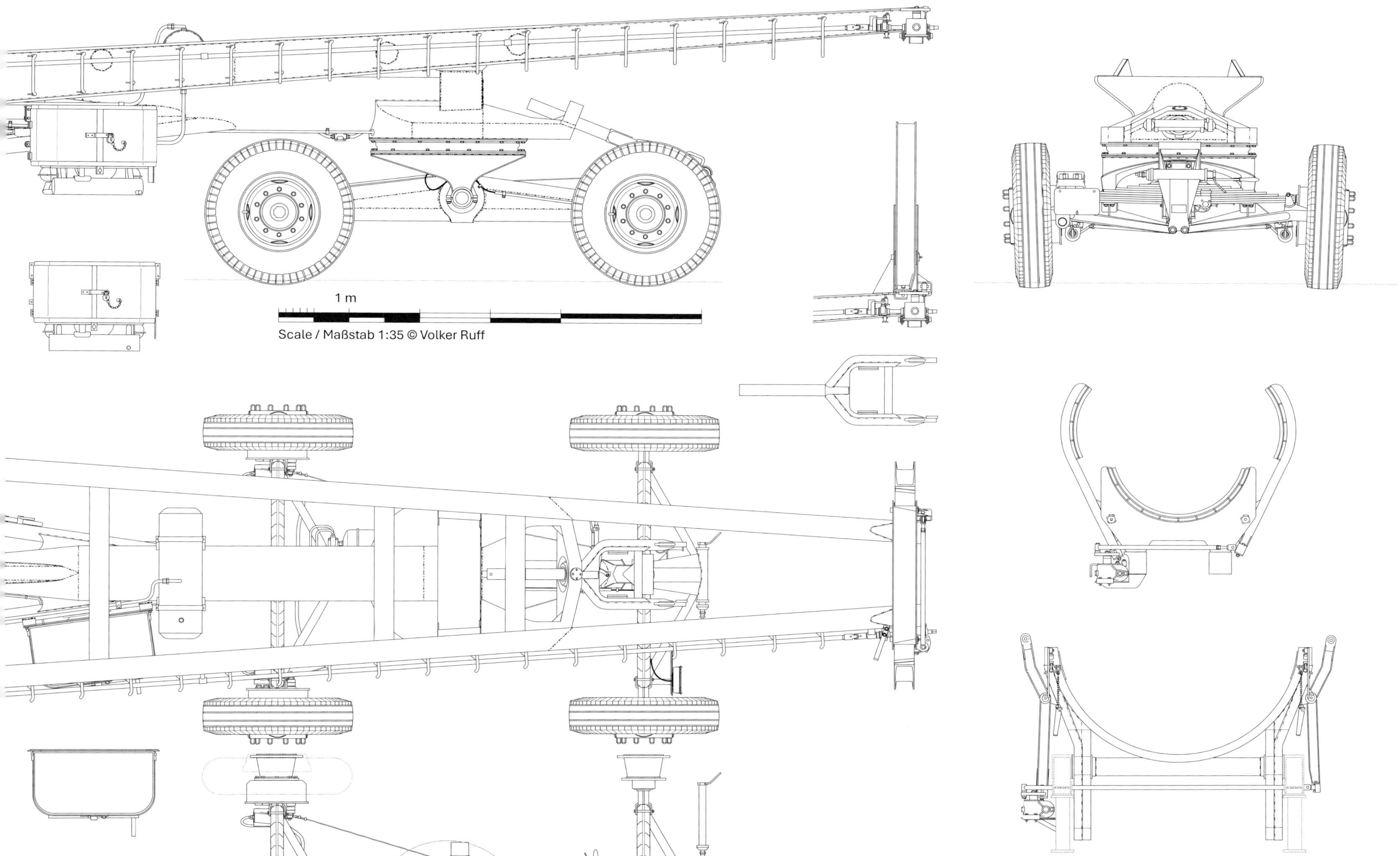

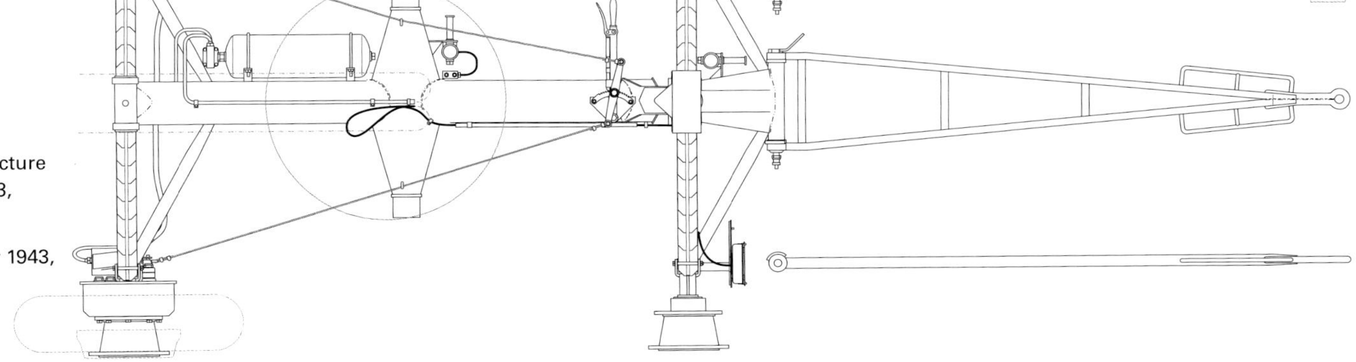

Meiller MA4-2, Manufacture year 1943, Fgst.Nr. 5353, without components

Meiller MA4-2, Baujahr 1943, Fgst.Nr. 5353, ohne Ausrüstung

Fahrzeug 102

FR-Anhänger (S) (3-achs.)

Drawing set SK7*(FXM) used for MA4-2

Für MA4-2 verwendeter Zeichnungssatz SK7*(FXM)

Meillerwagen MA4-3

The elaborately welded tilting frame on the FR trailer (S) MA4-2 was reworked again in mid-1943. Instead of a torsion frame, the tipping frame was assembled and welded like a double T-profile from two isosceles T-profiles and a web plate, giving it its shape. This construction method on the new tilting frame resulted in considerable time and material savings. Together with other minor changes to the chassis, the FR trailer (S) was further developed into the MA4-3 version.

From the end of 1943 this MA4-3 version was manufactured and delivered by the companies F.X.M., Lindner and Schenk.

Meillerwagen MA4-3

Der aufwendig geschweißte Kipprahmen am FR-Anhänger (S) MA4-2 wurde Mitte 1943 erneut überarbeitet. Anstelle eines Torsionsrahmens wurde der Kipprahmen ähnlich einem Doppel-T-Profil aus zwei gleichschenkeligen T-Profilen und einem Stegblech formgebend zusammengesetzt und verschweißt. Mit dieser Bauweise konnten am neuen Kipprahmen erhebliche Zeit- und Materialeinsparungen umgesetzt werden. Zusammen mit anderen kleineren Änderungen am Fahrgestell wurde der FR-Anhänger (S) zur Ausführung MA4-3 weiterentwickelt.

Ab Ende 1943 wurde diese Ausführung MA4-3 von den Firmen F.X.M., Lindner und Schenk gefertigt und ausgeliefert.

Deutsches Museum, München, Archiv- FA014/B105/44 TD5224 [CD74671]

61

DMA, GD639.5.11
T404
o.O, den 15.6.1944
Bd.Nr. E 2190/44 gK
Auszug
Teileliste Fz.102

Vehicles on order

Fahrzeugbestellungen

	Item Lfd. Nr.	Qty. ordered Best. Stückzahl	Qty. per vehicle Stück je Fahrzeug	Subject drawing number Gegenstand Zeichnungs-Nr.	Ordered by order number Besteller Bestell-Nr.	Supplier Lieferfirma
Fahrzeugbestellung	1a	78	1	FR. Anhänger (S) (3-achs.)	WaPrüf 11/III 361-0010/42	F.X.Meiller München 8 Lilienthalstr. 2/4
	1b	84	1	FR. Anhänger (S) (3-achs.)	WaPrüf 11/III 361-0007/43	Gottfried Lindner Ammendorf b. Halle
	1c	107	1	FR. Anhänger (S) (3-achs.)	WaPrüf 11/III 361-0026/43	Gottfried Lindner Ammendorf b. Halle
	1d	48	1	FR. Anhänger (S) (3-achs.)	WaPrüf 11/III 361-0034/43	Maschinen-Fabrik Karl Schenk
	1	*317*		*FR. Anhänger (S) (3-achs.)*	*Gesamtbestellung auf Fahrzeug 102*	

Customizing/outfitting

Ausrüstung

	Item Lfd. Nr.	Qty. ordered Best. Stückzahl	Qty. per vehicle Stück je Fahrzeug	Subject drawing number Gegenstand Zeichnungs-Nr.	Ordered by order number	Supplier Besteller Bestell-Nr. Lieferfirma
Ausrüstung für Fahrzeuge nach	1a			mechanische Ausrüstung	WaPrüf 11/III, WaPrüf 10/H, LR	Arbeitsgemeinschaft des Schlosserhandwerkes, München 2, Karlsplatz 24
				elektrische Ausrüstung	Gollnow & Sohn, Koblenz	Fa. Ertel, München
	1b			mechanische Ausrüstung	WaPrüf 11/III, WaPrüf 10/H, LR	Fa. Lindner, Ammendorf b. Halle
	1c			elektrische Ausrüstung	Gollnow & Sohn, Koblenz	Fa. Ertel, München
	1d			mechanische Ausrüstung	WaPrüf 11/III, WaPrüf 10/H, LR	Fa. Lindner, Ammendorf b. Halle
				elektrische Ausrüstung	Gollnow & Sohn, Koblenz	Fa. Ertel, München

Ergänzung
T66 (T404)
o.O, den 31.7.1944
Bd.Nr. 00761/44 gK
Auszug
Kabelmast
zu Teileliste Fz.102

	34	450	1	kompl. Kabelmast 21C115615/U1.U11 mit U ??	WaPrüf 10/H SS 4949-0338-6137/43	A. Rawie Fabrik für Eisenbahnbedarf, Osnabrück

DMA, GD639.5.21
OKW/Wa Prüf10
Karlshagen, 25.3.1944
Bd.Nr. 747/44 gK
Auszug
Besprechung der
Fahrzeuggruppe 1
vom 24./25.3.44
im Lager Rebstock

Production status: Vehicle 102
Delivered to date:
18 FR-trailers to Abt.836
6 FR-trailers to Abt.485
3 FR-trailers for Battery 444,
Further three trailers are finished and ready for shipment. Another 15 are expected to be shipped after the supply of still missing electrical parts expected by April 6. For shipping reasons they will be delivered in batches of 3 except the Stotz plugs, which will be delivered later. In addition there are currently 8 FR-trailers in the final stage of outfitting at the working group in Munich completion date and delivery readiness in Munich by April 15, 1944. Moreover, at Wa Prüf 10 there are 3 FR-trailers in the final stage and ready for shipment by April 2.

Current defects and their remedy
Internal illumination
The units' requirement for additional lighting equipment for internal illumination inside the *Gerät* will be met by Wa Prüf 10 by attaching two distribution boxes to the existing 27-V-line. This change of form will be advised to the units by Wa Prüf 10 and to the manufacturing companies by Wa Prüf 11/III. Evaluation as Case B.

Fertigungslage: Fahrzeug 102
Bisher sind ausgeliefert:
18 FR-Anhänger an Abt.836
6 FR-Anhänger an Abt.485
3 FR-Anhänger für Batterie 444
Weiter sind fertiggestellt und versandbereit 3 Stück. 15 weitere Stück kommen nach Anlieferung noch fehlender elektrischer Teile voraussichtlich zum 6.4. aus versandtechnischen Gründen in Raten à 3 Stück zum Versand, ausgenommen Stotz-Stecker, welche nachgeliefert werden. Außerdem befinden sich z.Zt. bei der Arbeitsgemeinschaft in München 8 FR-Anhänger in Endausrüstung; Fertigstellungstermin und Versandbereitschaft in München zum 15.4.44. Außerdem sind bei Wa Prüf10 3 FR-Anhänger in Endausrüstung und versandbereit zum 2.4.

Derzeitige Mängel und ihre Behebung
Innenbeleuchtung
Der Forderungen der Truppe nach Schaffung einer zusätzlichen Beleuchtungseinrichtung für Innenbeleuchtung im Gerät wird seitens Wa Prüf 10 durch Anbringung zweier Abzweigdosen an der vorhandenen 27-V-Leitung entsprochen. Formänderung geht seitens Wa Prüf 10 an die Truppe, über Wa Prüf 11/III an die Fertigungsfirmen. Bewertung als Fall B.

Deutsches Museum, München, Archiv- FA014/B104/44 TD5224 [CD74674]

62

BArch, RH8II Bild-B1152-44, o.Ang (Ausschnitt)

63

Changes to the tilting frame at MA4-3 version

Änderungen am Kipprahmen bei der Ausführung MA4-3

62, 63 If one compares the two tilting frames on the left and right with each other, the »Current defects and their remedy« described can be easily seen, e.g. the larger platform at the top, omission of the middle working platform, the higher T-Stoff filling container, the cable duct, and the cable mast are the conspicuous changes.

62, 63 Vergleicht man die beiden aufgestellten Kipprahmen links und rechts miteinander sind die beschriebenen »derzeitigen Mängel und ihre Behebung« gut nachvollziehbar. Die auffälligen Änderungen bestehen beispielsweise aus der größeren Bühne oben, dem Wegfall der mittleren Arbeitsbühne, dem höher gesetzten T-Stoff-Füllbehälter, dem Kabelschacht und dem Kabelmast.

Enlarged platforms

The general point of view is that it is necessary to enlarge the upper stage; in particular because of the A4 *Gerät* new folding doors. For Abt.836 this change is not urgent, because it will be equipped with Magirus ladders. For the other FR-trailers this change is considered urgent (Case A). The platforms can be built without interfering with current production and changed by the units. Drawings and data about required quantities will be passed by Wa Prüf 10 to Wa Prüf 11/III. c.

Vergrößerte Bühne

Es wird allgemein für erforderlich gehalten, dass die obere Bühne vergrößert wird; vor allem wegen der neuen Klapptüren am A4-Gerät. Für die Abt.836 ist diese Änderung nicht vordringlich, da sie mit Magirusleitern ausgerüstet wird. Für die weiteren FR-Anhänger wird die Änderung als dringlich (Fall A) angesehen. Die Bühnen können ohne Störung der augenblicklichen Fertigung erstellt und durch die Truppe ausgewechselt werden. Zeichnungsunterlagen und Angaben über erforderliche Stückzahlen gehen von Wa Prüf 10 an Wa Prüf 11/III.

T-Stoff tank

Because of the omitting of the middle platform the T-Stoff tank can be placed 150 mm higher, thus ensuring that the T-Stoff will run out of the tank and through the hoses. There is no further need for the middle platform for fuel emptying, because for this exceptional case the flap is accessible from the lower platform by using the engine ladder. If an immediate decision is made this improved installation of the T-Stoff tank can be made without delaying deliveries. Major Eberlein at the B.z.b.V. Heer will inform Wa Prüf 11/III directly about the decision by May 27.

T-Stoff-Behälter

Durch Fortfall der mittleren Bühne kann der T-Stoff-Behälter um 150 mm höher angeordnet werden, sodass dadurch der Ablauf des T-Stoffes aus dem Behälter und den Schläuchen gewährleistet wird. Die mittlere Bühne wird nicht mehr für die Brennstoffentleerung benötigt. Da für diesen Ausnahmefall eine Zugänglichkeit der Klappe mit Hilfe der Triebwerksleiter von der unteren Bühne aus möglich ist. Bei sofortigem Entscheid ist diese verbesserte Anbringung des T-Stoff-Behälters ohne Lieferungsverzug möglich. Entscheidung wird durch Major Eberlein beim B.z.b.V. Heer bis 27.5. unmittelbar an Wa Prüf11/III durchgegeben.

Telephone connection

The telephone connection distributors between the power supply vehicle and the BA vehicle currently in use in the Meillerwagen are not considered necessary at this point. It has to be checked whether their relocation to the power supply vehicle can take place without changing the length of the cables. Action by Wa Prüf 10.

Fernsprechanschluss

Die z.Zt. am Meillerwagen befindlichen Verteiler der Fernsprechanschlüsse zwischen Stromversorgungs- und BA-Wagen werden an dieser Stelle nicht für notwendig erachtet. Es muss überprüft werden, ob die Verlegung nach dem Stromversorgungswagen erfolgen kann, ohne dass die Kabellängen sich ändern. Veranlassung durch Wa Prüf 10.

Brace towards the strut

To improve the mounting of the cross bracing for the strut, the web plate of the coupling gear will be omitted from the vehicle's crosshead. Wa Prüf 11/III has issued a change notice for the vehicles already completed. The web plate will immediately be omitted from new vehicles to be delivered. The units' objection that the coupling pin fitting to this brace has insufficient tolerance cannot be confirmed. If distortions of the brace occur it must be straightened.

Verstrebung zur Spreize

Zur besseren Anbringung der Querverstrebung zur Spreize entfällt das Stegblech an der Kupplungsvorrichtung am Querhaupt des Wagen. Änderungsmitteilung für bereits fertiggestellte Fahrzeuge veranlasst Wa Prüf 11/III. Bei neu zur Auslieferung kommenden Wagen wird das Stegblech ab sofort weggelassen. Der Beanstandung der Truppe, dass die Passung des Kupplungsbolzens dieser Verstrebung zu gering toleriert ist, kann nicht entsprochen werden. Wenn Verbiegungen der Strebe auftreten, müssen diese ausgerichtet werden.

Bolt tolerance

For easier insertion of the reinforcing rod at the bracer, the web plate of the mounting at the front frame end is dispensed with. Wa Prüf 11/III will determine how far the bolt play can be increased, so that the bolt can be pushed through even if the pin eye is slightly off-centre.

Bolzenspiel

Zum leichteren Einlegen der Versteifungsstrebe an der Spreize entfällt das Stegblech der Halterung am vorderen Rahmenende. Wa Prüf11/III prüft, inwieweit das Bolzenspiel vergrößert werden darf, damit das Durchschieben des Bolzens auch bei leicht verkantetem Auge möglich ist.

BArch, RH8II Bild-B1143-44, o.Ang.

64

BArch, RH8II Bild-B1146-44, o.Ang.

65

64, 65 Camouflage device for the FR trailer (S) MA4-3 version. Two FR trailers were used in the EW to test the camouflage devices. The MA4-3 version shown here, manufactured by F.X. Meiller, has all electrical wiring and the P-battery missing. These were not part of the order at F.X. Meiller but were subcontracted. The camouflage scaffold represents a cannon. The camouflage arrangement consists of the framework (picture above) and the planning (picture below). Camouflage scaffolding to imitate airplanes and speedboats was also tried out.

64, 65 Tarnvorrichtung für den FR-Anhänger (S) Ausf. MA4-3. Für die Erprobung der Tarnvorrichtungen wurden im EW zwei FR-Anhänger benutzt. Dem hier gezeigten FR-Anhänger Ausf. MA4-3 aus der Fertigung von F.X. Meiller fehlen sämtliche elektrische Verkabelungen und auch die P-Stoffbatterie. Beide Ausrüstungen waren nicht Gegenstand der Beauftragung bei F.X. Meiller, sondern wurden im Unterauftrag ausgeführt. Das Tarngerüst stellt eine Kanone dar. Die Tarnvorrichtung besteht aus dem Gerüst (Bild oben) und aus der Verplanung (Bild unten). Auch wurden Tarngerüste zur Nachahmung von Flugzeugen und Schnellbooten erprobt.

Camouflage device

Two versions (airplane and gun) are currently under development at Wa-Prüf 10. In addition there is a camouflage canvas for the units. Wa Prüf 10 will pass the documentation about this to Wa Prüf 11 so that the required mountings can be included in future production. This is expected to take place in the second half of April. The current arrangement of the mountings will remain the same until further notice.

Lift gear

The raising of the lift gear caused by the intended implementation of the cable mast will be introduced after the drawings of Wa Prüf 10 arrive at the vehicle's manufacturing companies. The units will receive an appropriate notification about the changes in respect of vehicles already delivered and those currently being delivered.

Pressurized oil line

The pressurized oil line, which has repeatedly led to vibration fractures according to previous experience, now features an additional mounting on the vehicles delivered by the companies. At the same time the pipe connections leading to the control valve are secured against loosening. With regard to the vehicles already delivered, appropriate notifications of the changes will also be passed to the units through Wa Prüf 11/III.

Manometer

Because of their susceptibility the existing manometers (60 mm diameter) will be replaced by the proven manometers with 90 mm diameter (company Schäffer & Budenberg). Replacement by the units delivery arranged by Wa Prüf 11/III.

Ladder

The redesign of the fixed ladder on the lifting arm with a consistent rung distance will be considered in a new production run of FR-trailers (Case C). Altering delivered vehicles or those currently being delivered not to take place. The tread on the strut towards the ladder will be made skid-proof with welded additions. Change of form will be arranged by WaPrüf 11/III.

Name plates

Name plates for plugs, sockets etc. are undesirable for security reasons. Color coding and numbers are sufficient identification of them for the units. Decision by B.z.b.V. Heer requested by Wa Prüf.

Axle movements

Small movements of the rear axle are unimportant and related to its construction.

Monitoring on the move

Because of the need to monitor the FR-trailers' movements in respect of tire pressure loss in the rear wheels, etc. Wa Prüf 6 is asked to submit an appropriate proposal.

Tarnvorrichtung

Bei Wa Prüf 10 werden z.Zt. zwei Ausführungen entwickelt (Flugzeug und Geschütz). Bei der Truppe befindet sich außerdem eine Planentarnung. Wa Prüf 10 gibt die Unterlagen hierzu an Wa Prüf 11, damit die erforderlichen Halterungen bei der weiteren Fertigung vorgesehen werden. Eine Entscheidung über die Tarnart kann erst nach Erprobung erfolgen. Diese findet voraussichtlich in der zweiten Hälfte April statt. Die derzeitige Halterungsanordnung bleibt bis auf weiteres.

Aufzugvorrichtung

Das Höhersetzen der Aufzugvorrichtung, bedingt durch die beabsichtige Einführung des Kabelmastes, wird nach Eintreffen der Zeichnungen von Wa Prüf 10 bei den Herstellerfirmen der Wagen veranlasst. Bei bereits gelieferten und noch in Auslieferung befindlichen Fahrzeugen geht entsprechende Änderungsmitteilung an die Truppe.

Öldruckleitung

Die Öldruckleitung, die nach bisherigen Erfahrungen mehrfach zu Schwingungsbrüchen geführt hat, ist bei den von den Firmen zur Auslieferung kommenden Fahrzeugen zusätzlich gehaltert. Gleichzeitig sind die Rohrverbindungen zum Steuerventil gegen Lösen gesichert. Bei den bereits ausgelieferten Fahrzeugen gehen ebenfalls entsprechende Änderungsmitteilungen durch Wa Prüf 11/III an die Truppe.

Manometer

Die bisherigen Manometer (60 mm Dm) werden wegen Störanfälligkeit ausgewechselt gegen die erprobten Manometer 90 mm Durchmesser (Firma Schäffer & Budenberg). Austausch bei der Truppe, Zulieferung veranlasst Wa Prüf 11/III.

Leiter

Die Ausführung der Steigleiter am Hubarm in profilbestimmender Ausführung und gleichbleibendem Sprossenabstand wird bei einer neuen Auflage von FR-Anhängern berücksichtigt (Fall C). Eine Änderung bei ausgelieferten oder noch in Auslieferung befindlichen Fahrzeugen tritt nicht ein. Der Auftritt auf der Spreize zur Leiter wird durch Aufschweißungen gleitsicher gemacht, Formänderungsmitteilung veranlassen Wa Prüf 11/III.

Bezeichnungsschilder

Sämtliche Bezeichnungsschilder bei Steckern, Dosen usw. werden aus Geheimhaltungsgründen allgemein als unzweckmäßig bezeichnet. Der Truppe genügen Farbbezeichnungen und Zahlen. Entscheidung durch B.z.b.V. Heer nach Antrag Wa Prüf 1.

Schlingerbewegungen

Schlingerbewegungen der Hinterachse in geringem Ausmaß sind ohne Bedeutung und konstruktionsbedingt.

Marschüberwachung

Wegen der Marschüberwachung der FR-Anhänger bezüglich Reifenpannen an den hinteren Rädern usw. wird Wa Prüf 6 gebeten, geeigneten Vorschlag einzureichen.

BArch, RH8II Bild-B1153-44, o.Ang.

66

BArch, RH8II Bild-B11061-44, o.Ang.

67

66, 67 Details of the tongs, the upper working platform and cable mast. For launching with a cable mast fitted on the starting table, a transport device was attached to the tilting frame. The cable mast is secured with quick release fasteners. The upper working platform was considerably larger compared to the »old« version and allowed better movement and work on the opened equipment covers at the rocket head. The tongs consisted of the two movable clamps and a saddle piece. The saddle-piece and saddle were covered with a layer of felt.

66, 67 Details der Zange, der oberen Arbeitsbühne und des Kabelmasts. Für das Schießen mit Kabelmast, eingesetzt am Starttisch, wurde eine Transportvorrichtung am Kipprahmen angebracht. Der Kabelmast ist mit Schnellspannverschlüssen gesichert. Die obere Arbeitsbühne war gegenüber der »alten« Ausführung deutlich größer und erlaubte ein besseres Bewegen und Arbeiten an den geöffneten Gerätedeckeln der Raketenspitze. Die Zange bestand aus den beiden beweglichen Klammern und einem Formstück (Sattel). Formstück und Sattel waren mit einer Filzlage belegt.

Hydraulics

There is a danger that during an inaccurately guided movement of the pistons, the wear on the spring pins inside the smallest piston head can be so extreme in the final phase that they shear off (incident in K = Karlshagen). Wa Prüf 11 and the Meiller company believe that if operating instructions are correctly followed such a fault cannot occur. However, it is considered necessary that a finely threaded screw connection is provided instead of the current spring pins. Wa Prüf 11 is to arrange this construction executed by the Meiller company including testing by Wa Prüf 10. Completion by mid-April if possible. In order to avoid accidents the units must again be reminded that the manual handed out in January of 1944 must be followed exactly. This manual has to be amplified to state that nobody may stand in the direction of movement during the process of raising and lowering of the lifting arm. The piston's protection against dirt and dust in the extended position has to be provided by the placement of a protective cover. This can only be placed after extending the arm, so that the prescribed cycle of piston movement can be observed to be perfect.

To be introduced by Wa Prüf 11/III and followed exactly by the units.

General switch-through attempt

The units' need to execute a general switch-through test using electrical ground feeding is evaluated by Wa Prüf 10 as follows: Each regular board battery is sufficient for a general switch-through test with a maximum duration of 1,5 min. Additional switch-through tests would require changing the battery. Solution: A floor battery set including a flexible cable, which will be raised by a lifting device, will be connected for the case of necessary multiple general switch-through tests, and therefore the hatches will only be opened after a failed first general switch-through test. After all *Gerät* have passed through the tests at the interim facility, such disturbances are inherently less likely to occur. The placement of this cable inside the vehicle for the tests and use of firing devices I and III respectively is yet to be checked.

Winch

The altered winch placement for lowering the stage has already been introduced for new production. Wa Prüf 11/III has issued change notices for vehicles already delivered.

Cable mast

In future the bracing of the planned cable mast will be done using wire ropes instead of solid cable.

Hydraulik

Es besteht die Gefahr, dass bei nicht vorschriftsmäßigem Ausfahren der Kolben die Beanspruchung der Spannstifte im kleinsten Kolbenkopf in der Endphase so groß ist, dass diese abscheren (Vorfall in K = Karlshagen). Wa Prüf 11 und die Fa. Meiller sind der Ansicht, dass bei Einhaltung der Bedienungsvorschrift ein derartiger Fehler nicht auftreten kann. Trotzdem wird für erforderlich gehalten, dass an Stelle der bisherigen Spannstifte eine Feingewinde-Verschraubung vorgesehen wird. Wa Prüf 11 veranlasst Durchkonstruktion durch Fa. Meiller und Erprobung einschließlich gemeinsam mit dem Wa Prüf 10. Fertigstellung möglichst Mitte April. Zur Vermeidung von Unfällen muss die Truppe nochmals eingehenst darauf hingewiesen werden, dass die im Januar 1944 ausgegebene Bedienungsanleitung schärfstens beachtet wird. Diese Anleitung muss dahingehend erweitert werden, dass beim Aufrichten und Absenken des Hubarmes niemand in der Bewegungsrichtung steht. Der Schutz des Kolbens gegen Verschmutzung und Verstaubung im ausgefahrenen Zustand muss durch Anbringen eines Schutzüberzuges erfolgen, der erst nach dem Ausfahren angebracht werden darf, damit der vorschriftsmäßige Ablauf der Kolbenbewegung einwandfrei beobachtet werden kann.

Veranlassung durch Wa Prüf 11/III und Parallelvorschlag durch die Truppe.

Generaldurchschaltversuch

Die Forderung der Truppe, den Generaldurchschaltversuch mittels Bodenspeisung durchzuführen, wird von Wa Prüf 10 wie folgt beurteilt: Jede normale Bordbatterie ist für einen Generaldurchschaltversuch mit einer Maximaldauer von 1½ min ausreichend. Weitere Durchschaltversuche würden ein Auswechseln der Batterie erfordern. Lösung: Für evtl. erforderlichen mehrmaligen Generaldurchschaltversuch wird ein Satz Bodenbatterien mit flexiblen Kabel, welches mit Aufzugsvorrichtung hochgezogen wird, angeschlossen und damit die Klappen erst nach Misslingen des ersten Generaldurchschaltversuchs geöffnet. Nachdem sämtliche Geräte die Prüfung im Zwischenlager durchlaufen, sind derartige Störungen von Haus aus weniger zu erwarten. Unterbringung dieses Kabels im Lkw für Prüf- und Schiessgerät I bzw. III wird noch geprüft.

Seilwinde

Der geänderte Sitz der Seilwinde zum Herablassen der Bühne ist bei der Neufertigung bereits veranlasst. Diesbezügliche Änderungsmitteilungen für bereits ausgelieferte Fahrzeuge veranlasst Wa Prüf 11/III.

Kabelmast

Die Abspannung des geplanten Kabelmastes erfolgt in Zukunft mit Drahtseilen anstatt mit Volldraht.

Vehicle 103

Internal or former designation: Fz.-Pos.-Nr.103
Launching platform trailer (1-axle)
Meillertisch

DMA, GD639.5.18
HAP11/T 507
Karlhagen, 1.6.1944
Bd.Nr. E 3195/44 g

DMA, GD639.5.1
EW234
Fz.-Pos.-Nr. 103

DMA, GD639.51.7
Auszug
Bedienungsanweisung
Meillertisch

DMA, GD630.02
A4 Lehrbuch
der FR-Schule
Stand Mai 1944
Auszug aus
Technische Fahrzeuge

Intended use

The launching platform was originally part of the initial Meillerwagen (MA4). For handling reasons it was decided to separate the table from the Wagen and to develop it as an independent vehicle. The launching platform (Meillertisch) is used as a base for the *Sondergerät* during the preparations for firing and for launching. For transport the firing platform is mounted on a single-axle road trailer chassis with pneumatic-tired wheels. The accessories necessary for preparing and firing the *Sondergerät* are attached to the rotating ring of the final platform.

The firing accessories – refilling coupling, five way coupling

The valve box and rudder brackets are transported in the Fz.106 or 107 and only mounted on the launch table in the firing position. The mast required for firing with a cable mast is carried on the FR-trailer.

The Meillertisch has the following dimensions and weights:

Road trailer RSE-100 and launching table

Total length:	3,750 mm with table
Width:	2,650 mm with table
Height:	2,100 mm with table
Distance:	2,300 mm from axle center to eyelet center
Track width:	1,600 mm
Tare weight:	2,150 kg
Payload:	1,650 kg
Service weight:	2,260 kg
Wheels:	190-20

Launching table

Total length:	2,550 mm without cable mast
Width:	2,650 mm
Height:	1,300 mm with table
Tare weight:	1,650 kg
Payload:	12,700 kg A4 *Gerät* refueled and ready for launching
Service weight:	14,350 kg

The launching platform consists of the following main components:

- chassis with Chassis frame, wheels and tires, trailing arms and suspension springs, electrical system.
- The launch platform, rotating ring support and rotating ring supports, and deflector (Schurre).

Fahrzeug 103

Interne oder frühere Bezeichnung: Fz.-Pos.-Nr.103
Abschußplattform-Anhänger (1-achs.)
Meillertisch

Verwendungszweck

Ursprünglich war die Abschussplattform ein Bestandteil des ersten Meillerwagens (MA4). Aus Handhabungsgründen wurde entschieden, den Tisch vom Wagen zu trennen und als eigenständiges Fahrzeug weiter zu entwickeln. Die Abschussplattform (Meillertisch) dient dem Sondergerät als Untersatz bei den Schussvorbereitungen und beim Schießen. Für den Transport wird die Abschussplattform von einem luftbereiften einachsigen Straßenanhänger-Fahrgestell aufgenommen. An dem Drehring der Abschlussplattform wird das zum Fertigmachen des Sondergerätes und zum Schießen notwendige Zubehör befestigt.

Das Schiesszubehör – Nachtankkupplung, Fünffachkupplung

Ventilkasten, Ruderhalterungen werden im Fz.106 bzw. 107 befördert und erst in der Feuerstellung am Starttisch montiert. Der für das Schießen mit Kabelmast benötigte Mast wird am FR-Wagen mitgeführt.

Der Meillertisch hat folgende Maße und Gewichte:

Fahrgestell RSE-100 mit Starttisch

Gesamtlänge:	3.750 mm mit Tisch
Breite:	2.650 mm mit Tisch
Höhe:	2.100 mm mit Tisch
Abstand:	2.300 mm von Achsmitte bis Ösenmitte
Spurweite:	1.600 mm
Leergewicht:	2.150 kg
Nutzlast:	1.650 kg
Gefechtsgewicht:	2.260 kg
Räder:	190-20

Starttisch

Länge:	2.550 mm ohne Mastantrieb
Breite:	2.650 mm
Höhe:	1.300 mm
Leergewicht:	1.650 kg
Nutzlast:	12.700 kg A4-Gerät betankt und startklar
Gefechtsgewicht:	14.350 kg

Die Hauptbauteile des Abschussplattform-Anhängers sind:

- Fahrgestell mit Fahrgestellrahmen, Rädern und Bereifung, Längslenkern und Tragfedern, elektrische Anlage.
- Abschussplattform mit Drehringauflage, Drehring, Stützen und Schurre.

Meillertisch »Version MA4«
Part of the Meillerwagen MA4 manufactured according to drawing SK 537 of Company F.X. Meiller.

Meillertisch »Ausführung MA4«
Als Bauteil des Meillerwagen MA4 gefertigt nach Zeichnung SK 537 der Firma F.X. Meiller

Barch, RH8 Bild-B0720-42BSM

68

68 The original version of the Meillertisch (MA4) was simpler in design than the subsequent redesign of the Meillertisch with single-axle trailer. The deflector (Schurre), a main component of the table, had the task of keeping the enormous forces of about 15 t/m^2 of the propulsion jet, with temperatures of over 2.500 °C, from the ground and diverting them laterally. The deflector (Schurre) was repaired several times: burst sheet metal parts were welded together again, annealed sheets were replaced. Various cone tips were tested.

68 Der Meillertisch in der Urausführung (MA4) war in der Konstruktion einfacher ausgeführt als die nachfolgenden Neukonstruktionen des Meillertisches mit Ein-Achs-Anhänger. Die Schurre, ein Hauptbauteil des Tisches, hatte die Aufgabe, die gewaltigen Kräfte des Antriebstrahles von ca. 15 t/m^2 mit Temperaturen von über 2.500 °C vom Boden abzuhalten und seitlich abzuleiten. Mehrmals wurde die Schurre repariert, zerplatzte Blechteile wieder zusammengeschweißt und ausgeglühte Bleche ersetzt. Zusätzlich wurden verschiedene Kegelspitzen erprobt.

Meillertisch MA4-2 »early version«
manufactured according to drawing SK 774 with chassis RSE-100 and drawing B8760 by Company F.X. Meiller.

Meillertisch MA4-2 »frühe Ausführung«
gefertigt nach Zeichnung SK 774 mit Fahrgestell RSE-100 und Zeichnung B8760 der Firma F.X. Meiller.

Deutsches Museum, München, Archiv- FA014/B696/43 BSM [CD81798]

69

69 On the »early version« of the Meiller table MA4-2, the crankcases were fitted further down on the support legs. However, the design was still too weak for the enormous forces, which could lead to bending of the jacks. As an interim solution, the support legs with crankcases were kept at a distance with a cross brace (EW modification). In this version, the rotating ring was made of a tube and was stiffened by a U-profile in the later MA4-3 version. The deflector was welded together from individual rounded sheets to form a cone.

69 An der »frühen Ausführung« des Meillertisches MA4-2 waren die Kurbelgehäuse an den Stützfüßen weiter unten eingepasst. Allerdings war die Konstruktion immer noch zu schwach für die enormen Kräfte. Dies konnte zum Verbiegen der Stützen führen. Als Zwischenlösung wurden die Stützen mit integrierten Kurbelgehäuse von einer Querverstrebung auf Abstand gehalten (EW-Modifikation). Der Drehring war in dieser Ausführung aus einem Rohr gefertigt und wurde bei der späteren Version MA4-3 durch ein U-Profil versteift. Die Schurre war aus einzelnen gerundeten Blechen zu einem Kegel zusammengeschweißt.

Deutsches Museum, München, Archiv- FA014/B1269/43 TD5224 [BN55698]

70

70 The insufficient ground clearance objected to by the units is clearly visible in the old Meillertisch version. It is just under 20 cm, which could lead to damage to the support legs and the base plate (deflector- Schurre) in the field. A ground clearance of about 50 cm was required. These requirements resulted in a new design of the single-axle trailer RSE 100.

The table already matches to the middle version MA4-3, here the crankcases of the supports were placed further up

70 Die von der Truppe beanstandete zu geringe Bodenfreiheit ist bei der alten Meillertisch-Ausführung deutlich zu erkennen. Sie liegt bei knappen 20 cm, was im Gelände durchaus zu Beschädigungen an den Stützfüßen und der Bodenplatte (Schurre) führen konnte. Es wurde eine Bodenfreiheit von ca. 50 cm gefordert. Diese Vorgaben hatten eine Neukonstruktion des Ein-Achs-Anhängers RSE 100 zur Folge. Der Tisch entspricht bereits der mittleren Ausführung MA4-3. Hier wurden die Kurbelgehäuse der Stützen weiter oben angebracht.

Archiv F.X.M
Fabrikationsbuch
Anhängerbuch 10 + 11

Development variants

The launching table, separated from the Meillerwagen MA4, was used for experiments until the end of 1943. The mountings for the A4 *Gerät*, the design and position of the attachments on the launching table for the rudder mountings, the refuelling device, the five-way coupling and the valve box were tested. The deflector (Schurre) posed a special challenge for the engineers. The brute forces acting on the chute during launch had to be sensibly diverted and the overall design of the launch table had to be stable enough to withstand the enormous pressure. During the first firing attempts in mid-1942 there were considerable problems with the deflector (Schurre) – bursting of the welded or partly riveted plates, or deformations at the cone made the deflector useless. Under the heat and enormous pressure of the launch flame, some of the spindles of the support legswere bent. This endangered the stability of the table construction.

With the experience gained from the first firing tests the entire construction was redesigned by F.X. Meiller.

The table (stands and deflector) was completely redesigned. A single-axle trailer of type RSE-100 was developed for its transport and was designated with the table as Fz.103. The ground clearance of the table on the attached single-axle trailer, which was considered too low, led to further changes. The tubular frame of the trailer was modified by offsetting and the wheels were occasionally replaced by larger ones. Smaller changes were incorporated into the ongoing production. For firing with a cable mast the table was extended by a cable mast winch device.

To differentiate between the manufacturer FXM divided the firing table for the Fz.103 into the versions MA4-2 and MA4-3. This categorization was amended by the author with the term »early version«, »middle version« and »late version«.

As with the Meillerwagen, the first Fz.103 were manufactured exclusively by F.X. Meiller. With the introduction of the cranked trailer variant production by the Lindner and Schenk companies was also commissioned by Wa Prüf 11/II . The Lindner company was designated as the leading firms for the Fz.103.

Entwicklungsvarianten

Der vom Meillerwagen MA4 getrennte Starttisch wurde bis Ende 1943 für Versuche verwendet. Erprobt wurden die Aufnahmen für das A4-Gerät, die Art und Lage der Befestigungen am Abschusstisch der Ruderhalterungen, der Nachtankvorrichtung, der Fünffachkupplung und dem Ventilkasten. Eine besondere Herausforderung stellte die Schurre für die Ingenieure dar. Die brachialen Kräfte, die beim Start auf dieses Bauteil einwirkten, mussten sinnvoll umgelenkt und die Gesamtkonstruktion des Starttisches so stabil ausgelegt sein, dass sie dieser Belastung standhielt. Bei den ersten Startversuchen Mitte 1942 gab es erhebliche Probleme. Zerplatzer der geschweißten bzw. zum Teil auch genieteten Bleche und Verformungen am Kegel machten die Schurre unbrauchbar. Unter der Hitze und dem enormen Druck des Antriebsstahls verbogen sich teilweise die Spindeln der Standfüße. Die Standsicherheit der Tischkonstruktion wurde dadurch gefährdet.

Mit den gewonnenen Erfahrungen aus den ersten Startversuchen wurde die gesamte Konstruktion von F.X. Meiller neu ausgelegt.

Der Tisch (Stützen und Schurre) wurden komplett neu definiert. Für den Transport wurde ein Einachsanhänger Type RSE-100 entwickelt und mit dem Tisch als Fz.103 bezeichnet. Die als zu gering erachtete Bodenfreiheit des Tisches am eingehängten Ein-Achs-Anhänger führten zu weiteren Änderungen. Der Rohrrahmen des Anhängers wurde durch Abkröpfen modifiziert und die Räder gelegentlich durch größere Räder getauscht. Kleinere Änderungen flossen in die laufende Produktion ein. Für das Schießen mit Kabelmast wurde der Tisch um eine Kabelmast-Aufwindevorrichtung erweitert.

Zur Unterscheidung teilte der Hersteller FXM den Abschußtisch des Fz.103 in die Versionen MA4-2 und MA4-3 ein. Diese Einteilung wurde vom Autor um die Begriffe »frühe Version«, »mittlere Version« und »späte Version« ergänzt.

Wie beim Meillerwagen wurden die ersten Fz.103 exklusiv von F.X. Meiller gefertigt. Mit Einführung der abgekröpften Anhängervariante wurden auch die Firmen Lindner und Schenk als Nachbaufirmen durch Wa Prüf 11/III beauftragt. Die Firma Lindner wurde als Leitfirma für das Fz.103 bestimmt.

DMA, GD639.5.11
T404
o.O., den 26.5.1944
T404 o.O., den 26.5.1944
Bd.Nr. E 2191/44 gK
Auszug
Teileliste Fz.103

Vehicles on order

Fahrzeugbestellungen

	Item Lfd. Nr.	Qty. ordered Best. Stückzahl	Qty. per vehicle Stück je Fahrzeug	Subject drawing number Gegenstand Zeichnungs-Nr.	Ordered by order number Besteller Bestell-Nr.	Supplier Lieferfirma
Fahrzeugbestellung	1a	78	1	Abschußplattformanhg. Nr.103	WaPrüf 11/III 361-0010/42	F.X. Meiller München 8 Lilienthalstr. 2/4
	1b	84	1	Abschußplattformanhg. Nr.103	WaPrüf 11/III 361-0007/43	Gottfried Lindner Ammendorf b. Halle
	1c	107	1	Abschußplattformanhg. Nr.103	WaPrüf 11/III 361-0026/43	Gottfried Lindner Ammendorf b. Halle
	1d	48	1	Abschußplattformanhg. Nr.103	WaPrüf 11/III 361-0034/43	Maschinen-Fabrik Karl Schenk
Ergänzung Teileliste 103	2	238	1	Halterung mit Antrieb zum Kabelmast Fz.103	WaPrüf 10/H SS4949-0338-6137/43	A. Rawie Fabrik für Eisenbahnbedarf Osnabrück

Meillertisch MA4-3 »mid version«

With modified chassis RSE-100 manufactured according to drawing B9020 by Company F.X. Meiller

Meillertisch MA4-3 »mittlere Ausführung«

Mit modifiziertem Fahrgestell RSE-100, gefertigt nach Zeichnung B9020 der Firma F.X. Meiller

71

71 The tubular frame of the single-axle trailer RSE-100 was offset between the chassis winch and the cross tube of the fork. This resulted in a considerable increase in ground clearance. The deflector was lifted with the chassis winch to the tubular frame and staked. Turntable support and turntable ring were protected with a tarpaulin, which was folded inwards almost around the turntable ring and held together between the supports with push buttons.

71 Am Ein-Achs-Anhänger RSE-100 wurde der Rohrrahmen zwischen Fahrgestellwinde und dem Querrohr der Gabel gekröpft. Dadurch erreichte man eine erhebliche Vergrößerung der Bodenfreiheit. Mit der Fahrgestellwinde wurde die Schurre zum Rohrrahmen gekurbelt und abgesteckt. Drehringauflage und Drehring waren mit einer Plane geschützt, die nach innen eingeschlagen war, um den Drehring fasste und zwischen den Stützen mit Druckknöpfen zusammengehalten wurde.

DMA, GD639.5.21
OKW/Wa Prüf10
Karlshagen, 25.03.1944
Bd.Nr. 747/44 gK
Auszug
Besprechung der
Fahrzeuggruppe 1
vom 24./25.3.44
im Lager Rebstock

Production status: Vehicle 103
So far 9 units have been delivered to Abt.836. Of these 3 have been handed over to Battery 444 by order of B.z.b.V. Heer and will remain in HL; 1 unit was destroyed. As a result, Abt.836 currently possesses 5 launch platform trailers.
Six units have been delivered to Abt.485 to date, namely 3 units directly to the units and 3 pieces for prior testing of the ground system in live fire to Wa Prüf 10.

Further deliveries will be.
7 units by April 2 from Wa Prüf 10 (already designated for 836),
6 units by April 5 from LR,
5 units from LR, as soon as they are delivered from Wa Prüf 11/III.

Current defects and their remedy:
The units' modification requests of the launch platforms described above have been considered differently depending on the delivery dates. The delivery of 36 launch platforms between early April and May has those improvements including greater ground clearance as demanded by the units. Until the receipt of these vehicles, the units will make do with interim solutions, which have already been partly executed.

These are: A trunnion lock consisting of a round iron bar with a diameter of 10 mm is designated instead of the locking pin between cone and table (units' version). For the guide bushing mounting inside the table base, 4 cylinder screws (M10) are arranged instead of the two stud screws. Strengthening of the limber's bent towing eye by welding on a 10 mm flat iron bar sideways. The removal of the lateral brace is rejected; during tire change the carriage has to be extended as before.

Ratchets: Adjustment of the rotating ring is not currently possible even with the simultaneous operation of both ratchets. The units make do by moving the rotating ring by hand, which needs 8 men. Things to be checked
a) whether this rough adjustment can be made with the help of levers and
b) whether fine adjustment is possible by simultaneous use of the ratchets.

If an accurate adjustment by these means is not possible, then the worm gear drive designed by Wa Prüf 10 will be put into effect in such a way that the rough adjustment can be done by movement with levers and the fine adjustment by using the worm gear drive. Likewise, in a troop trial whether the insufficient ground clearance of the old launch platform trailers will be improved by removing the foot plates, with the plates placed in suitable holders at the table.

Manufacturing documents: Are to be compiled by Wa Prüf 11.

Fertigungslage: Fahrzeug 103
An Abt.836 wurden bisher ausgeliefert 9 Stck. Von diesen wurden über Befehl B.z.b.V. Heer 3 Stck. an Batterie 444 übergeben und verbleiben in HL., 1 Stck. wurde zerstört. Abt.836 besitzt daher z.Zt. 5 Stck. Abschussplattform-Anhänger.
An Abt.485 wurden bisher ausgeliefert 6 Stück und zwar 3 Stück unmittelbar an die Truppe und 3 Stück zur vorherigen Erprobung der Bodenanlage im scharfen Schuss über Wa Prüf 10.

Zur weiteren Auslieferung kommen:
7 Stück zum 2.4 von Wa Prüf 10 (bereits für 836 vorgesehen),
6 Stück zum 5.4 von LR,
5 Stück von LR, sobald sie von Wa Prüf11/III zur Auslieferung kommen.

Derzeitige Mängel und ihre Behebung:
Bei obigen Abschussplattformen sind Änderungswünsche der Truppe je nach dem Auslieferungstermin unterschiedlich berücksichtigt. Die Auslieferung von 36 Abschussplattformen in der Zeit von Anfang April bis Mai hat die von der Truppe geforderten Verbesserungen einschl. größerer Bodenfreiheit. Bis zum Erhalt dieser Fahrzeuge behilft sich die Truppe mit Zwischenlösungen, welche zum Teil schon durchgeführt sind.

Diese sind: Statt der Zapfenverriegelung zwischen Kegel und Tisch wird ein Überfallriegel aus Rundeisen mit 10 mm Durchmesser vorgesehen (Truppenausführung). Statt der beiden Stiftschrauben zur Halterung der Führungsbuchse im Tischfuß werden 4 Zylinderschrauben M10 angeordnet. Verstärkung der abgekröpften Protzöse durch seitliches Aufschweißen von 10 mm Flacheisen. Ein Herausnehmen der seitlichen Strebe wird abgelehnt, bei Reifenwechsel muss das Fahrgestell wie bisher ausgefahren werden.

Ratschen: z.Zt. ist die Verstellung des Drehringes auch bei gleichzeitiger Betätigung beider Ratschen nicht möglich. Die Truppe behilft sich durch Bewegen des Drehringes von Hand und stellt hierzu 8 Mann an. Es muss überprüft werden, ob es
a) bei dieser Grobeinstellung verbleiben kann unter Zuhilfenahme von Hebebäumen und
b) die Feineinstellung durch gleichzeitige Benutzung der Ratschen möglich ist.

Ist ein genaues Einrichten mit diesen Mitteln nicht möglich, dann wird der von Wa Prüf 10 entwickelte Schneckenantrieb so zur Ausführung gebracht, dass für Grobeinstellung eine Bewegung mit Hebebäumen und die Feineinstellung durch Einklinken des Schneckenantriebes vorgenommen werden kann. Desgleichen wird in einem Truppenversuch die ungenügende Bodenfreiheit der alten Abschussplattformanhänger durch Abnehmen der Fußplatten verbessert, wobei die Platten in entsprechende Halterungen am Tisch aufgebracht werden.

Fertigungsunterlagen: Sind durch Wa Prüf 11 erstellt.

Chassis and launch platform

The trailer RSE-100 is made from tubular frame with swing axle, wheels, manuel brake system and electrical system. The photographs on the left show the structural differences between the individual chassis frames.

The launch platform MA4-3 consists of the deflector, supports, rotating ring support and rotating ring. The launching platforms on page 111 show the differences between their individual constructions.

Fahrgestell und Abschussplattform

Der Anhänger RSE-100 besteht aus einem Rohrrahmen mit Schwingachse, Rädern, Handbremse und elektrischer Anlage. Die Fotografien links zeigen die konstruktiven Unterschiede der einzelnen Fahrgestellrahmen.

Die Abschußplattform MA4-3 besteht aus Schurre, Stützen, Drehringauflage und Drehring. Die Abbildungen der Abschussplattformen auf Seite 111 zeigen die Unterschiede der einzelnen Konstruktionen.

Deutsches Museum, München, Archiv- FA014/B1270/43 TD5224 [CD74704]

72

Deutsches Museum, München, Archiv- FA014/B469/44 TD5224 [CD74704]

73

72 The tubular frame of th early version RSE-100 is designed like an A when viewed from above. The two tube ends are designed as fork, they engage in the holders on the deflector, which is lifted up to the chassis frame with a pole winch. A tail light with night driving light and hand brake complete the early single-axle trailer. The wheels are size 6.00-20

73 The angled contour of the tubular frame of the late version RSE-100 can be seen clearly in the side profile. The type of design as A-tube frame was retained and instead of the pole winch a chassis winch with lateral crank access was mounted on the tubular frame. The brake system is omitted, only a tail light with night driving light is attached. The wheels are now size 190-20.

72 Der Rohrrahmen hat von oben gesehen bei der frühen Ausführung RSE-100 die Form eines A. Die beiden Rohrenden sind als Gabeln ausgeführt. Sie greifen in die Halter an der Schurre, die mit einer Stockwinde an den Fahrgestellrahmen herangekurbelt wird. Ein Rücklicht mit Nachtmarschgerät und Handbremse ergänzen den frühen Ein-Achs-Anhänger. Die Räder haben die Größe 6,00-20.

73 Die abgekröpfte Kontur des Rohrrahmens der späten Ausführung RSE-100 wird im Seitenprofil durch die vorliegende Abbildung sehr gut dargestellt. Die Art der Ausführung als A-Rohrrahmen wurde beibehalten und anstelle der Stockwinde eine Fahrgestellwinde mit seitlichem Kurbelzugang am Rohrrohrrahmen montiert. Die Bremsanlage ist entfallen, nur ein Rücklicht mit Nachtmarschgerät ist angebracht. Die Räder sind jetzt in der Größe 190-20 ausgeführt.

74 While the deflector of the early versions MA4 and MA4-2 was even smaller and was welded together from individual rounded plates to form a cone tip, it was now manufactured as a pyramid from rounded 5 mm thick special steel plates and provided with a cone tip. Small ramps for the trailer wheels and holders for the lock bearings on the chassis are also welded to the deflector. On the left and right side two flaps are worked into the contour of the deflector and have to be opened to retract the trailer. The deflector is bolted to the protective ring of the support at four corners. The MA4-3 table supports carry the rotating ring support, and struts stiffen the construction. The gear part of a pole winch is incorporated in all four supports; the table is aligned in the firing position via the cranks. The rotating ring is reinforced on its four corners, where the plates are mounted on which the tail fins of the rocket rest. On the ring are mounted clamping holders forthe internal control vanes, valve box, as well as clamping sleeves for the refuelling device and five way coupling. A holder (erected) for the igniter flask is mounted next to the retainer tubes for the valve box. Two bracket for fixing it to the Meiller trailer are welded to the rotating ring frame. The ring lies on the ring support. When the rocket is set up, the rotating ring is aligned in the firing direction by means of a rotating device (ratchet) via a link chain and locked with four locking devices.

75 The table MA4-3 of the late version remained almost unchanged and was only supplemented by a cable mast winding device on the rotating ring.

74 War die Schurre in den frühen Ausführungen MA4 und MA4-2 noch kleiner und aus einzelnen gerundeten Blechen zu einer Kegelspitze zusammengeschweißt, wurde diese nun als Pyramide aus gerundeten, 5 mm dicken Sonderstahlblechen gefertigt und mit einer Kegelspitze versehen. Kleine Auffahrrampen für die Anhängerräder und Halter für die Aufnahme der Verschlusslager am Fahrgestell sind ebenfalls an der Schurre angeschweißt. Links und rechts sind zwei Klappen in die Kontur der Schurre eingearbeitet. Diese müssen zum Einfahren des Anhängers geöffnet werden. Die Schurre ist an vier Ecken am Schutzring der Stützen verschraubt. Die Stützen am Tisch MA4-3 tragen die Drehringauflage, Streben versteifen die Konstruktion. In allen vier Stützen ist der Getriebeteil einer Stockwinde eingearbeitet; über die Kurbeln wird der Tisch in der Feuerstellung ausgerichtet. Der Drehring ist an vier Ecken verstärkt. Dort sind die Teller angebracht, auf der die Heckflossen der Rakete aufliegen. Auf dem Ring sind Klemmhülsen für Strahlruder-Halter, Ventilkasten, sowie Klemmbüchsen für die Nachtankvorrichtung und Fünffachkupplung montiert. Eine Halterung (aufgestellt) für die Zündflasche ist neben den Halterohren für den Ventilkasten befestigt. Zur Fixierung am Meillerwagen sind zwei Haltelager am Drehringrahmen angeschweißt. Der Drehring liegt auf der Drehringauflage. Mit aufgestellter Rakete wird der Drehring mittels einer Drehvorrichtung (Ratsche) über eine Gelenkkette in die Schussrichtung ausgerichtet und mit vier Feststellern blockiert.

75 Der Tisch MA4-3 der späten Ausführung blieb nahezu unverändert und wurde nur um eine Kabelmast-Aufwindevorrichtung am Drehring ergänzt.

Deutsches Museum, München, Archiv- FA014/B472/44 TD5224 [CD74754]

74

BArch, RH8II Bild-B1574-44, o.Ang.

75

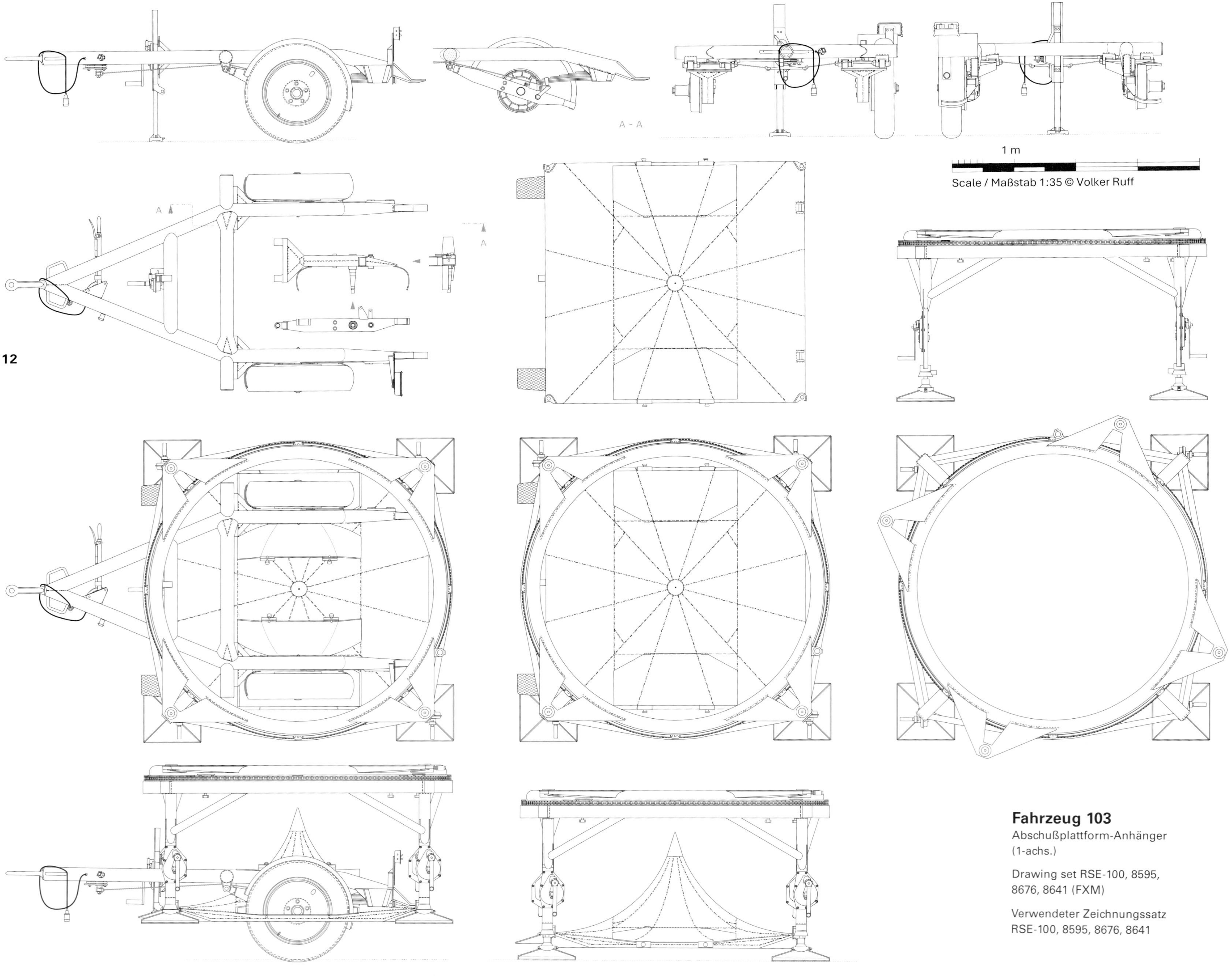

Fahrzeug 103

Abschußplattform-Anhänger
(1-achs.)

Drawing set RSE-100, 8595, 8676, 8641 (FXM)

Verwendeter Zeichnungssatz
RSE-100, 8595, 8676, 8641

Meillertisch MA4-3 »late version« (final version)
Manufactured according to the drawing B9020 of Company F.X. Meiller

Meillertisch MA4-3 »späte Ausführung« (Abschlussausführung)
Gefertigt nach der Zeichnung B9020 der Firma F.X. Meiller

BArch, RH8II Bild-B0107-45, o.Ang.

76

76 Since fewer and fewer towing vehicles were available an additional beam with coupling claw was welded between the two pick-up forks on the final version of the single-axle trailer RSE-100. Thus it was possible to couple further single-axle launch platform trailers one behind the other.

76 Da immer weniger Zugmittel zur Verfügung standen, wurde an der Abschlussausführung des Ein-Achs-Anhängers RSE-100 noch ein zusätzlicher Träger mit Kupplungsklaue zwischen die beiden Aufnahmegabeln angeschweißt. Somit war es möglich, weitere Ein-Achs-Abschussplattform-Anhänger hintereinander zu koppeln.

DMA, GD639.5.18
HAP11/T 507
Karlshagen, 1.6.1944
Bd.Nr. E 3195/44 g

DMA, GD639.51.3
Auszug
Bedienungsanweisung
Maschinensatz 11

Vehicle 104

Internal or former designation: Fz.-Pos.-Nr.104
Machine set 11 (Strüver)
Strüver

Intended use
The machine set 11 is used to supply power for ground network (BN) to vehicles and equipment:
- Fire control vehicle
- B.A. cars
- A4 *Gerät* (Meillerwagen / Meillertisch)

Initially, only the Strüver machine set 11 was available for power supply. The vehicles to be supplied with power were adapted to the Strüver circuitry:

Fz.101 (Fz.101A with Strüver circuit) and the B.A. cars Fz.106 or 107 (Fz.106A or 107A) with Strüver circuit.

The FF connection between the vehicles of the firing position and combustion cut-off position were also switched via the Citomat system in Strüver. The firing position of the Abteilungen equipped with a Strüver was called operation mot.I.

Only Abt.836 was designed and equipped for operation with machine set 11 »Strüver«. An exchange of the vehicles of mot.II (without Strüver circuit) with vehicles of mot.I (Fz.101A and Fz.106A or 107A) was not possible.

The Strüver has the following dimensions and weights:

Total length:	~ 7,450 mm
Width:	~ 2,400 mm
Height:	~ 3,000 mm
Wheelbase:	~ 3,700 mm
Track width:	~ 1,840 mm distance between dual tires
Tare weight:	n.a.
Payload:	n.a.
Service weight:	n.a.
Wheels:	8.00-20 double tires

The machine set 11 consists of the following main components:
- Chassis with turntable, overrun and Knorr air brake
- Sheet metal body with »Strüver« diesel direct current aggregate 48 kw, 230 V, four-field switchgear, Citomat*, direct current converter, relay switch box, TFb1

* Citomat = Telephone- Self-connection central

Fahrzeug 104

Interne oder frühere Bezeichnung: Fz.-Pos.-Nr.104
Machinensatz 11 (Strüver)
Strüver

Verwendungszweck
Der Maschinensatz 11 dient zur Stromversorgung im Bodennetz (BN) für folgende Fahrzeuge und Geräte:
- Feuerleitwagen
- B.A. Wagen
- Gerät A4 (Meillerwagen / Meillertisch)

Für die Stromversorgung stand zunächst nur der Maschinensatz 11 Strüver zur Verfügung. Die mit Strom zu versorgenden Fahrzeuge sind schaltungstechnisch an die Strüver Schaltung angepasst.

Es sind: Fz.101 (Fz.101A mit Strüver-Schaltung) und der B.A.-Wagen Fz.106 bzw. 107 (Fz.106 bzw. 107A mit Strüver-Schaltung).

Auch die FF-Verbindung zwischen den Fahrzeugen der Feuerstellung und BS-Stellung wurden über die Citomat-Anlage im Strüver geschaltet. Die Feuerstellung der mit einem Strüver ausgerüsteten Abteilungen wurde als Betrieb mot.I bezeichnet.

Nur die Abteilung Abt.836 wurde für den Einsatz mit dem Maschinensatz 11 »Strüver« vorgesehen und ausgerüstet. Ein Austauschen der Fahrzeuge mot.II (ohne Strüver-Schaltung) in Fahrzeuge der mot.I (Fz.101A und Fz.106A bzw. 107A) ist nicht möglich.

Der Strüver hat folgende Maße und Gewichte:

Gesamtlänge:	~ 7.450 mm
Breite:	~ 2.400 mm
Höhe:	~ 3.000 mm
Radstand:	~ 3.700 mm
Spurweite:	~ 1.840 mm Mittenabstand zwischen Doppelreifen
Leergewicht:	k.A.
Nutzlast:	k.A.
Gefechtsgewicht:	k.A.
Räder:	8,00-20 Doppelräder

Der Maschinensatz 11 besteht aus folgenden Hauptbauteilen:
- Fahrgestell mit Drehschemel, Auflauf- und Knorr-Druckluftbremse
- Blechaufbau mit »Strüver« Diesel-Gleichstrom-Aggregat 48 kw, 230 V, Vierfeldrige Schaltanlage, Citomat*, Gleichstrom-Umformer, Relais-Umschaltkasten, TFb1

* Citomat = Fernsprech- Selbstanchluss-Zentrale

Deutsches Museum, München, Archiv- FA014/B694/43 BSM [CD79126]

77

Strüver 11 with test equipment
Strüver 11 mit Testausrüstung

77 This Fz.104 with test equipment belongs to the HAP group c training battery and is set up for a firing test on test stand X. Mounted on both sides are 3 flaps each. In the back side is the outlet for the honeycomb radiator, and behind it is the diesel engine with coupled direct current generator and switch box 220 V. The accumulator for the TF1b-unit is located in the middle of the metal box, behind it is the switch box with three converters 220/24 V.
The FL cable connection for the firing position are mounted in one of the metal boxes on the chassis.

77 Dieses Fz.104 mit Testausrüstung gehört zur Gruppe c Lehrbatterie HAP und ist für einen Schießversuch auf dem Prüfstand X aufgebaut. Beidseitig sind die drei Klappen aufgestellt. An der Rückwand ist der Austritt für den Wabenkühler, dahinter sitzt der Dieselmotor mit gekoppeltem Gleichstromgenerator und Schaltkasten 220 V. Mittig im Blechkasten ist der Akkumulator für das TFb1-Gerät untergebracht, dahinter sitzt der Umschaltkasten mit drei Stück Umformern 220/24 V.
Die FL-Kabelverbindungen für die Schussstellung sind in einem der Blechkästen am Chassis eingerüstet.

Chassis and body

The machine set 11 is a two-axle trailer vehicle that can be driven on the road, with a chassis frame made of electrically welded and rigidly ribbed profil irons that supports the sheet metal protective housing with a curved roof. There are three opening flaps on each side with individually lockable locks. On the rear wall are flaps that can be opened to allow the cooling air for the honeycomb cooler to exit. A flap is installed on the front wall at the top and doors for access to the power access panel at the bottom.

Fahrgestell und Aufbau

Der Maschinen-Satz 11 ist ein zweiachsiges, straßenfahrbares Anhänger-Fahrzeug mit einem Fahrgestellrahmen aus elektrisch verschweißten und starr verrippten Profileisen, die das Blechschutzgehäuse mit gewölbtem Dach tragen. Auf jeder Seite befinden sich drei aufstellbare Klappen mit einzeln verschließbarer Verriegelung. An der hinteren Stirnseite sind aufstellbare Klappen zum Austritt der Kühlluft für den Wabenkühler und an der vorderseitigen Stirnwand oben eine Klappe und unten Türen für den Zugang zur Stromzugangstafel verbaut.

78

Final version
Abschlussausführung

78 This »Strüver« Fz.104 is painted is in a striking camouflage pattern. A 12 t Zgkw. Sd.Kfz.8 serves as the tractor. Two sheet steel boxes for holding tools and normal spare parts for the diesel dynamo unit and chassis are attached on both sides below the chassis frame. The sheet steel boxes are lockable and equipped with a hinged flap on the front. A spare wheel holder is mounted in the center of the chassis frame. The photo was taken in March 1944 at the Groß-Born military training area, Art.Abt.836 training area.

78 In einem auffälligen Tarnmuster ist dieser »Strüver« Fz.104 lackiert. Als Zugmaschine dient ein 12 t Zgkw. Sd.Kfz.8. Zur Aufnahme von Werkzeugen und normalen Reserveteilen für das Diesel-Dynamo-Aggregat und Fahrgestell sind auf beiden Seiten zwei Stahlblechkästen unterhalb des Fahrgestellrahmens angebracht. Die Stahlblechkästen sind verschließbar und mit vorderseitiger Scharnierklappe ausgerüstet. Mittig ist am Fahrgestellrahmen eine Reserverad-Halterung montiert. Das Foto entstand im März 1944 auf dem Truppenübungsplatz Groß-Born, Ausbildungsgelände der Art.Abt.836.

The fixtures

The required current is generated by a »Strüver« diesel direct current generator with 48 KW and is regulated by a converter unit consisting of three direct current converters to 24/36 V with 335/223 A.
In sheet steel cabinets are housed :
1 four-field switchgear
1 Telephone self-connection centre (Citomat),
1 relay switch box,
1 carrier frequency device and 1 additional device
The entry panel for the *Feuerleitpanzer* and exit panel to the firing sites I, II, III and a quick controller for the DC motors are housed in the front part of the sheet metal structure.

* Citomat = Is a fully automatic dial system for 20 subscribers and 1 exchange line (OB/ZB/W operation).

Die Einbauten

Der benötigte Strom wird von einem »Strüver« Diesel-Gleichstrom-Aggregat mit 48 KW erzeugt und über eine Umformer-Einheit aus drei Gleichstrom Gleichstrom-Umformer auf 24/36 V mit 335/223 A geregelt.
In Stahlblechschränken sind untergebracht :
1 vierfeldrige Schaltanlage
1 Fernsprech-Selbstanschluss-Zentrale (Citomat),
1 Relais-Umschaltkasten,
1 Träger-Frequenz-Gerät und 1 Zusatz-Gerät
Die Zugangstafel für den Feuerleitpanzer und die Abgangstafel zu den Schießstellen I, II, III und ein Schnellregler für die Gleichstrommotoren sind in Gehäusen im vorderen Teil des Blechaufbaus montiert.

* Citomat = Ist eine vollautomatische Wählscheibenanlage für 20 Teilnehmer und 1 Amtsleitung (OB/ZB/W-Betrieb).

BArch, RH8II Bild-B0772-42_BSM

79

Deutsches Museum, München, Archiv- FA014/B542/44 TD5224 [CD74788]

80

79, 80 Behind the open flap is the control panel for the power supply system. The current produced was controlled by handwheels and monitored by ammeters; a multitude of fuses of different strengths protected the system from overload.

79, 80 Hinter der geöffneten Klappe befindet sich die Bedientafel für die Stromversorgungsanlage.
Die produzierte Stromstärke wurde mit Handrädern gesteuert und von Amperemetern überwacht, eine Vielzahl verschieden starker Sicherungen schützt die Anlage vor Überlastung.

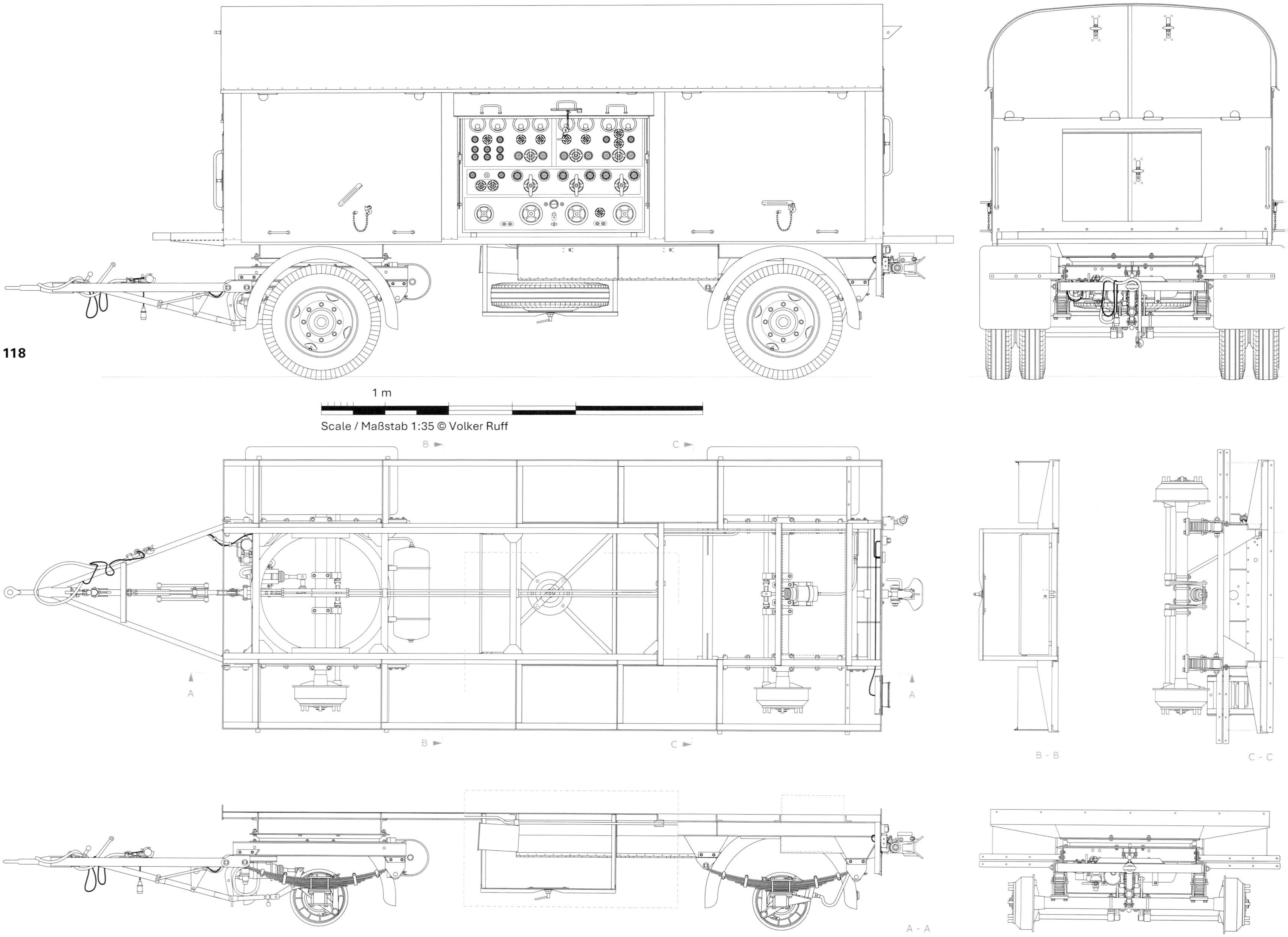

Scale / Maßstab 1:35 © Volker Ruff

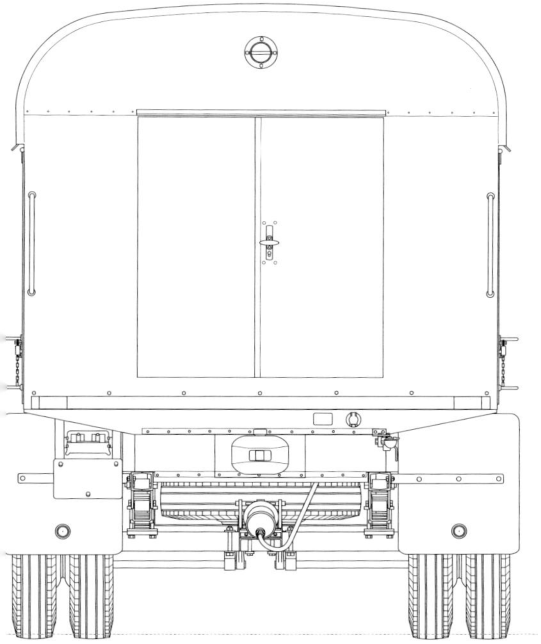

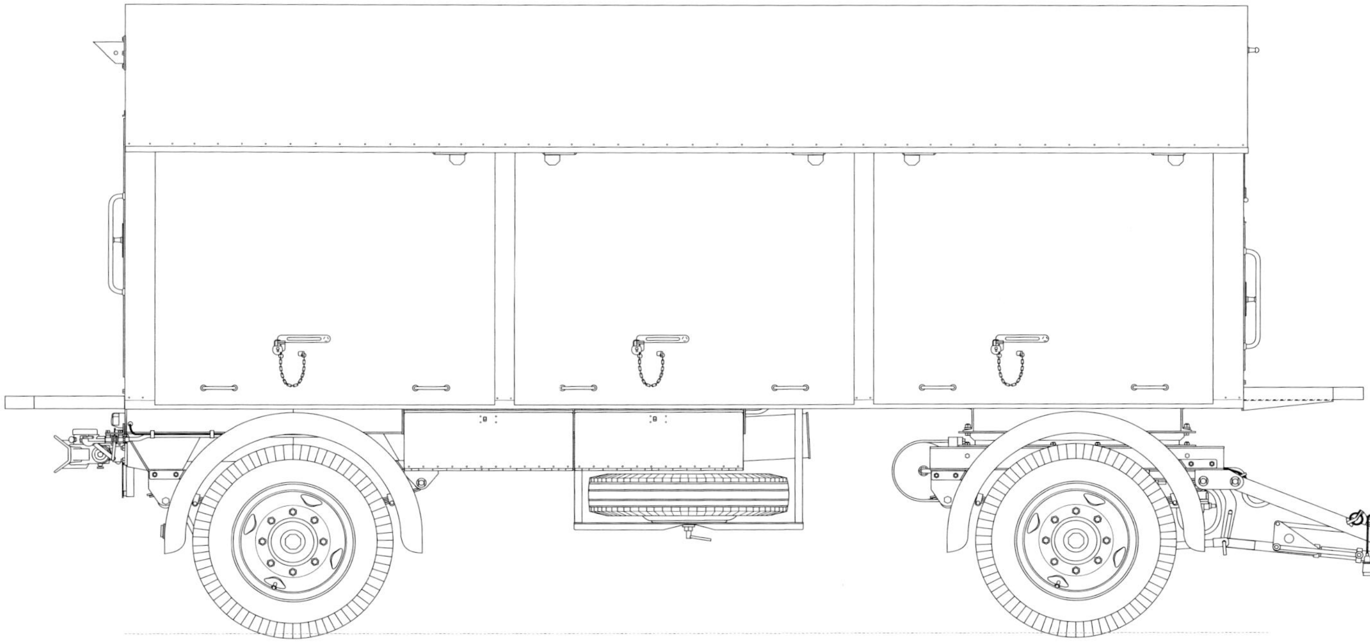

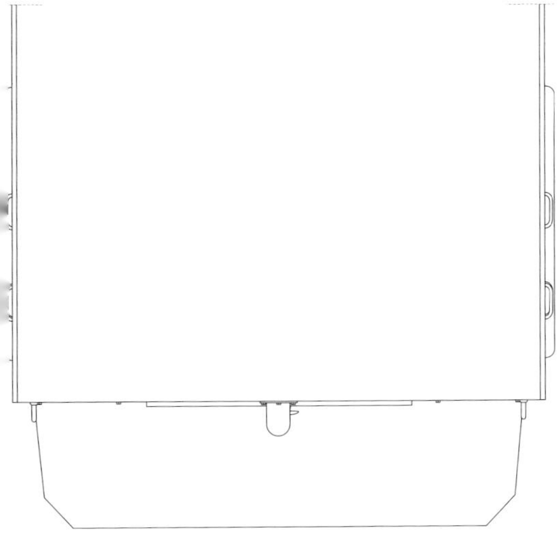

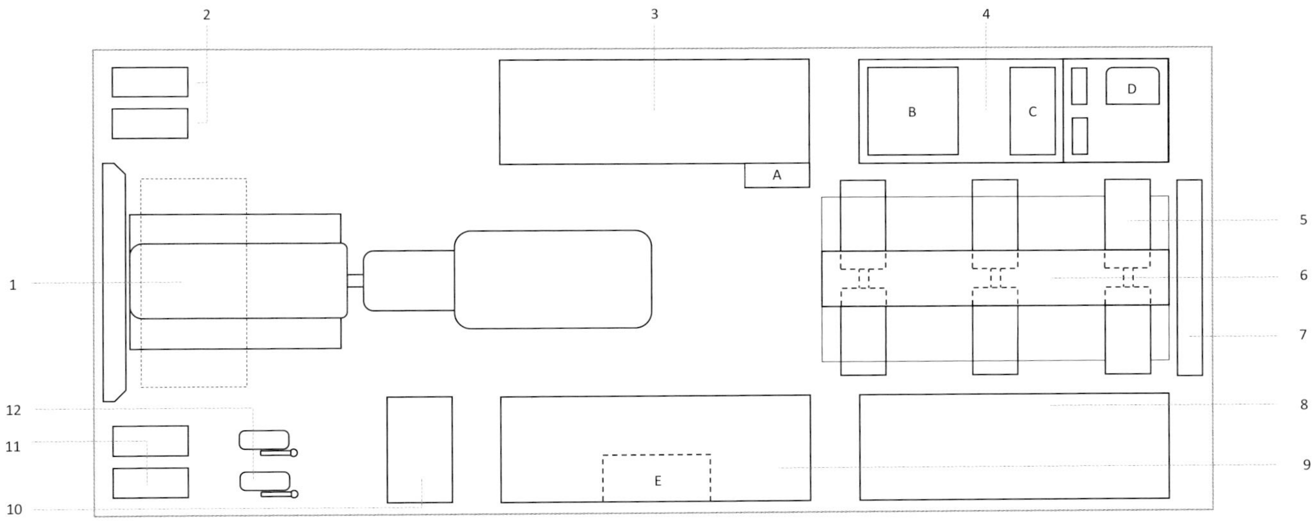

1 Diesel engine with flanged 220V DC generator
2 Accumulators (starter)
3 Control panel for power supply system and inverter (A)
4 Rack with TFb (B), device (C) and quick-acting controller (D)
5 3 converters 220/24 V with motor and shunt generator
6 Switch box
7 Output panel for fire control vehicle
8 Output panel for launching sites I, II, and III
9 Accumulator for TFb device and panel for Citomat (E)
10 220 V switch box
11 Pumps for fuel filling and cooling water
12 Command transmitter battery and plug battery (Stotz)

1 Dieselmotor mit angeflanschten Gleichstromgenerator 220V
2 Akkumulatoren (Starter)
3 Schalttafel für Stromversorgungsanlage und Wechselrichter (A)
4 Gestell mit TFb-Gerät (B), Zusatzgerät (C) und Schnellregler (D)
5 3 Stück Umformer 220/24 V mit Motor und Nebenschlussgenerator
6 Umschaltkasten
7 Abgangstafel zum Feuerleitpanzer
8 Abgangstafel Startstelle I, II und III
9 Akkumulator für TFb-Gerät und Anschlusstafel für Citomat (E)
10 Schaltkasten 220 V
11 Pumpen für Brennstoffeinfüllung und Kühlwasser
12 Kommandogeber- und Stotzstecker-Batterie

Fahrzeug 104
Machinensatz 11 (Strüver)

Drawing set used n.a.
Verwendeter Zeichnungssatz k.A.

DMA, GD639.5.18
HAP11/T 507
Karlshagen, 1.6.1944
Bd.Nr. E 3195/44 g

DMA, GD639.5.1
EW234
Fz.-Pos.-Nr.105

DMA, GD639.51.2
Bd.Nr. 23/240/44 gK
August 1944
Auszug
Bedienungsanweisung
Fz.-Nr.105

DMA, GD630.02
A4 Lehrbuch
der FR-Schule
Stand Mai 1944
Auszug aus
Technische Fahrzeuge

Vehicles 105

Internal or former designation: Fz.-Pos.-Nr.105
Power supply car on 2 t truck (Steyr)
Steyr

Intended use

In order not to weaken the on-board battery in the *Sondergerät* prematurely by tests and through-connection tests, the on-board power supply is supplied with electrical energy by a power supply car before the shot is fired. In addition, all switching processes and test messages coming from the *Sondergerät* on the one hand and from the individual ground stations on the other hand converge in a control centre of the same vehicle.

For the operation of mot.II each firing point has a power supply vehicle. Within vehicle group 1 this is the Fz.105. It is a 1.5 t Steyr vehicle and tows the cable reel trailer (Fz.110) needed for operation of the entire electrical on-board network.

The power supply truck contains 2 sets of machines and a multi-part electrical switchgear, which in connection with the 5/+1 reserve reels on the Steyr itself and the 8 cable reels loaded on the cable trailer, are used to ensure that the launching required tests and firing commands can be carried out by the operating personnel.

The power supply car has the following dimensions and weights:

Total length:	5,540 mm
Width:	2,140 mm
Height:	2,250 mm
Wheelbase:	3,250 mm
Track width:	1,458/1.572 mm front/rear
Tare weight:	2,810 kg
Payload:	1,500 kg, 2,000 kg started in mid 1944
Combat weight:	4,290 kg
Wheels:	190-20

The power supply car consists of the following main components:

- Chassis
- Built-in relay box, two-armature converter, electrical machine set, control panel and accessories

Chassis

The chassis used is 1,5 t Steyr truck of type 1500A/02 (equipment description according to D 665/11). Built as type 2000A onward mid 1944 with stronger springs for 2 t payload.

Fahrzeug 105

Interne oder frühere Bezeichnung: Fz.-Pos.-Nr.105
Stromversorgungswagen auf Lkw 2 t (Steyr)
Steyr

Verwendungszweck

Um die Bordbatterie im Sondergerät durch Prüfungen und Durchschaltversuche nicht vorzeitig zu schwächen, wird das Bordnetz vor dem Schuss von einem Stromversorgungswagen mit elektrischer Energie gespeist. Außerdem laufen in einer Schaltzentrale des gleichen Fahrzeuges alle Schaltvorgänge und Prüfmeldungen zusammen, die einerseits vom Sondergerät andererseits von den einzelnen Bodenstellen kommen.

Für den Betrieb mot.II hat jede Abschuss-Stelle einen Stromversorgungswagen. Dieser hat innerhalb der Fahrzeuggruppe I die Pos.105, ist ein Steyr Fahrzeug 1,5 t und zieht den zum Gesamtbetrieb des elektrischen Bordnetzes notwendigen Kabeltrommelanhänger (Fz.110).

Im Stromversorgungswagen befinden sich 2 Maschinensätze und eine mehrteilige elektrische Schaltanlage, die in Verbindung mit den 5/+1 Reservetrommeln auf dem Steyr selbst und den 8 auf dem Kabelanhänger verladenen Kabeltrommeln dafür sorgen, dass die für den Abschuss erforderlichen Tests und Feuerbefehle durch das Bedienpersonal durchgeführt werden können.

Der Stromversorgungswagen hat folgende Maße und Gewichte:

Gesamtlänge:	5.540 mm
Breite:	2.140 mm
Höhe:	2.225 mm
Radstand:	3.250 mm
Spurweite:	1.458/1.572 mm vorne/hinten
Leergewicht:	2.810 kg
Nutzlast:	1.500 kg, 2.000 kg ab Mitte 1944
Gefechtsgewicht:	4.290 kg
Räder:	190-20

Der Stromversorgungswagen besteht aus folgenden Hauptbauteilen:

- Fahrgestell
- Einbauten-Relaiskasten, Zweianker-Umformer, Elektromaschinensatz, Schalttafel und Zubehör

Fahrgestell

Als Fahrgestell dient der Lastkraftwagen 1,5 t Steyr Baumuster 1500A/02 (Gerätebeschreibung siehe D 665/11). Ab 1944 als Baumuster 2000A mit verstärktem Federpaket für 2 t Zuladung gebaut.

Mechanical structure of the chassis

The Fz.105 was a 1.5 t Steyr truck model based on the Steyr V8-3.5 l version. The Steyr had an air-cooled V8 eight-cylinder engine and has a front-wheel-rear-whee axle drive, which gave it good off-road capability. Its top speed on the motorway was 70 km/h, and its permissible total weight 4.160 kg.

Mechanischer Aufbau des Fahrgestells

Das Fahrzeug Nr.105 ist auf dem Lastkraftwagen 1,5 t Steyr und auf dem Baumuster Steyr V8-3,5 l luftgekühlt aufgebaut. Der Steyr hat einen luftgekühlten V8-Zylinder-Motor und besitzt einen Vorderrad-Hinterrad-Achsenantrieb, der ihm eine gute Geländegängigkeit verleiht. Seine Höchstgeschwindigkeit auf der Autobahn beträgt 70 km/h, das zulässige Gesamtgewicht 4.160 kg.

BArch, RH8II Bild-D13-44, o.Ang.

BArch, RH8II_Bild-B1451-44, o.Ang.

82

DMA, GD639.5.21
OKW/Wa Prüf10
Karlshagen, den 25.3.1944
Bd.Nr. 747/44 gK
Auszug
Besprechung der
Fahrzeuggruppe 1
vom 24./25.3.44
im Lager Rebstock

Production status: Vehicle 105
So far, 6 units have been delivered to Abt.485 in the version which are still susceptible to faults (relay box), namely:
3 units directly to the units and
3 units for prior live-firing testing of the ground devices through Wa Prüf 10.
The following units with already improved relay boxes will be delivered in exchange for the 6 already delivered to units:
3 units by April 1, 1944 and
3 units by April 17, 1944, each of them without the cable reels and batteries which had already been delivered along with the first 6 units.

Current defects and their remedy
Dust proofing of the relay box by using a cover version with angle iron and a rubber seal; fastening of the cover with »swing stopper« clips. The rear cover will be provided with a stronger flat rubber and additional screws. The alternating current generator and the converter will be provided with a dust protection canvas, which is only intended for use during a march. It was ascertained by Prof. Wierer that 3 aggregates ran irregularly during the first 30 minutes of the startup. As part of the functionality testing it has to be examined by a half-hour run whether or not the unit will run regularly under load. For the testing of the guide beam part inside the *Gerät*, the supply voltage for the B-rack must have a frequency of 0 Hz +/- 1 Hz.

At LR it will be checked whether the frequency speed can be kept constant by manual corrections of the frequency changer set fitted in the Steyr power supply vehicle as requested. For this check a vibrating reed instrument must be installed in the power supply vehicle. In case of supply difficulties, the built-in frequency meter can be taken from the control vehicle and used for the test. In the case of insufficient numbers of frequency meters, they must be arranged to be easily exchangeable and issued only to the power supply vehicles of the guide beam battery. The necessary connection box must be provided in all power supply vehicles.

Fertigungslage: Fahrzeug 105
An Abt.485 wurden bisher geliefert 6 Stück in der Ausführung mit noch störanfälligem Relaiskasten und zwar:
3 Stck. unmittelbar an die Truppe und
3 Stck. zur vorherigen Erprobung der Bodenanlagen im scharfen Schuss über Wa Prüf 10.
Es kommen zur Auslieferung mit bereits verbessertem Relaiskasten gegen Austausch der bisher gelieferten 6 Stück:
3 Stck. zum 1.4.44 und
3 Stck. zum 17.4.44, jedesmal ohne Kabeltrommeln und Batterie, welche bereits mit den ersten 6 Fahrzeugen mitgeliefert wurden.

Derzeitige Mängel und ihre Behebung
Staubdichtigkeit des Relaiskastens durch Deckelausführung mit Winkeleisen und Wulstgummis sowie Befestigung des Deckels mit Bierflaschenverschlüssen. Der rückwärtige Deckel wird mit stärkerem Flachgummi und mehr Schrauben versehen. Der Drehstromgenerator sowie der Umformer werden mit einem Segeltuch-Staubschutz versehen, der jedoch nur während des Marsches vorgesehen ist. Durch Prof. Wierer wurde festgestellt, dass bei 3 Aggregaten unregelmäßiger Lauf innerhalb der ersten 30 Minuten des Einlaufens eintrat. Bei der Funktionsprüfung ist durch einen halbstündigen Lauf zu untersuchen, ob das Aggregat nach dieser Zeit unter Belastung regelmäßig läuft oder nicht. Für die Prüfung des LS-Teiles im Gerät muss die Speisespannung für das B-Gestell eine Frequenz von 0 Hz +/- 1 Hz. haben.

In LR wird geprüft, ob bei dem im Steyr-Stromversorgungswagen eingebauten Maschinensatz die Drehzahl in der geforderten Weise durch Regelung von Hand aus konstant gehalten werden kann. Zur Kontrolle muss ein Zungen-Frequenzmesser im Stromversorgungswagen eingebaut werden. Falls bei der Beschaffung Schwierigkeiten auftreten, muss zunächst der im Prüfwagen eingebaute Frequenzmesser dort herausgenommen und für diesen Zweck verwandt werden. Für den Fall, dass die Frequenzmesser nicht in genügender Zahl beschafft werden können, müssen sie leicht auswechselbar angeordnet werden und werden dann jeweils nur den Stromversorgungswagen mit LS-Batterie beigegeben. Die Anschlussenden hingegen sind bei allen Stromversorgungswagen vorzusehen.

DMA, GD639.5.11
T 66 (T 404)
o.O., den 31.7.1944
Bd.Nr. E 2835/44 gK
Auszug
Teileliste Fz.105

Vehicles on order | Fahrzeugbestellungen

	Item Lfd. Nr.	Qty. ordered Best. Stückzahl	Qty. per vehicle Stück je Fahrzeug	Subject drawing number Gegenstand Zeichnungs-Nr.	Ordered by order number Besteller Bestell-Nr.	Supplier Lieferfirma
Fahrzeugbestellung	1	100	1	Lkw 2 t (Steyr) Fahrgestell	WaPrüf 11/III	Steyr
Ausrüstung	2	50	1	Stromversorgungswagen	WaPrüf 10/H0112-3769/43	Gollnow & Sohn, Koblenz
		50	1	mit vollständiger Ausrüstung und Schaltung	WaPrüf 10/H0112-6100/43	A. Strüver, Hamburg

BArch, RH8II_Bild-B1452-44, o.Ang.

83

83 The frame of an cable lying cart is lashed down in brackets on the rear doors, the associated wheels are mounted in bearings on the left and right of the rear. Through the open door you look into the crew's compartment, where a two-armature converter, empty reel, scaffolding with cable reel and relay and switch box are installed.

83 Der Rahmen eines Verlegewagens ist an den hinteren Türen in Halterungen festgezurrt. Die zugehörigen Räder sind links und rechts am Heck aufgesteckt. Durch die offene Tür blickt man in den Mannschaftsraum. Dort ist ein Zweianker-Umformer, eine Leertrommel, ein Gerüst mit Kabeltrommel sowie ein Relais- und Schaltkasten eingebaut.

Partial views from right and left
With the top folded back and the doors open, you can look into the control centre with its installations in the former crew compartment.

Teilansicht von rechts und links
Mit zurückgeschlagenem Verdeck und geöffneten Türen blickt man in die Schaltzentrale mit den Einbauten im ehemaligen Mannschaftsraum.

BArch, RH8II_Bild-B1453-44, o.Ang.

84

BArch, RH8II_Bild-B1454-44, o.Ang.

85

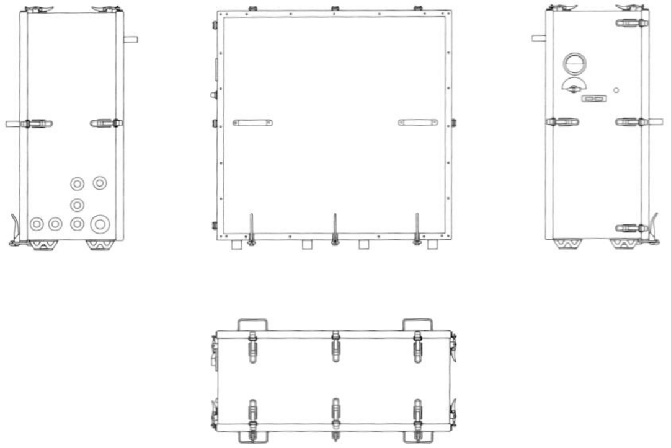

84 The relay box is mounted instead of the rear seat. Next to it is a switch cabinet with a sheet canopy that shows, facing to the interior the switchboard. In front of it there is a set of electric machines and a two-armature converter for 380 V/27-32 V 3 kVA mounted on a floor assembly.

85 The switch cabinet is equipped with the connector panel on the outside. The individual connections are accessible when the car door is open. For differentiation the Flaksteckdosen are marked in red, blue, green and with position marks* for the socket toggles. A plate with cooling fins and bearing for the crank handle closes the spare wheel housing, behind it is the electric machine set which is started by a crank handle.

* The socket of the anti aircraft cablemust be inserted with the toggles aligned with the position marks and then turned until it engages.

84 Anstelle der Rücksitzbank ist ein Relaiskasten montiert. Daneben steht ein Schaltschrank mit Schutzdach. Die darauf befindliche Schalttafel zeigt in den Innenraum. Davor ist ein Elektromaschinensatz und ein Zweianker-Umformer für 380 V/27-32 V 3 kVA auf einer Bodengruppe montiert.

85 Die nach außen gerichtete Rückseite des Schaltschranks ist mit einer Steckertafel ausgerüstet. Die einzelnen Anschlüsse sind bei geöffneter Wagentüre zugänglich. Die Flaksteckdosen tragen zur Unterscheidung eine in rot, blau und grün gehaltene Farbkennzeichnung und sind mit Lagemarken* für die Steckerknebel versehen. Ein Blech mit Kühlrippen und Lager für die Anwurfkurbel verschließt den Reserve-Radkasten. Dahinter sitzt der Elektromaschinensatz, der über eine Anwurfkurbel gestartet wird.

* Der Stecker muss so eingesteckt werden, dass die Knebel an den Positionsmarkierungen ausgerichtet sind und dann gedreht werden, bis er einrastet.

Rear view of the installations in the rear and rear wall of the vehicle
The tarpaulin is folded together or opened at the rear and folded upwards, you have a view of the shelf with the stowage boxes for spare batteries.

Rückansicht auf die Einbauten im Heck und an der Wagen-Rückwand
Wird das Verdeck zusammengefaltet oder hinten geöffnet und nach oben geklappt, blickt man auf das Ablagefach mit den Staukästen für Reserve-Batterien.

BArch, RH8II_Bild-B1456-44, o.Ang.
86

BArch, RH8II_Bild-B1455-44, o.Ang.
87

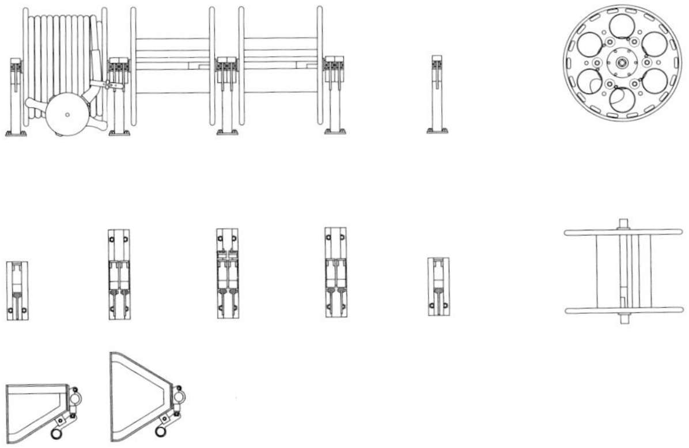

86 The Picture left shows the attachment of the inserted rewinding stand on the rear wall of the vehicle. For transport, the rewinding stand was inserted on the frame above the converter unit.

87 The rear doors are open. 4 cable reels are inserted in a storage rack in the luggage compartment. The Flakstecker 104-polig are equipped with a strain relief made of leather and secured with small leather straps to the cable reel disc for transport. The crank handle for turning the cable reel is mounted in the rewinding stand on the inside wall of the right rear door. On the shelf two boxes for spare batteries are mounted, these include: Ignition battery, Stotz-Stecker battery and command transmitter battery.

86 Das Bild links zeigt die Anbringung des eingesetzten Aufspulbockes an der Wagenrückwand. Zum Transport wurde der Aufspulbock am Gerüst oberhalb der Umformer-Einheit eingesteckt.

87 Die Hecktüren sind geöffnet. Vier Kabeltrommeln sind in einem Lagergerüst im Gepäckraum eingesetzt. Die 104-poligen Flakstecker sind mit einer Zugentlastung aus Leder versehen und mit kleinen Lederriemen zur Transportsicherung an der Kabelrollenscheibe gesichert. An der Innenwand der rechten Hecktüre ist die Handkurbel zum Drehen der Kabelrolle im Aufspulbock gehaltert. Auf der Ablage sind zwei Kästen für Reserve-Batterien montiert. Sie enthalten die Zünd-, Stotz-Stecker- und Kommandogeber-Batterie.

The internals – view from the front top

The driver and passenger seats have been removed to improve visibility of the installations. The inclined position of the steering column with the steering wheel is clearly visible, an otherwise rather inconspicuous detail.

Die Einbauten – Ansicht von vorn oben

Zur besseren Sicht auf die Einbauten wurde der Fahrer- und Beifahrersitz ausgebaut. Gut zu erkennen ist die Schrägstellung der Lenksäule mit dem Lenkrad. Ein ansonsten eher unscheinbares Detail.

BArch, RH8II_Bild-B1458-44, o.Ang

88

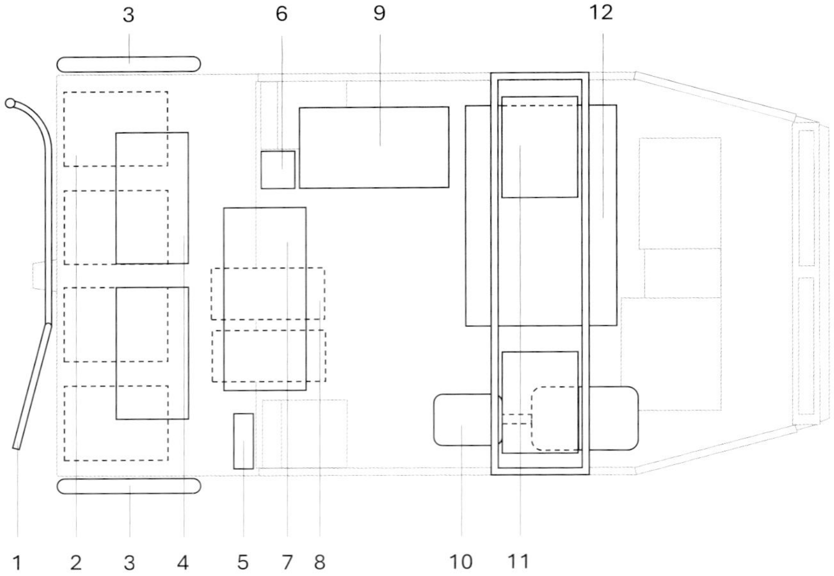

1 Cable cart frame
2 Cable reel 4x
3 Wheel for cable laying cart
4 Spare battery for stotz, command and ignition
5 Field phone
6 Pintch (carbon pressure regulator)
7 Relay cabinet
8 Stotz plug holding, command and ignition battery
9 Switch and connector cabinet
10 Anchor converter 220/380 V and 27 V rectifier
11 Frame with cable reels
12 Electric machine set EVZ 6d

1 Kabelwagenrahmen
2 Kabeltrommel 4x
3 Rad für Kabelauslegewagen
4 Reserve für Stotz-, Kommando- und Zündbatterie
5 Feldfernsprecher
6 Kohledruckregler (Pintch)
7 Relaiskasten
8 Stotzsteckerhalte-, Kommando- und Zündbatterie
9 Schalt- und Steckerkasten
10 Ankerumformer 220/380 V und 27 V Gleichrichter
11 Gestell mit Kabelrollen
12 Elektromaschinensatz EVZ 6d

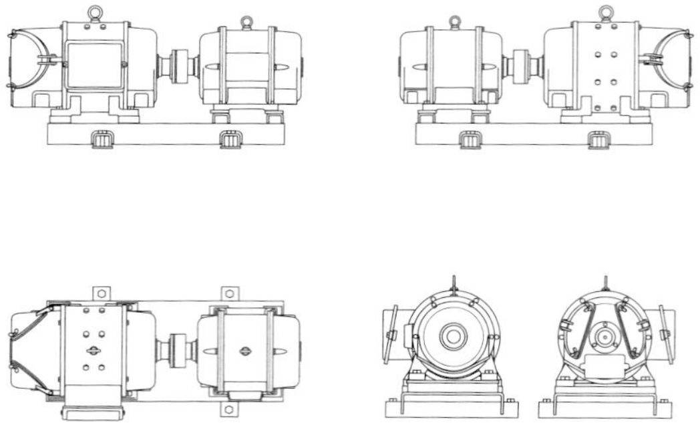

88-92 The electric machine set (EVZ 6 d) with three-phase alternator and petrol drive engine – Zündapp brand (right) – sits on a base plate. On the left of it the two-armature converter for 380 V/27-32 V 3 kVA is also mounted on a base plate. On the frame above it there is space for two cable reels or a roll-up stand with an empty reel. On the left in the direction of travel the switch cabinet with protective roof is mounted. The control panel of the control cabinet is located on the inside, the connector panel on the outside. In the middle is the relay box. On the shelf behind it are the field telephone 33 and the boxes for spare batteries. The pictures on the following page are showing other details of the fixtures in different designs.

88-92 Der Elektromaschinensatz (EVZ 6 d) mit Drehstromgenerator und Benzin-Antriebsmotor der Marke Zündapp (rechts) sitzt auf einer Bodenplatte. Links davon befindet sich der Zweianker-Umformer für 380 V/27-32 V 3 kVA. Er ist ebenfalls auf einer Bodenplatte montiert. Das darüber installierte Gestell ist zur Ausrüstung mit zwei Kabeltrommeln oder einem Aufrollbock mit einer Leertrommel vorgesehen. In Fahrtrichtung links ist der Schaltschrank mit Schutzdach montiert. Die Schalttafel des Schaltschranks befindet sich auf dessen Innenseite, die Steckertafel auf der Außenseite. Mittig ist der Relaiskasten, auf der Ablage dahinter der Feldfernsprecher 33 und die Kästen für Reserve-Batterien angebracht. Die Bilder der Folgeseite zeigen weitere Einbauten in unterschiedlichen Ausführungen.

Deutsches Museum, München, Archiv- FA014/B969/44 T5224 [CD68761]

89

Deutsches Museum, München, Archiv- FA014/B968/44 T5224 [CD79124]

90

BArch, RH8II_Bild-B1457-44, o.Ang.

91

BArch, RH8II_Bild-B1459-44, o.Ang.

92

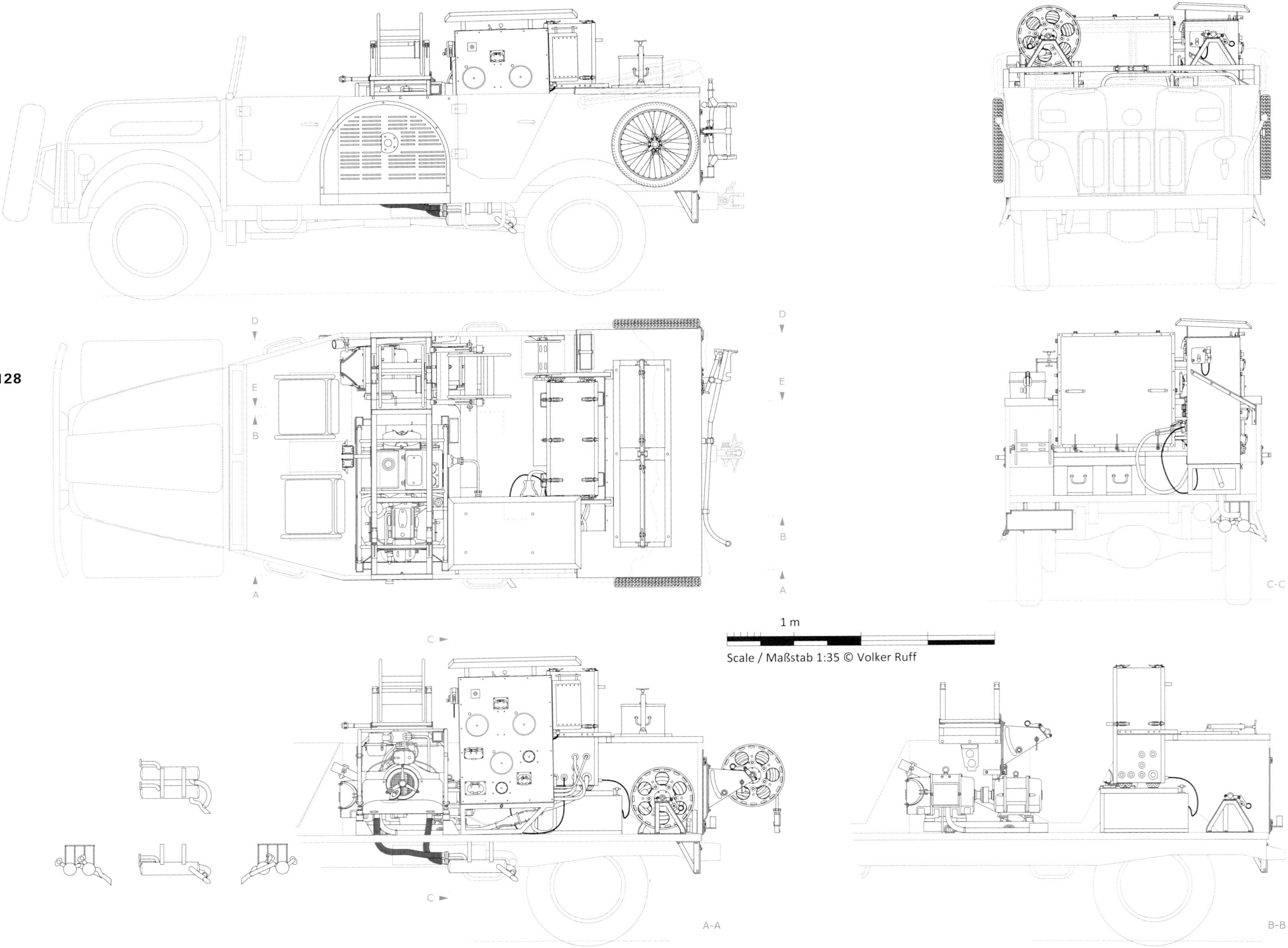

Scale / Maßstab 1:35 © Volker Ruff

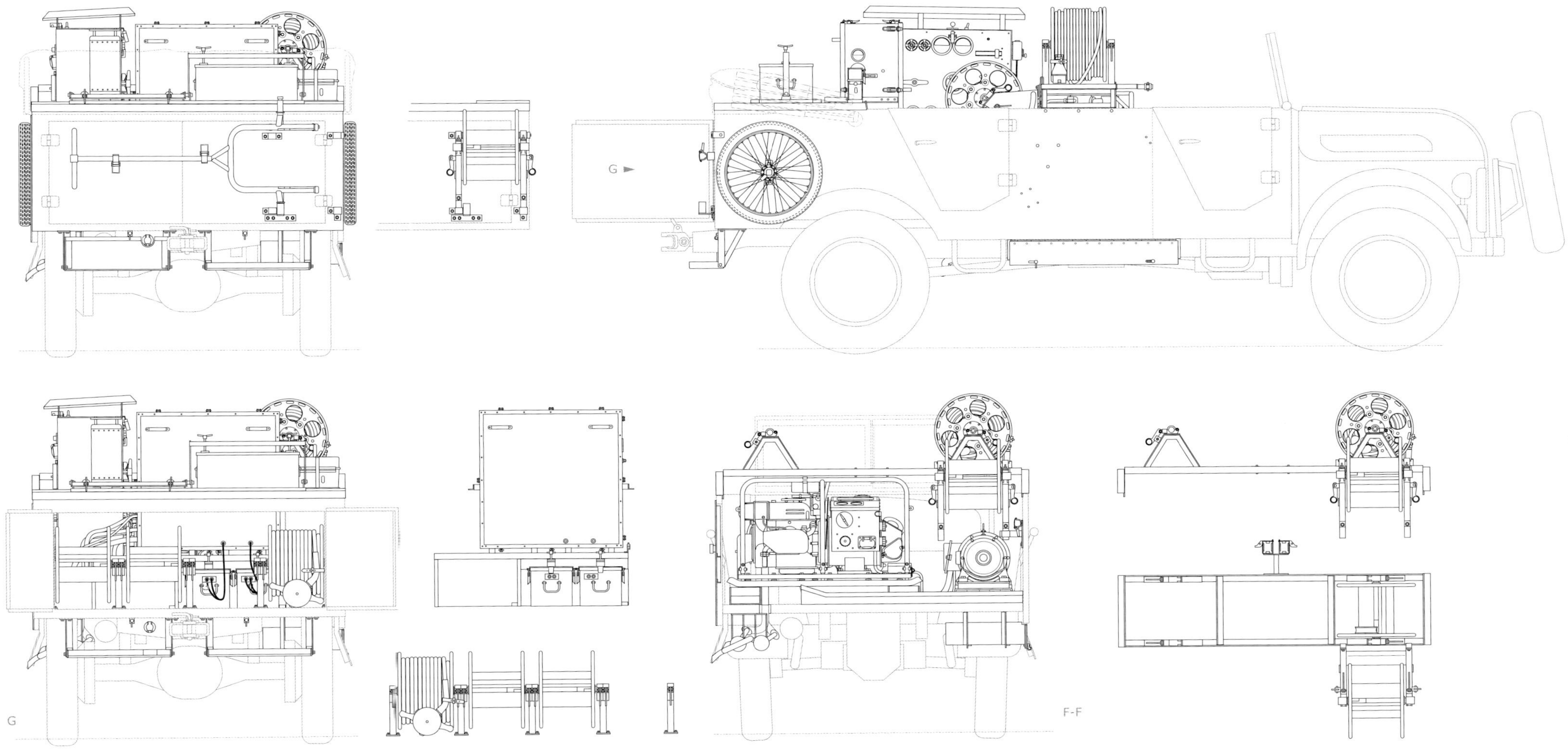

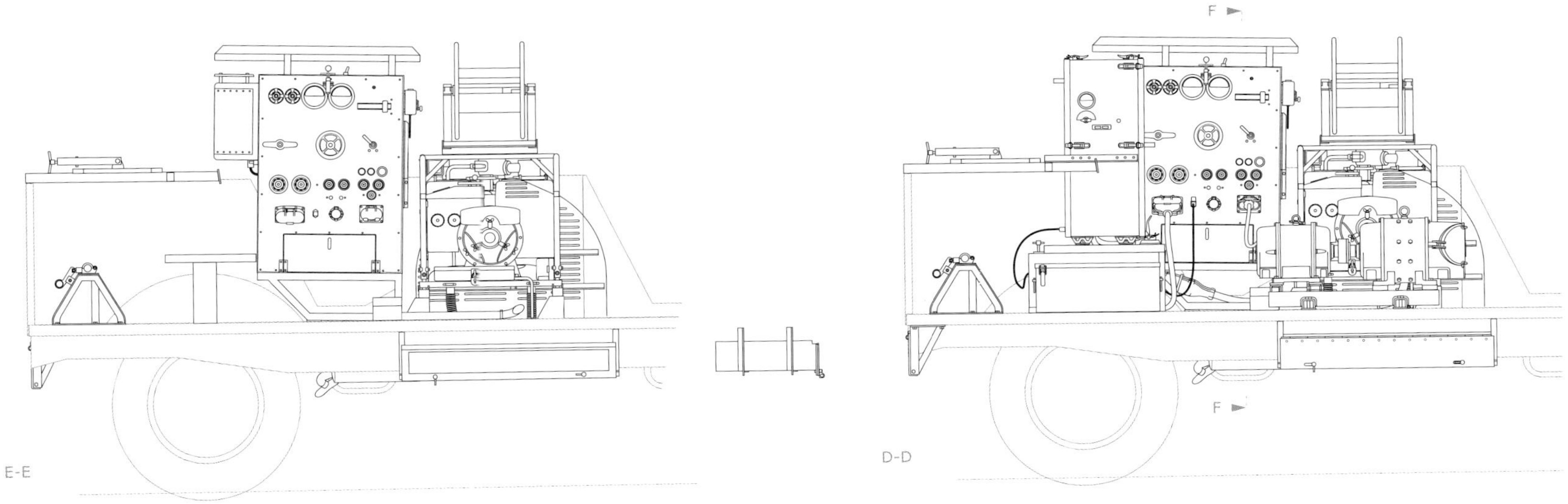

Fahrzeug 105

Stromversorgungswagen
auf Lkw 2 t (Steyr)

Partially used drawings acc. to group list 21Gr226

Teilweise verwendete Zeichnungen nach Gruppenliste 21Gr226

BArch, RH8II_Bild-B0068-45, o.Ang.

93

Fz.105 Steyr in the final version
Fz.105 Steyr in der Abschlussausführung

93 This side view from the left shows all major changes to the chassis and body of the Fz.105 Steyr final version. The wheels for the cable-laying cart are mounted on both sides of the rear side wall. The sheet metal cover with ventilation slots in the spare wheel cutout and the alternative mounting of the spare wheel on the bumper frame in front of the hood are also seen.

93 Folgende Seitenansichten zeigen alle wesentlichen Änderungen am Fahrgestell und Aufbau des Fz.105 Steyr der Abschlussausführung. Die Laufräder für den Kabelverlegewagen sind beidseitig an der Heckseitenwand montiert. Die Blechabdeckung mit Lüftungsschlitzen im Ersatzradausschnitt und die alternative Montage des Ersatzrades am Stossfängerrahmen vor der Motorhaube sind gut sichtbar.

BArch, RH8II_Bild-B0069-45, o.Ang.

94

BArch, RH8II_Bild-B0062-45, o.Ang.

95

BArch, RH8II_Bild-B0073-45, o.Ang.

96

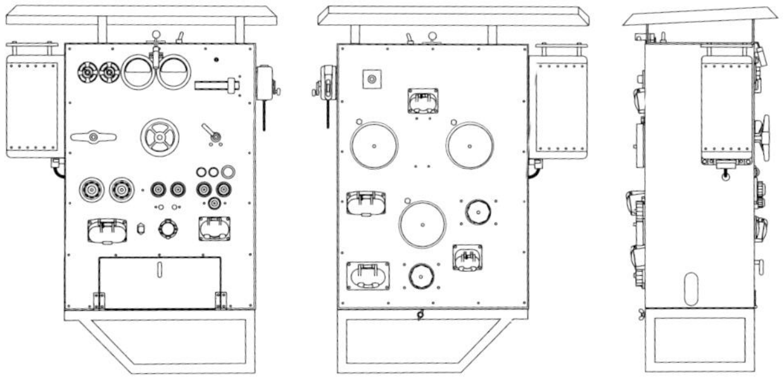

94-96 A 27 V Pintsch regulator is mounted on the rear wall of the switch cabinet and a small box with a hand lamp is mounted on its front. The plug panel and switch panel seen here show the instruments and socket arrangement of the final version.

In the inside are the relay box and a box with switch and plug panel. The vehicle has new paintwork in dark yellow »nach Muster«, but the lettering is cut out so appears on a darker background.

94-96 An der Rückwand des Schaltschrankes ist ein Pintsch Regler 27 V und an der Stirnseite ein Kästchen mit Handlampe montiert. Die Stecker- und Schalttafel zeigen die Instrumente- und Steckeranordnung in der Abschlussausführung.

Im Innenraum befindet sich der Relaiskasten und Schaltschrank mit Schalt-und Steckertafel. Das Fahrzeug hat einen neuen Anstrich in Dunkelgelb nach Muster erhalten. Die Beschriftung ist ausgespart und erscheint auf dunklerem Grund.

DMA, GD639.5.18
HAP11/T 507
Karlshagen, 1.6.1944
Bd.Nr. E 3195/44 g

DMA, GD639.5.1
EW234
Fz.-Pos.-Nr. 106/107

DMA, GD630.02
A4 Lehrbuch
der FR-Schule
Mai 1944
Auszug aus
Technische Fahrzeuge

Vehicle 106/107

Internal or former designation: Fz.-Pos.-Nr.106/107
Test vehicle I/III on 3t truck closed (305)
Test vehicle I
Test vehicle III

Intended use

The truck is a Kfz 305 Opel Blitz with closed superstructure (LC box body). The body interior is divided into two halves by a partition wall. In the front half is the B-rack used for testing the combustion cut-off and guide beam on-board equipment.

The rear part of the test vehicle is used to accommodate equipment for the engine crew, which is stored in shelves, drawers or brackets on the walls

The test vehicle I/III has the following dimensions and weights:

Total length:	5,900 mm
Width:	2,100 mm
Height:	2,800 mm
Wheelbase:	3,450 mm
Track width:	1,630/1,642 mm front/rear
Tare weight:	3,275 kg
Maximum weight	6,100 kg
Combat weight:	n.a.
Wheels:	190-20

The main parts of test vehicles I/III are:

- Chassis
- Closed superstructure (305) LC box body
- Fixtures
- B-rack for combustion cut-off and guide beam test equipments

The chassis

The chassis is the chassis of the 3 t Opel 6700 Type A truck (equipment description according to D 666/5) but the normal chassis of the Opel 3 t truck Type 3.6/36 S (equipment description according to D 669/19B) was also used.

Fahrzeug 106/107

Interne oder frühere Bezeichnung: Fz.-Pos.-Nr.106/107
Prüfwagen I/III auf Lkw 3 t geschl. (305)
Prüfwagen I
Prüfwagen III

Verwendungszweck

Der Wagen ist ein Kfz 305 Opel-Blitz mit geschlossenem Aufbau. Das Wageninnere ist durch eine Zwischenwand in zwei Hälften aufgeteilt. In der vorderen Wagenhälfte ist das B-Gestell unterbracht, das zur Prüfung der BS- und LS-Bordgeräte dient.

Der hintere Teil des Prüfwagens dient zur Aufnahme der Geräte für den Triebwerkstrupp, die in Regalen, Schubfächern oder Halterungen an den Wänden untergebracht sind

Der Prüfwagen I/III hat folgende Maße und Gewichte:

Gesamtlänge:	5.900 mm
Breite:	2.100 mm
Höhe:	2.800 mm
Radstand:	3.450 mm
Spurweite:	1.630/1.642 mm vorne/hinten
Leergewicht:	3.275 kg
Zul.Höchstgew.	6.100 kg
Gefechtsgewicht:	k.A.
Räder:	190-20

Die Hauptteile des Prüfwagen I/III sind:

- Fahrgestell
- Geschlossener Aufbau (305) LC-Koffer
- Einbauten
- B-Gestell für BS- und BS/LS-Prüfgeräte

Das Fahrgestell

Als Fahrgestell dient das Fahrgestell des Lastkraftwagen 3 t Opel 6700 Typ A (Gerätebeschreibung nach D 666/5) aber auch das normale Fahrgestell des Opel 3 t Lastkraftwagens Typ 3,6/36 S (Gerätebeschreibung nach D 669/19B) wurde verbaut.

DMA, GD639.5.11
ABW
o.O., den 27.10.1944
Bd.Nr. ABW 13/44 gK
T 404
o.O., den 15.6.1944
Bd.Nr. E 2153/44 gK
Auszug
Teileliste Fz.106 und 107

Vehicles on order

Fahrzeugbestellungen

	Item Lfd. Nr.	Qty. ordered Best. Stückzahl	Qty. per vehicle Stück je Fahrzeug	Subject drawing number Gegenstand Zeichnungs-Nr.	Ordered by order number Besteller Bestell-Nr.	Supplier Lieferfirma
Fahrzeugbestellung	1	115	1	Lkw 3 t Chassis	WaPrüf 11/III	Opel, Brandenburg
		115	1	Aufbau geschlossen (305)	WaPrüf 11/III	Gaubschat, Berlin-Neukölln
Ausrüstung	1	115	1	Prüfwagen I und III, kompl. Ausrüstung	WaPrüf 11/III	Gollnow & Sohn, Koblenz

97 For test vehicles I and III, the LC box body (305) was placed on a chassis of the Opel 3 t truck type 3.6/36 S, but chassis of the 3 t Opel 6700 type A were also used. Only minor changes were made to the box body. In the front flap an opening for a socket was cut, and behind it a marine socket with holder was mounted on the door sill. Behind the opened flap one looks at the B-rack side wall. One half of a two-part ladder is lashed down on the left side of the roof.

97 Für den Prüfwagen I und III wurde der LC-Koffer (305) auf ein Fahrgestell des Opel 3 t Lastkraftwagens Typ 3,6/36 S gesetzt, aber auch Fahrgestelle des 3 t Opel 6700 Typ A wurden verwenden. Der Kofferaufbau wies nur geringfügige Änderungen auf. In der vorderen Klappe wurde ein Durchbruch für eine Steckdose geschnitten und dahinter eine Marinesteckdose mit Halter am Türschweller montiert. Hinter der geöffneten Klappe sieht man auf die B-Gestell-Seitenwand. Eine Hälfte einer zweiteiligen Leiter ist auf der linken Dachseite verzurrt.

Production status: Vehicle 106, Vehicle 107

A distinction must be made between test vehicle I [1] (106 without guide beam (LS)) and test vehicle III (107 with guide beam (LS)). Furthermore, the test vehicles are different, whether they work with Strüver or Steyr power supply. The former are equipped with converters* and switched accordingly in the B-rack with Strüver power supplies. Such vehicles are used exclusively, initially at 836. For all other departments, the test vehicle without converter is used. Test vehicles 2.-10*. and 14.-17*. are equipped with converters.

Delivery status: So far first (1) test vehicle has been delivered in October 1943, converted at HL. The B-rack of this is currently in use in HL.

Second (2) test vehicles (107A) on 2.2.44 to 444 – destroyed by accident. The B-rack of this is conditionally usable in the first test vehicle. The test vehicle returns to HAP 11. Test vehicle has *Heidegerät***.

3.-7. Test vehicles with *Heidegerät* (1 of which 107A, 4 of which 106A). On 2.2. to 836, of which one temporarily to 444.

8.-9. Test vehicle (107A) On 29.3. to 836 without *Heidegerät*. This is delivered by HAP 11. Responsible: Dipl. Ing. Decker.

10. test vehicle (107A) currently in HAP 11 (damaged on PR.X) arrives on 4.4 at 836 without *Heidegerät*. Ing. Schuran immediately checks engine equipment thereof.

11.-13. Test vehicle (106) Currently in HAP 11 to complete on engine equipment, electrical equipment. Electrial equipment completed except *Heidegerät*.

14.-16. test vehicle (106A) as replacement vehicle for 836. At present Rebstock is to be delivered to the unit up to 15.4. without *Heidegerät*.The test transmitters for this are being sent by courier according to HAP 11. The vehicles are to be completed by the Gollnow company except for the B-rack and converter and will leave on 3.4. (2 pcs) and on 4.4. from Rebstock with escort and full run number according to HAP 11. Reason: Fa. Gollnow, or for the provision of accompanying personnel: Major Kleidorfer. The vehicles will be made ready for collection by the units by 15.4. The pick-up squad of the unit is requested in good time by Wa Prüf 10, Gruppe IV.

Trucks for the test vehicles 17.-21. are already in Lager Rebstock. For test vehicles 22.-26. they are to be immediately delivered by Wa test 11/III.

17. test vehicle (107A) is intended as replacement vehicle for test vehicle III no. 2.

Test vehicles 18.-20. (107) and 21.-23. (106) are intended for Battery 485.

Test vehicles 24. and 25. (106) and 26. (107) are intended as replacement vehicles for 485.

* Test vehicle I und Test Vehicle II with converters / Strüver circuit were named as Fz.106A and Fz.107 A

** Heidegerät = Modulator and encoder for operating and cloaking frequency.

Fertigungslage: Fahrzeug 106, Fahrzeug 107

Es ist zu unterscheiden zwischen Prüfwagen I [1] (106 ohne Leitstrahl (LS)) und Prüfwagen III (107 mit Leitstrahl (LS)). Ferner sind die Prüfwagen unterschiedlich, ob sie mit Strüver oder Steyr-Stromversorgung zusammenarbeiten. Erstere sind mit Umformer* ausgestattet und im B-Gestell entsprechend geschaltet bei Strüver-Stromversorgung. Derartige Fahrzeuge kommen ausschließlich, zunächst bei 836, zur Verwendung. Für alle weiteren Abteilungen wird der Prüfwagen ohne Umformer verwendet. Prüfwagen 2-10* und 14-17* sind mit Umformer ausgerüstet.

Ausgeliefert sind bisher: Erster (1.) Versuchswagen im Oktober 1943, in HL umgebaut. Das B-Gestell hiervon ist stationär zurzeit in HL in Verwendung. Zweiter (2.) Prüfwägen (107A) am 2.2.44 zu 444 – durch Unfall zerstört. Das B-Gestell hiervon ist bedingt brauchbar und im ersten Versuchswagen verwendet. Der Prüfwagen kommt zu HAP 11 zurück. Prüfwagen hat Heidegerät**.

3.-7. Prüfwagen mit Heidegerät (1 davon 107A, 4 davon 106A). Am 2.2. zu 836, davon einer vorübergehend zu 444.

8.-9. Prüfwagen (107A) Am 29.3. zu 836 ohne Heidegerät. Dieses wird durch HAP 11 nachgeliefert. Verantwortlich: Dipl. Ing. Decker.

10. Prüfwagen (107A) z.Zt. im HAP 11 (beschädigt auf PR.X) Kommt am 4.4 zu 836 ohne Heidegerät. Ing.Schuran überprüft sofort triebwerkseitige Ausrüstung hiervon.

11.-13. Prüfwagen (106) zur Zeit im HAP 11 zur triebwerkseitigen und elektrischen Ausrüstung, elektrische Ausrüstung vollständig bis auf Heidegerät.

14.-16. Prüfwagen (106A) als Ersatzfahrzeug für 836. Zur Zeit sind Rebstock bis 15.4. ohne Heidegerät an die Truppe auszuliefern. Die Prüfsender hierfür sind mit Kurier nach HAP 11 unterwegs. Die Fahrzeuge werden bis auf das B-Gestell und Umformer von der Firma Gollnow fertiggestellt und gehen am 3.4. (2 Stück) und am 4.4. von Rebstock mit Begleitmann und Vollfahrtnummer nach HAP 11 ab. Veranlassung: Fa. Gollnow bzw. für Gestellung des Begleitpersonals: Major Kleidorfer. Die Fahrzeuge werden elektrisch und triebwerksseitig bis 15.4. für die Truppe abholbereit gemacht. Abholkommando der Truppe wird rechtszeitig angefordert durch Wa Prüf 10, Gruppe IV.

Für 17.-21. Prüfwagen befinden sind die Fahrzeuge bereits in Rebstock. Für 22.-26. Prüfwagen werden sie umgehend durch Wa Prüf 11/III dorthin zugestellt.

17. Prüfwagen (107A) ist als Ersatzfahrzeug für Prüfwagen III Nr. 2 vorgesehen.

18.-20. Prüfwagen (107) und 21.-23. (106) sind für Abteilung 485 vorgesehen.

24. und 25. (106) sowie 26. Prüfwagen (107) sind als Ersatzfahrzeuge für 485 vorgesehen.

* Prüfwagen I und Prüfwagen III mit Umformer / Strüverschaltung werden als Fz.106A und Fz.107A bezeichnet.

** Heidegerät = Modulator und Geber für Arbeits- und Tarnfrequenz

Current defects and their remedy

Labelling is partially incomplete on the B-rack assemblies. Oberleutnant Eisner, Kommandostelle S hands over all necessary documents to HAP 11, TD5, Dipl.Ing. Decker, after a drawing and necessary instructions for the 10 copies are to be delivered to B.z.b.V. Heer by 10.4.. Uniform execution by the units must be ensured by B.z.b.V. Heer, so that the operating instructions drawn up by Kommandostelle S can be used generally.

The damage to the B-rack that happened at the first Battery 836 (wire breaks and instrument damage) is to be regarded as one-off. Such defects have not been detected by other bodies despite the driving stress on the test vehicle. The B-rack is checked for shaking resistance as part of the generally shake tests.

In contrast, the previously used »Baugruppe 37« was able to monitor the doppler until the end, but only with a certain position of the test carriage in relation to the *Gerät* that had to be determined by the units. Both solutions are therefore unsatisfactory for the units. Documents for Baugruppe 37 are to immediately be given to Gollnow for the one-off accelerated production of 20 sub-assemblies 37. WUG 10 is providing the necessary tubes. A better solution for the doppler test must be prepared, tested and reported to B.f.E. by HAP 11, TD5 by 1.6.44. Some frequency meters are still missing in the B-rack, which are necessary for the adjustment of the converter set in the test vehicle III (107). Battery 836 already has two test vehicles with frequency meters and Leutnant Westermann has delivered a frequency meter for a 3rd test vehicle. The necessary distribution of the vehicles within the Battery is a matter of course. Another frequency meter is to be provided as soon as possible for test vehicle 17 – responsible: HAP 11, TD 5, Dipl. Ing. Decker. TD 5 will soon examine the need for of the general use of the frequency meter for test vehicles with (LS) guide beam. Notification of this to B.f.E. to 1.6.

The glass plate of a voltmeter in the B-rack is fogging up. HAP 11 will check whether it is possible to replace this voltmeter with another type or what changes are necessary to the current voltmeters (change case B). In order to control the frequency of the alternating current supplied by the converter it is necessary to install an additional controller on test vehicle 107. This modification is to be carried out by HAP 11 on the vehicles already delivered. The date depends on the delivery of the necessary sliding resistance by the company. (Change case A) WuG 10 supports the immediate delivery of the resistors using order form Z.

In order to connect the test vehicle in command position 2 (100 m offside position), some reworking (cable connection etc.) has become necessary and has already been initiated by HAP 11. The work has been carried out up to 7.4. (Change case A).

Derzeitige Mängel und ihre Behebung

Auf den Baugruppen des B-Gestelles ist die Beschriftung teilweise unvollständig. Oberleutnant Eisner, Kommandostelle S übergibt während der Tagung alle erforderlichen Unterlagen an HAP 11, TD5, Dipl.Ing. Decker, nach denen eine Zeichnung und erforderliche Anleitung für die Truppe in 10-facher Ausfertigung bis 10.4. an B.z.b.V. Heer geliefert werden. Einheitliche Ausführung durch die Truppe muss durch B.z.b.V. Heer sichergestellt werden, damit die von Kommandostelle S aufgestellte Bedienungsanweisung allgemein verwendbar wird.

Die bei der ersten Batterie 836 aufgetretenen Beschädigungen des B-Gestelles (Drahtbrüche und Instrumentenbeschädigungen) sind als einmalig anzusehen. Von anderen Stellen sind derartige Mängel trotz Fahrbeanspruchung des Prüfwagens nicht festgestellt worden. Das B-Gestell wird im Rahmen der allgemein vorgesehenen Schüttelprüfungen auf Schüttelfestigkeit überprüft.

Dem gegenüber war die vorher verwendetet »Baugruppe 37« bis zum Schluss in der Lage, den Verdoppler zu überwachen, aber nur bei einer bestimmten von der Truppe erst zu ermittelnden Stellung des Prüfwagens zum Gerät anzeigesicher. Beide Lösungen sind also für die Truppe unbefriedigend. Es werden sofort Unterlagen für die Baugruppe 37 an die Firma Gollnow gegeben zur einmalig beschleunigten Fertigung von 20 Baugruppen 37. WUG 10 sorgt für die Bereitstellung der erforderlichen Röhren. Eine bessere Lösung der Verdopplerprüfung ist durch HAP 11, TD5 vorzubereiten, zu erproben und bis 1.6.44 darüber B.f.E. zu melden.

Im B-Gestell fehlen zum Teil noch Frequenzmesser, die beim Prüfwagen III (107) zur Einregelung des Umformersatzes notwendig sind. Die Abteilung 836 hat bereits zwei Prüfwagen mit Frequenzmesser und bekommt durch Leutnant Westermann ein Frequenzmesser für einen 3. Prüfwagen zugestellt. Die entsprechende Aufteilung der Fahrzeuge innerhalb der Abteilung ist selbstverständlich. Ein weiterer Frequenzmesser ist baldigst für den 17. Prüfwagen sicherzustellen – verantwortlich: HAP 11, TD 5, Dipl. Ing. Decker. TD 5 überprüft baldigst die Notwendigkeit der allgemeinen Verwendung des Frequenzmessers bei Prüfwagen mit Leitstrahl. Meldung hierüber an B.f.E. zum 1.6.

Bei einem Spannungsmesser im B-Gestell beschlägt die Glasplatte. HAP 11 überprüft, ob Austausch dieses Spannungsmessers gegen andere Bauart möglich ist. bzw. welche Änderung an den derzeitigen Spannungsmessern notwendig ist (Änderungsfall B).

Zur Regelung der Frequenz des vom Umformer gelieferten Wechselstromes ist es beim Prüfwagen 107 notwendig, einen zusätzlichen Regler einzubauen. An den bereits ausgelieferten Wagen wird diese Änderung durch HAP 11 ausgeführt. Der Termin hängt von der Lieferung des dazu notwendigen Schiebewiderstades durch die Firma ab. (Änderungsfall A) WuG 10 unterstützt die umgehende Lieferung der Widerstände an Hand des Bestellscheines Z.

Zum Anschluss des Prüfwagens in der befohlenen Stellung 2 (100 m Abseitsstellung) ist die Ausführung einiger Nacharbeiten (Kabelanschlüsse usw.) notwendig geworden und durch HAP 11 bereits veranlasst. Die Arbeiten sind bis zum 7.4. durchgeführt. (Änderungsfall A)

98

Devices in B-rack for combustion cut-off (BS)
Test transmitter for doppler and radio command-receiver testing.
Command transmitter with mixer stage for modulating the radio command test transmitter.
Circuit and signal field for monitoring the operating voltages and commands.
The doppler signal is tested by radio by illuminating the door antennas with the test transmitter and measuring the illuminated energy of the transmitter directly at the door antennas with an aperiodic measuring instrument.
The command receiver receives its RF energy modulated with the working frequencies via an RF cable and returns the commands executed during the test as current pulses via cable on display signals.

Devices in the B-rack for guide beam (LS)
model transmitter for testing the LS receiver with RF cable
RF keying, LF keying, modulation stage and measurement field
The command voltage is fed to the on-board receiver via HF cable. The resulting control current is displayed in the *Feuerleitpanzer* on the control console.

Geräte im B-Gestell für BS (Brennschluss)
Prüfsender für Verdoppler- und Funkkommando-Empfänger-Prüfung.
Kommandogeber mit Mischstufe zur Modulation des Funkkommando-Prüfsenders.
Schaltung- und Signalfeld zur Überwachung der Betriebsspannungen und Kommandos.
Der Verdoppler wird auf dem Funkwege geprüft, indem der Prüfsender die Türantennen anstrahlt und mit einem aperiodischen Messinstrument direkt an den Türantennen die angestrahlte Energie des Verdopplersenders gemessen wird.
Der Kommando-Empfänger erhält seine mit den Arbeits-Frequenzen modulierte HF-Energie über ein HF-Kabel und gibt die ausgeführten Kommandos bei der Prüfung als Stromimpulse über Kabel auf Schauzeichen zurück.

Geräte im B-Gestell für LS (Leitstrahl)
Modellsender zur Prüfung des LS-Empfängers mit HF-Kabel
HF-Tastung, NF-Tastung, Modulationsstufe und Messfeld
Die Kommandospannung wird dem Bordempfänger über HF-Kabel zugeführt. Der dabei entstehende Steuerstrom wird im Feuerleitpanzer am Steuerungspult angezeigt.

98 The converter on the left side of the picture is only available in Fz.106A or Fz.107A (with Strüver circuit). The converter is not required for test vehicles I and III, which work together with Steyr power supply vehicle 105. On the right side the B-rack with the test transmitters is installed only for BS or BS and LS test transmitters.

98 Der links im Bild vorhanden Umformer ist nur im Fz.106A bzw. Fz.107A (mit Strüverschaltung) vorhanden. Der Umformer entfällt bei den Prüfwagen I und III, die mit dem Steyr-Stromversorgungswagen Fahrzeug 105 zusammenarbeiten. Rechts ist das B-Gestell nur für BS- oder BS- und LS-Prüfsendern eingebaut.

Deutsches Museum, München, Archiv- FA014/B912/44T5224 [CD79120]

99

99 When the access door on the LC box body is open, the B rack with the empty equipment racks for test transmitter B red and test transmitter B green for the alternating frequencies is visible. In front of the door, a power socket for the internally mounted power distribution panel is mounted on the outside wall. An additional equipment box is mounted on the driver's roof. The second half of the split ladder is lashed to the right on the roof.

99 Bei geöffneter Zugangstür am LC-Koffer blickt man auf das B-Gestell mit den leeren Geräteeinschüben für die Prüfsender B rot und Prüfsender B grün für die Wechselfrequenzen. Vor der Tür ist an der Außenwand eine Kraftsteckdose für die innen angebrachte Stromverteilungstafel montiert. Auf dem Fahrerdach ist zusätzlich ein Gerätekasten aufgesetzt. Die zweite Hälfte der geteilten Leiter ist rechts auf dem Dach verzurrt.

Deutsches Museum, München, Archiv- FA014/B987/44 T5224 [CD83222]

100

Deutsches Museum, München, Archiv- FA014/B988/44 T5224 [CD83223]

101

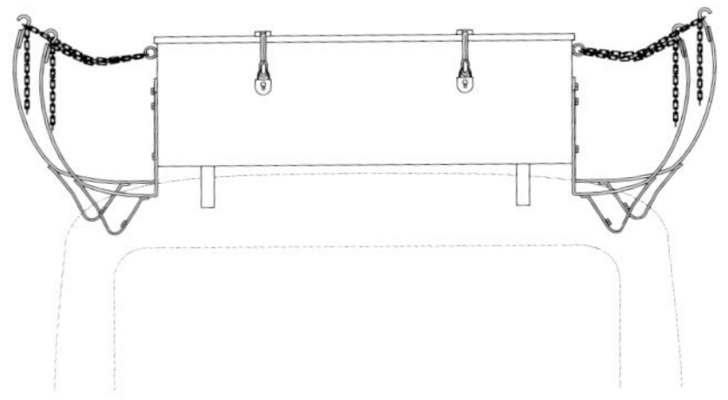

1 m

Scale / Maßstab 1:35 © Volker Ruff

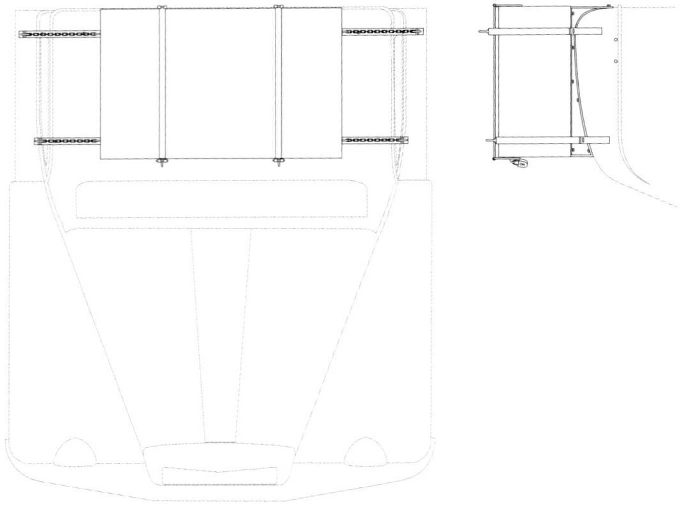

100, 101 Another test vehicle I/III Fz.106/107. Only a few modifications were made to the LC box body and were mainly limited to cable connections and counter brackets for installations in the interior, which also differed slightly from vehicle to vehicle. To accommodate camouflage mats, a pair of brackets was attached to the left and right of the large stowage box and supported to the roof of the vehicle cabin.

100, 101 Ein weiterer Prüfwagen I/III Fz.106/107. Nur wenige Änderungen wurden am LC-Koffer durchgeführt. Sie beschränkten sich überwiegend auf Kabelanschlüsse und Gegenhalter für Einbauten im Innenraum. Diese wichen auch von Fahrzeug zu Fahrzeug geringfügig voneinander ab. Zur Aufnahme von Tarnmatten wurde je ein Bügelpaar links und rechts der großen Staukiste befestigt und zum Dach der Fahrzeugkabine abgestützt.

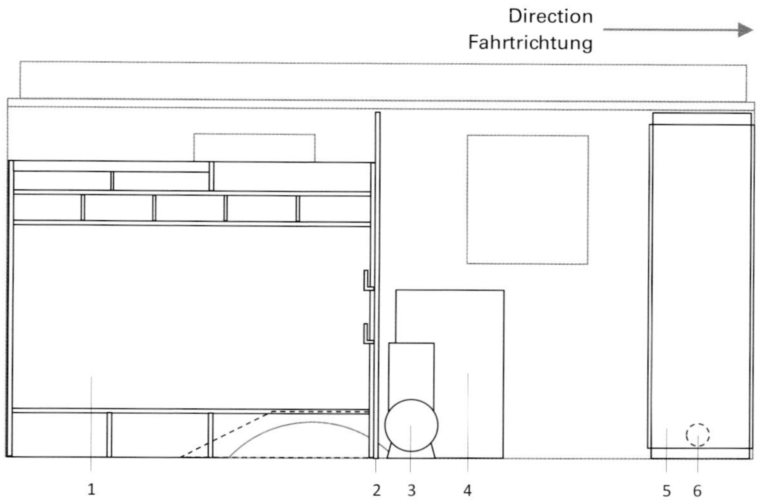

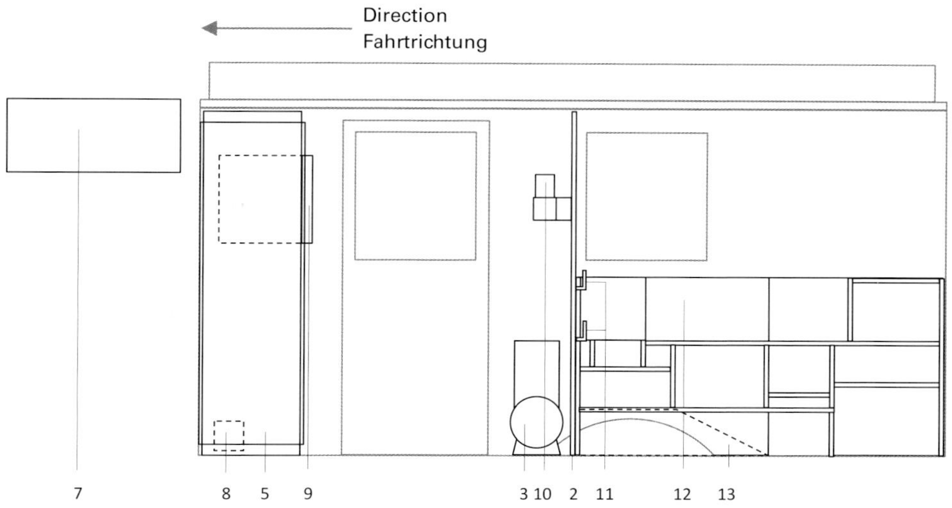

View right side wall
Ansicht rechte Seitenwand

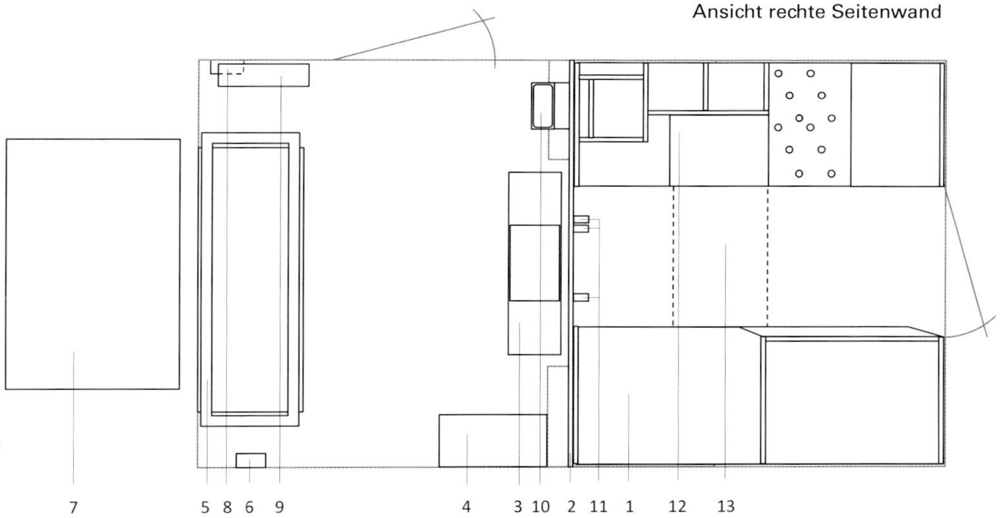

Top view
Ansicht von oben

1 Rack LH to store HB50 hot air blower, valve box and drawers for accessories
2 Partition wall
3 Converter (only for design Fz.106/107A)
4 Cabinet
5 B-rack
6 Naval Plug
7 Storage
8 Power plug
9 Switch panel
10 Field phone
11 Bracket fiveway- and refueling device
12 Rack RH to store Z-Stoff pre heater, Bracket rudder and fueling hose
13 Ramp optional

5 B-rack
- 1 Power supply unit
- 2 Mains switch and signal field
- 3 Keying stage*
- 4 Empty drawer*
- 5 Mains stage*
- 6 Control stage*
- 7 NF stage*
- 8 Empty drawer*
- 9 Test transmitter B yellow (operating frequency*
- 10 Preheating and remote keying stage*
- 11-18 Test transmitter B yellow (alternating frequency)*
- 19 Mixing stage
- 20 Test transmitter B red (operating frequency)
- 21-28 Test transmitter B red (alternating frequency)
- 29 Command transmitter
- 30 Test transmitter B green (operating frequency)
- 31 Doubler test receiver
- 32-39 Test transmitter B green (alternating frequency)

* only for Fz.107

1 Gestell L Ablage für HB50 Heißluftbläser, Ventilkasten und Schubladen für Zubehör
2 Trennwand
3 Umformer (nur bei Ausführung Fz.106/107A)
4 Schubladenschrank
5 B-Gestell
6 Marinesteckdose
7 Staukasten
8 Kraftsteckdose
9 Sicherungstafel
10 FF Feldtelefon
11 Halter Fünffachkupplung und Nachtankvorrichtung
12 Gestell R Ablage für Z-Stoff Anwärmgerät, Halterung für Strahlruder, Betankungsschlauch
13 Rampe optional

5 B-Gestell
- 1 Netzgerät
- 2 Netzschalt- und Signalfeld
- 3 Taststufe*
- 4 Leerfach*
- 5 Netzstufe*
- 6 Kontrollstufe*
- 7 NF-Stufe*
- 8 Leerfach*
- 9 Prüfsender B gelb (Betriebsfrequenz)*
- 10 Vorheiz- und Ferntaststufe*
- 11-18 Prüfsender B gelb (Wechselfrequenz)*
- 19 Mischstufe
- 20 Prüfsender B rot (Betriebsfrequenz)
- 21-28 Prüfsender B rot (Wechselfrequenz)
- 29 Kommandogeber
- 30 Prüfsender B grün (Betriebsfrequenz)
- 31 Verdopplerprüfempfänger
- 32-39 Prüfsender B grün (Wechselfrequenz)

* nur für Fz.107

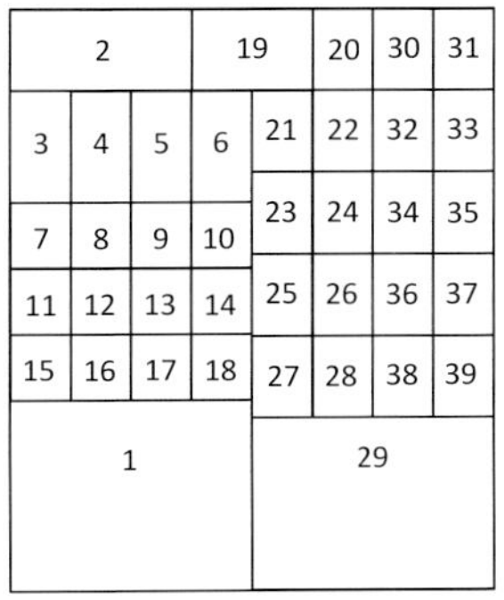

Fahrzeug 106/107
Prüfwagen I/III
auf Lkw 3 t geschl. (305)

Loading plan
Beladeplan

Equipment for the engine bay

The rear part of the test vehicle is used to accommodate the equipment for the engine crew, which is stored in shelves, drawers or brackets on the walls.

Geräte für den Triebwerksraum

Der hintere Teil des Prüfwagens dient zur Aufnahme der Geräte für den Triebwerkstrupp, die in Regalen, Schubfächern oder Halterungen an den Wänden untergebracht sind.

102

103

102, 103 Left side wall – Brackets and equipment
- Valve box
- Hot-air blower HB50 with accessories.
- Carriage for hot air blower
- Device for five way coupling
- Device for oxygen topping up connection
- Lift cage

102, 103 Seitenwand links – Halterungen und Geräte
- Ventilkasten
- Heissluft-Gebläse HB50 mit Zubehör
- Schlitten zum Heizluftbläser
- Vorrichtung zur Fünffachkupplung
- Vorrichtung zur Nachtankkupplung
- Aufzugskorb

Partition wall and right side wall for holders and equipment

- Z-Stoff heating device
- Fire extinguisher
- Brackets for thrusters
- A-Stoff refueling hose

Trennwand und Seitenwand rechts für Halterungen und Ausrüstung

- Z-Stoff-Anwärmgerät
- Löschgerät
- Halterungen für Strahlruder
- A-Stoff-Betankungsschlauch

Deutsches Museum, München, Archiv- FA014/B447/44 TD5224 [CD74802]

104

Deutsches Museum, München, Archiv- FA014/B444/44 TD5224 [CD79129]

105

104, 105

- Test board with the fine measuring manometer for adjusting the pressure reducer
- Lift-off contacts and associated device
- Switch and test box for the general circuit test
- Exhaust pipe with holding device for the A-tank vent
- A- and B-Stoff hoses with coupling
- T-Stoff refuelling hose with refuelling tube
- B-Stoff and T-Stoff defuelling hose with coupling
- Pipe – NW 4 and NW 6 for connecting ground devives with the *Gerät*
- Tools and toolboxes

104, 105

- Prüfbrett mit dem Feinmessmanometer zum Einstellen des Druckminderers
- Abhebekontakte und zugehörige Vorrichtung
- Schalt- und Prüfkästchen für den Generaldurchschaltversuch
- Abgasrohr mit Haltevorrichtung für den A-Tankentlüfter
- A- und B-Stoffschläuche mit Kupplung
- T-Stoff-Betankungsschlauch mit Betankungsrohr
- B-Stoff und T-Stoff-Enttankungsschlauch mit Kupplung
- Rohrleitung – NW 4 und NW 6 zur Verbindung der bodenseitigen Ausrüstungsteile mit dem Gerät
- Werkzeuge und Werkzeugkästen

BArch, RH8II Bild-B1940-44 o. Ang.

106

106 This test vehicle I or III with a large equipment box on the roof of the driver's cab is mounted on a 3 t Opel 6700 Type A chassis. The Fz.106 or 107 vehicle belongs to the 2./836 and was used during the »Schiesserprobung Schiene« (firing tests rail) in November 1944. At the front flap a navy plug is installed and with a NSH cable one is connected to the Fz.105. The vehicle is painted in standard dark yellow »nach Muster« overall and has a light spot camouflage.

106 Dieser Prüfwagen I oder III mit dem großen Gerätekasten auf dem Dach des Führerhaus ist auf einem 3 t Opel 6700 Typ A Fahrgestell aufgebaut. Das Fahrzeug Fz.106 oder 107 gehört zur 2./836 und wurde bei der Erprobung »Schießerprobung Schiene« im November 1944 eingesetzt. An der vorderen Klappe befindet sich ein Marinestecker. An dieser Stelle ist ein NSH-Kabel eingesteckt, welches mit dem Fz.105 verbunden ist. Das Fahrzeug ist »nach Muster« komplett in Standard dunkelgelb lackiert und mit einer leichten Flecktarnung versehen.

107

107 A test car I or III with LC box body (305) on a chassis of the Opel 3 t truck type 3.6/36 S. The windows in the rear access door and in the side wall are closed by the fixtures in the interior or with a panel from the inside. The vehicle is painted in dark yellow 'nach Muster' over all. Note the single tyres on the otherwise double-tyred rear axle, signs of beginning supply shortages.

107 Ein Prüfwagen I bzw. III mit LC-Koffer (305) auf einem Fahrgestell des Opel 3 t Lastkraftwagens Typ 3,6/36 S.
Die Fenster in der hinteren Zugangstüre und in der Seitenwand sind durch die Einbauten im Innenraum bzw. mit einer Füllung von innen verschlossen. Das Fahrzeug ist komplett in dunkelgelb »nach Muster« lackiert. Beachte die Einfachbereifung der ansonsten doppelbereiften Hinterachse, ein Anzeichen beginnender Versorgungslücken.

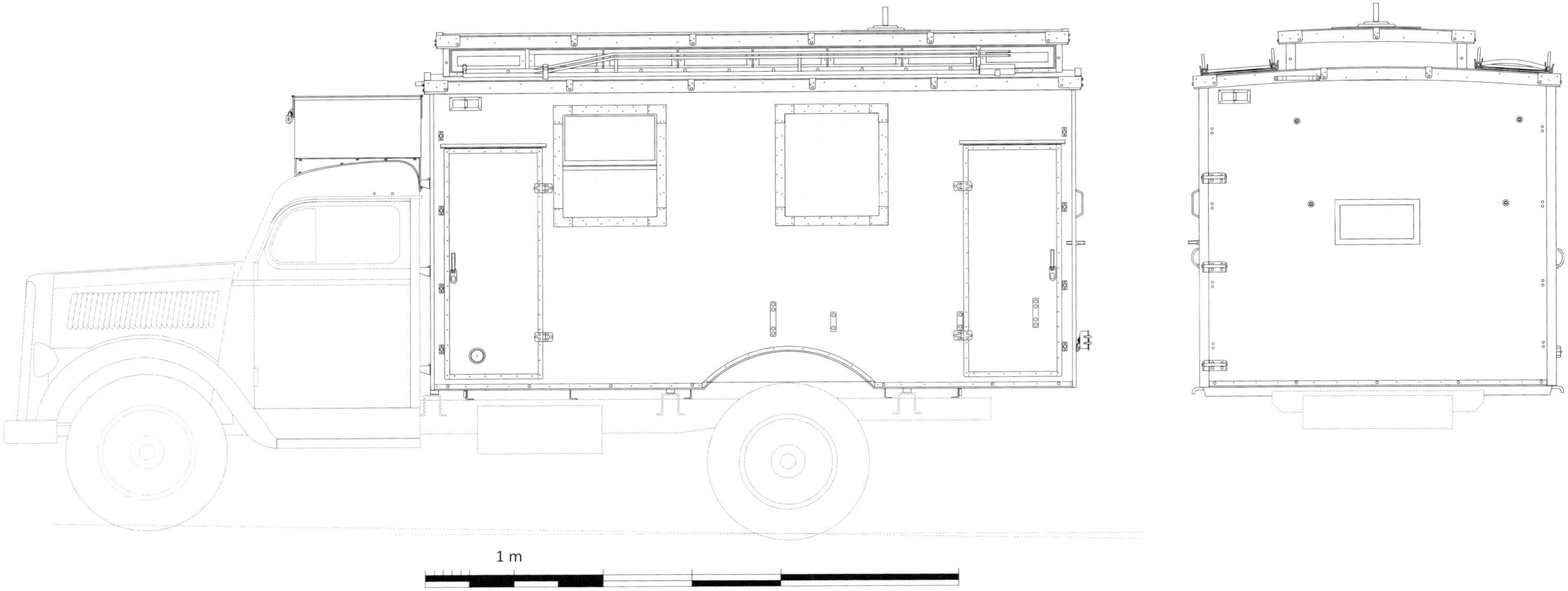

Scale / Maßstab 1:35 © Volker Ruff

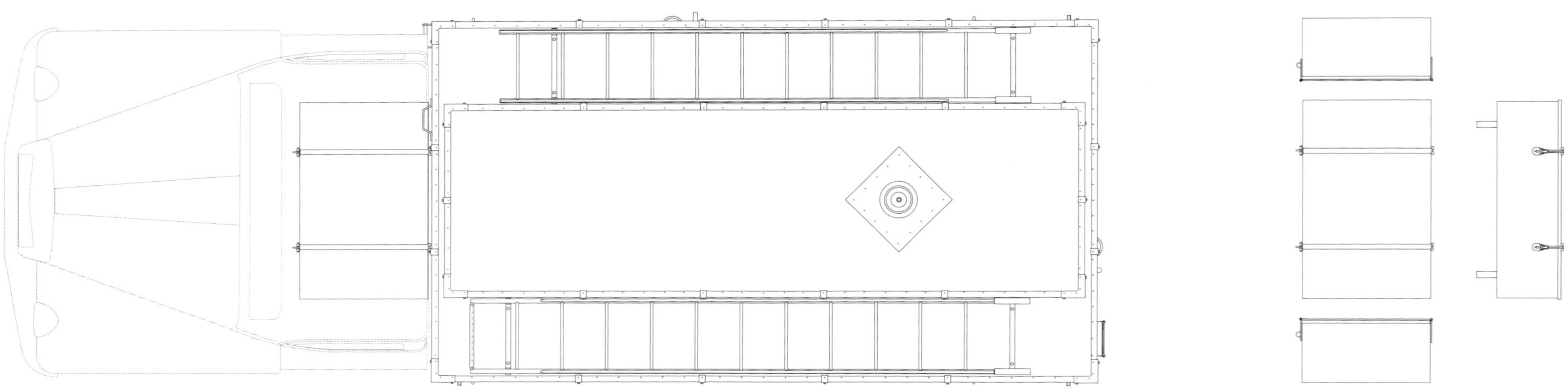

Fahrzeug 106/107

Prüfwagen I/III
auf Lkw 3 t geschl. (305)

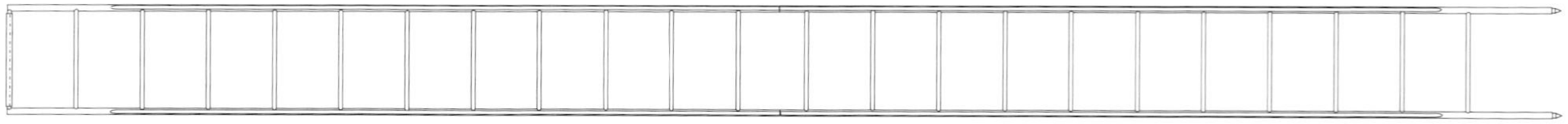

Partially used drawings acc. to group list 21Gr357, E2734

Teilweise verwendete Zeichnungen nach Gruppenliste 21Gr357

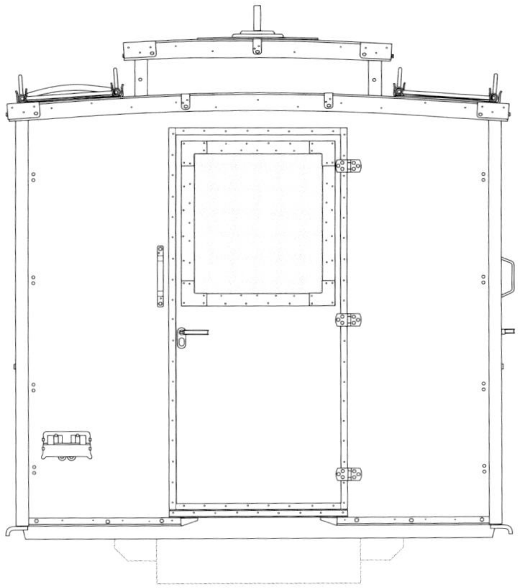

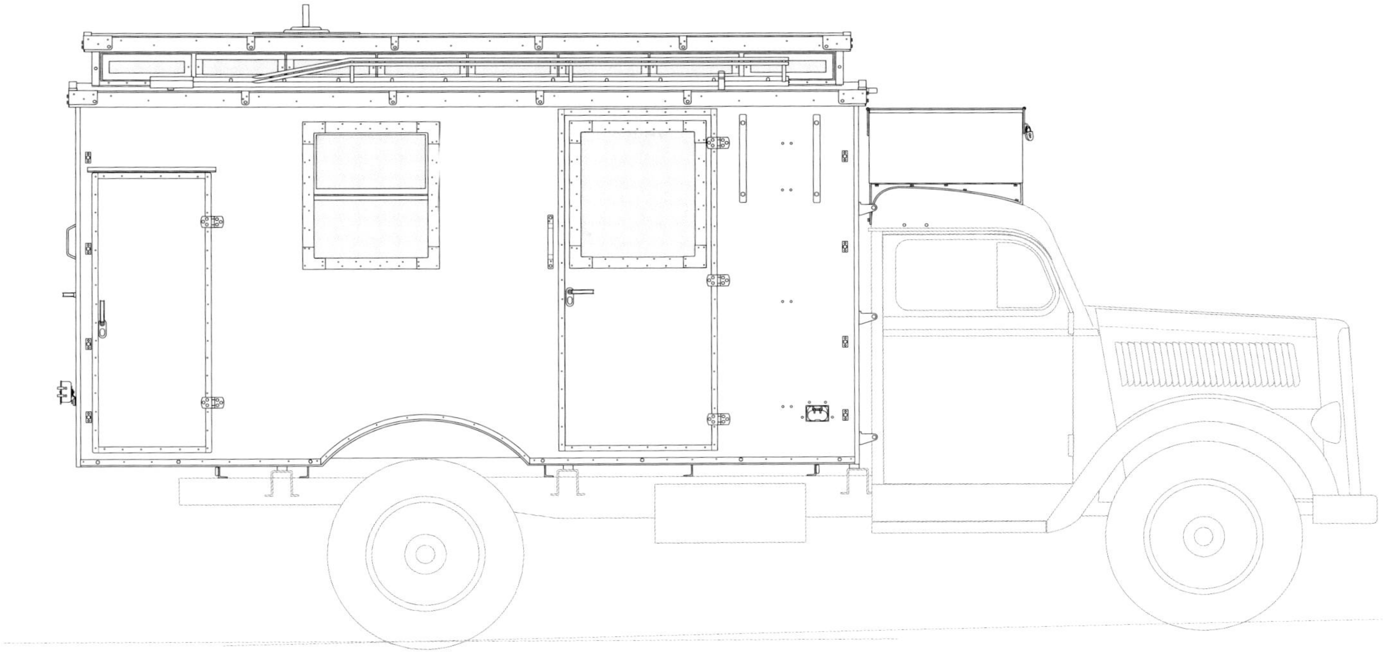

1 m

Scale / Maßstab 1:35 © Volker Ruff

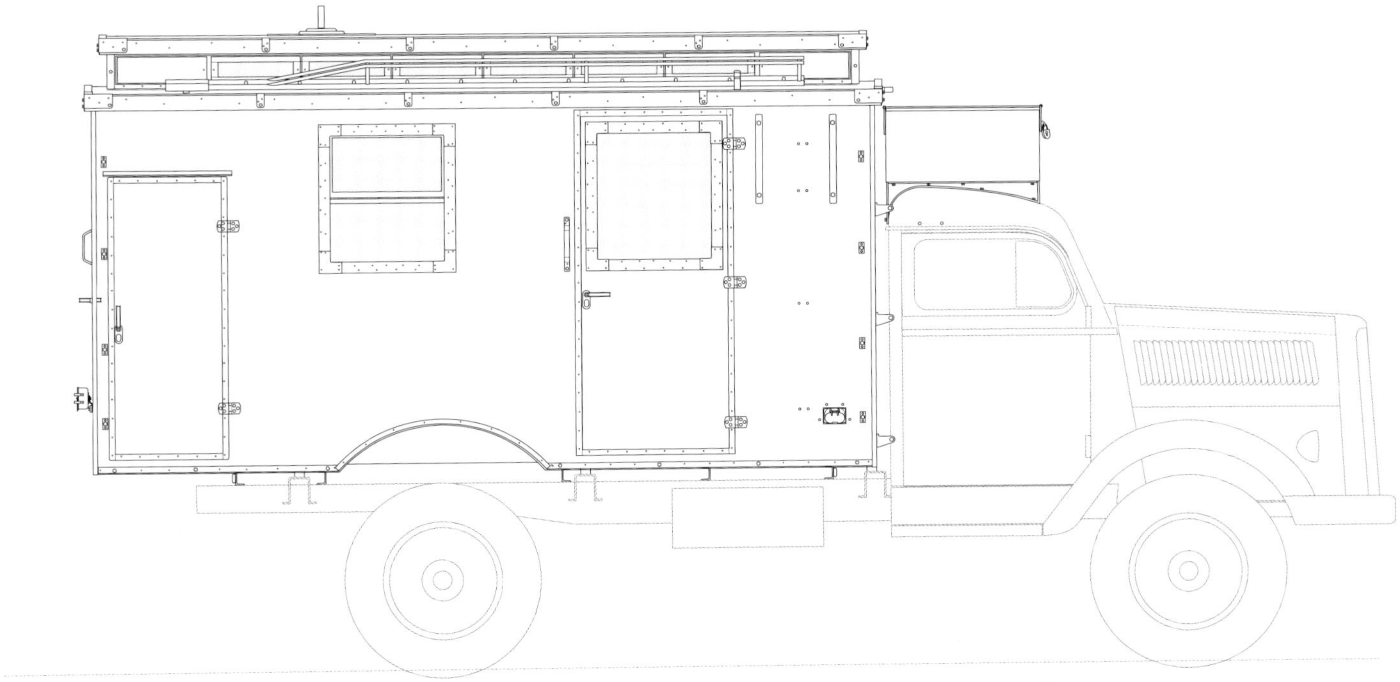

DMA, GD639.5.18
HAP11/T 507
Karlshagen, 1.6.1944
Bd.Nr. E 3195/44 g

DMA, GD639.5.1
EW234
Fz.-Pos.-Nr. 108

DMA, GD639.51.1
Auszug
Beladeplan

DMA, GD630.02
A4 Lehrbuch
der FR-Schule
Auszug aus
Technische Fahrzeuge

DMA, GD639.5.21
OKW/Wa Prüf10
Karlshagen, 25.03.1944
Bd.Nr. 747/44 gK
Auszug
Besprechung der
Fahrzeuggruppe 1
vom 24./25.3.44
im Lager Rebstock

Vehicle 108

Internal or former designation: Fz.-Pos.-Nr.108
Cable reel trailer (2 axle)
Cable reel trailer

Intended use
The necessary cable reel with NSH, NMH, HF, SF, FL, FF and FL cables connecting firing position I, II and III to the vehicles of the firing range were stowed and transported in a cable reel trailer. For the cabling according to mot.I with »Strüver« power generator there was a »small« cable reel trailer based on the 3 t trailer E3. For mot.II operations with the »Steyr« power supply trailer the cable reel trailer was mounted on the 5 t trailer and was designed for 50 cable reels + 10 empty cable reels.

The cable reel trailer has the following dimensions and weights:

Total length:	6,550 mm
Width:	2,200 mm
Height:	2,750 mm
Wheelbase:	3,000 mm
Track wight:	1,735 mm
Tare weight:	2,000 kg
Payload:	5,000 kg
Service weight:	6,920 kg
Wheels:	7.00-20

The cable reel trailer consists of the following main components:
- Chassis to 5 t unit trailer adapted
- Closed body with loading- and false floor with cable reel holders and flaps that can be opened on both sides

Chassis
The basis for the adapted chassis is the chassis of the unit trailer 5 t, type E5 (type 03) (equipment description according to D 692/41)

Production status: Vehicle 108
Supplied so far:
3 to Abt.836
1 to Battery 444
Further production is stopped, vehicle is no longer delivered *

* This refers to the »small« version of the cable reel trailer with 36 cable reels for operation according to mot.I. For operation according to mot.II, the cable reel trailer was continued in the »large« version for 50 + 10 cable reels.

Fahrzeug 108

Interne oder frühere Bezeichnung: Fz.-Pos.-Nr.108
Kabeltrommelanhänger (2-achs.)
Kabeltrommelanhänger

Verwendungszweck
Die notwendigen Kabeltrommeln mit NSH-, NMH-, HF-, SF-, FL-, FF und FL-Kabeln zur Verbindung der Schiessstelle I, II und III mit den Fahrzeugen der Feuerstellung wurden in einem Kabeltrommelanhänger verstaut und transportiert. Für die Verkabelung nach mot.I mit »Strüver« Stromerzeuger gab es einen »kleinen« Kabeltrommelanhänger auf Basis des 3 t Anhängers E3. Für den Betrieb mot.II mit »Steyr« Stromversorgungswagen wurde der Kabeltrommelanhänger auf den 5 t Anhänger aufgebaut und war für 50 Kabelrollen + 10 Leerkabelrollen ausgelegt.

Der Kabeltrommelanhänger hat folgende Maße und Gewichte:

Gesamtlänge:	6.550 mm
Breite:	2.200 mm
Höhe:	2.750 mm
Radstand:	3.000 mm
Spurweite:	1.735 mm
Leergewicht:	2.000 kg
Nutzlast:	5.000 kg
Gefechtsgewicht:	6.920 kg
Räder:	7,00-20

Der Kabeltrommelanhänger besteht aus folgenden Hauptbauteilen:
- Fahrgestell 5 t Einheits-Anhänger angepasst
- Geschlossener Aufbau mit Lade- und Zwischenboden mit Kabeltrommelaufnahmen und beidseitig zu öffnenden Klappen

Fahrgestell
Als Basis für das angepasste Fahrgestell dient das Fahrgestell des Einheitsanhängers 5 t, Baumuster E5 (Typ 03) (Gerätebeschreibung nach D 692/41)

Fertigungslage: Fahrzeug 108
Bisher geliefert:
3 Stck. an Abt.836
1 Stck. an Batterie 444
Weitere Fertigung ist eingestellt, Fahrzeug wird nicht mehr geliefert *

* Das bezieht sich auf die »kleine« Ausführung des Kabeltrommelanhängers mit 36 Kabeltrommeln für Betrieb nach mot.I. Für den Betrieb mot.II wurde der Kabeltrommelanhänger in der »großen« Ausführung für 50 + 10 Kabeltrommeln weitergebaut.

Chassis and body

A body with loading- and false floor, lockable with loading walls and with rails for removable cable reel racks, was placed on a modified 5 t trailer chassis. The lower loading floor and upper false floor are accessible on the left and right via 4 flaps each.

Current defects and their remedy

The units wish to have the side loading walls made of sheet metal instead of wood is therefore no longer fulfilled.

Fahrgestell und Aufbau

Auf ein modifiziertes 5 t Anhänger-Fahrgestell wurde ein mit Ladewänden verschließbarer Aufbau aus Lade- und Zwischenboden und Aufnahmeschienen für herausnehmbare Kabelrollengestelle gesetzt. Der untere Ladeboden und obere Zwischenboden ist links und rechts über jeweils vier Klappen zugänglich.

Derzeitige Mängel und ihre Behebung

Auf den Wunsch der Truppe, die seitlichen Ladewände nicht aus Holz, sondern aus Blech auszuführen, wird daher auch nicht entsprochen.

108

Fz.108 Cable reel trailer (early version)
Fz.108 Kabeltrommelwagen (frühe Ausführung)

108 On the early cable reel trailer the upper loading floor is set somewhat lower, so the wooden hinged flaps are different sizes for each loading level. On the later version of Fz.108, the upper loading floor was installed higher, the hinged flaps were adapted and covered with sheet metal. Both early and late versions of the cable reel trailer were used by Abteilung 836.

108 Am frühen Kabeltrommelwagen ist der obere Zwischenboden etwas tiefer angesetzt. Dadurch sind die Klapptüren (mit Holzfüllung) pro Ladeebene unterschiedlich groß. An der späteren Ausführung von Fz.108 wurde der obere Ladeboden höher eingesetzt, die Klapptüren angepasst und mit Blech verkleidet. Die frühe wie auch die späte Ausführung des Kabeltrommelwagens wurden bei der Abteilung 836 eingesetzt.

109

109 In the spring of 1944, in the inventory of Abt.836 was this cable reel trailer Fz.108 of an earlier version. The side view taken from the left, shows very clearly the details of the early hinged flaps, in the different sizes for the lower and upper loading areas.

109 Im Bestand der Abt.836 befand sich im Frühjahr 1944 dieser Kabeltrommelanhänger Fz.108 früherer Ausführung. Die Seitenansicht von links zeigt sehr deutlich die Details und unterschiedlichen Größen der frühen Klapptüren der unteren und obere Ladefläche.

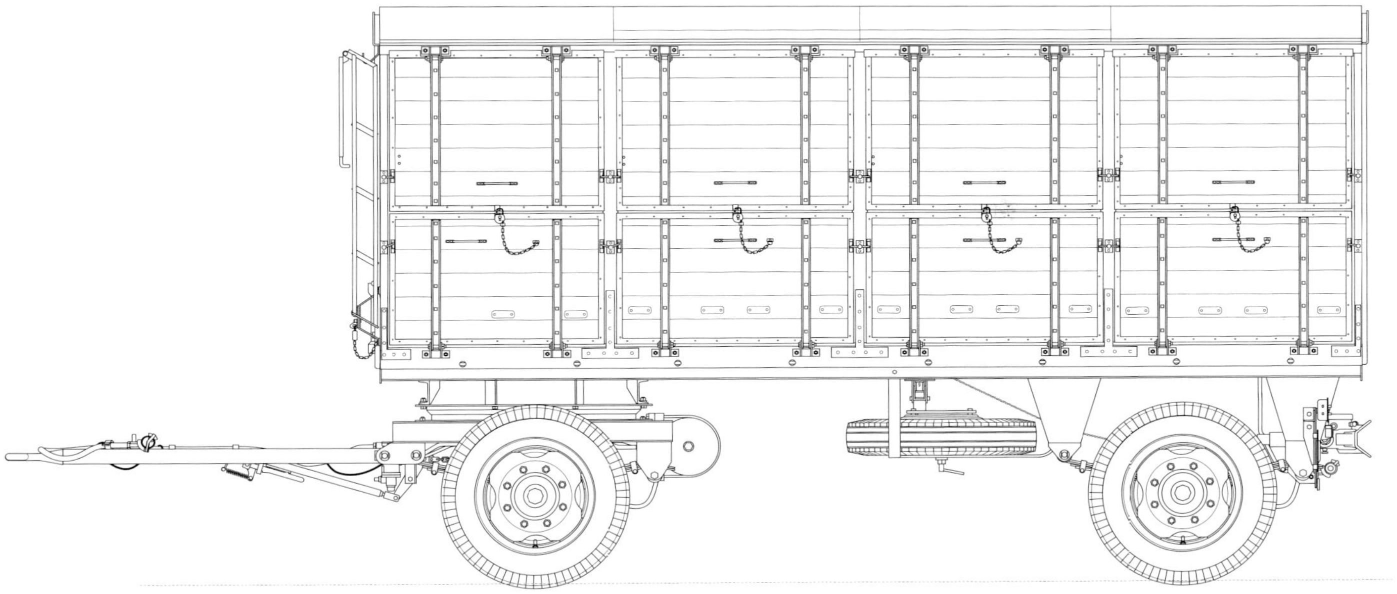

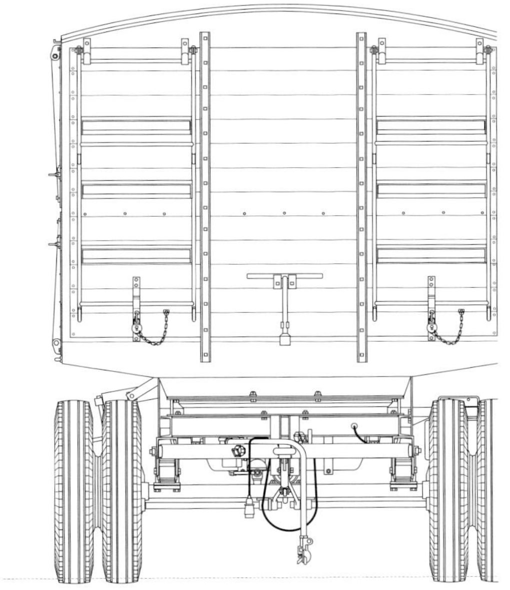

1 m

Scale / Maßstab 1:35 © Volker Ruff

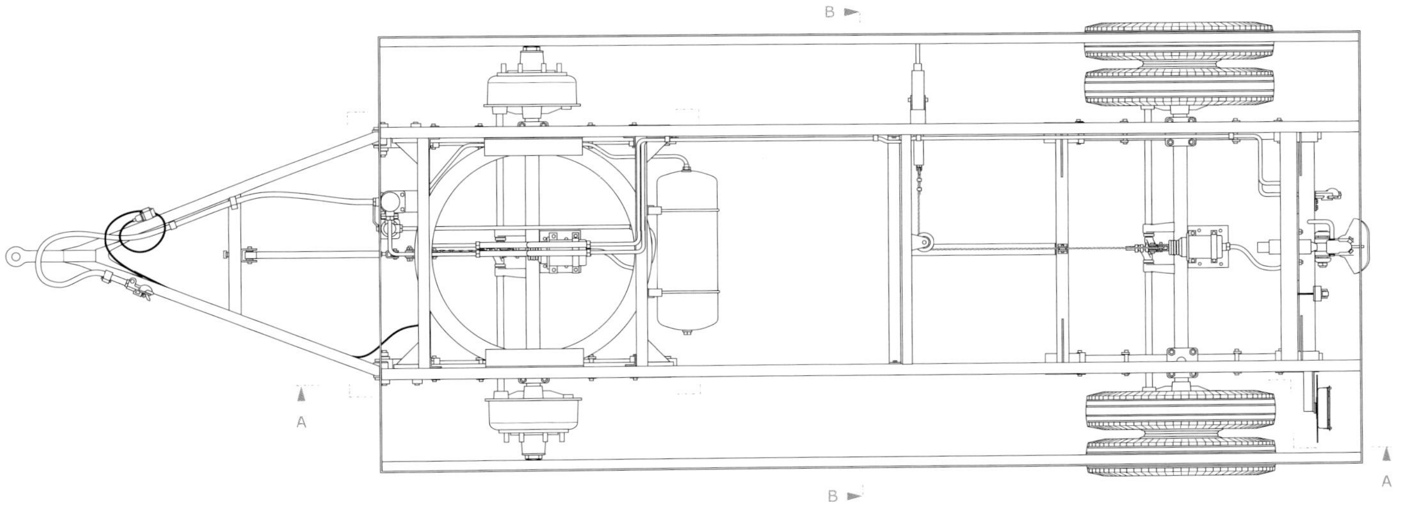

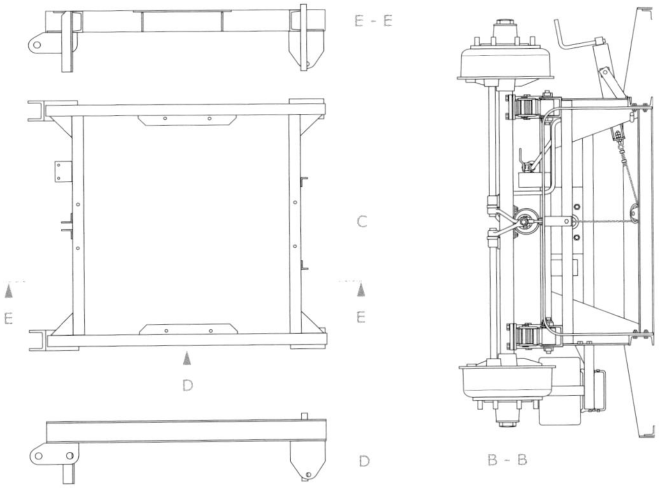

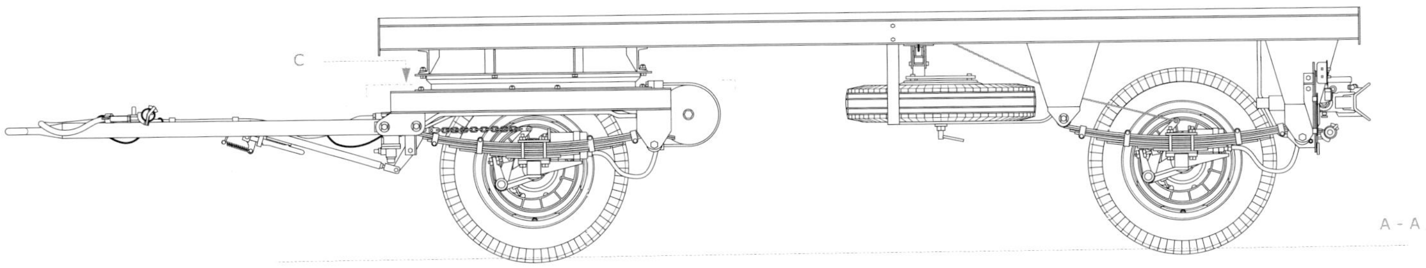

Early version 1943
Frühe Ausführung 1943

Fahrzeug 108

Kabeltrommelanhänger
(2-achs.)

Partially used drawings acc. to
group list 21Gr157

Teilweise verwendete Zeichnungen
nach Gruppenliste 21Gr157

BArch, RH8II Bild-B1443-44, o.Ang.

110

Fz.108 Cable reel trailer (late version)
Fz.108 Kabeltrommelwagen (späte Ausführung)

110 Lager Rebstock in spring 1944 – a cable reel trailer is prepared for acceptance. To facilitate access to the loading floors, two foldable platform ladders are carried along the front of the superstructure, with a wheel-nut socket wrench resting in a holder between them.

110 Lager Rebstock im Frühjahr 1944 – ein Kabeltrommelwagen ist für die Abnahme vorbereitet. Zur Erleichterung für den Zugang zu den Ladeböden wurden an der Stirnseite des Aufbaus zwei klappbare Plattform-Leitern mitgeführt. Dazwischen ruht in einer Halterung ein Radmuttersteckschlüssel.

BArch, RH8II Bild-B1444-44, o.Ang.

111

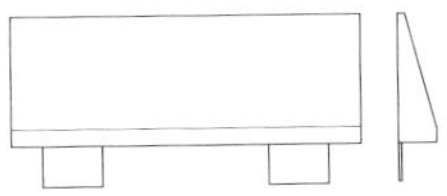

111 At the rear, the chassis for two cable laying carts are suspended in brackets. The corresponding rubber-tyred spoked wheels are stowed in the front cable compartments on the top. Cable bridges, simple bevelled wooden planks to protect the cables when driving over them, are inserted between the cable laying carts, further cable bridges have been stored in the front cable compartments in the bottom.

111 Auf der Rückseite sind die Fahrgestelle zweier Verlegewagen in deren Halterungen eingehängt. Die gummibereiften Speichenräder der Verlegewagen befinden sich in den vorderen oberen Kabelfächern. Kabelbrücken und einfache, abgeschrägte Holzbohlen zum Schutz der Kabel beim Überfahren sind zwischen die Kabelverlegewagen eingesteckt. Weitere Kabelbrücken wurden in den vorderen unteren Kabelfächern aufbewahrt.

The fixtures

The body, which is divided into two loading levels, is divided into eight chambers on the left and right and accessible via flaps. Additional columns in the lower chambers supports the upper floor. Each chamber is equipped with two rail tracks on which two winding blocks run, which can be pulled out. The spoke wheels are stowed in the front chambers. A total of 54 cable reels (for FL-, SF- and FF cables) and 12 cable bridges can be stored.

Die Einbauten

Der in zwei Ladeebenen unterteilte Aufbau ist links und rechts in acht Kammern aufgeteilt und über Klappen zugänglich. Zusätzliche Säulen stützen die oberen Boden in den unteren Kammern. Je Kammer sind zwei Schienenbahnen montiert, auf der zwei Aufspulböcke laufen, die ausziehbar sind. In den vorderen Kammern sind die Speichenräder verstaut. Insgesamt können 54 Kabelrollen (für FL-, SF- und FF-Kabel) und 12 Kabelbrücken untergebracht werden.

BArch, RH8II Bild-B1445-44, o.Ang.

112

BArch, RH8II Bild-B1446-44, o.Ang.

113

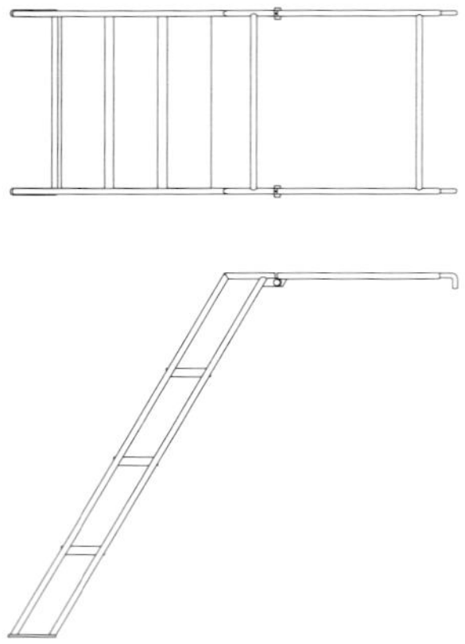

112 The lower flaps are opened and the rewinding stands with the cable reels are pulled out and tipped out, now the cable reels are ready to be taken over. The bearing shell on the rewinding stand is opened, a shaft is inserted into the cable reel hole and the cable reels are prepared for transfer to the laying cart.

113 To open the upper flaps, a platform ladder is inserted into the frame. The lower flap is now opened so that it rests on the frame of the ladder and forms a platform. The upper flap can be reached via the platform; it is now hinged upwards and locked in place. Next the cable reels are pulled out of the upper chamber and tipped out.

112 Die unteren Klappen sind geöffnet und vier der vorderen Aufspulböcke mit den Kabelrollen herausgezogen und ausgekippt. Nun stehen die Kabelrollen bereit. Die Lagerschale am Aufspulbock wird geöffnet, eine Welle in die Kabelrollenbohrung gesteckt und die Kabelrollen zur Übernahme auf den Verlegewagen vorbereitet.

113 Für den Zugang zu den oberen Kammern werden die Klappen nach oben ausgeschwenkt und von einem Abstandshalter an der Seitenwandtrennung gehalten. Eine Plattformleiter wird am Rahmen eingesetzt und darauf die untere Klappe abgelegt. Über diese Plattform erreicht man nun die obenen Kammern. Jetzt werden die Kabelrollen aus den oberen Kammern herausgezogen und ausgekippt.

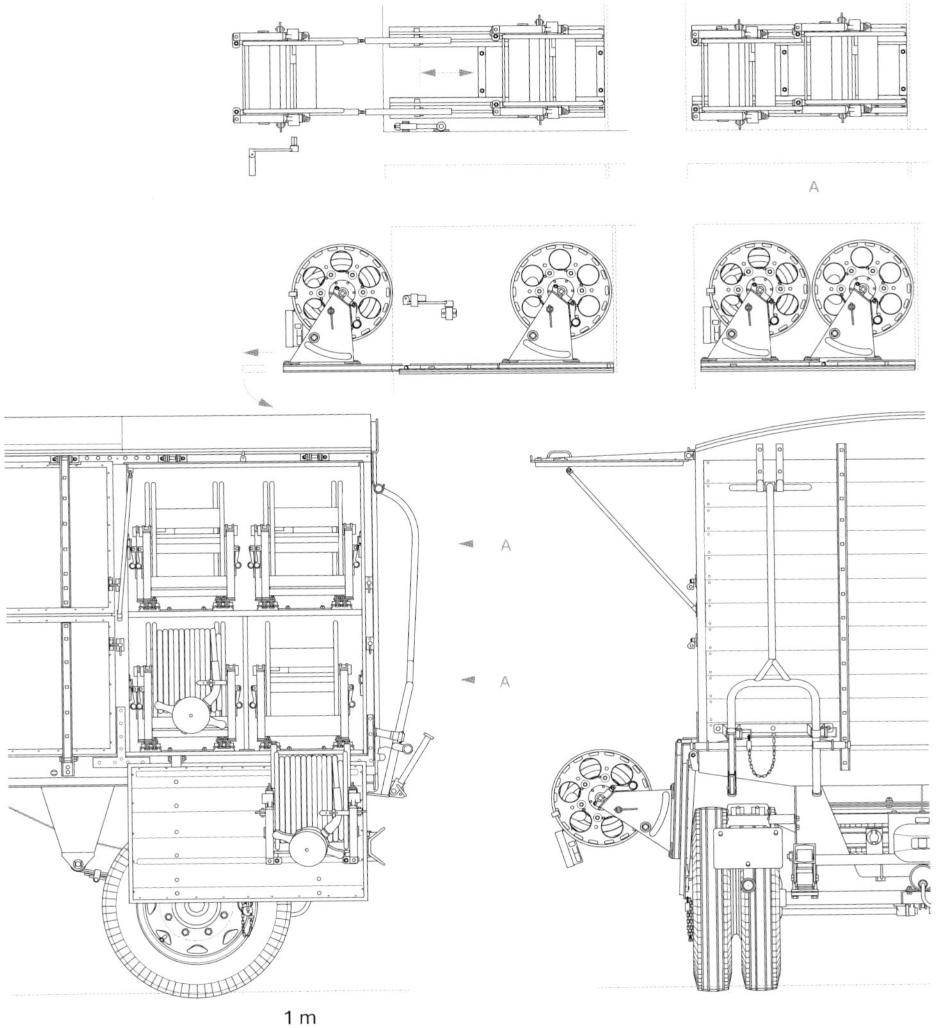

Scale / Maßstab 1:35 © Volker Ruff

BArch, RH8II Bild-B1447-44, o.Ang.

114

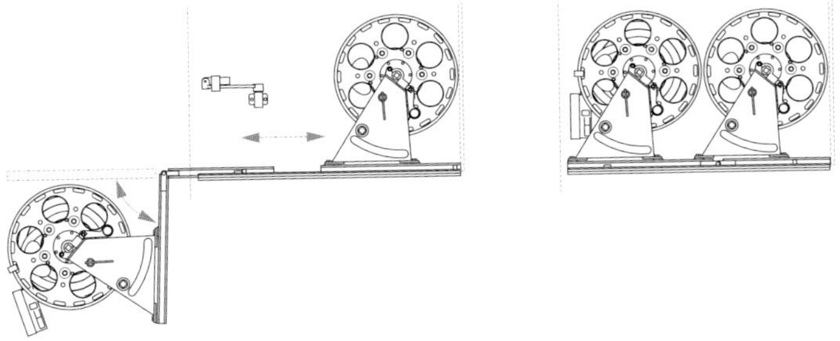

114 Two pairs of double rails are mounted in each chamber. The rewinding stands with the cable reels are pushed onto a pair of rails offset to each other. A protective strip on the flap and a partition inserted in the center serve as transport safety feature. In the lower chambers the heavy cables with the 104-pin Flaksteckdose or Flakstecker are stowed away. Only the front row of cable reels can be pulled out all the way and tilled down for removal. The spare cable reels are located in the rear row of the cable reels and can only be pulled out as far as they will go.

114 Je Kammer sind zwei Doppelschienenpaare montiert. Die Aufspulböcke mit den Kabeltrommeln werden auf ein Schienenpaar versetzt zueinander geschoben. Eine Stoßleiste an den Klappen und eine mittig eingesetzte Trennwand dienen als Transportsicherung. In den unteren Kammern sind die schweren Kabel mit der 104-poligen Flaksteckdose oder Flakstecker verstaut. Nur die vordere Reihe der Aufspulböcke mit Kabelrollen können bis zum Anschlag herausgezogen und zur Entnahme nach unten gekippt werden. In der hinteren Reihe sind die Reservekabelrollen in den Aufspulböcken eingesetzt. Diese lassen sich nur bis zum Anschlag herausziehen.

BArch, RH8II Bild-B1448-44, o.Ang.

115

BArch, RH8II Bild-B1449-44, o.Ang.

116

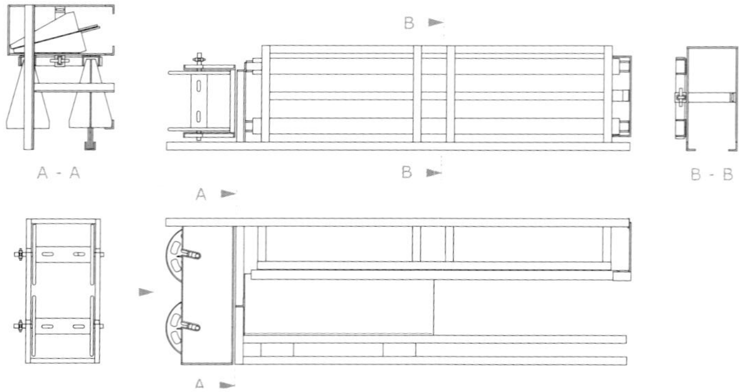

115 In the two front compartments, one above the other, only one double rail track is mounted and equipped with two rewinding stands with empty reels for FL and SF cable. The remaining space is used to store the rubber-tired spoked wheels for the laying cart and for 12 cable bridges.

116 The platform ladder inserted at the frame is used to the access to the upper cable compartments.

115 In den beiden vorderen übereinanderliegenden Kammern ist nur je eine doppelte Schienenbahn montiert und mit zwei Aufspulböcken mit Leertrommeln für FL- und SF-Kabel bestückt. Der verbliebene Raum wird zum Lagern der gummibereiften Speichenräder für den Verlegewagen und für 12 Kabelbrücken verwendet.

116 Die am Rahmen eingesteckte Plattformleiter dient dem Zugang zu den oberen Kabelkammern.

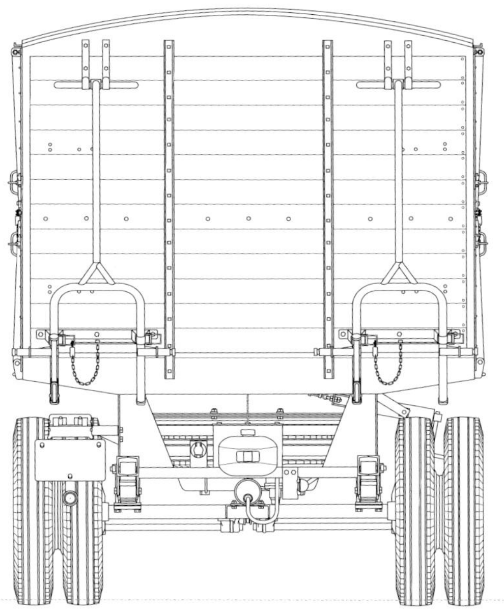

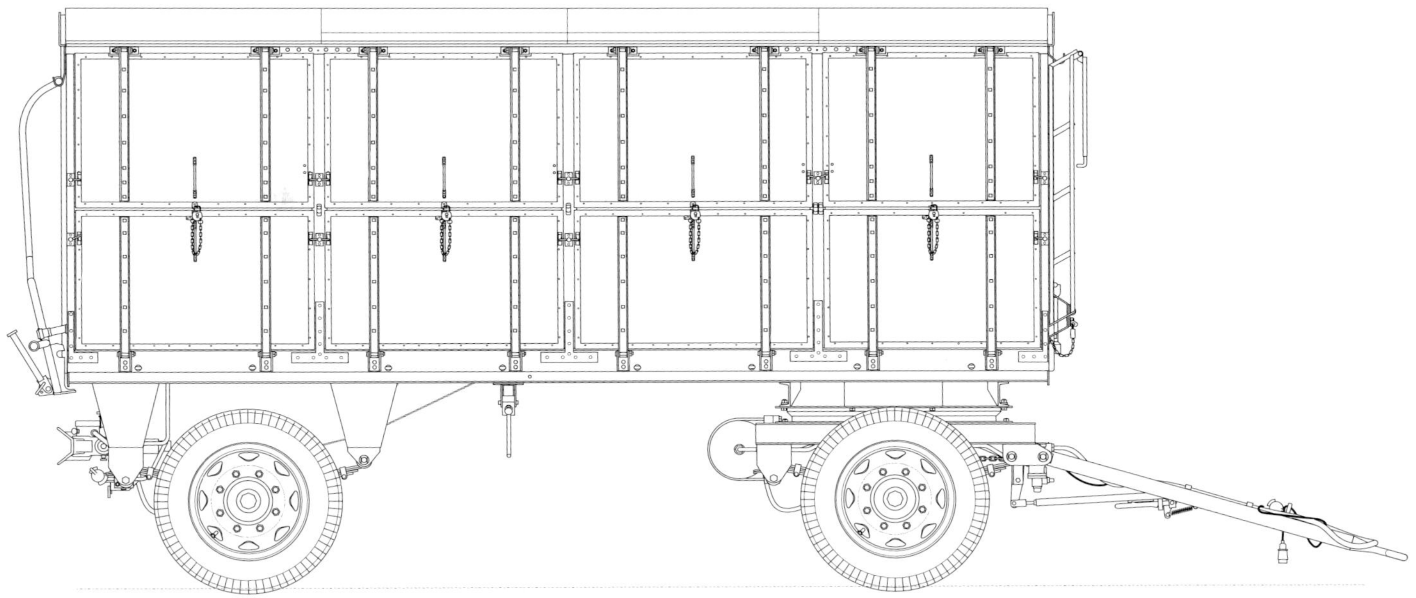

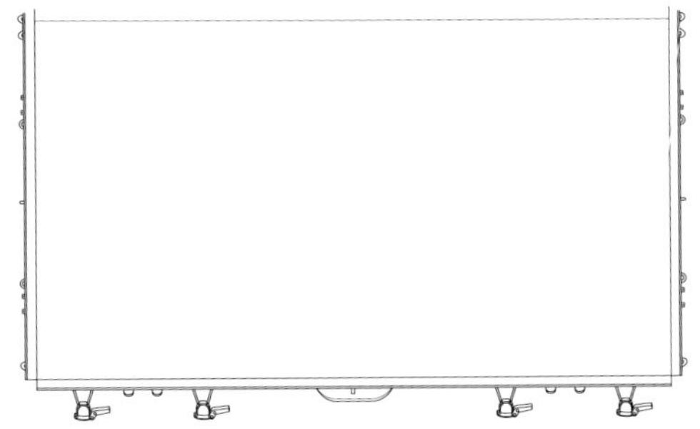

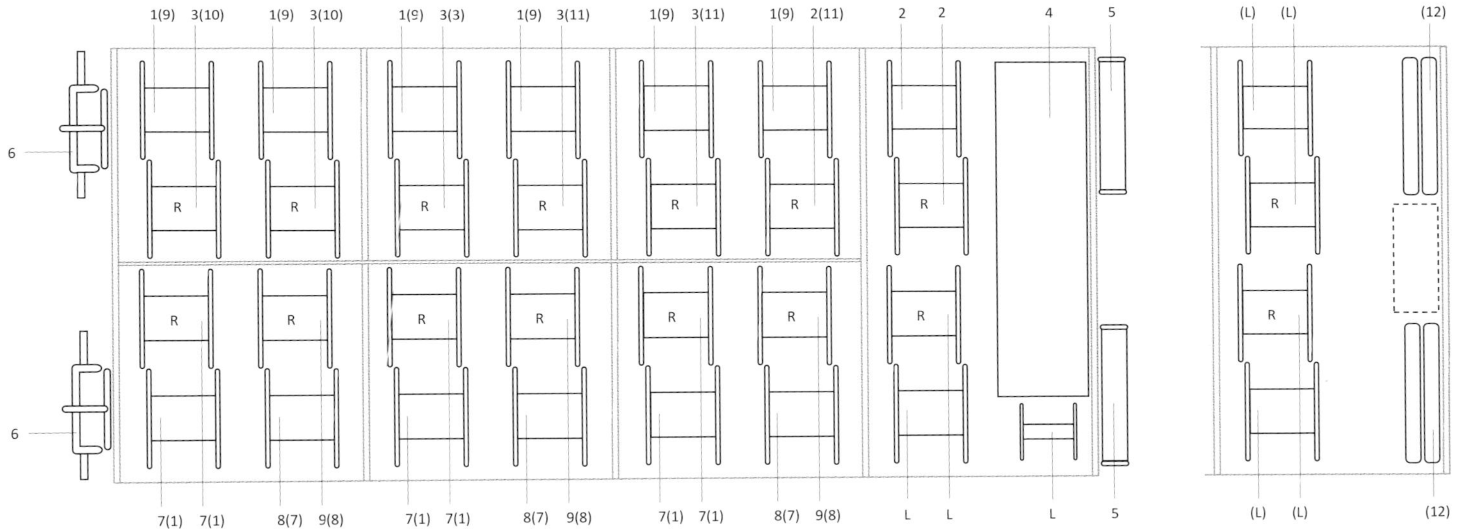

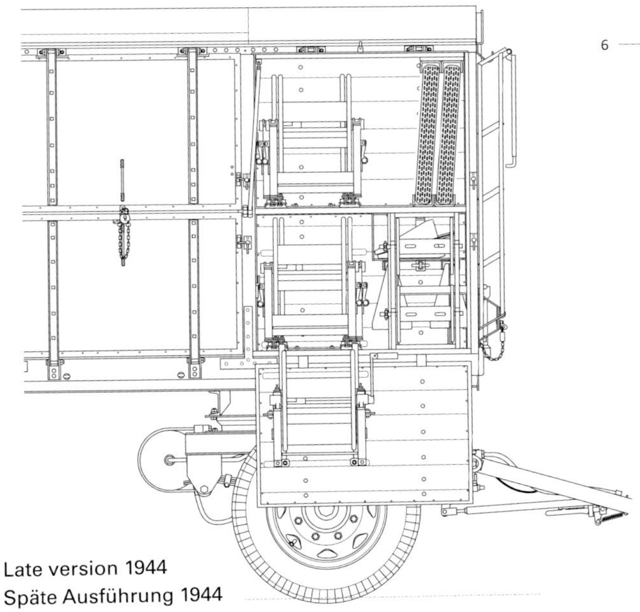

Late version 1944
Späte Ausführung 1944

1 Cable reel, cable FL 108 x 0.5 ea 70 m
2 Cable reel, cable NMH 4 x 4 ea 120 m
3 Cable reel, cable HF
4 Storage rack with 12 cable bridges
5 Platform ladder
6 Laying cart
7 Cable reel, cable NSH 4 x 25 ea 30 m
8 Cable reel, cable NMH 21 x 1.5 ea 70 m
9 Cable reel, cable FF ea 180 m
10 Cable reel, cable SF ea 300 m
11 Cable reel, cable NSH 3 x 2,5 m ea 30 m
12 Pair of wheels for laying cart
13 Spare part box for plugs
() upper floor + numbering
L Cable reel empty
R Backup

1 Kabelrolle, Kabel FL 108 x 0,5 je 70 m
2 Kabelrolle, Kabel NMH 4 x 4 je 120 m
3 Kabelrolle, Kabel HF
4 Gestell mit 12 Kabelbrücken
5 Plattformleiter
6 Kabelverlegewagen
7 Kabelrolle, Kabel NSH 4 x 25 je 30 m
8 Kabelrolle, Kabel NMH 21 x 1.5 je 70 m
9 Kabelrolle, Kabel FF je 180 m
10 Kabelrolle, Kabel SF je 300 m
11 Kabelrolle, Kabel NSH 3 x 2,5 m je 30 m
12 Räderpaar für Kabelverlegewagen
13 Ersatzteilkiste für Stecker
() oberer Boden + Nummer
L Kabelrolle leer
R Reserve

1 m

Scale / Maßstab 1:35 © Volker Ruff

Fahrzeug 108

Kabeltrommelanhänger
(2-achs.)

Partially used drawings acc. to group list 21Gr157

Teilweise verwendete Zeichnungen nach Gruppenliste 21Gr157

DMA, GD639.5.18
HAP11/T 507
Karlshagen, 1.6.1944
Bd.Nr. E 3195/44 g

DMA, GD639.5.1
EW234
Fz.-Pos.-Nr. 110

DMA, GD630.02
A4 Lehrbuch
der FR-Schule
Mai 1944
Auszug aus
Technische Fahrzeuge

DMA, GD639.5.21
OKW/Wa Prüf10
Karlshagen, 25.03.1944
Bd.Nr. 747/44 gK
Auszug
Besprechung der
Fahrzeuggruppe 1
vom 24./25.3.44
im Lager Rebstock

Vehicle 110

Internal or former designation: Fz.-Pos.-Nr.110
Small cable reel trailer (1-axle)
Cable reel trailer, small 900 kg

Intended use
The small cable reel trailer is towed by the power supply trailer Fz.105. Apart from the cable reels accommodated in the power supply car, it safely holds all the other power supply cables and FF cables used for the cabling of the firing position for single shot firing.

The cable reel trailer has the following dimensions and weights:

Total length:	3,440 mm
Width:	1,800 mm
Height:	1,440 mm
Distance:	2,265 mm from axle center to eyelet center
Track width:	1,560 mm
Tare weight:	410 kg
Payload:	900 kg
Combat weight:	1,170 kg
Wheels:	6.00-20

The cable reel trailer consists of the following main components:
- Chassis
- Body – sheet metal box with hood covers and two cable reel racks for a total of 8 cable reels

Chassis
The vehicle for the small cable reel trailer is the pneumatic-tired single-axle trailer type »A« for 900 kg load capacity. The equipment description is in manual D691/21 B.

Production status: Vehicle 110
6 units have been delivered to Abt. 485. Further deliveries are always possible in conjunction with power supply vehicles (Steyr).

Current defects and their remedy:
From now on the sheet metal side wall has to be rounded off to protect the cable while it is pulled out. Design changes to be notified by Wa Prüf 10. The last 5 m of the connecting cable from the power supply vehicle to the valve box on the launch pad has to be fire protected.
Manufacturing documents: created by Wa Prüf 10 in cooperation with Fa. Gollnow.

Fahrzeug 110

Interne oder frühere Bezeichnung: Fz.-Pos.-Nr.110
Kleiner Kabeltrommel-Anhänger (1-achs.)
Kabeltrommel-Anhänger, klein 900 kg

Verwendungszweck
Der kleine Kabeltrommel-Anhänger wird vom Stromversorgungswagen Fz.105 gezogen. In ihm befinden sich außer den im Stromversorgungswagen untergebrachten Kabeltrommeln sämtliche weiteren Stromversorgungs- und FF-Kabel, die für die Verkabelung der Feuerstellung beim Einzelschuss gebraucht werden.

Der Kabeltrommel-Anhänger hat folgende Maße und Gewichte:

Gesamtlänge:	3.440 mm
Breite:	1.800 mm
Höhe:	1.440 mm
Abstand:	2.265 mm von Achsmitte bis Ösenmitte
Spurweite:	1.560 mm
Leergewicht:	410 kg
Nutzlast:	900 kg
Gefechtsgewicht:	1.170 kg
Räder:	6,00-20

Der Kabeltrommel-Anhänger besteht aus folgenden Hauptbauteilen:
- Fahrgestell
- Aufbau – Blechkasten mit Deckelhauben und zwei Kabelrollengestelle für insgesamt für 8 Kabelrollen

Fahrgestell
Das Fahrgestell für den Kabeltrommel-Anhänger, klein ist der luftbereifter Einachs-Anhänger Typ »A« für 900 kg Tragfähigkeit.
Gerätebeschreibung siehe D691/21 B.

Fertigungslage: Fahrzeug 110
Bisher geliefert für Abt. 485 6 Stück. Weitere Lieferungen termingemäß mit den dazugehörigen Steyr-Stromversorgungswagen stets möglich.

Derzeitige Mängel und ihre Behebung:
Ab sofort muss die seitliche Blechwand zwecks Schonung der Kabel beim Herausziehen abgerundet werden. Formänderungsmitteilung veranlasst Wa Prüf 10. Das Verbindungskabel zwischen Anschluss am Stromversorgungswagen zum Ventilkasten der Abschussplattform muss auf den letzten 5 m mit einem Feuerschutz versehen werden.
Fertigungsunterlagen: erstellt von Wa Prüf 10 in Zusammenarbeit mit Fa. Gollnow und Sohn.

Structure

A sheet metal box with dtwo-winged sheet metal lids was mounted on the base frame of the 900 kg chassis. Two racks for cable reels for four cable reels each were inserted in the sheet metal box and bolted to the reinforced base plate.

Aufbau

Auf den Grundrahmen des Fahrgestells 900 kg war ein Blechkasten mit zweiflügeligen Deckelhauben montiert. Im Blechkasten wurden zwei Gestelle mit Leitungstrommeln für je vier Kabelrollen eingesetzt und mit dem verstärkten Bodenblech verschraubt.

BArch, RH8II Bild-B1465-44, o.Ang.

117

Stenciling of the cable designations in the cover hoods of the Fz.110

(As of 9/1/44)

Beschriftung der Kabelbezeichnungen in den Deckelhauben des Fz.110

(Stand vom 1.9.44)

BArch, RH8II Bild-B1764-44, o.Ang.

118

BArch, RH8II Bild-B1763-44, o.Ang.

119

118, 119 Labeling of the cable fittings are used and the two plates in the cover are fixed one above the other or painted on.

118, 119 Es werden die Bezeichnungsschilder der Kabelarmaturen verwendet und jeweils die beiden Schilder im Deckel übereinander befestigt oder auflackiert.

Simplified representation

Image at left – from left to right

250 m FF cable	Cable 51 Fz.101-102	
70 m NSH 4x2.5	Fz.105 220/380 V	to Fz.106/107 220/380 V
60 m NSH 3x10	Fz.105 Converter	at firing point Converter
120 m NMH 4x4	Fz.105 27V-220 V	to FL tank 27V-220 V

Image at right – from left to right

60 m NMH 21x1,5	Fz.105 Valve box	at firing point Valve box
30 m NSH 4x25	separation point weft point Fz.105 27 V	at weft point 27 V
30 m NSH 4x25	Fz.105 27 V	Fz.105 27 V seperation point weft point
250 m FF cable	Cable 52 Fz. 101-101	

Vereinfachte Darstellung

Bild links – von links nach rechts

250 m FF-Kabel	Kabel 51 Fz.101-102	
70 m NSH 4x2,5	Fz.105 220/380 V	an Fz.106/107 220/380 V
60 m NSH 3x10	Fz.105 Umformer	an Schussstelle Umformer
120 m NMH 4x4	Fz.105 27V-220 V	an FL-Panzer 27V-220 V

Bild rechts – von links nach rechts

60 m NMH 21x1,5	Fz.105 Ventilkasten	an Schussstelle Ventilkasten
30 m NSH 4x25	Trennstelle Schussstelle Fz.105 27 V	an Schussstelle 27 V
30 m NSH 4x25	Fz.105 27 V	an Trennstelle Schussstelle Fz.105 27 V
250 m FF-Kabel	Kabel 52 Fz.101-101	

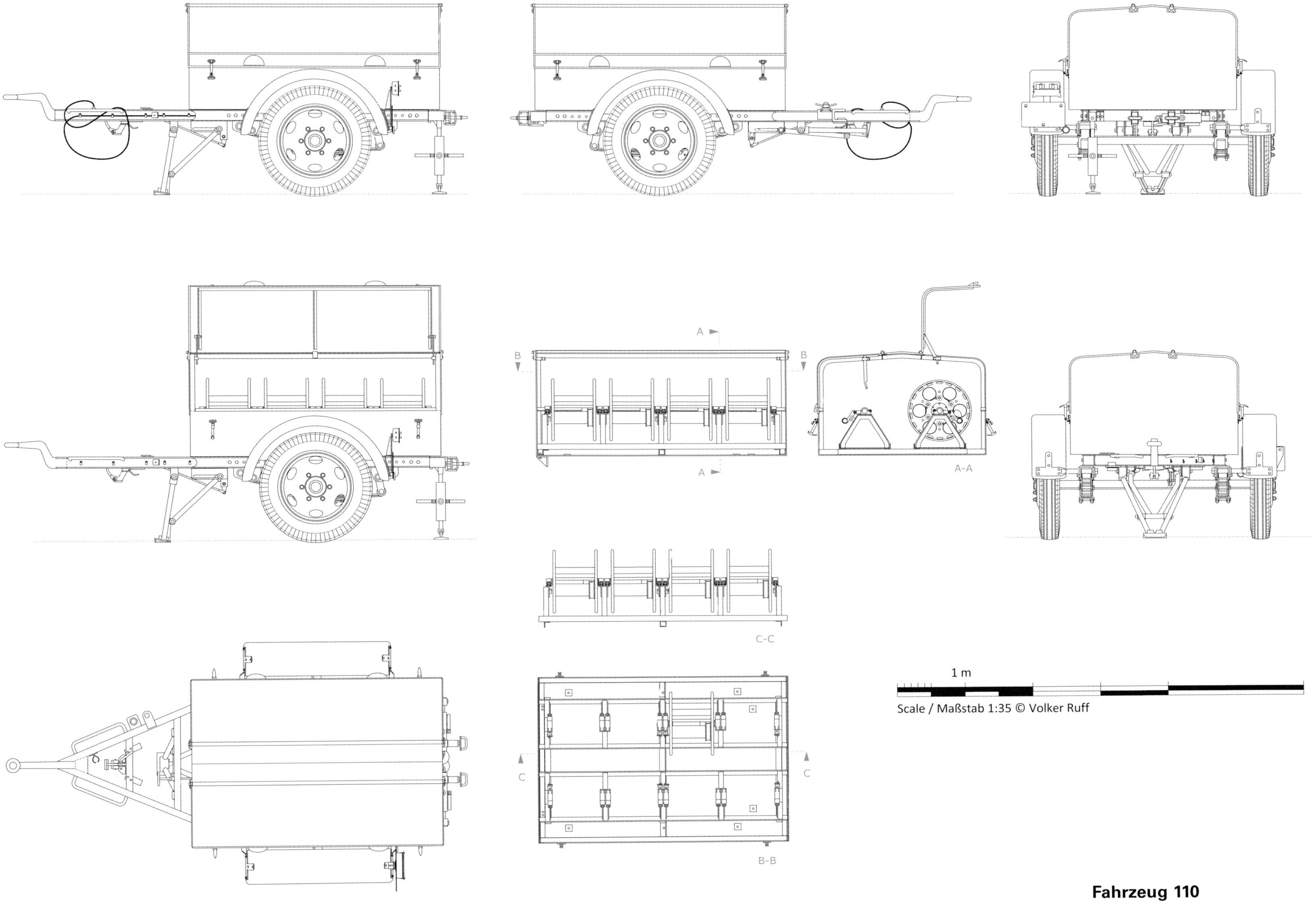

Fahrzeug 110

Kleiner Kabeltrommel-Anhänger (1-achs.)

Drawing set used D691/2

Verwendeter Zeichnungssatz D691/2

VEHICLE GROUP II
FAHRZEUGGRUPPE II

Remote control position

The remote control position is divided into a combustion cut-off ground installation for combustion cut-off and a guide beam ground installation for precise control of the aggregate on the trajectory [1].

Combustion cut-off ground installation

DMA, GD630.02
A4 Lehrbuch
der FR-Schule
Mai 1944
Auszug aus Kapitel III
Technische Fahrzeuge
in der Brennschluss-
stellung
Ing. Gassmann

DMA, GD632.41.12
EW22/463/44gK
Stand Sept.44
Auszug
Gerätebeschreibung

Purpose and use
The task of the combustion cut-off installation is to measure the speed of the flying A4 *Gerät* by radio transmission using the Doppler principle and to switch off its propulsion at the ordered V_0 speed. The *Gerät* is subdivided into two on-board devices, the Doppler and the signal control receiver, the most important of which are the V_0 transmitter, the V_0 surveying transmitter, the V_0 receiver and the signal transmitter. In addition there are the necessary antennas, a power supply unit and other accessories.

The combustion cut-off ground installation consists of:
- two vehicles in which all equipment is accommodated
- the transmitter vehicle
- and the receiver vehicle
- as well as in two trailers, the dipole trailer
- and the power supply trailer

BAMA, RH8-1310
Spezialfahrzeuge
Baugruppen und
wesentliche Bauteile
der Bodenanlage FR

Special vehicles in the combustion cut-off position
Fz.-Pos.-Nr.201 Transmitter vehicle on 3 t truck closed (305)
Fz.-Pos.-Nr.202 Cut-off receiver vehicle on 3 t truck closed (305)
Fz.-Pos.-Nr.204 Antenna trailer (1-axle)

Fernlenkstellung

Die Fernlenkstellung ist aufgeteilt in eine BS-Bodenanlage für den Brennschluss und einer LS-Bodenanlage zur genauen Steuerung des Aggregates auf der Flugbahn [1].

BS-Bodenanlage

Zweck und Aufgabe
Die Brennschlussanlage hat die Aufgabe, auf funktechnischem Wege unter Anwendung des Doppler'schen Prinzips die Geschwindigkeit des fliegenden A4-Gerätes zu messen und bei der befohlenen V_0-Geschwindigkeit den Antrieb abzuschalten. Die Anlage gliedert sich in zwei bordseitige Geräte: dem Verdoppler und dem Kommandoempfänger, deren wichtigster der V_0-Sender, der V_0-Vermessungssender, der V_0-Empfänger und der Kommandosender sind. Hinzu kommen die erforderlichen Antennen, ein Stromversorgungs-Aggregat und sonstiges Zubehör.

Die BS-Bodenanlage besteht aus:
- Zwei Kraftfahrzeugen in denen sämtliche Geräte untergebracht sind
- dem Sendewagen
- Empfangswagen
- sowie in zwei Anhängern, dem Dipolanhänger
- und einem Stromversorgungs-Anhänger

Spezialfahrzeuge in der Brennschlussstellung
Fz.-Pos.-Nr.201 BS-Sendewagen auf Lkw 3 t geschl. (305)
Fz.-Pos.-Nr.202 BS-Empfangswagen auf Lkw 3 t gl. geschl. (305)
Fz.-Pos.-Nr.204 BS-Antennen-Anhänger (1-achs.)

Location of vehicles in the combustion cut-off position
The combustion cut-off position was about 8-12 km behind the launching site in the rear extension of the firing direction. The drawing on page 163 shows the prescribed position of the vehicles Fz.201, Fz.202 and in particular the antennas like Fz.204 dipol and rhombus.

Aufstellung der Fahrzeuge in der BS-Stellung
Die BS-Stellung befand sich etwa 8-12 km hinter der Feuerstellung in rückwärtiger Verlängerung der Schussrichtung. Die Zeichnung auf Seite 163 zeigt die vorgeschriebene Position der Fahrzeuge Fz.201, Fz.202 und insbesondere der Antennen, wie Fz.204 Dipol und Rhombus.

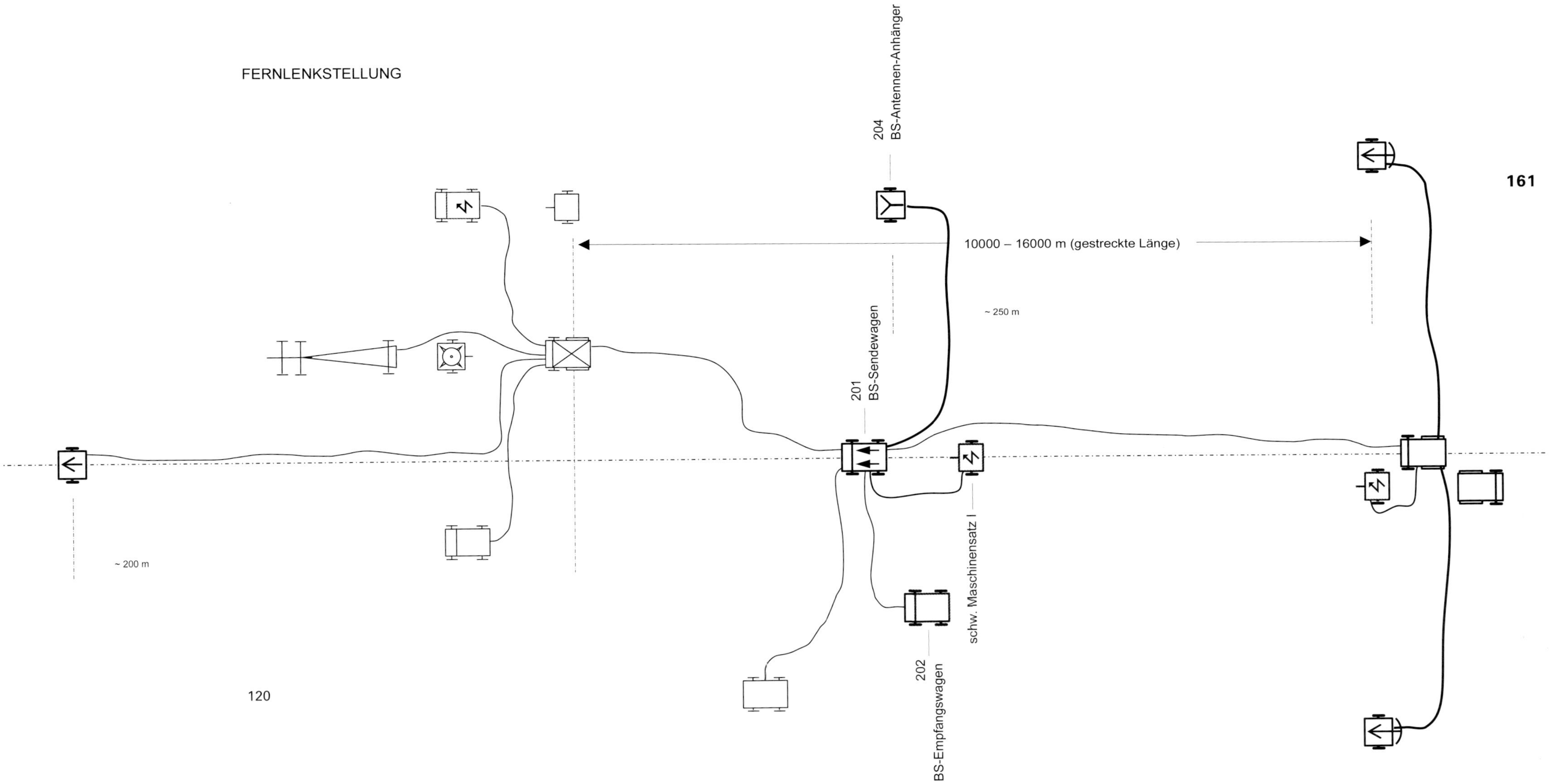

BS-Bodenanlage
in der Fernlenkstellung

DM

DMA, GD630.02
A4 Lehrbuch
der FR-Schule
Mai 1944
Auszug aus Kapitel III
Technische Fahrzeuge
in der Brennschluss-
stellung
Ing. Gassmann

Principle of the combustion cut-off

The shooting distance of a projectile depends on the starting angle and the starting velocity V_0.

With the A4 *Gerät*, the deflection program ensures that the exit angle at the end of the drive path is always the same, so that the firing distance is governed by the initial velocity V_0. The longer propulsion remained in operation, the greater the speed achieved when propulsion was stopped.

For a rocket, the initial speed V_0 is the speed when the propulsion system is switched off (combustion cut-off).

The speed of the A4 *Gerät* was determined by a radio system that used the Doppler effect. As soon as the correct speed was reached a radio command was given for »combustion cut-off«. The accuracy was +/- 5 km in firing distance of 275 km.

Prinzip des Brennschlusses

Die Schussweite eines Geschosses ist abhängig vom Anfangswinkel und der Anfangsgeschwindigkeit V_0.

Beim A4-Gerät war durch das Umlenkprogramm der Abgangswinkel am Ende der Antriebsbahn stets gleich groß, sodass die Schussweite durch die Größe der Anfangsgeschwindigkeit V_0 verändert wurde. Die bei Stillsetzen des Antriebes erreichte Geschwindigkeit war umso grösser, je länger der Antrieb in Betrieb blieb.

Prinzipiell versteht man bei einer Rakete unter der Anfangsgeschwindigkeit V_0 die Geschwindigkeit bei Abschalten des Antriebes (Brennschluss).

Durch eine Funkanlage, die den Doppler-Effekt ausnutzte, wurde die Geschwindigkeit des A4-Gerätes ermittelt. Sobald die richtige Geschwindigkeit erreicht war, wurde durch Funkkommando »Brennschluss« gegeben. Die Genauigkeit betrug +/- 5 km in der Schussweite von 275 km.

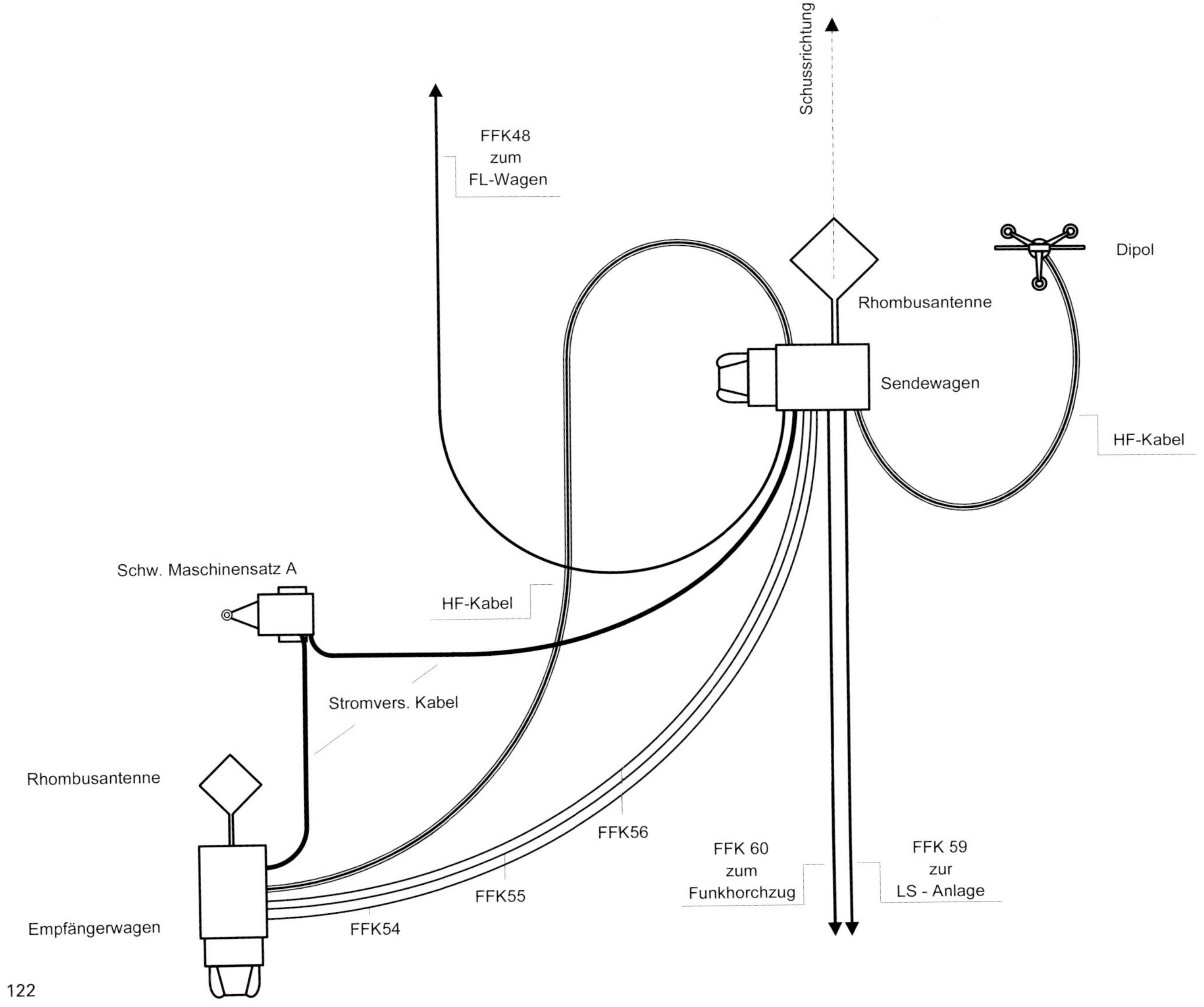

Aufbau der BS-Bodenanlage einer FR.-Batterie

DMA, GD639.5.18
HAP11/T 507
Karlshagen, 1.6.1944
Bd.Nr. E 3195/44 g
Interne oder frühere Bezeichnung

DMA, GD639.5.1
T507
Fz.-Pos.-Nr. 201

DMA, GD630.02
A4 Lehrbuch
der FR-Schule
Mai 1944
Auszug aus Kapitel III
Technische Fahrzeuge
in der Brennschluss-stellung
Ing. Gassmann

Vehicle 201

Internal or former designation: Fz.-Pos.-Nr.201
Combustion cut-off transmitter vehicle on 3 t truck closed (305)
Campina I

Intended use
In the transmitter vehicle of the type Opel-Blitz Kfz.305 LC box body different transmitters and signal modulators are installed for the encryption and transmission of the combustion cut-off command via the dipole antenna.

In the LC box body of the vehicle are
- The V_0-test transmitter, power 1 KW
- The corresponding rhombus transmitting antenna is packed in a box in folded form. The four wooden antenna masts (two pieces) have a length of 3,50 m, the four antenna sides have a length of 10 m*.
- The V_0-calibrating transmitter with normal frequency auxilary transmitter, power 100 W
- The signal modulator and phantom signal modulator for generating the audio frequency or encrypting it.
- The signal transmitter, power 800 W, which sends the encrypted combustion cut-off signal to the A4 *Gerät*. The corresponding dipole antenna is located in the dipole trailer.

The transmitter vehicle has the following dimensions and weights:

Total length:	6,020 mm
Total width:	2,100 mm
Total height:	2,930 mm
Wheelbase	3,450 mm
Track width:	1,630/1,642 mm front/rear
Tare weight:	3,275 kg
Payload:	2,825 kg
Maximum weight :	6,100 kg
Service weight:	5,030 kg
Wheels:	190-20

The main parts of the transmitter vehicle are:
- Chassis
- Closed superstructure (305) LC box body
- Installations – transmitter for combustion cut-off

The chassis
The chassis used is that of the 3 t Opel 6700 Type A truck (equipment description according to D 666/5) but the normal chassis of the Opel 3 t Type 3.6/36 S truck (equipment description according to D 669/19B) was also used.

* BS = combustion cut-off

Fahrzeug 201

Interne oder frühere Bezeichnung: Fz.-Pos.-Nr.201
BS-Sendewagen auf Lkw 3 t geschl. (305)
Campina I

Verwendungszweck
Im Sendewagen vom Typ Opel-Blitz Kfz.305 LC-Koffer sind verschiedene Sende- und Kommandogeber zur Verschlüsselung und Übertragung des Brennschlusskommandos über die Dipolantenne installiert.

Im LC-Koffer des Fahrzeuges befinden sich:
- Der V_0-Maß-Sender, Leistung 1 KW.
- Die zugehörige Rhombus-Sende-Antenne. Sie wird in zusammengelegter Form in einer Kiste verpackt mitgeführt. Die vier hölzernen Antennenmaste (2teilig) haben eine Länge von 3,50 m, die vier Antennenseiten haben eine Länge von 10 m*.
- Der V_0-Vermessungssender, Leistung 100 W.
- Der Kommandogeber und Tarngeber zur Erzeugung der Tonfrequenz bzw. deren Verschlüsselung.
- Der Kommandosender, Leistung 800 W, der das verschlüsselte Brennschlusskommando an das A4-Gerät sendet. Die dazu gehörige Dipolantenne befindet sich im Dipolanhänger.

Der Sendewagen hat folgende Maße und Gewichte:

Gesamtlänge:	6.020 mm
Gesamtbreite:	2.100 mm
Gesamthöhe:	2.930 mm
Radstand	3.450 mm
Spurweite:	1.630/1.642 mm vorne/hinten
Leergewicht:	3.275 kg
Nutzlast:	2.825 kg
Zul. Höchstgew.	6.100 kg
Gefechtsgewicht:	5.030 kg gewogen
Räder:	190-20

Die Hauptteile des Sendewagens sind:
- Fahrgestell
- Geschlossener Aufbau (305) LC-Koffer
- Einbauten/Sender für Brennschluss-Kommando

Das Fahrgestell
Als Fahrgestell dient das Fahrgestell des Lastkraftwagen 3 t Opel 6700 Typ A (Gerätebeschreibung nach D 666/5). Aber auch das normale Fahrgestell des Opel 3 t Lastkraftwagens Typ 3,6/36 S (Gerätebeschreibung nach D 669/19B) wurde verbaut.

* BS = Brennschluss

The superstructure – LC box body (305)

A Kfz.305 was used as the base vehicle for the transmitter vehicle. The LC box body had been slightly modified, so in addition to the transmitter and signal modulator, an EGO III oven for heating the interior had been installed. For heating operation the smoke exhaust was introduced and connected through the roof. For interior ventilation, ventilation hoods were mounted on the roof and ventilation grills in the side walls.

Der Aufbau – LC-Koffer (305)

Als Basisfahrzeug für den Sendewagen wurde ein Kfz.305 verwendet. Am LC-Koffer wurden geringfügige Änderungen vorgenommen. So wurde neben den Einbauten für Sende- und Kommandogeber auch ein Ofen der Bauart EGO III zur Innenraumbeheizung eingebaut. Für den Heizbetrieb wurde der Rauchabzug über das Dach eingeführt und angeschlossen. Zur Innenraumbelüftung waren Entlüftungshauben auf dem Dach und Lüftungsgitter in den Seitenwänden montiert.

BArch, RH8II Bild-B1402-44, o.Ang.

123

Deutsches Museum, München, Archiv- FA014/B376/44 TD5224 [CD74744]

124

124 Behind the front flap there is an access with a cover for power, FF and grounding cable connections in the lower part of the side wall. Above the wheel arch on both sides are ventilation grills for interior ventilation. Above, between the two window frames, there is an inlet for the dipol cable(s), below that the fixing screws for cable holders. Ventilation hoods are mounted on the roof to the left and right side of the gallery.
.

124 Hinter der vorderen Klappe befindet sich im unteren Teil der Seitenwand ein Zugang mit Deckel für Energie-, FF- und Erdungskabelanschlüsse. Oberhalb des Radkastens ist beidseitig ein Lüftungsgitter zur Innenraumbelüftung eingesetzt. Zwischen den beiden Fensterrahmen befindet sich oben ein Einlass für die Dipolkabel, darunter die Befestigungsschrauben für Kabelhalter. Auf dem Dach links und rechts der Galerie ist jeweils eine Lüftungshaube montiert.

125

125 At the front of the closed superstructure are mounted porcelain insulators as connections for the auxiliary dipole and for for surveying antennas. The standing area on the roof of the cabin can be reached via the access steps on the front wall. Between the lower and middle step there is a support for the antenna rods for the Rhombus antenna.

125 An der Vorderseite des geschlossenen Aufbaus sind Porzellanisolatoren als Anschlüsse für den Hilfsdipol und für Vermessungsantennen angebracht. Der Stehbereich auf dem Kabinendach ist über die klappbare Trittstufen an der Frontwand erreichbar. Zwischen der unteren und mittleren Trittstufe befindet sich eine Halterung für die Antennenstäbe der Rhombus-Antenne.

The installations – Transmitter for combustion cut-off signal

Fz.201 Side view in driving direction left. The open flaps on the LC box body give a view of the installations. Behind the front flap are the installation (from bottom to top) for battery box, phantom modulator and charging rectifier. The opened rear flap shows the rack with the power supplies for the signal transmitter.

Die Einbauten – Sender für Brennschluss Kommando

Fz.201 Seitenansicht in Fahrtrichtung links. Die geöffneten Klappen am LC-Koffer geben einen Blick auf die Einbauten. Hinter der vorderen Klappe befinden sich die Einbauten (von unten nach oben) für Batteriekasten, Tarngeber und Ladegleichrichter. Die geöffnete hintere Klappe zeigt das Gestell mit den Netzteilen für den Kommandosender.

Deutsches Museum, München, Archiv- FA014/B911/44 T5224 [CD74723]

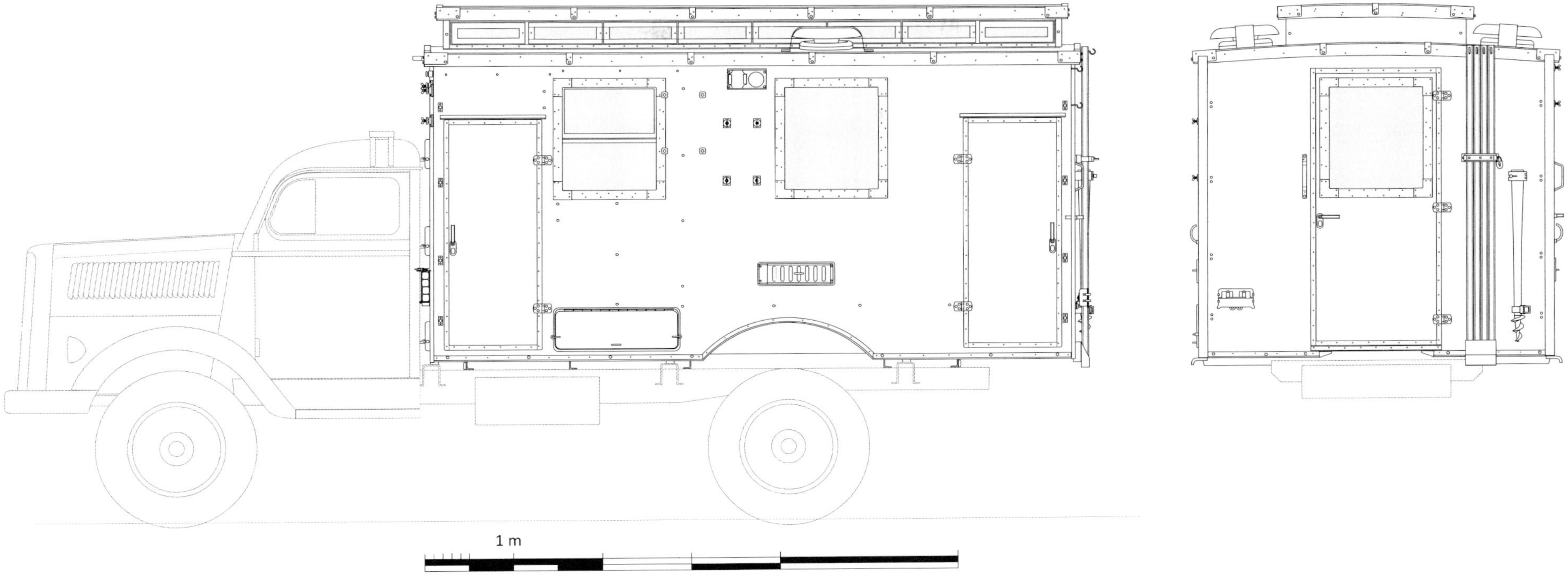

Scale / Maßstab 1:35 © Volker Ruff

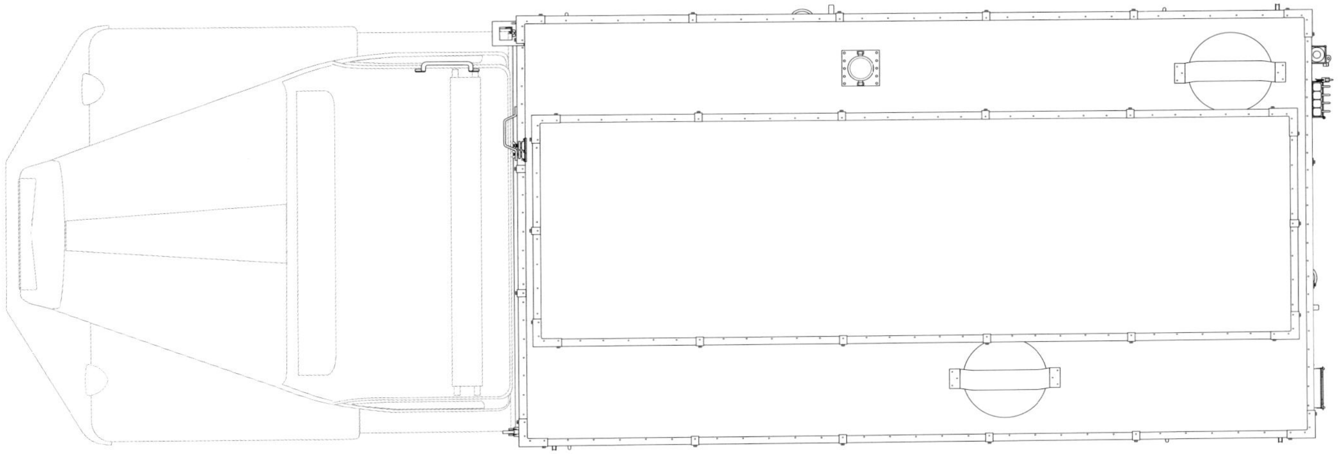

Fahrzeug 201

BS-Sendewagen auf Lkw 3 t geschl. (305)

Partially used drawings acc. to group list 21Gr362

Teilweise verwendete Zeichnungen nach Gruppenliste 21Gr362

Deutsches Museum, München, Archiv- FA014/B418/44 TD5224 [CD74763]

Interior view through the rear door

- In front of the rear wall is a V_0 transmitter first cabinet mounted, consisting of: Control panel, auxiliary transmitter; switch and control panel, power stage, empty compartment, control stage and fuse panel.
- Relay box to the right, signal modulator.
- To the left of it is the rectifier
  On the right side wall above the wheel arch is a frame, the V_0 calibrating transmitter mounted, consisting of:
- Power supply unit, with automatic circuit breaker for mains voltage and control unit with indicator lamps, special power supply unit and above it the output stage.
- In front of it is an EGO III stove with an, access door
  On the left side wall above the wheel arch is a rack for the signal transmitter, consisting of:
- Power supply unit, with automatic circuit breaker for main voltage, and control unit with pilot lamps, control stage with thermostat, modulation and power stage and above it the power amplifier.
  Above this is followed by a box for frequency means and key means.
- Cut-off signal control receiver, Oemig converter, TF devices b (TF1b), transmission device b, flap cabinet with 10 lines, field telephone, and phantom modulator.

Innenansicht durch die hintere Tür

Vor der Rückwand ist ein V_0-Meßsender-Vorgestell montiert, bestehend aus:

- Bedienungstafel, Hilfssender, Schalt- und Kontrollfeld, Leistungsstufe, Leerfach, Steuerstufe und Sicherungsfeld.
- Rechts daneben Relaiskasten, Kommandogeber.
- Links daneben der Ladegleichrichter.
  An der rechten Seitenwand über den Radkasten ist ein Gestell, das V_0-Meßsender-Endgestell, montiert. Es besteht aus:
- Netzanschlussgerät, mit Sicherungsautomaten für Netzspannung und Bedienungsgerät mit Kontrolllampen, Sondernetzgerät und darüber die Endstufe.
- Davor ein Ofen Bauart EGO III, Zugangstüre.

An der Linken Seitenwand über dem Radkasten ist ein Gestell für den Kommando-Sender montiert, bestehend aus:

- Netzanschlussgerät, mit Sicherungsautomaten für Netzspannung und Bedienungsgerät mit Kontrolllampen, Steuerstufe mit Thermostat, Modulations- und Leistungsstufe und darüber die Endstufe.

Es folgt ein Kasten für Frequenzmittel und Schlüsselmittel, darauf aufgebaut:

- Kontroll-Kommandoempfänger, Oemig-Umformer, TF-Geräte b (TF1b), Übertragungsgerät b, Klappenschrank mit zehn Leitungen, Feldtelefon, und Tarngeber.

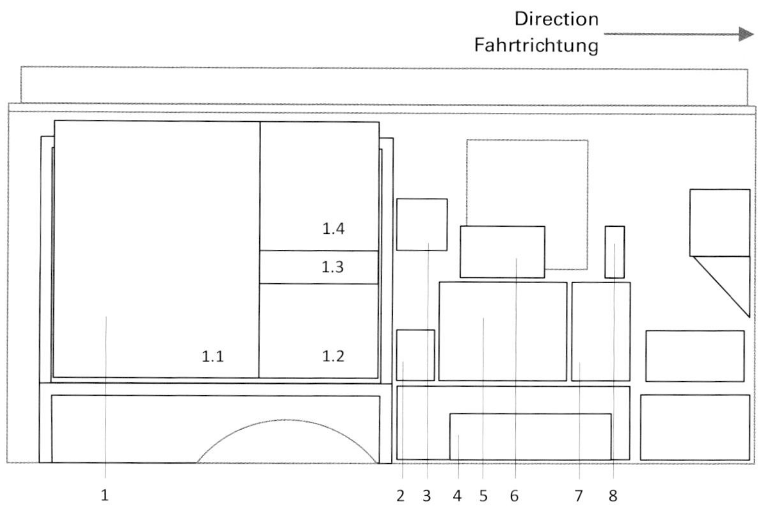

View left side wall
Ansicht linke Seitenwand

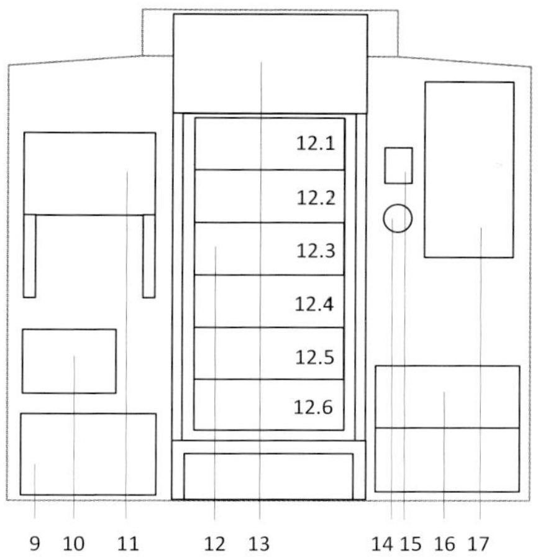

View on back wall
Ansicht auf Rückwand

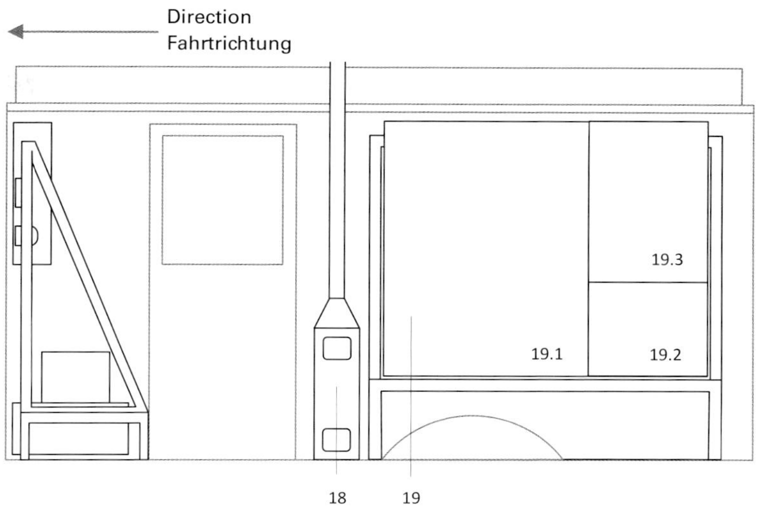

View right side wall
Ansicht rechte Seitenwand

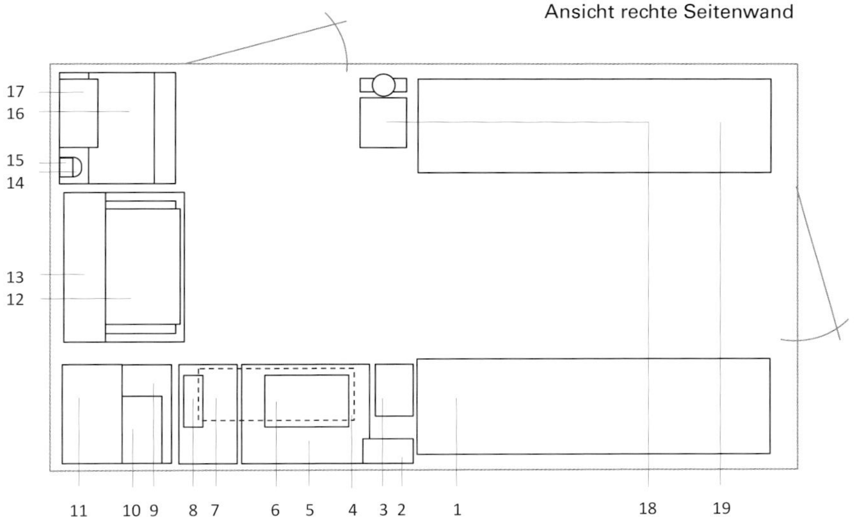

Top view
Ansicht von oben

1 Rack for the signal transmitter
2 Control command receiver
3 Oemig-converter
4 Box for frequency and key means
5 TF device TF1b
6 Hinged cabinet with 10 lines
7 Transmission device B
8 Field phone
9 Battery box
10 Phantom modulator
11 Charging rectifier
12 V_0-Transmitter first cabinet
13 Control panel
14 Antenna modulator
15 Antenna switch for auxiliary transmitter
16 Signal modulator
17 Relay box
18 Stove
19 V_0-Calibrating transmitter

1 Rack for the signal transmitter I
- 1.1 Power supply unit
- 1.2 Control stage with Thermostat
- 1.3 Modulation + power stage
- 1.4 Power amplifier

12 V_0-Transmitter first cabinet
- 12.1 Auxiliary transmitter
- 12.2 Switch and control panel
- 12.3 Power stage
- 12.4 Empty drawer
- 12.5 Control stage
- 12.6 Fuse panel

19 V_0-Calibrating transmitter
- 19.1 Power supply unit
- 19.2 Special power supply unit
- 19.3 Output stage

1 Kommandosender I mit Gestell
2 Kontroll-Kommandoempfänger
3 Oemig-Umformer
4 Kasten für Frequenz- / Schlüsselmittel
5 TF-Gerät b
6 Klappenschrank für 10 Leitungen
7 Übertragungsgerät B
8 FF-Telefon
9 Batteriekasten
10 Tarngeber
11 Ladegleichrichter
12 V_0-Messsender Vorgestell
13 Bedienungstafel
14 Antennenanpassung
15 Antennenumschalter für Hilfssender
16 Kommandogeber
17 Relaiskasten
18 Heizofen
19 V_0-Messsender Endgestell

1 Kommandosender I mit Gestell
- 1.1 Netzanschlussgerät
- 1.2 Steuerstufe mit Thermostat
- 1.3 Modulation + Leistungsstufe
- 1.4 Endstufe

12 V_0-Messsender Vorgestell
- 12.1 Hilfssender
- 12.2 Schalt- Kontrollfeld
- 12.3 Leitungsstufe
- 12.4 Leerfach
- 12.5 Steuerstufe
- 12.6 Sicherungsfeld

19 V_0-Messsender Endgestell
- 19.1 Netzanschlussgerät
- 19.2 Sondernetzgerät
- 19.3 Endstufe

Fahrzeug 201
BS-Sendewagen
auf Lkw 3 t geschl. (305)

Loading plan
Beladeplan

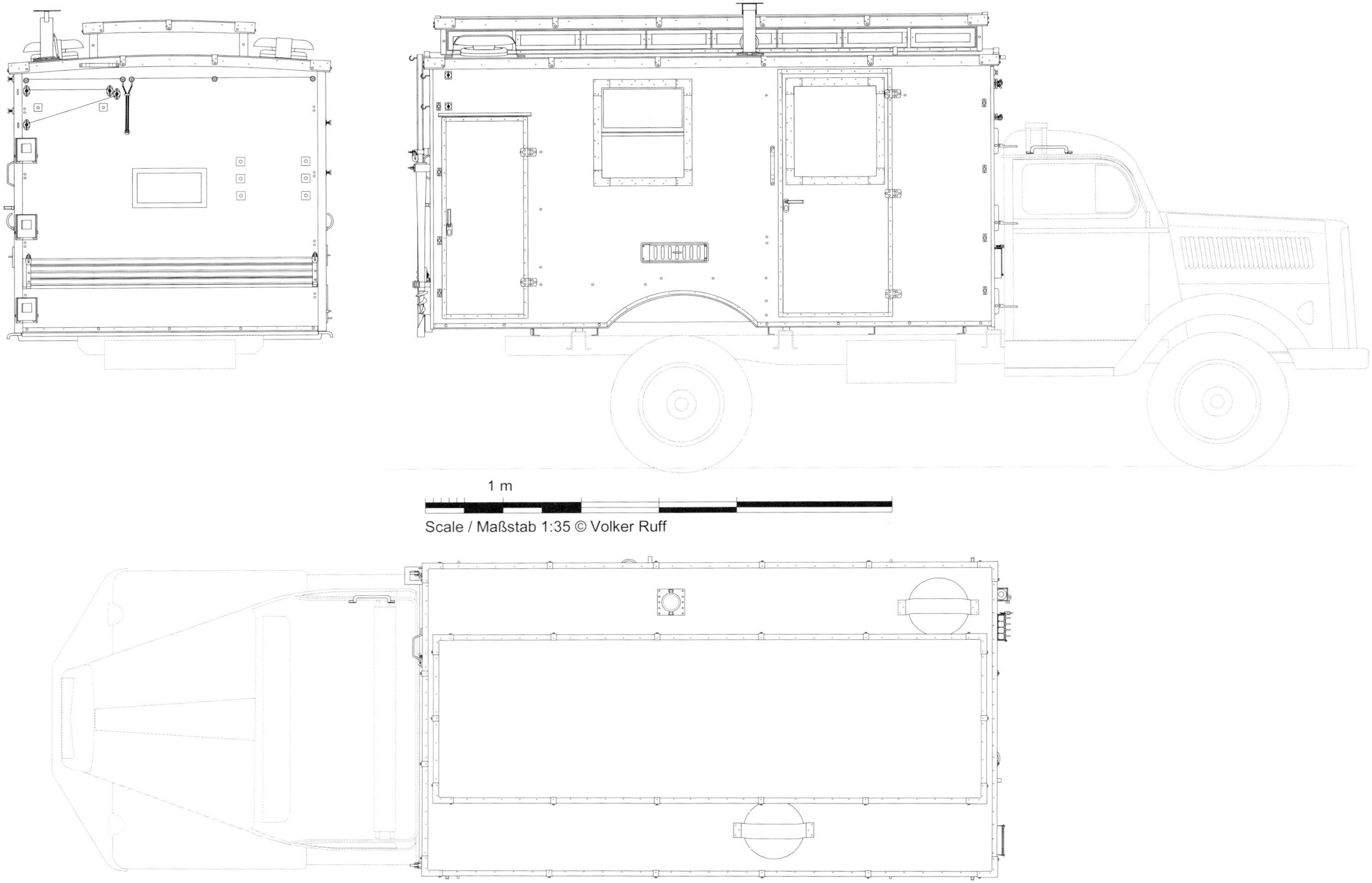

Fahrzeug 201
BS-Sendewagen
auf Lkw 3 t geschl. (305)

Partially used drawings acc. to group list 21Gr362

Teilweise verwendete Zeichnungen nach Gruppenliste 21Gr362

Deutsches Museum, München, Archiv- FA014/B945/44 T5224 [CD74713] Ausschnitt

128

128 The side view clearly shows the changes made, they are well visible: additional ventilation sliders on the side, insulators for the auxiliary antenna are mounted on the LC box body, front, side and rear wall. The color scheme is dark yellow according to the pattern. It has no camouflage. Weathering marks and paint peeling are pronounced to varying effect, e.g., on the frame rails around the windows and door.

128 In der Seitenansicht sind die ausgeführten Änderungen gut zu erkennen: Zusätzliche Lüftungsschieber sind seitlich, Isolatoren für die Hilfsantenne am LC-Koffer, vorne, seitlich und an der Rückwand montiert. Die Farbgebung ist dunkelgelb nach Muster. Es besitzt keine Tarnung. Unterschiedlich stark sind Verwitterungsspuren und Farbablösungen ausgeprägt, z.B. an den Rahmenleisten um Fenster und Tür.

DMA, GD639.5.18
HAP11/T 507
Karlshagen, 1. 6.1944

DMA, GD639.5.1
T507
Fz.-Pos.-Nr. 202

DMA, GD630.02
A4 Lehrbuch
der FR-Schule
Mai 1944
Auszug aus Kapitel III
Technische Fahrzeuge
in der Brennschluss-
stellung
Ing. Gassmann

Vehicle 202

Internal or former designation: Fz.-Pos.-Nr.202
Combustion cut-off receiver vehicle on 3 t truck closed (305)
Campina II

Intended use
The combustion cut-off receiver vehicle of the Opel Blitz Kfz.305 LC box body is equipped with a V_0-measuring rack which contains the V_0-calibrating transmitter and other special instruments. The radio signals of the *Gerät* are acquired via the Rhombus antenna with connected receiver and passed on to the transmitter vehicle for processing. The power supply cable reel and FF cable reel are stowed in a rack and used for communication and power supply with Fz.201.

In the LC box body of the vehicle are
- V_0-calibrating transmitter with corresponding resonance amplifier, Wien-Robinson Bridge, zero gain, control gain ring modulator, bridge relay, audio frequency oscilliator for testing purposes and main connecting unit.
- The associated rhombus receiving antenna is carried folded in a box.
- Signal panel with control loudspeaker for checking the entire system and a rack for power supply and FF cable reel.

The receiver vehicle has the following dimensions and weights:

Total length:	6,020 mm
Total width:	2,100 mm
Total height:	2,930 mm
Wheelbase	3,450 mm
Track width:	1,630/1,642 mm front/rear
Tare weight:	3,275 kg
Payload:	2,825 kg
Maximum weight :	6,100 kg
Service weight:	4,780 kg
Wheels:	190-20

The main parts of the evaluation vehicle are:
- Chassis
- Closed superstructure (305) LC box body
- Installations – reveiver, power supply and FF cable reel

The chassis
The chassis used is that of the 3 t Opel 6700 Type A truck (equipment description according to D 666/5) but the normal chassis of the Opel 3 t Type 3.6/36 S truck (equipment description according to D 669/19B) was also used.

* BS = combustion cut-off

Fahrzeug 202

Interne oder frühere Bezeichnung: Fz.-Pos.-Nr.202
BS-Empfangswagen auf Lkw 3 t geschl. (305)
Campina II

Verwendungszweck
Der Empfangswagen vom Typ Opel-Blitz Kfz.305 LC-Koffer ist ausgerüstet mit einem V_0-Meß-Gestell, das den V_0-Meß-Empfänger und weitere Sondergeräte enthält. Über die Rhombusantenne mit angeschlossenem Empfänger werden die Funksignale des Gerätes erfasst und an den BS-Sendewagen zur Verarbeitung weitergegeben. In einem Gestell sind Stromversorgungskabeltrommel und FF-Kabeltrommel verstaut, die zur Kommunikation und Energieversorgung mit Fz.201 Verwendung finden.

Im LC-Koffer des Fahrzeuges befinden sich:
- V_0-Meß-Empfänger mit dazugehörigen Resonanzverstärker, Wien-Robinson-Brücke, Nullverstärker, Steuerverstärker Ringmodulator, Brückenrelais, Tongenerator für Prüfzwecke und Netzteil.
- Zugehörige Rhombus-Empfangsantenne, die zusammengelegt in einer Kiste mitgeführt wird.
- Signaltafel mit Kontrolllautsprecher zum Überprüfen der gesamten Anlage, und Gestell für Stromversorgung- und FF-Kabeltrommel.

Der Empfangswagen hat folgende Maße und Gewichte:

Gesamtlänge:	6.020 mm
Gesamtbreite:	2.100 mm
Gesamthöhe:	2.930 mm
Radstand	3.450 mm
Spurweite:	1.630/1.642 mm vorne/hinten
Leergewicht:	3.275 kg
Nutzlast:	2.825 kg
Zul.Höchstgew.	6.100 kg
Gefechtsgewicht:	4.780 kg gewogen
Räder:	190-20

Die Hauptteile des Empfangswagens sind:
- Fahrgestell
- Geschlossener Aufbau (305) LC-Koffer
- Einbauten/Empfänger, Stromversorgung- und FF-Kabeltrommel

Das Fahrgestell
Als Fahrgestell dient das Fahrgestelle des Lastkraftwagen 3 t Opel 6700 Typ A (Gerätebeschreibung nach D 666/5). Aber auch das normale Fahrgestell des Opel 3 t Lastkraftwagens Typ 3,6/36 S (Gerätebeschreibung nach D 669/19B) wurde verbaut.

* BS = Brennschluss

The closed superstructure (305) – LC box body

A Kfz.305 was used as the basic vehicle for the receiver vehicle. The vehicle shown here had a LC box body without gallery roof, but box bodies with gallery roof were also used. The LC box body had been slightly modified. Beside the installations for the measuring device and the rack for cable reels there was also a EGO III stove for heating the interior. For heating operation, the smoke outlet was connected through the roof.

Der geschlossene Aufbau (305) – LC-Koffer

Als Basisfahrzeug für den Empfangswagen wurde ein Kfz.305 verwendet. Das hier gezeigte Fahrzeug hatte einen LC-Koffer ohne Oberlichter. Es wurden aber auch geschl. Aufbauten mit Laufgangdach verwendet. Der LC-Aufbau wurde leicht modifiziert. Neben den Einbauten für das Meß-Gestell und dem Gestell für Kabeltrommeln gab es auch einen EGO III Ofen zur Beheizung des Innenraums. Für den Heizbetrieb wurde der Rauchabzug über das Dach angeschlossen.

BArch, RH8II Bild-B1403-44, o.Ang.

129

Built-in/receiver, power supply and FF cable reel

A luggage box was mounted on the roof of the Fz.202. Behind the opened flap, a battery box with battery was installed and, as with the Fz.201, there was also a connection box for the power, FF cable in the lower part in front of the wheel house of the side wall.

Einbauten/Empfänger, Stromversorgung- und FF-Kabeltrommel

Auf dem Dach des Fz.202 war eine Gepäckkasten montiert. Hinter der geöffneten Klappe war ein Batteriekasten mit Batterie eingebaut, und wie beim Fz.201 befand sich auch ein Anschlusskasten für das Strom-, FF-Kabel vor dem Radkasten im unteren Teil der Seitenwand.

130

130 The racks for the installations were secured through the outer wall with carriage bolts and repainted in color. The chimney is part of a EGO III heating stove.

130 Die Gestelle für die Installationen wurden durch die Außenwand mit Wagenschrauben gesichert und farblich neu lackiert. Der Schornstein war Teil eines EGO III-Heizofens.

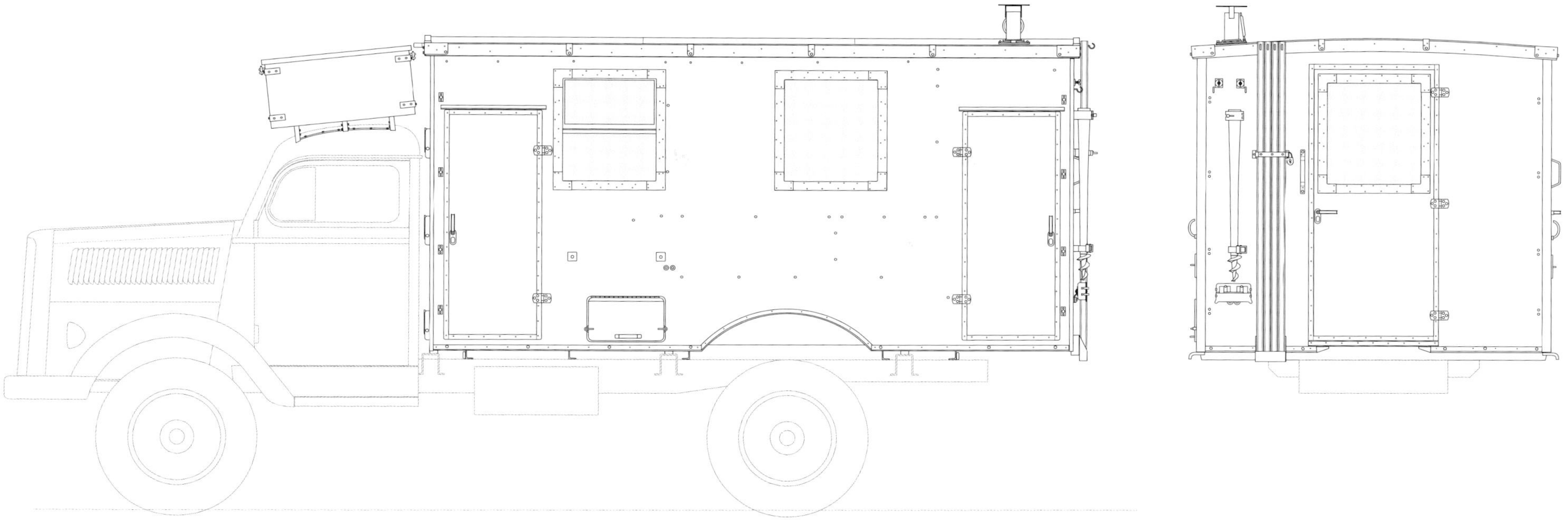

1 m

Scale / Maßstab 1:35 © Volker Ruff

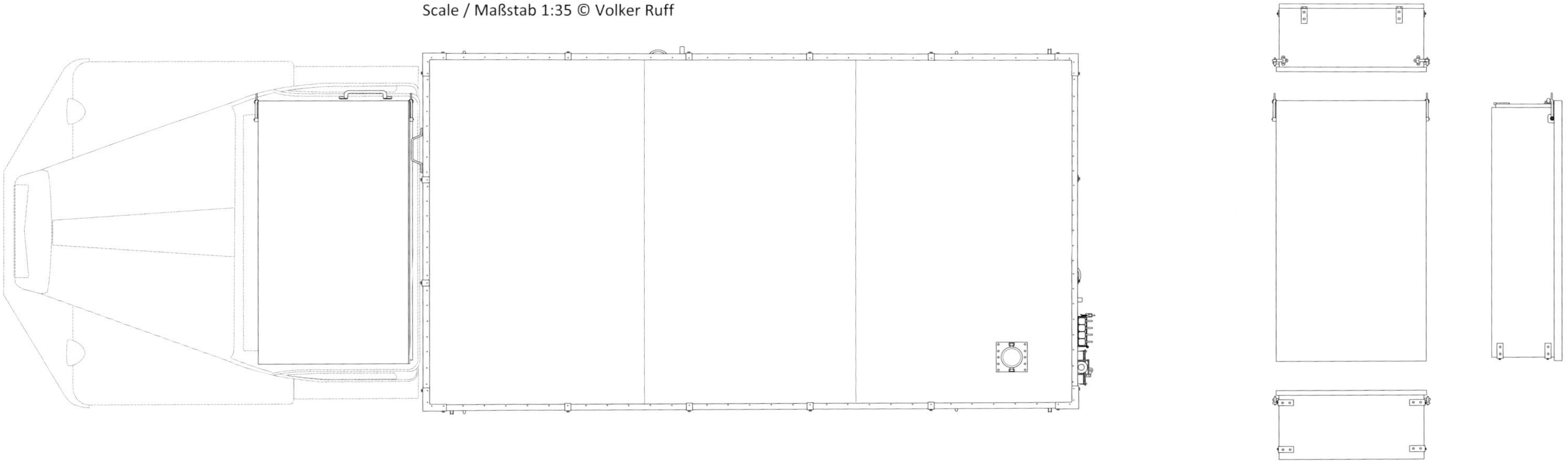

Fahrzeug 202

BS-Empfangswagen
auf Lkw 3 t geschl. (305)

Partially used drawings acc. to group list 21Gr363

Teilweise verwendete Zeichnungen nach Gruppenliste 21Gr363

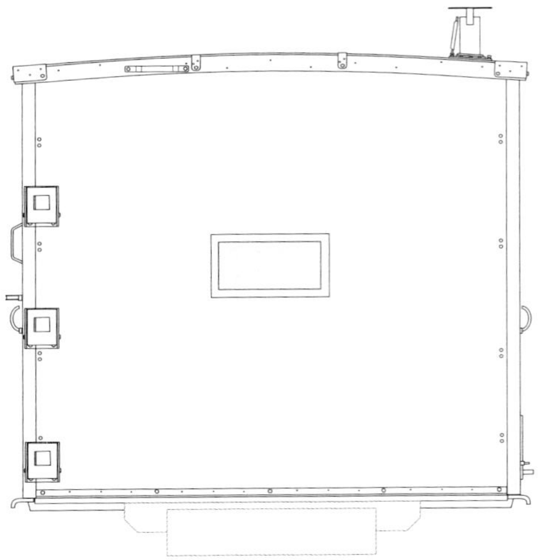

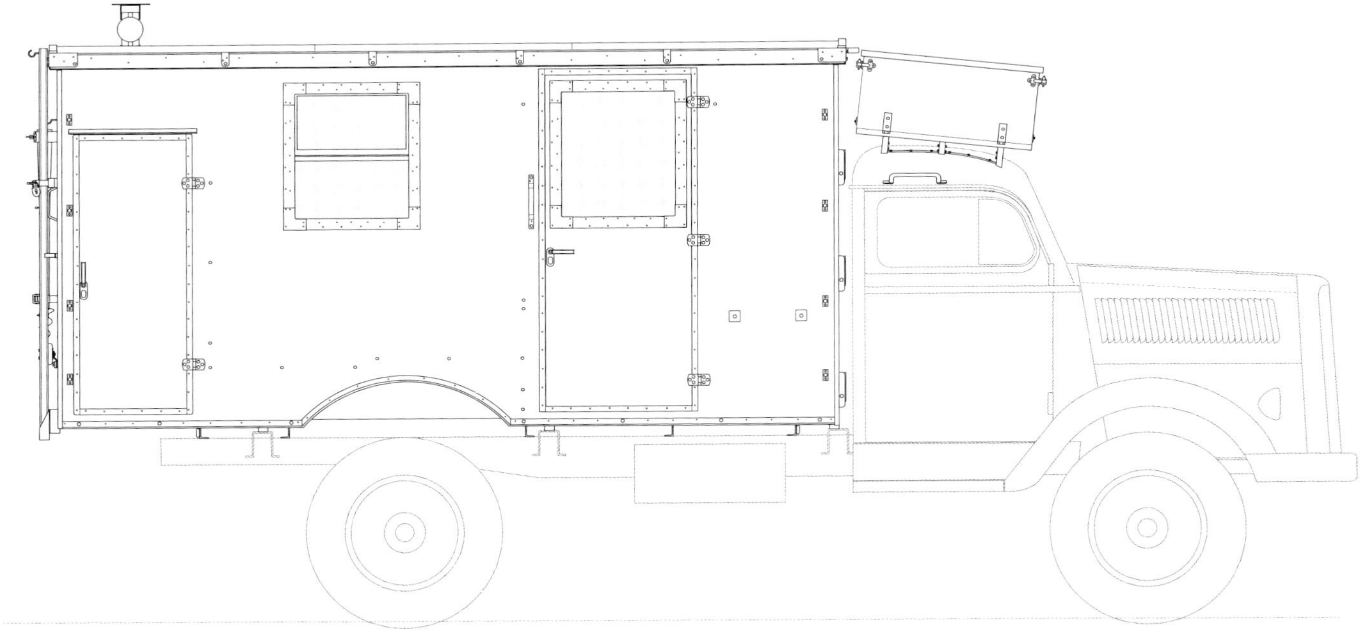

1 m

Scale / Maßstab 1:35 © Volker Ruff

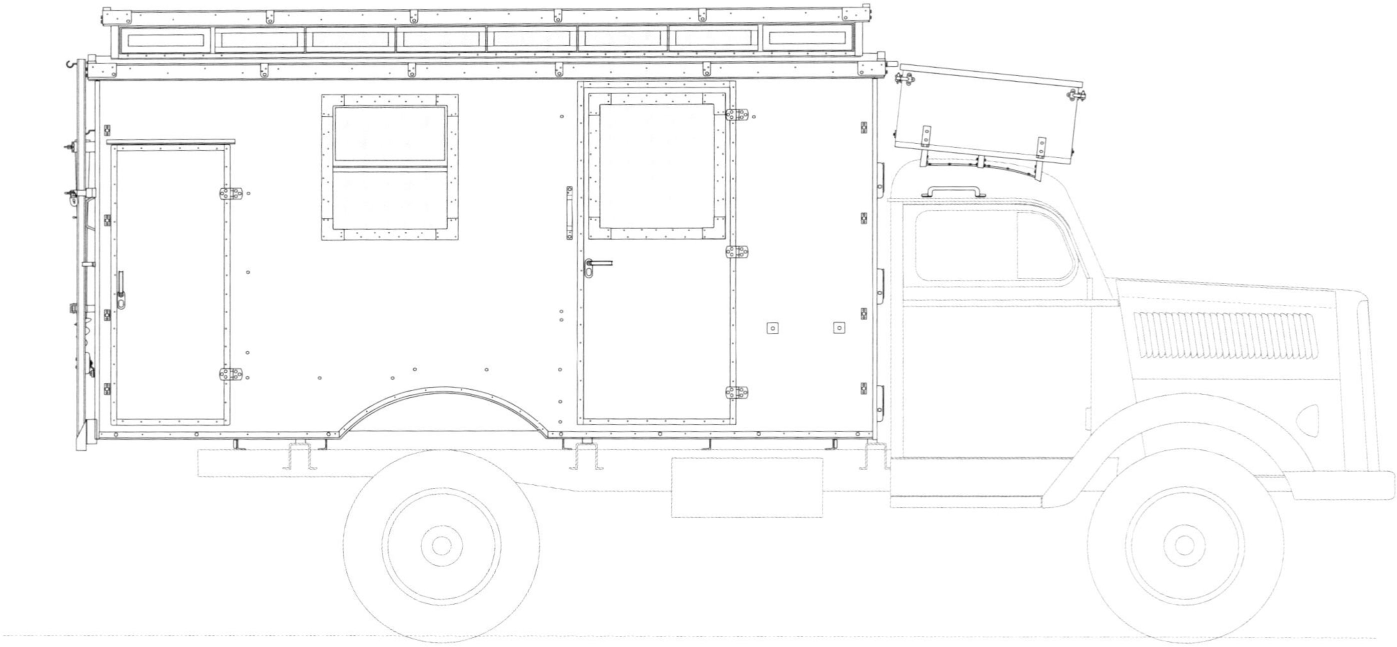

Fahrzeug 202

BS-Empfangswagen auf Lkw 3 t geschl. (305)

Partially used drawings acc. to group list 21Gr363

Teilweise verwendete Zeichnungen nach Gruppenliste 21Gr363

131

131 The LC box body on this vehicle 202 has a gallery roof and is mounted on a 3 t Opel 6700 type A chassis. The lettering on the driver's door for loading weights is conspicuously large compared to the usual loading class lettering. When looking at the different vehicle cabs, one will notice the different curved and different heights of the roofs; accordingly, the mounted storage boxes were adapted to the superstructure.

131 Der LC-Koffer an diesem Fahrzeug 202 hat ein Laufgangdach und ist auf einen 3 t Opel 6700 Typ A Fahrgestell aufgesetzt. Die Beschriftung an der Fahrertür zu den Beladegewichten ist verglichen mit der sonst üblichen Beladeklassenbeschriftung auffällig groß. Bei Betrachtung der verschiedenen Fahrzeugkabinen fallen die unterschiedlich gewölbten und ungleichen Höhen der Dächer auf; entsprechend wurden bei aufgesetzten Staukisten die Unterkonstruktionen angepasst.

Deutsches Museum, München, Archiv- FA014/B926/44 T5224 (CD74719)

132

BArch, RH8II Bild-B1409-44, o.Ang

133

Interior view

On the right, above the wheel arch, there is a rack for cable reels and a laying cart. Here are stowed:

- Three FF cable reels of 250 m each for the connections between the receiver vehicle and transmitter vehicle. Via these cables one call, three signals and one measuring tone are sent for checking.
- Three power supply cables. They are routed between the power supply trailer and the two vehicles in the BS position.
- Cable laying cart

On the left is an EGO III stove and above the wheel arch is a rack. The rack contains:

- The V_0-measuring rack is mounted on dampers in the equipment rack. On top a field telephone was stored for the connection via FF cable to the hinged cabinet in the BS transmitter vehicle.
- Integrated in the measuring rack are: Control panel, bridge amplifier, resonance amplifier with frequency indicator V_0-calibrating transmitter, bridge, test generator, main connected unit and two empty compartments V_0.
- On the rear wall two material crates are mounted over corner.

Innenansicht

Rechts ist über dem Radkasten ein Gestell für Kabelrollen und einem Verlegewagen vorhanden. Es sind verstaut:

- Drei FF-Kabelrollen zu je 250 m zur Verbindung zwischen Empfangswagen und Sendewagen. Über diese Kabel gehen ein Gespräch, drei Signale und ein Messton zur Überprüfung.
- Drei Stromversorgungskabel. Sie werden zwischen Stromversorgungs-Anhänger und den beiden Kraftfahrzeugen der BS-Stellung verlegt.
- Kabelverlegewagen

Links ist ein Ofen der Bauart EGO III und über dem Radkasten ein Gerätegestell montiert. Im Gestell sind enthalten:

- Das V_0-Messgestell ist im Gerätegestell auf Dämpfern montiert. Darauf ist ein Feldfernsprecher gelagert, der über FF-Kabel an den Klappenschrank im BS-Sender-KW angeschlossen ist.
- Im Meßgestell intergriert sind:
  Signalverteiler, Brückenverstärker, Resonanz-Verstärker mit Frequenzanzeiger, V_0-Messempfänger, Brücke, Prüfgenerator, Netzgerät und zwei Leerfächer V_0.
- An der Rückwand sind zwei Materialkisten über Eck montiert.

View left side wall
Ansicht linke Seitenwand

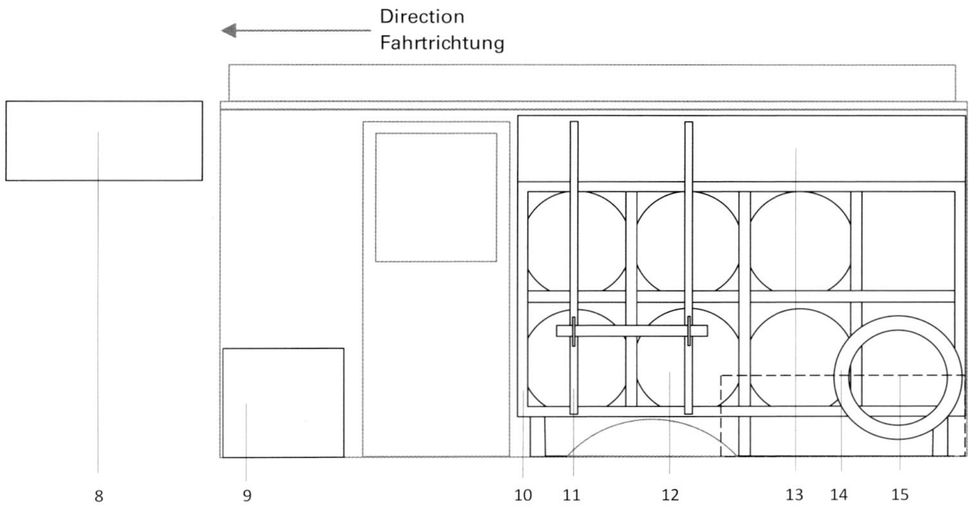

View right side wall
Ansicht rechte Seitenwand

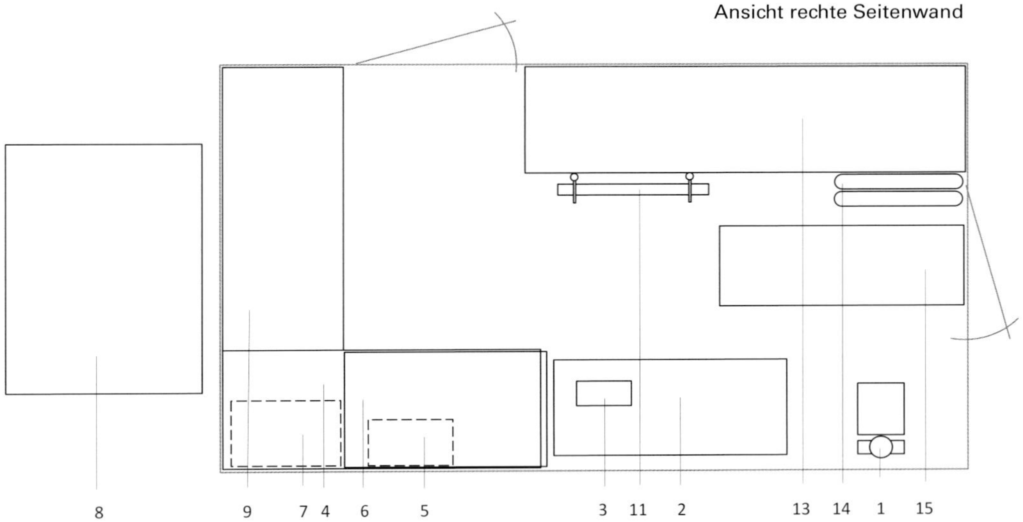

Top view
Ansicht von oben

1	Stove	1	Ofen
2	V_0-Measuring rack	2	V_0-Messgestell
3	Field phone	3	FF-Telefon
4	Storage box	4	Staukasten
5	Tabletop	5	Tischplatte
6	Terminal box	6	Anschlusskasten
7	Battery box	7	Batteriekasten

2	V_0-Measuring rack	2	V_0-Messgestell
2.1	empty	2.1	leer
2.2	V_0-Measuring receiver	2.2	V_0-Messempfänger
2.3	Resonance amplifier	2.3	Resonanzverstärker
2.4	Bridge amplifier	2.4	Brückenverstärker
2.5	Signal distributor	2.5	Signalverteiler
2.6	Bridge	2.6	Brücke
2.7	empty	2.7	leer
2.8	Test generator	2.8	Prüfgenerator
2.9	Power supply unit	2.9	Netzgerät

8	Luggage box	8	Gepäckkiste
9	Material box	9	Materialkasten
10	Rack for cable reels	10	Gestell für Kabelrollen
11	Laying cart	11	Verlegewagen
12	Cable reels	12	Kabeltrommeln
13	Stowage box	13	Staukasten
14	Wheels for laying cart	14	Räder für Verlegewagen
15	Storage box aisle	15	Staukasten Gang

Fahrzeug 202
BS-Empfangswagen
auf Lkw 3 t geschl. (305)

Loading plan
Beladeplan

Vehicle 204

Internal or former designation: Fz.-Pos.-Nr.204
Combustion cut-off antenna trailer (1-axle)
Dipole trailer

DMA, GD639.5.18
HAP11/T 507
Karlshagen, 1. 6.1944
Bd.Nr. E 3195/44 g

DMA, GD639.5.1
T507
Fz.-Pos.-Nr. 204

DMA, GD630.02
A4 Lehrbuch
der FR-Schule
Mai 1944
Auszug aus Kapitel III
Technische Fahrzeuge
in der Brennschluss-
stellung
Ing. Gassmann

Intended use

The dipole trailer is pulled by the transmitter vehicle Fz.201. It is designed as a single-axle trailer and carries the dipole antenna for the combustion cut-off signal transmitter.

The dipole antenna consists of an:

- antenna mast
- dipole head
- blocking pot (condenser)
- and two dipole arms.

Fahrzeug 204

Interne oder frühere Bezeichnung: Fz.-Pos.-Nr.204
BS-Antennen-Anhänger (1-achs.)
Dipol-Anhänger

Verwendungszweck

Der Dipol-Anhänger wird vom BS-Sender-KW gezogen. Er ist als Einachsanhänger ausgebildet und befördert die Dipolantenne für den BS-Kommandosender.

Die Dipolantenne besteht aus:

- Antennenmast
- Dipolkopf
- Sperrtopf (Kondensator)
- und zwei Dipolarmen.

134

135 For transport, the dipole is tilted over two joints onto the trailer and the dipole mast is secured to a bracket. The head is fitted with a dustcover and is also supported on the trailer. The outriggers are tilted inwards or towards the dipole base. The cable reel is protected by a barrel-shaped cylindrical cover. The dipole trailer Fz.204 is pulled by the transmitter vehicle Fz.201. The pictures shown represent the construction status of the 5th system and are essentially valid for systems 4 to 14.

135 Für den Transport wird der Dipol über zwei Gelenke auf den Anhänger gekippt und der Dipolmast an einer Halterung befestigt. Der Kopf ist mit einer Staubschutzhaube versehen und wird ebenfalls auf dem Anhänger abgestützt. Die Ausleger werden nach innen oder zur Dipolbasis hin gekippt. Die Kabeltrommel ist durch eine tonnenförmige zylindrische Abdeckung geschützt. Der Dipol-Anhänger Fz.204 wird vom Sendewagen Fz.201 gezogen. Die wiedergegebenen Bilder stellen den Bauzustand der 5. Anlage dar und sind im Wesentlichen gültig für Anlage 4 bis 14.

The dipole trailer

The dipole trailer is manufactured as a triangular-shaped frame made of U-profiles and sheet metal and is equipped with a towing eye. Large handles facilitate mobility. On the rear of the frame there is a cross beam with two mountings to accommodate the dipole base. The 5.50-16 wheels are mounted on a swing axle and bolted to the frame. Transport boxes for the highly sensitive dipole arms are mounted on the left and right of the trailer frame. The antenna cable is rolled up in the middle on the cable reel and closed with a protective cover, not shown here. The dipole mast is secured in the holder in front of the cable dreel for transport.

The dipole base is folded from sheet metal and bolted to the base assembly of the carriage. The dipole is aligned with adjusting spindles in the arms of the cradle until the plumb line and level indicate its vertical and horizontal alignment.

Der Dipol-Anhänger

Der Dipolanhänger wird als dreieckiger Rahmen aus U-Profilen und Stahlblechen gefertigt und ist mit einer Zugöse ausgestattet. Große Handgriffe erleichtern die Beweglichkeit. An der Rückseite des Rahmens befindet sich ein Querträger mit zwei Lafettenlagern zur Aufnahme des Dipolsockels. Die Räder der Größe 5,50-16 sind auf einer Schwingachse montiert und mit dem Rahmen verschraubt. Für die hochempfindlichen Dipolarme sind Transportkisten links und rechts am Anhängerrahmen montiert. Das Antennenkabel wird in der Mitte auf der Kabeltrommel aufgerollt und mit einer Schutzhaube verschlossen (hier nicht dargestellt). In der Halterung vor der Kabeltrommel wird der Dipolmast zum Transport gesichert.

Der Dipolsockel ist aus Blechen gekantet und auf der Bodengruppe der Lafette verschraubt. Der Dipol wird mit Justierspindeln in den Auslegern der Lafette ausgerichtet bis Lot und Libelle seine vertikale und horizontale Ausrichtung anzeigen.

Deutsches Museum, München, Archiv- FA014/B919/44 T5224 [CD74721]

136

Deutsches Museum, München, Archiv- FA014/B918/44 T5224 [CD74722]

137

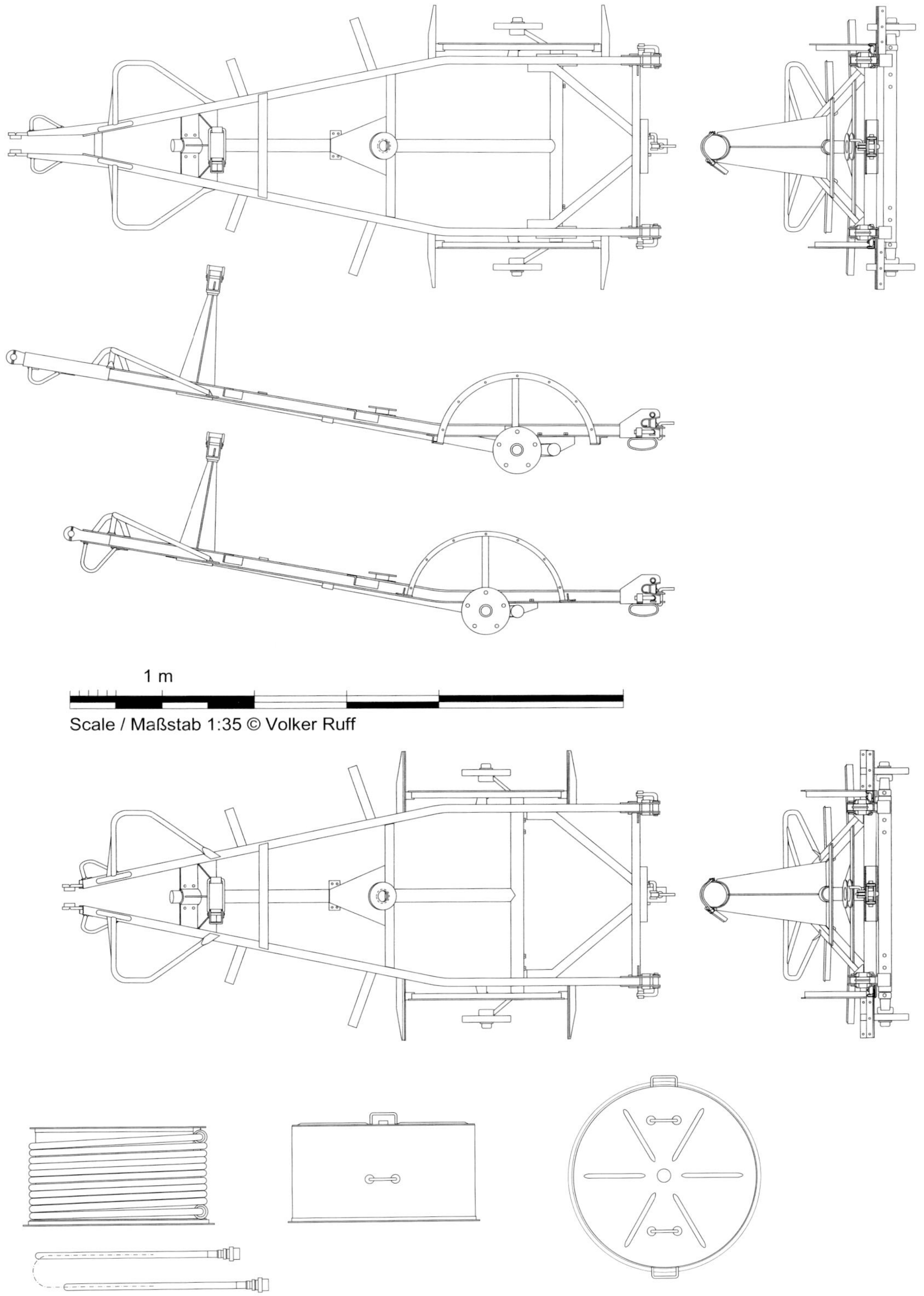

Trailer for Fz.204
Anhänger zu Fz.204

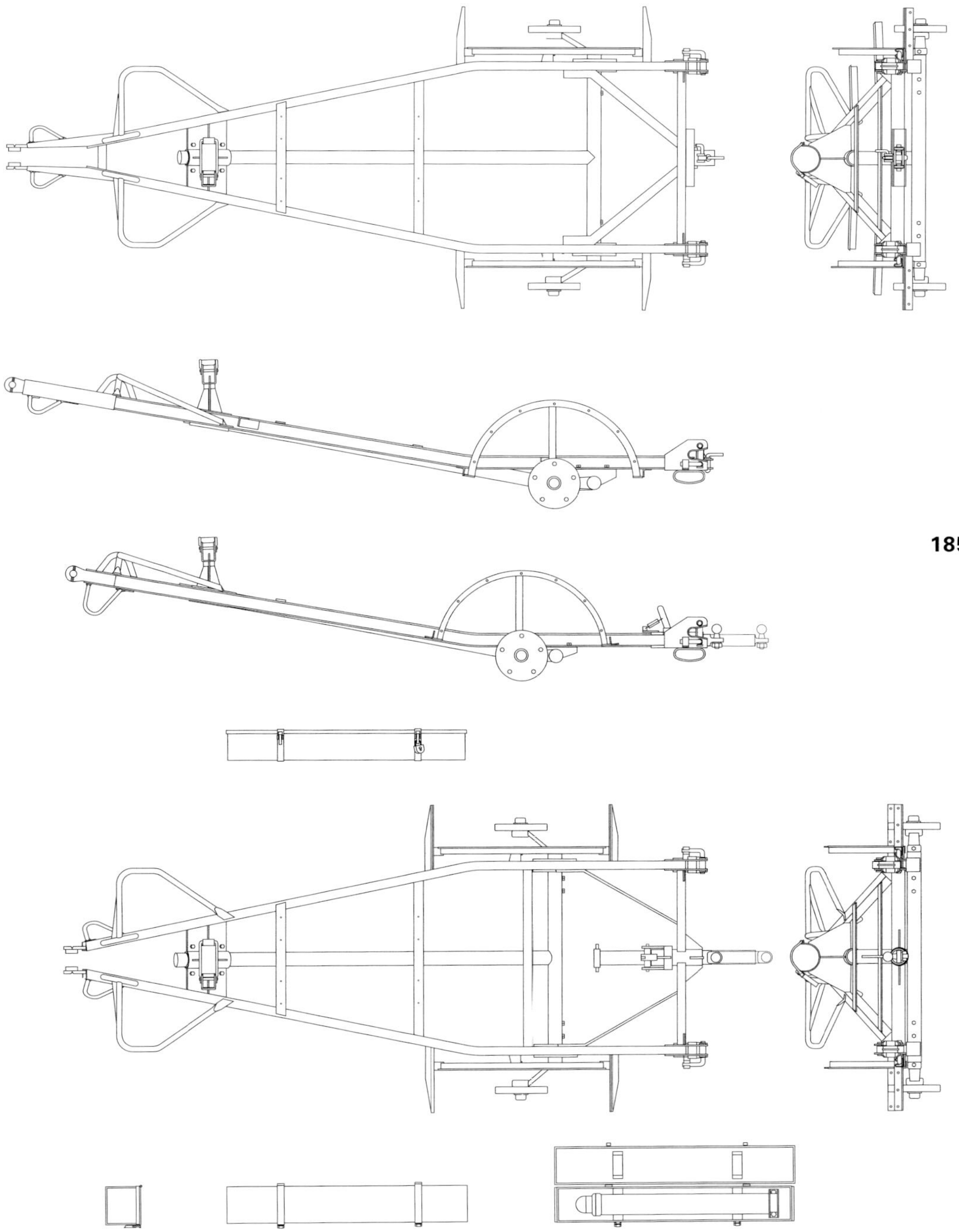

Trailer for Fz.207
Anhänger zu Fz.207

Fahrzeug 204

BS-Antennen-Anhänger
(1-achs.)

Partially used drawings acc. to group list 21Gr169

Teilweise verwendete Zeichnungen nach Gruppenliste 21Gr169

186

Deutsches Museum, München, Archiv- FA014/B917/44T5224 [CD66931]

The dipole – assembled and connected

The two dipole arms are inserted in the dipole head. The head is fixed to the dipole mast with a connection plate. The mast with swivel arm passes through the dipole base and is connected to the blocking pot. The alignment of the dipole to the firing direction is done by a swivel arm. The symmetry pot (blocking pot), which is movably mounted transversely on the front side, is equipped with a connection socket. At the front side a connection socket is connected to the HF antenna cable. The condenser at the opposite side is adjusted according to a calibration table. The dipole base is screwed to the base assembly (cradle). Three outriggers are swivel-mounted on the cradle and are lashed down with a locking wheel. Adjustment spindles with support plates enable its alignment. For transport, the dismantled dipole (without dipole arms) is picked up on the cradle bearings, turned over and lashed via the swivel bearings on the antenna trailer. The dipole from Fz.204 is identical in construction with the dipole from Fz.207.

Der Dipol – aufgebaut und angeschlossen

Die beiden Dipolarme sind im Dipolkopf eingesetzt. Das Kopfteil ist auf einer Anschlussplatte am Dipolmast befestigt. Der Mast mit Schwenkarm führt durch den Dipolsockel und ist mit dem Sperrtopf verbunden. Die Ausrichtung des Dipols zur Schussrichtung erfolgt über einen Richtschwenkarm. Der quer im Sockel beweglich gelagerte Symmetrietopf (Sperrtopf) ist an der Stirnseite mit einer Anschlussbuchse ausgerüstet. Hier wird das Antennenkabel angeschlossen. Der Kondensator an der gegenüberliegenden Stirnseite wird nach einer Kalibriertabelle eingestellt. Der Dipolsockel ist mit der Bodengruppe (Lafette) verschraubt. Drei Ausleger sind an der Lafette schwenkbar gelagert und werden mit einem Feststellrad verzurrt. Justierspindeln mit Auflagetellern ermöglichen die Ausrichtung. Für den Transport wird der abgebaute Dipol (ohne Dipolarme) an den Lafetten-Lagern aufgenommen und über die Schwenklager am Antennen-Anhänger umgelegt und verzurrt. Der Dipol des Fz.204 ist baugleich mit dem Dipol des Fz.207.

The BS-antenna trailer has the following dimensions and weights:

Total length:	~ 3,830 mm w/o dipole
Width:	1,670 mm
Height:	~ 1,375 mm, 3,200 mm dipole
Track width:	1,400 mm, 3,000 mm dia dipole lafette
Service weight:	740 kg
Wheels:	5,50-16

Der BS-Antennen-Anhänger hat folgende Maße und Gewichte:

Gesamtlänge:	~ 3.830 mm ohne Dipol
Breite:	1.670 mm
Höhe:	~ 1.375 mm, 3.200 mm Dipol
Spurweite:	1.400 mm, 3.000 mm ø Dipol-Lafette
Gefechtsgewicht:	740 kg
Räder:	5,50-16

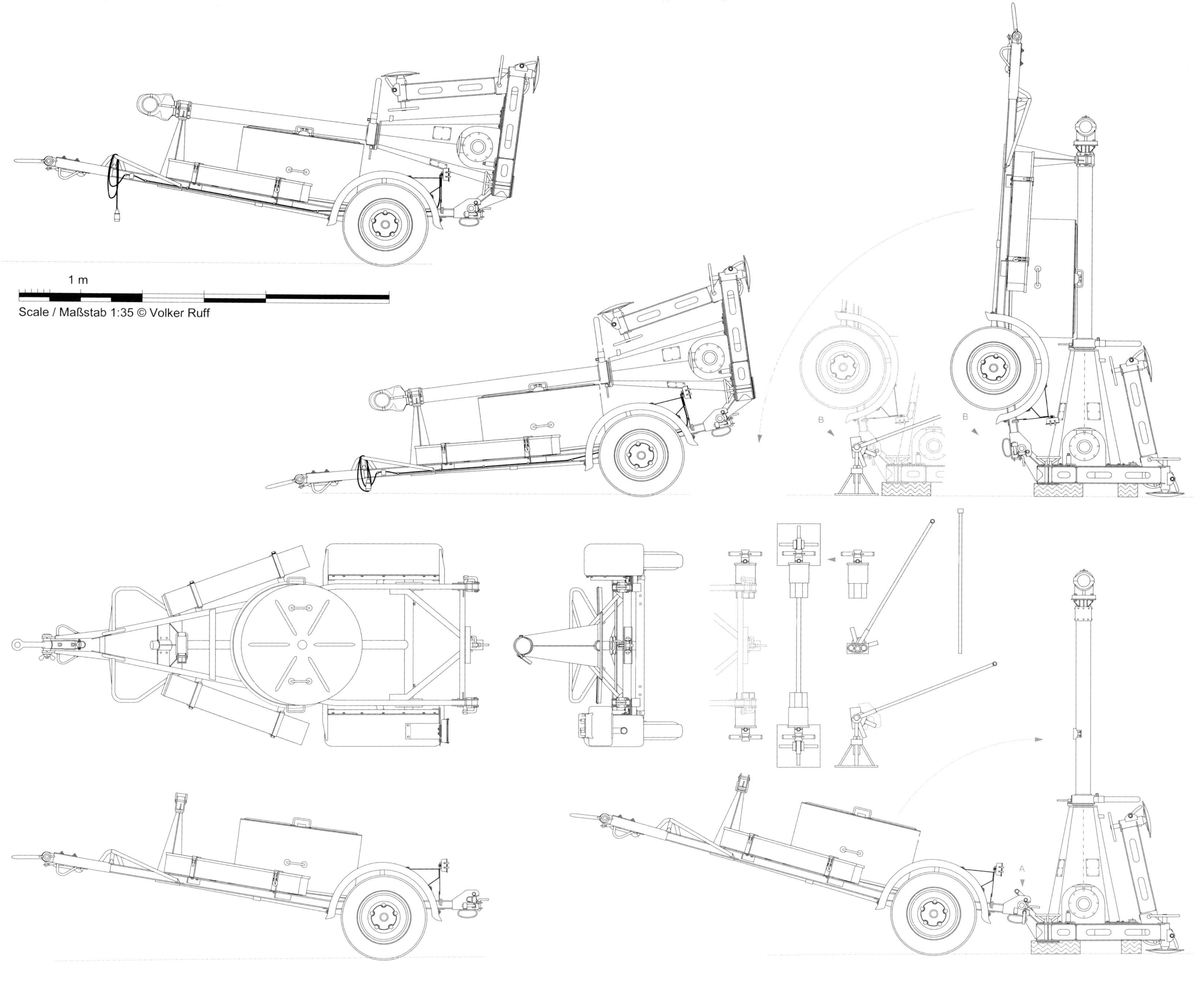

Scale / Maßstab 1:35 © Volker Ruff

Guide beam ground installation

Purpose and task
The guide beam ground installation is set up about 10 to 16 km behind the launch position. The lateral deviation of the flying A4 *Gerät* caused by wind forces is recorded with the guide beam method, modulated and transmitted to the corresponding signals for precise control to the Aggregat.

The guide beam ground installation consists of:
The transmission vehicle with LS-transmitter and lobe switching device. The equipment vehicle with HF supply cable, the two dipole trailers that are attached to the equipment vehicle and a power supply unit a trailer behind transmission vehicle.

BAMA, RH8-1310
Spezialfahrzeuge
Baugruppen und
wesentliche Bauteile
der Bodenanlage FR

Special vehicles in guide beam site
Fz.-Pos.-Nr.205 transmitter vehicle on 3 t truck closed (Maultier)
Fz.-Pos.-Nr.206 equipment vehicle on 3 t truck closed (Maultier)
Fz.-Pos.-Nr.207 antenna trailer (1 axle)
Fz.-Pos.-Nr.208 controll trailer (1 axle)

DMA, GD630.02
A4 Lehrbuch
der FR-Schule
Mai 1944
Auszug aus Kapitel II
Technische Fahrzeuge
in der LS-Stellung
Ing. Gassmann

Principle of the guide beam control
When firing, the task is to hit a target point with a projectile from a firing point. Deviations in the firing range of the shot are called length scattepoint. Deviations in the firing range of the shot are called length scattering, and those across it are called lateral scattering. Compared to an artillery projectile, the lateral deviation is about twice as large as the length deviation. In order to reduce the lateral deviation from the firing lane, a radio-technical method has been developed, the beam guiding method. Its task is to measure the lateral deviation of the missile, which is mainly caused by external forces so is not detected by the gyroscopes, and to convert the measured values into a corresponding control command.

The two dipoles (VHF transmitter system) are used to generate a guide beam plane. If the *Gerät* flies to the left or right of this LS lane, the deviation can be measured via the field strength of the transmitting dipoles. The rod antennas on fins 2 and 4 receive this pulse. These pulses are converted by on-board units and fed as control current to the steering control unit (*Mischgerät*) of the control system. In the steering control unit the control current of the guide beam system is converted into an additional control command for the jet and air rudders I and III – and the correction is initiated.

* Mischgerät = a ampflifier of five independent panels mounted on a common base plate. It tooks signals from the gyroscope and then translated into a signal that directed the servo steering mechanism and steering vans to correct for attitude deviations in powered flight.

LS-Bodenanlage

Zweck und Aufgabe
Die LS-Bodenanlage wird etwa 10 bis 16 km hinter der Abschussstellung aufgebaut. Die durch Windkräfte hervorgerufene Seitenabweichung des fliegenden A4-Gerätes wird mit dem Leitstrahlverfahren erfasst, moduliert und entsprechende Signale zur genauen Steuerung an das Aggregat übermittelt.

Die LS-Bodenanlage besteht aus:
Dem Sendewagen mit LS-Sender und Umtastgerät, dem Gerätewagen mit HF-Zuführungskabel, den beiden Dipolanhängern, die an den Gerätewagen angehängt werden und einem Stromversorgungsaggregat als Anhänger an den Sendewagen.

Spezialfahrzeuge in der LS-Stellung
Fz.-Pos.-Nr.205 LS-Sende-Kw auf Lkw 3 t geschl. (Maultier)
Fz.-Pos.-Nr.206 LS-Geräte-Kw auf Lkw 3 t geschl. (Maultier)
Fz.-Pos.-Nr.207 LS-Antennen-Anhänger (1-achs.)
Fz.-Pos.-Nr.208 LS-Kontroll-Anhänger (1-achs.)

Prinzip der LS-Steuerung
Bei einem Schuss ist die Aufgabe gestellt, mit einem Geschoss von einem Abschusspunkt aus einen Zielpunkt zu treffen. Abweichungen in der Schussweite nennt man Längenstreuung, die senkrecht dazu liegenden Abweichungen Seitenstreuung. Gegenüber einem Artilleriegeschoss ist die Seitenabweichung etwa doppelt so groß als die Längenabweichung. Um nun die Seitenabweichung von der Schussebene zu verringern hat man ein funktechnisches Verfahren entwickelt: das Leitstrahlverfahren. Es hat die Aufgabe die Seitenabweichung des Gerätes, die vorwiegend durch äußere Kräfte entstehen und durch die Kreisel nicht erfasst werden, zu messen und die Messwerte in ein entsprechendes Steuerkommando umzuwandeln.

Mit den beiden Dipolen (UKW-Sende-Anlage) wird eine Leitstrahlebene (LS-Ebene) erzeugt. Fliegt das Gerät links oder rechts von dieser LS-Ebene wird die Abweichung über die Feldstärke der sendenden Dipole messbar. Die Stabantenne an der Flosse 2 und 4 empfangen diesen Impuls. Über die Bordgeräte werden diese Impulse umgeformt und als Steuerstrom dem Mischgerät der Steuerungsanlage zugeführt. Im Mischgerät wird der Steuerstrom der LS-Anlage in ein zusätzliches Steuerkommando für die Strahl- und Luftruder I und III umgewandelt – und die Korrektur eingeleitet.

* Mischgerät = ein Verstärker aus fünf getrennten Schalttafeln, die auf einer gemeinsamen Grundplatte montiert sind. Er nimmt die Signale des Gyroskops auf und setzt sie in ein Steuersignal um, das den Servolenkungsmechanismus der Ruder (Segel) zur Korrektur der Fluglage bei Abweichungen ansteuert.

Setup of the guide beam position

The guide beam signal apparatus is installed approx. 200 m in front of or behind the firing position. The guide beam ground installation is set up 10 to 16 km behind the firing position and aligned in the firing direction.

Aufbau der LS-Stellung

Die LS-Kontrollanlage wird ca. 200 m vor oder hinter der Feuerstellung aufgebaut. Die LS-Bodenanlage wird zwischen 10 bis 16 km hinter der Feuerstellung in Schussrichtung aufgebaut und ausgerichtet.

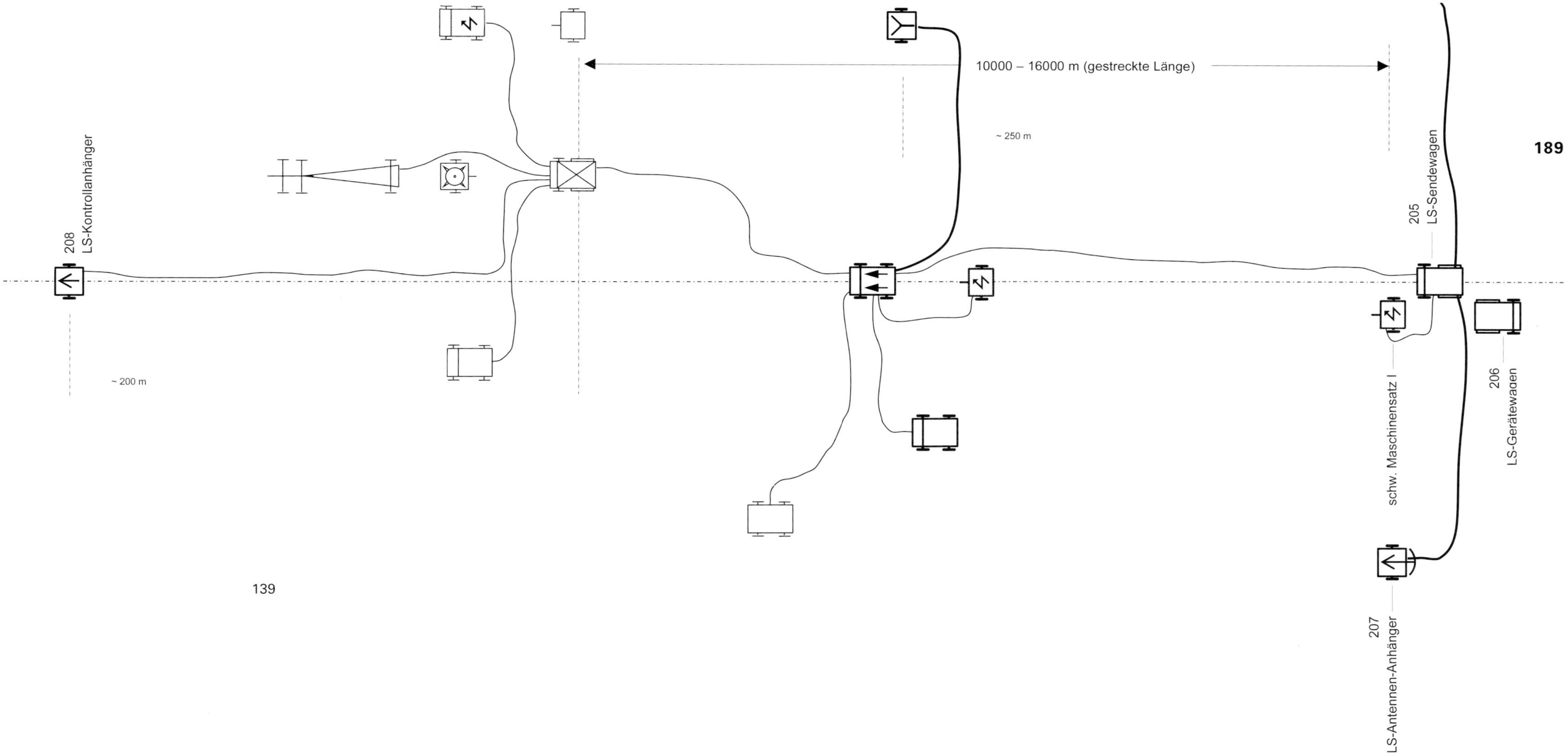

139

LS-Bodenanlage
in der Fernlenkstellung

Deutsches Museum, München, Archiv- FA014/B934/44 T5224 [CD74716]

The guide beam controlling apparatus

Before firing, the guide beam controlling apparatus checks whether the control plane passes through the firing position and thus coincides with the firing direction. A 25 m high mast with dipole antenna is positioned on a surveyed point exactly in the firing direction, about 200 m behind or in front of the firing position.

The energy absorbed by the dipole antenna from the guide beam transmitter is fed via an antenna amplifier to the guide beam control receiver. It works in the same way as the guide beam on-board unit and indicates on an output instrument whether the guiding beam deviates to the left or right, or whether it coincides with the firing direction. This output instrument is connected in parallel to an identical measuring instrument in the Fz.205 transmitter vehicle and an FF cable that connects the firing position with the guide beam position. The tuning devices in the LS-transmission trolley must now be adjusted until the measuring instrument has zero deflection. The entire guide beam signal apparatus is housed in a single-axle trailer.

Die LS-Kontrollanlage

Mit der LS-Kontrollanlage wird vor dem Abschuss geprüft, ob die Leitebene durch die Feuerstellung geht und somit mit der Schussrichtung zusammenfällt. Ein 25 m hoher Kurbelmast mit Dipolantenne steht auf einem vermessenen Punkt genau in der Schussrichtung rund 200 m hinter oder vor der Feuerstellung.

Die von der Dipolantenne aufgenommene Energie des LS-Senders wird über einen Antennen-Verstärker dem LS-Kontrollempfänger zugeführt. Er arbeitet genauso wie das LS-Bordgerät und zeigt an einem Ausgangs-Instrument an, ob der Leitstrahl nach links oder rechts abweicht oder ob er mit der Schussrichtung zusammenfällt. Dieses Ausgangsinstrument ist parallel zu einem identischen Messinstrument im LS-Sendewagen und einem FF-Kabel angeschlossen, das die Feuerstellung mit der LS-Stellung verbindet. Die Abstimmmittel im LS-Sendewagen müssen nun so lange verstellt werden bis das Messgerät Nullausschlag hat. Die gesamte LS-Kontrollanlage ist in einem einachsigen Anhänger untergebracht.

DMA, A4 Fibel Nr.34
Auszug

Setup of the guide beam ground installation

The sketch shows the installation of the vehicles and the dipoles in the guide beam position. The transmitter vehicle drives about 5 m in front of the surveyed point. The two dipoles are moved with the equipment vehicle to the points surveyed 100 m to the right and left of the central point perpendicular to the firing direction, in which they are set up and aligned. After the antenna cable has been laid out with the equipment trolley, it moves alongside the transmission trolley.

Aufbau der LS-Bodenanlage

Die Skizze zeigt die Aufstellung der Fahrzeuge und der Dipole in der LS-Stellung. Der Sendewagen fährt etwa 5 m vor den vermessenen Punkt. Die beiden Dipole werden mit dem Gerätewagen auf die 100 m rechts und links vom Zentralpunkt senkrecht zur Schussrichtung vermessenen Punkte gefahren, dort werden sie aufgestellt und ausgerichtet. Nachdem mithilfe des Gerätewagens das Antennenkabel ausgelegt ist, fährt dieser neben den Sendewagen.

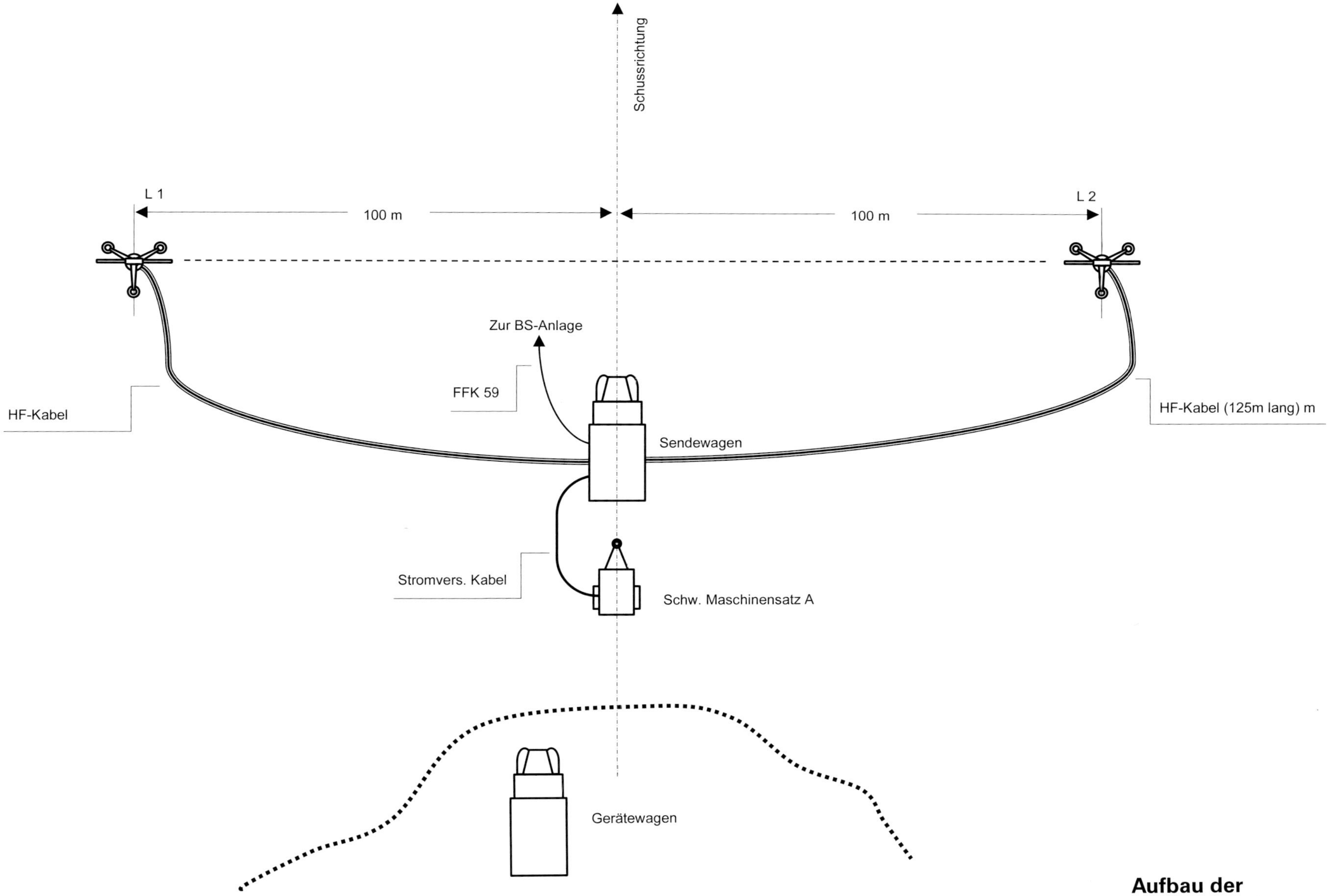

Aufbau der LS-Bodenanlage

DMA, GD639.5.18
HAP11/T 507
Karlshagen, 1.6.1944

DMA, GD639.5.1
T507
Fz.-Pos.-Nr. 205

DMA, GD630.02
A4 Lehrbuch
der FR-Schule
Mai 1944
Auszug aus Kapitel III
Technische Fahrzeuge
in der LS-Stellung
Ing. Gassmann

Vehicle 205

Internal or former designation: Fz.-Pos.-Nr.205
Guide beam transmitter vehicle on 3 t truck closed (Maultier)
Guide beam transmitter vehicle

Intended use

An Opel Blitz Gleisketten Lastkraftwagen 2 t, geschlossen (Maultier) (Sd.Kfz.3/4) with closed superstructure (LC box body) and gallery roof is used for the transmitter vehicle, in which the

- LS-transmitter,
- lobe switching device,
- the display unit for the guide beam control,
- the clear alarm device and the
- communication devices are located.

The deviation of the rocket from the guiding beam plane is measured with the built-in lobe switching device. For direction correction the transmitted impulse is received via the antennas on the *Gerät*, transmitted to the steering control unit (*Mischgerät**) where it is converted into a control correction and transmitted to the rudders on fins I and III.

The LS Transmitter vehicle has the following dimensions and weights:

Total length:	6,380 mm
Width:	2,160 mm
Height:	2,950 mm
Track width:	1,542/1,790 mm rear center distance between the tracks
Tare weight:	4,830 kg
Payload:	2,000 kg
Service weight:	6,040 kg
wheels:	190-20
Track:	Kgs 67 280/90

The LS-transmitter vehicle consists of the following main components:

- Chassis
- Closed superstructure (305) LC box body and attachments, blowers
- Installations – transmitters, control-, lobe switching, alarm and message units for the guide beam system

Chassis

The chassis is the chassis Gleisketten-Lastkraftwagen 2 t, (Maultier) Opel Typ 3,6-3 (see technical manual D 669/19 and D669/20)

* Mischgerät = a ampflifier of five independent panels mounted on a common base plate. It tooks signals from the gyroscope and then translated into a signal that directed the servo steering mechanism and steering vans to correct for attitude deviations in powered flight.

Gleisketten Lastkraftwagen = tracked 2 t truck (Maultier)

Fahrzeug 205

Interne oder frühere Bezeichnung: Fz.-Pos.-Nr.205
LS-Sende-Lkw auf Lkw 3 t geschl. (Maultier)
LS-Sendewagen

Verwendungszweck

Für den Sendewagen wird ein Gleisketten Lastkraftwagen 2 t, geschlossen (Maultier) (Sd.Kfz.3/4) mit geschlossenem Aufbau (LC-Koffer) und Laufgangdach (Gallerie) verwendet, in dem sich der

- LS-Sender,
- das LS-Umtastgerät,
- das Anzeigegerät für die LS-Kontrolle,
- die Klarmeldeeinrichtung und die
- Nachrichtengeräte befinden.

Mit den eingebauten LS-Umtastgerät wird die Abweichung der Rakete von der Leitstrahlebene gemessen. Zur Richtungskorrektur wird der gesendete Impuls über die Antennen am Gerät empfangen, an das Mischgerät* übermittelt und dort in eine Steuer-Korrektur umgewandelt und an die Ruder an Flosse I und III übermittelt.

Der LS-Sende-Kw hat folgende Maße und Gewichte:

Gesamtlänge:	6.380 mm
Breite:	2.160 mm
Höhe:	2.950 mm
Spurweite:	1.542/1.790 mm hinten Mittenabstand zwischen Kettenlaufwerk
Leergewicht:	4.830 kg
Nutzlast:	2.000 kg
Gefechtsgewicht:	6.040 kg
Räder:	190-20
Kette:	Kgs 67 280/90

Der LS-Sende-Kw besteht aus folgenden Hauptbauteilen:

- Fahrgestell
- Geschlossener Aufbau (305) LC-Koffer und Anbauten, Kühlgebläse
- Einbauten – LS-Sender, LS-Kontrolle-, Umtast-, Klarmelde- und Nachrichtengeräte

Fahrgestell

Als Fahrgestell dient das Fahrgestell des Gleisketten-Lastkraftwagen 2 t, (Maultier) Opel Typ 3,6-3 (Gerätebeschreibung siehe D 669/19 und D 669/420).

* Mischgerät = ein Verstärker aus fünf getrennten Schalttafeln, die auf einer gemeinsamen Grundplatte montiert sind. Er nimmt die Signale des Gyroskops auf und setzt sie in ein Steuersignal um, das den Servolenkungsmechanismus der Ruder (Segel) zur Korrektur der Fluglage bei Abweichungen ansteuert.

BArch, RH8II Bild-B1405-44, o. Ang.

142

142 The Sd.Kfz.3/4 tracked truck 2 t, (Maultier) with LC box body was converted into a guide beam transmission vehicle and designated as Fz.205. The vehicle is painted with a striking camouflage paint, which was also sprayed over the windows of the LC-box body. Afterwards the windows were only partly cleaned of paint. On the driver's door is the loading class, which is already heavily faded here, and 50 km/h is written as driving speed. That 50 km/h speed can never be reached with a maximum speed of 33 km/h for the tracked vehicle. The trailer is a Fz.504 generator trailer.

142 Das Sd.Kfz.3/4, ein Gleisketten-Lastkraftwagen 2 t (Maultier) mit LC-Koffer, wurde in einen LS-Sendekraftwagen umgebaut und als Fz.205 bezeichnet. Das Fahrzeug ist mit einem auffälligen Tarnanstrich versehen, der auch über die Fenster am LC-Koffer aufgespritzt wurde. Anschließend wurde die Farbe nur zum Teil von den Fenstern entfernt. Auf der Fahrertüre ist die bereits stark verblasste Beschriftung der Verladeklasse und der Fahrgeschwindigkeit zu sehen. Die in der Beschriftung angegebene Geschwindigkeit von 50 km/h wurden bei einem zulässigen Höchstwert von 33 km/h bei diesem Gleisskettenfahrzeug nie erreicht. Der Anhänger ist ein Fz.504 Generator-Anhänger.

The closed superstructure

The two hatches in the side wall of the tried-and-tested LC box body were blocked. An antenna connection with cover was integrated in the rear door, and a power- and FF cable connection was inserted in the side wall. The many screw heads in the side wall were fasteners for interior installations. In late 1944, the LC box body was mounted on a frame to protect the delicate equipment, set in four springs and connected to the chassis frame with shock absorbers.

Der geschlossene Aufbau

In der Seitenwand des bewährten LC-Koffers waren die beiden Klappen blockiert. Ein Antennenanschluss mit Deckel in der hinteren Klappe und je ein Energie- und FF-Kabelanschluss wurden eingesetzt. Die zahlreichen Schraubenköpfe in der Seitenwand waren Befestigungen für Einbauten im Innenraum. Ende 1944 wurde der LC-Koffer zum Schutz der empfindlichen Geräte auf einen Rahmen montiert, auf vier Federn gesetzt und mit Stoßdämpfern mit dem Fahrgestellrahmen verbunden.

194

Deutsches Museum, München, Archiv- FA014/B907/44 T5224 [CD74726]

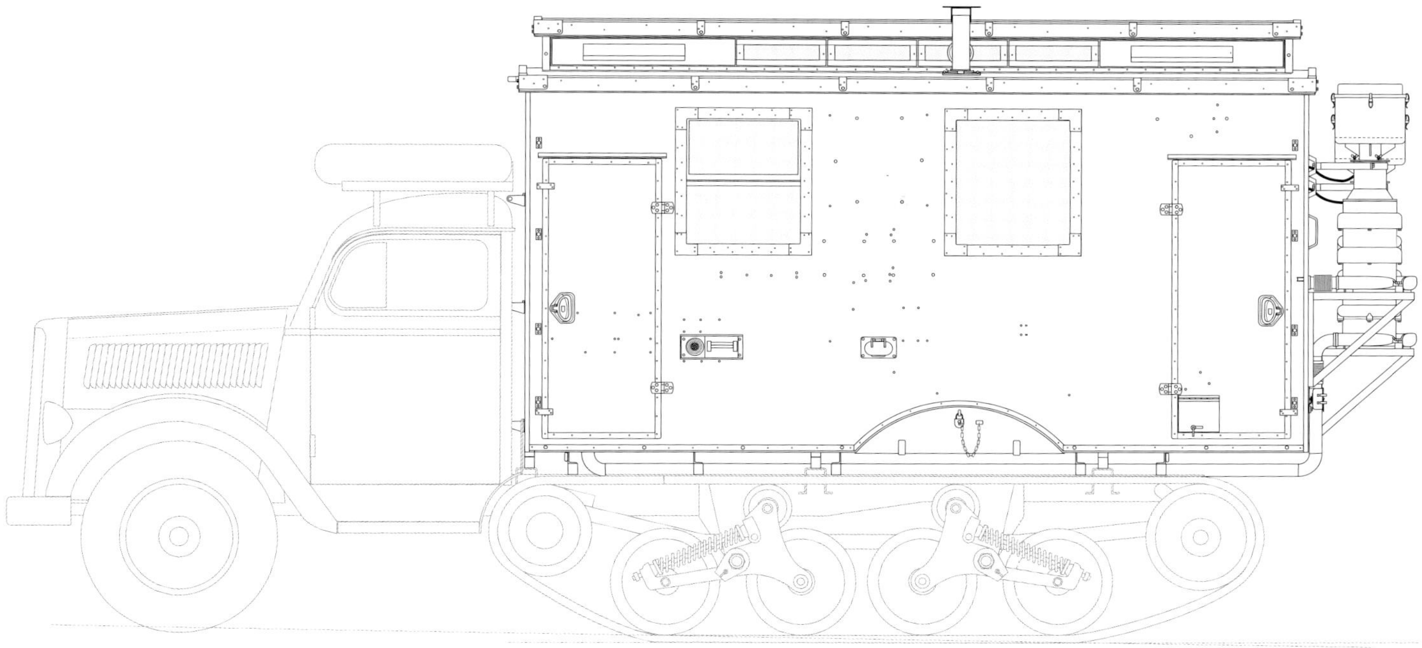

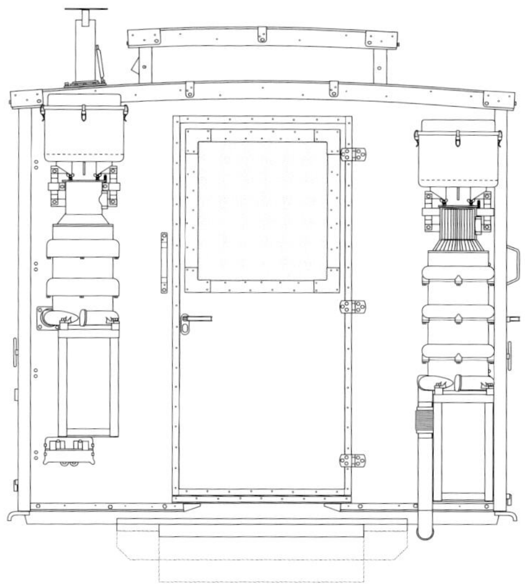

Scale / Maßstab 1:35 © Volker Ruff

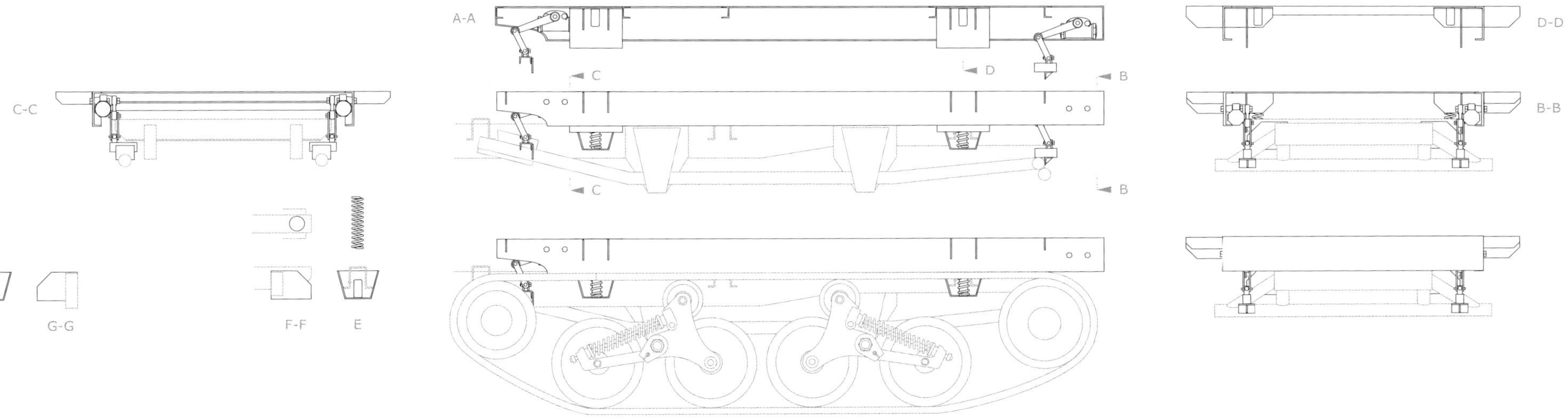

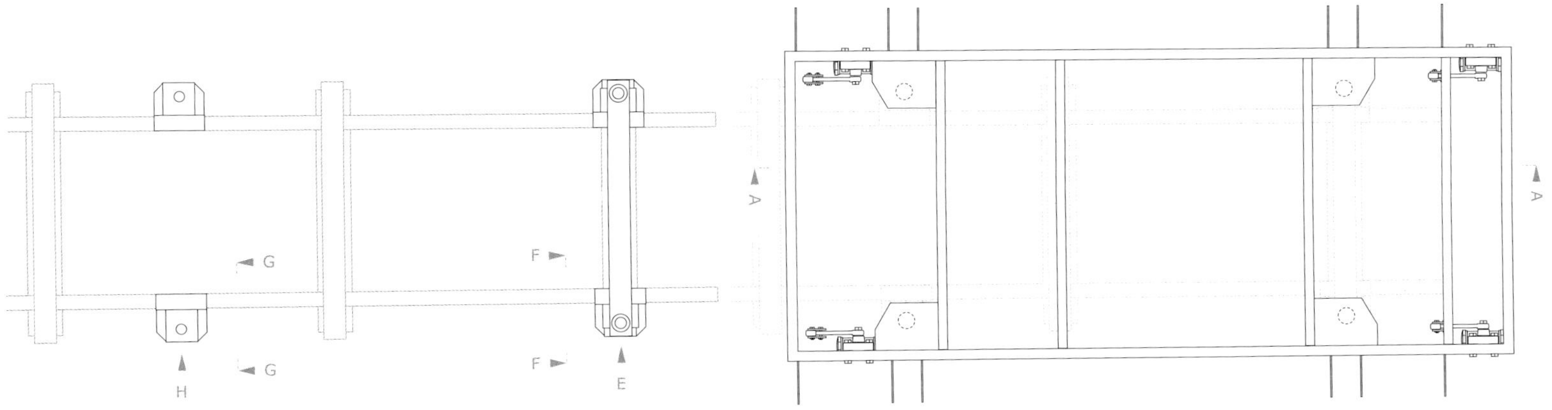

Fahrzeug 205

LS-Sende-Lkw auf Lkw 3 t geschl. (Maultier)

Partially used drawings acc. to group list 21Gr364

Teilweise verwendete Zeichnungen nach Gruppenliste 21Gr364

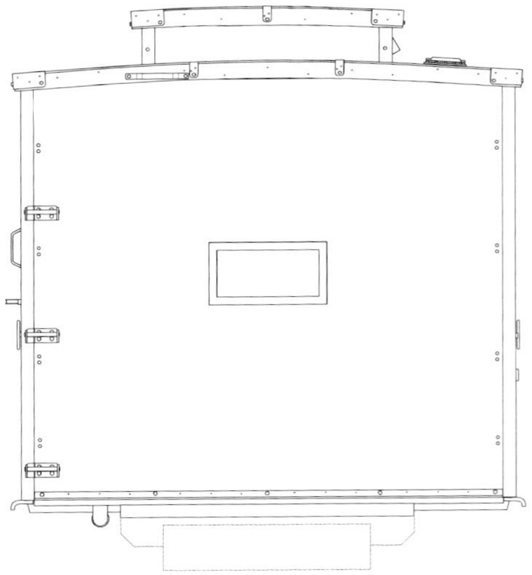

1 m

Scale / Maßstab 1:35 © Volker Ruff

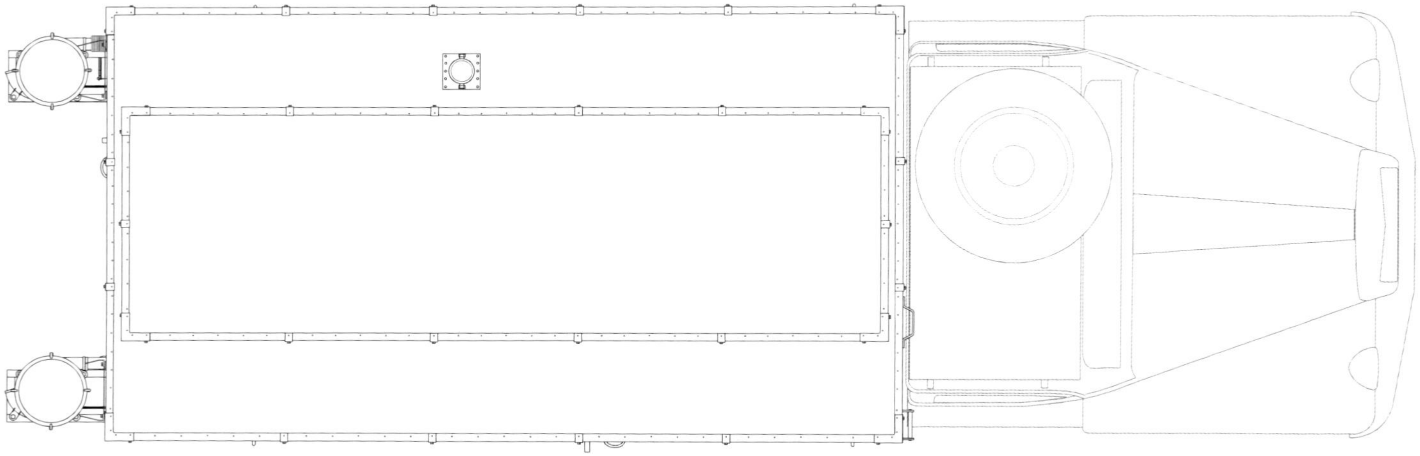

Deutsches Museum, München, Archiv- FA014/B948/44 T5224 [CD75224] Ausschnitt

144

144 A large lid for the second antenna connection is fitted in the right-hand side wall between the flap and the tool compartment (in the wheel cut-out). The interior can be reached by the access door with a boarding ladder. Fresh air is supplied to the interior by a fan between the door and the window to dissipate the heat generated during operation. Another fan removes the warm air from the guide beam transmitter rack to the outside.

In addition, the two front and rear hinged windows in the gallery roof on the left side have been replaced by an air outlet. See picture on page 193.

144 In der rechten Seitenwand ist zwischen Klappe und dem Werkzeugfach (im Radausschnitt) ein großer Deckel für den zweiten Antennenanschluss eingesetzt. Der Innenraum ist über die Zugangstür mit einer Einstiegsleiter erreichbar. Frischluft wird dem Innenraum über einen Ventilator zwischen Tür und Fenster zugeführt bzw. die entstandene Betriebswärme abgeführt. Über einen weiteren Ventilator wird die warme Luft aus dem LS-Sendergestell nach außen abgeführt.

Zusätzlich sind im Laufgangdach an der linken Seite die beiden vorderen und hinteren Ausstellfenster durch einen Luftauslass ersetzt. Siehe Bild auf Seite 193.

Deutsches Museum, München, Archiv FA014/D377/44 TD5224 [CD/4/46]

Body – rear view

A blower for the mock antenna is mounted on angle brackets on the left rear wall and to the right a blower for the transmitter. The blowers are intended to reduce the high operating temperatures of the highly sensitive devices in the lobe switching unit and the guide beam transmitter frame with HF and LF tray.

Aufbau – Rückansicht

An der Rückwand links sind auf Winkelkonsolen ein Gebläse für die künstliche Antenne und rechts ein Gebläse für den Sender montiert. Die Gebläse sollen die hohen Betriebstemperaturen der hochempfindlichen Geräte im Umtastgerät und LS-Sendergestell mit HF- und NF-Wanne absenken.

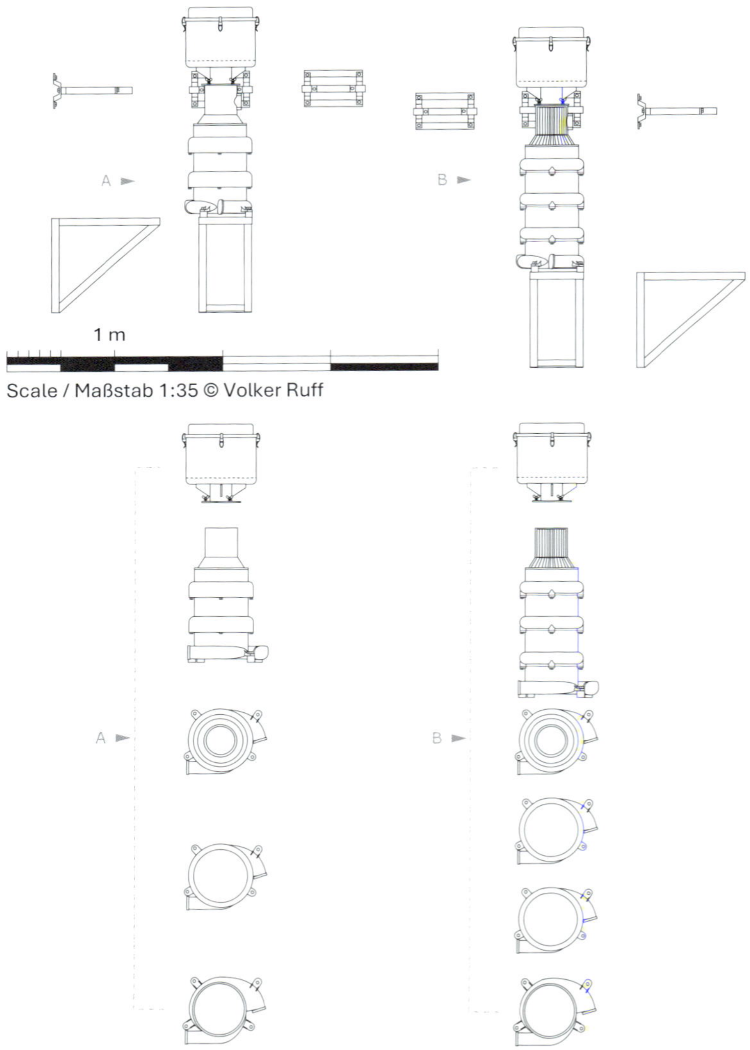

Scale / Maßstab 1:35 © Volker Ruff

The fixtures

Rear access to the lobe switching unit

With the rear door open you can see the back of the LS-keying unit with components: an resistor of 39 Ohm with shielded housing (top) , resistor of 78 Ohm with shielded housing (middle) a shielded housing (middle right) with the instrument column named »Posaune« (bottom). The »Posaune«, turns at 3000 rpm and transmit the signals to both antennas. The opening of the rear door is stopped by a door limiter, and a canvas cover is clipped onto the window frame.

Die Einbauten

Rückwärtiger Zugang zum Umtastgerät

Bei geöffneter Hecktür blickt man auf die Rückseite des LS-Umtastgerätes mit Komponenten: einem Widerstand von 39 Ohm im geschirmten Gehäuse (oben), einem Widerstand von 78 Ohm mit geschirmtem Gehäuse (Mitte) und einem geschirmten Gehäuse (Mitte rechts) mit der Instrumentensäule namens »Posaune« (unten). Die »Posaune« dreht sich mit 3000 U/min und überträgt die Signale an beide Antennen. Die Öffnung der Hecktür wird von einem Türbegrenzer gestoppt. Eine Segeltuchabdeckung ist auf dem Fensterrahmen befestigt.

146 The lobe switching unit also contains the cabins (shielded housing) in which the power cables are build in. One at the power line input and two at the power line output just before the connections for the antenna cables. Furthermore, there are two (mock) antennas on the right side which consume the energy emitted by the transmitter when the dipoles are cut off. The heat emitted is compensated by a cooling fan.

146 Das Umtastgerät enthält ferner die Kabinen (geschirmte Gehäuse), die in die Energieleitungen eingebaut sind. Ein Gehäuse befindet sich am Eingang und zwei am Ausgang der Energieleitung kurz vor den Anschlüssen für die Antennenkabel. Desweiteren befinden sich rechtsseitig zwei Scheinantennen, die bei abgeschalteten Dipolen die vom Sender abgegebene Energie verbrauchen. Die dabei abgegebene Wärme wird durch die Kühlung eines Lüfters kompensiert.

BArch, RH8II Bild-B1407-44, o.Ang.

147

Deutsches Museum, München, Archiv- FA014/B386/44 TD5224 (CD79127)

148

In the interior

On the wheel arch, which is accessible from the outside and is used as a toolbox, is an EGO III type oven. To the right of it is a switch panel with circuit breakers, fuses and cable ducts mounted on the inner wall. On the left is the front panel for the lobe switching unit.

In a box on the right wall of the LC box body is the alarm system, consisting of a switch that lights up on the control panel of the *Feuerleitpanzer* when the system is »clear«.

In addition to the communication system on the right wall, there is also the support frame for a field telephone; long distance calls can be made via this frame, using a crossover network to switch the combustion cut-off ground installation.

Im Innenraum

Auf dem Radkasten, der von außen zugänglich ist und als Werkzeugkasten verwendet wird, ist ein Ofen vom Typ EGO III montiert. Rechts davon ist an der Innenwand eine Schaltertafel mit Schutzschaltern, Sicherungen und Kabelkanälen angebracht. Links ist die Frontplatte für das LS-Umtastgerät zu sehen.

In einem Kästchen an der rechten Wand des LC-Koffers befindet sich die Klarmeldeeinrichtung, bestehend aus einem Schalter, der bei »Anlage klar« auf dem Steuerungspult des Feuerleitpanzers aufleuchtet.

Neben der Nachrichtenanlage an der rechten Wand befindet sich ferner ein Tragegestell für einen Feldfernsprecher. Über diese Verbindung können über eine Frequenzweiche Ferngespräche zur Vermittlung der BS-Bodenanlage geführt werden.

147, 148 The guide beam transmitter consists of 4 main components:
HF power meter and HF tray (top), LF tray (middle) and mains power and rectifier section (bottom).

The tilted HF tray in the picture shows the complex structure of the various devices. From left to right: quartz container with thermostat, A-stage (LV1), next to it the B-stage and C-stage, partition wall, box with toggle handles and tube LS 1500, box with HF contact (D-stage), partition wall, tube RS384 (K-stage)

147, 148 Der LS-Sender besteht aus 4 Hauptbaugruppen:
HF-Leistungsmesser und HF-Wanne (oben), NF-Wanne (Mitte) und Netz- und Gleichrichterteil (unten).

Die im Bild gekippte HF-Wanne zeigt den komplexen Aufbau der unterschiedlichsten Geräte. Von links nach rechts: Quarzbehälter mit Thermostat, A-Stufe (LV1), daneben die B-Stufe und C-Stufe, Trennwand, ein Kasten mit Knebelgriffen Röhre LS 1500 und ein Kasten mit HF-Kontakt (D-Stufe), Trennwand, Röhre RS384 (K-Stufe).

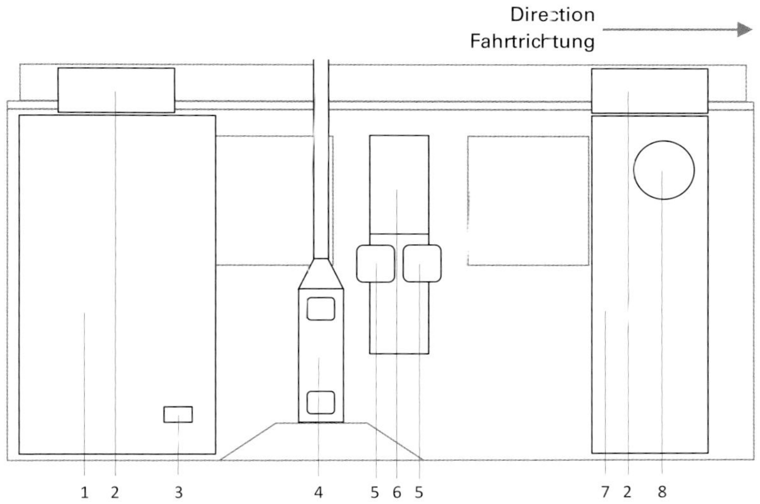

View left side wall
Ansicht linke Seitenwand

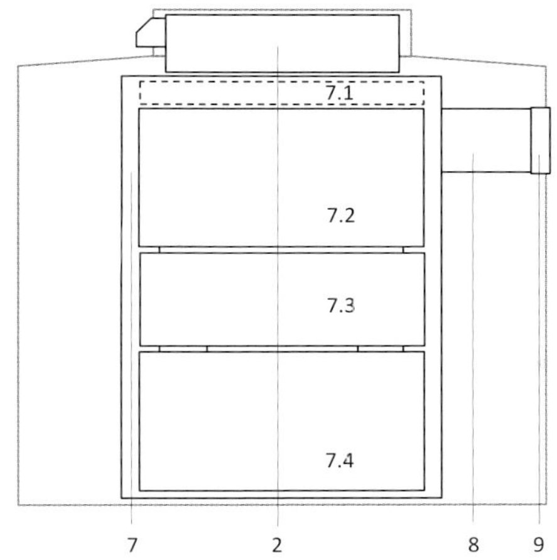

View on back wall
Ansicht auf Rückwand

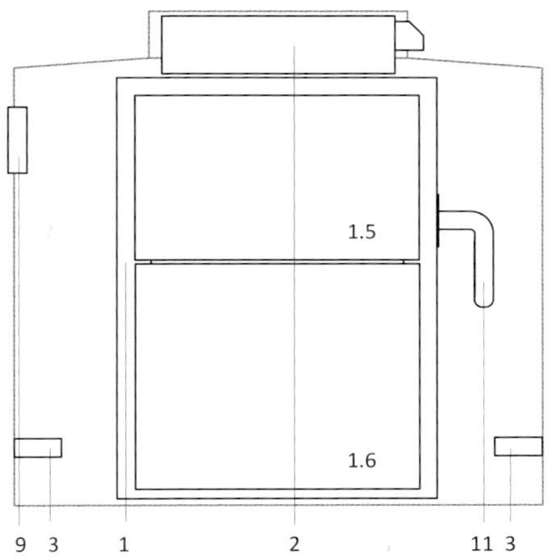

View on back wall (entrance)
Ansicht auf Rückwand (Zugang)

1	Lobe-switching-unit (transducer)	1	Umtastgerät
2	Exhaust air panel with outlet	2	Abluftverkleidung mit Auslass
3	Antenna connection	3	Antennenanschluß
4	Stove	4	Heizofen
5	Circuit breaker	5	Schutzschalter
6	Switchboard	6	Schaltbrett
7	LS transmitter	7	LS-Sender
8	Exhaust air (heat)	8	Abluft (Wärme)
9	Fan (exhaust air)	9	Lüfter (Abluft)
10	Storage box with bench	10	Staukasten mit Sitzbank
11	Cooling air air	11	Zuluft (Kühlung)

7	LS-transmitter	7	LS-Sender
7.1	Zero instrument power supply unit	7.1	Nullinstrumente Netzanschlussgerät
7.2	HF tray	7.2	HF-Wanne
7.3	LF tray	7.3	LF-Wanne
7.4	Mains and rectifier section	7.4	Netz- und Gleichrichterteil

1	Lobe-switching-unit – back side	1	Umtastgerät – Rückseite
1.1	78 Ohm cabin (housing)	1.1	78-Ohm-Kabine (Gehäuse)
1.2	39 Ohm cabin (housing)	1.2	39-Ohm-Kabine (Gehäuse)
1.3	Shields	1.3	Abschirmungen
1.4	Large horn	1.4	Große Posaune
	Lobe-switching-unit – front panel		Umtastgerät – Frontseite
1.5	Front panel top with hand wheels	1.5	Frontplatte oben mit Handrädern
1.6	Front panel bottom with calibration	1.6	Frontplatte unten mit Skalierung

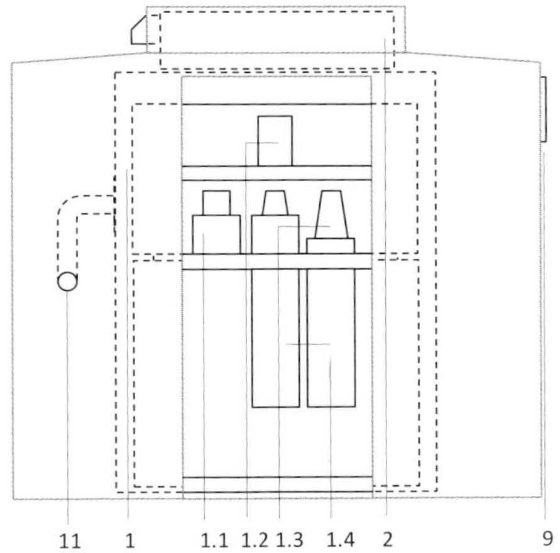

Rear view to the Lobe-Switching-Device by open entrance door
Sicht von hinten auf das Umtastgerät bei offener Zugangstür

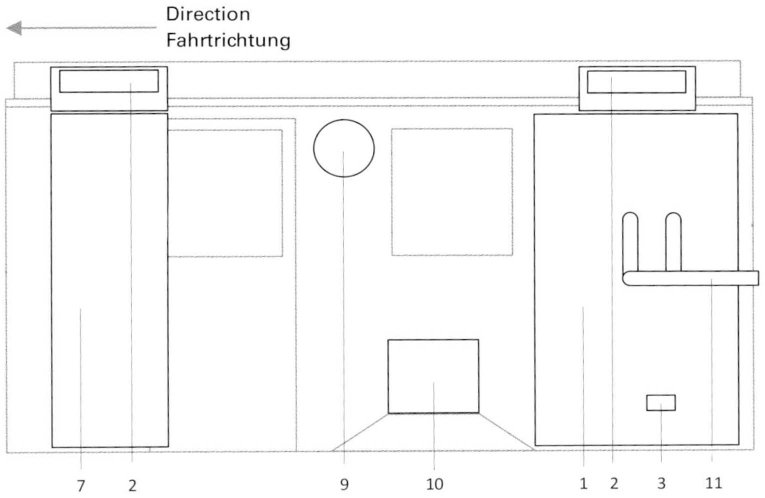

View right side wall
Ansicht rechte Seitenwand

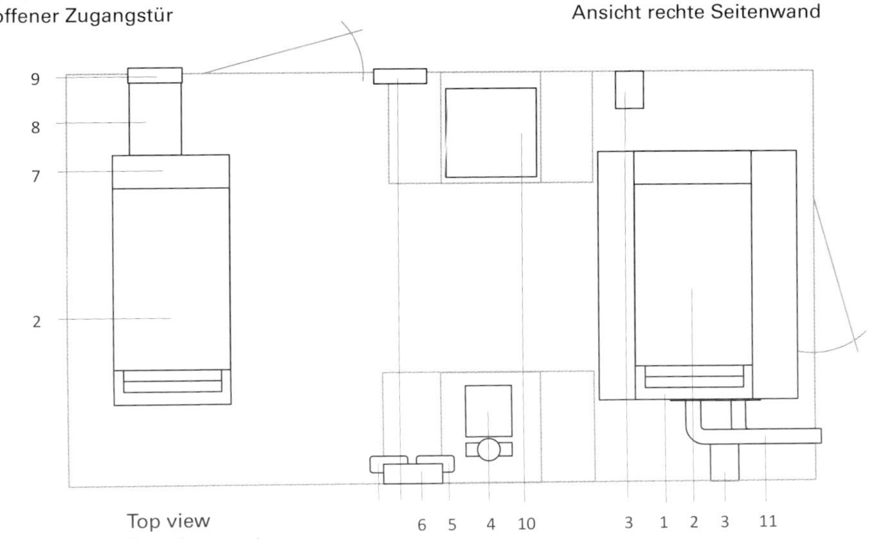

Top view
Ansicht von oben

Fahrzeug 205

LS-Sende-Lkw
auf Lkw 3 t geschl. (Maultier)

Loading plan
Beladeplan

DMA, GD639.5.18
HAP11/T 507
Karlshagen, 1.6.1944
Bd.Nr. E 3195/44 g

DMA, GD639.5.1
T507
Fz.-Pos.-Nr. 206

DMA, GD630.02
A4 Lehrbuch
der FR-Schule
Mai 1944
Auszug aus Kapitel III
Technische Fahrzeuge
in der LS-Stellung
Ing. Gassmann

Vehicle 206

Internal or former designation: Fz.-Pos.-Nr.206
Guide beam equipment vehicle on 3 t truck closed (Maultier)
Guide beam equipment vehicle

Intended use

A tracked 2 t truck (Gleisketten Lastkraftwagen), closed body (Maultier) (Sd.Kfz.3/4) with closed superstructure (LC box body) and gallery roof is used as the equipment vehicle.

Two cable reels, each with 125 m high-frequency cable, are installed in its rear. The cable reels are mounted on a shaft and can be rolled up or unrolled during travel using cable rollers through two flaps located on the rear body. A workbench with vice, a tool cabinet, a wall cabinet, and two boxes for storing spare tubes are installed in the front part of the LC box body.

The equipment truck has the following dimensions and weights:

Total length:	6,200 mm
Width:	2,160 mm
Height:	2,960 mm
Track width:	1,542/1,790 mm rear center distance between the tracks
Tare weight:	4,500 kg
Payload:	2,000 kg
Service weight:	5,285 kg
Wheels:	190-20
Tracks:	Kgs 67 280/90

The equipment vehicle consists of the following main components:

- Chassis
- Closed superstructure (305) LC box body
- Fittings – HF cable reel, table, storage and tool box

Chassis

The chassis of Gleisketten-Lastkraftwagen 2 t, (Maultier) Opel Typ 3,6-3 (see technical manual D 669/19 and D 669/20) is used.

Fahrzeug 206

Interne oder frühere Bezeichnung: Fz.-Pos.-Nr.206
LS-Geräte-Lkw auf Lkw 3 t geschl. (Maultier)
LS-Gerätewagen

Verwendungszweck

Für den Gerätewagen wird ein Gleisketten-Lastkraftwagen 2 t, geschlossen (Maultier) (Sd.Kfz.3/4) mit geschl. Aufbau(LC-Koffer) und Laufgangdach verwendet. Im hinteren Teil des Gerätewagens sind zwei Kabeltrommeln mit je 125 m Hochfrequenzkabel eingebaut. Die Kabeltrommeln sind auf einer Welle gelagert. Sie lassen sich während der Fahrt durch zwei am hinteren Aufbau des Wagens befindliche Klappen über Kabelroller auf- bzw. abrollen. Ferner sind eine Werkbank mit Schraubstock, ein Werkzeugschrank, ein Wandschrank und zwei Kästen zur Aufbewahrung der Ersatzröhren im vorderen Teil des LC-Koffers eingebaut.

Der LS-Geräte-Kw hat folgende Maße und Gewichte:

Gesamtlänge:	6.200 mm
Breite:	2.160 mm
Höhe:	2.960 mm
Spurweite:	1.542/1.790 mm hinten Mittenabstand zwischen Kettenlaufwerk
Leergewicht:	4.500 kg
Nutzlast:	2.000 kg
Gefechtsgewicht:	5.285 kg
Räder:	190-20
Kette:	Kgs 67 280/90

Der LS-Geräte-Kw besteht aus folgenden Hauptbauteilen:

- Fahrgestell
- Geschlossener Aufbau (305) LC-Koffer
- Einbauten – HF-Kabeltrommel, Tisch, Vorrats- und Werkzeugkasten

Fahrgestell

Es wird das Fahrgestell des Gleisketten-Lastkraftwagen 2 t, (Maultier) Opel Typ 3,6-3 (Gerätebeschreibung siehe D 669/19 und D 669/420) verwendet.

BArch, RH8II Bild-B1404-44, o.Ang.

149

149 For the Fz.206 the Sd.Kfz. 3/4 tracked 2 t truck (Maultier) with LC box body, was converted to the guide beam-equipement vehicles. Like the Fz.205, this vehicle is also provided with a stricking camouflage paint which was also sprayed generously over the windows. Afterwards the windows were only partly cleared of paint. Also here the driving speed is indicated only 50 km/h. A speed which is not reached with the tracked truck.

The Fz.206 is also a towing vehicle for the two Fz.207 dipol trailers coupled one behind the other.

149 Für das Fz.206 wurde das Sd.Kfz.3/4, ein Gleisketten-Lastkraftwagen 2 t (Maultier) mit LC-Koffer, zum LS-Gerätewagen umgebaut. Auch dieses Fahrzeug ist wie das Fz.205 mit einem auffälligen Tarnanstrich versehen, der ebenfalls großzügig über die Fenster aufgesprüht wurde. Anschließend sind die Fenster nur zum Teil wieder von der Farbe gereinigt worden. Auch hier ist die Fahrgeschwindikeit mit 50 km/h angegeben. Eine Geschwindigkeit die mit dem Gleisketten-Lastkraftwagen nicht erreicht wird.
Das Fz.206 ist Zugmittel für die beiden hintereinander gekoppelten Fz.207 Dipol-Anhänger.

Superstructure – rear and side view

Two sets of grounding tubes (grounding screws) are mounted on the rear wall to the left and right of the access door. In the lower corners of the rear wall are cable outlets, behind which are cable rollers for rolling up and unrolling the HF cables. Two cable reels for the approx. 125 m HF cables are mounted inside. A connection for the power cable is in front of the side access door.

Aufbau – Rück und Seitenansicht

Zwei Sätze Erdungsrohre (Erdungsschrauben) sind an der Rückwand links und rechts von der Zugangstür angebracht. In den unteren Ecken der Rückwand sind Kabeldurchlässe eingepasst, dahinter sind Kabelroller für das Auf- und Abrollen der HF-Kabel vorhanden. Zwei Kabeltrommeln für ca. 125 m HF-Kabel sind im Innenraum montiert. Ein Anschluss für ein Energiekabel ist vor der seitlichen Zugangstüre eingesetzt.

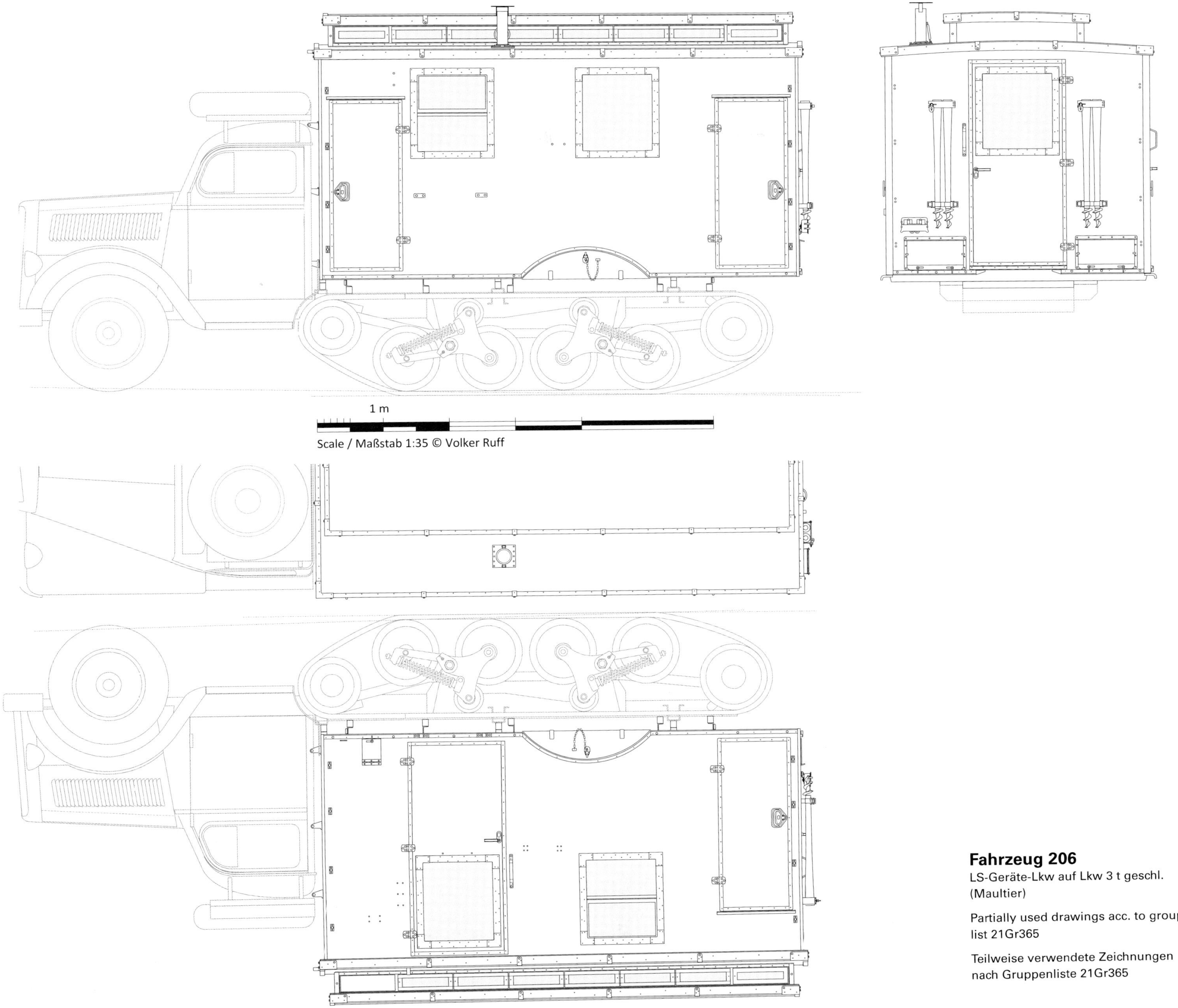

Fahrzeug 206

LS-Geräte-Lkw auf Lkw 3 t geschl. (Maultier)

Partially used drawings acc. to group list 21Gr365

Teilweise verwendete Zeichnungen nach Gruppenliste 21Gr365

BArch, RH8II Bild-B1408-44, o.Ang.

151

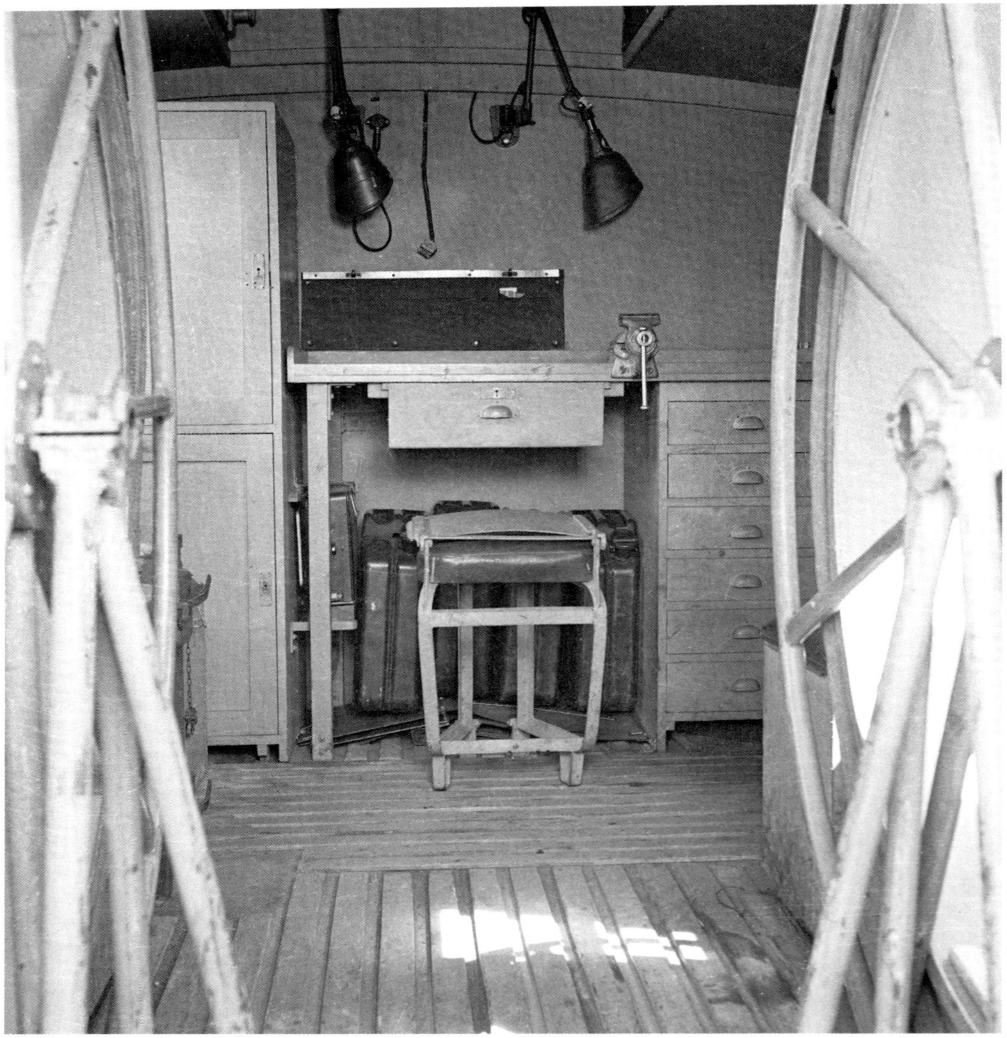

Deutsches Museum, München, Archiv- FA014/B927/44 T5224 [CD74718]

152

Interieur
Innenraum

151, 150 The view through the right-hand access door shows an EGO III stove on the opposite wall, with a box seat mounted on the right. On the wheel arch and side wall are brackets for two rifles.

Through the rear door you can see the rear wall, with from left to right: a cupboard with storage boxes, a workbench with vice, and toolboxes. The stool can be folded up and stowed under the workbench if no canisters are stored there. Two work lights provide sufficient lighting.

A further box seat is mounted on the right wheel arch. The cable reel racks are mounted to the left and right. Each holds about 125 m of HF cable for the connections from the dipoles Fz.207 to the guide beam transmitter vehicle Fz.205.

151, 152 Der Blick durch die rechte Zugangstüre zeigt an der gegenüberliegenden Wand einen Ofen vom Typ EGO III, rechts davon ist ein Sitzkasten montiert. Auf dem Radkasten sind Halterungen für zwei Gewehre angebracht.

Durch die Hecktür blickt man auf die Rückwand. Von links nach rechts: ein Schrank mit Vorratskästen, eine Werkbank mit Schraubstock und Werkzeugkästen. Der Hocker ist zusammenfaltbar und kann, wenn keine Kanister unter der Werkbank gelagert werden, dort hineingeklappt werden. Zwei Arbeitsleuchten sorgen für ausreichende Beleuchtung.

Ein weiterer Sitzkasten ist auf dem rechten Radkasten befestigt. Die Kabelrollengestelle sind links und rechts davon montiert. Jeweils ca. 125 m HF-Kabel für die Verbindung von den Dipolen Fz.207 zum LS-Sendewagen Fz.205 sind dort aufgerollt.

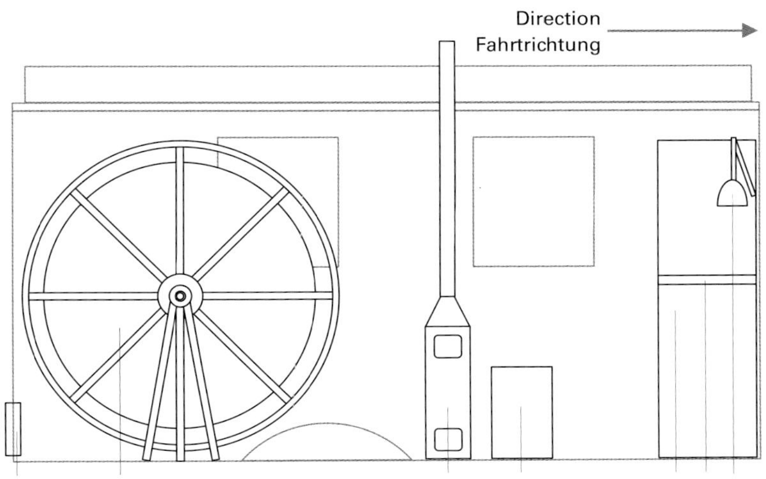

View left side wall
Ansicht linke Seitenwand

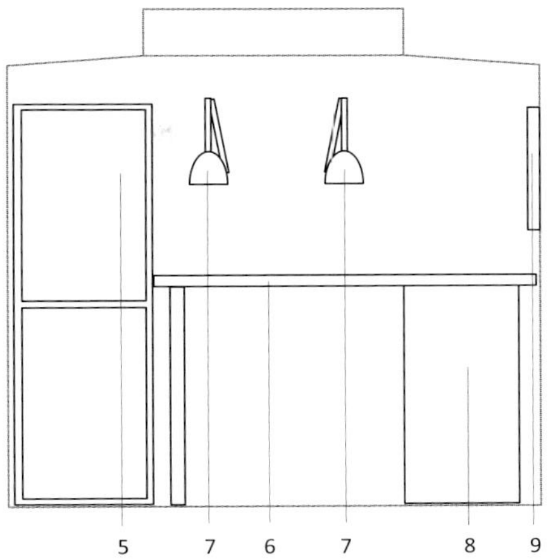

View on back wall
Ansicht auf Rückwand

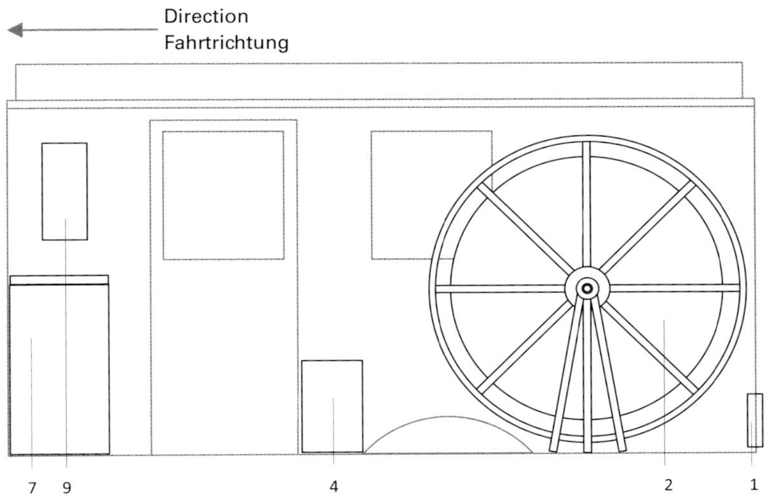

View right side wall
Ansicht rechte Seitenwand

1 Flap	1 Klappe
2 HF cable reel	2 HF-Trommel
3 Stove	3 Heizofen
4 Storage box with bench	4 Staukasten mit Sitzbank
5 Cabinet	5 Schränke
6 Workbench	6 Werkbank
7 Work lamp	7 Arbeitsleuchte
8 Tool drawers	8 Werkzeugschubladen
9 Switch board	9 Sicherungskasten

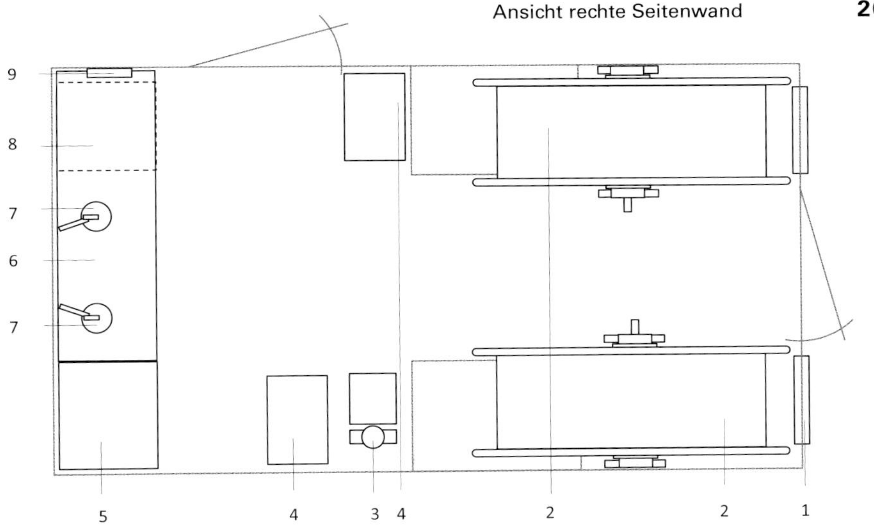

Top view
Ansicht von oben

Fahrzeug 206
LS-Geräte-Lkw
auf Lkw 3 t geschl. (Maultier)

Loading plan
Beladeplan

DMA, GD639.5.1
T507
Fz.-Pos.-Nr. 207

DMA, GD630.02
A4 Lehrbuch
der FR-Schule
Mai 1944
Auszug aus
Kapitel III
Technische Fahrzeuge
in der LS-Stellung
Ing. Gassmann

Vehicle 207

Internal or former designation: Fz.-Pos.-Nr.207
Antenna trailer (1-axle)
Dipol trailer

Intended use

These contain the two dipoles of the LS ground transmitter. The antenna mast of the two dipoles is jacked up while driving and is erected on the corresponding tripods during operation. The two dipole arms are mounted on the antenna mast. They are housed in special boxes during the trip.

Fahrzeug 207

Interne oder frühere Bezeichnung: Fz.-Pos.-Nr.207
LS-Antennen-Anhänger (1-achs.)
Dipol-Anhänger

Verwendungszweck

Sie enthalten die beiden Dipole des LS-Bodensenders. Der Antennenmast der beiden Dipole ist während der Fahrt aufgebockt und wird bei Betrieb auf die dazugehörigen Dreibeine aufgerichtet. Die beiden Dipolarme werden am Antennenmast aufgebaut. Sie sind während der Fahrt in besonderen Kästen untergebracht.

153

153 For transport, the dipole and its cradle are hinged by its supports on the trailer, tipped over towards the towing eyelet and secured in a bracket in front of it. The sensitive dipoles are stowed in their own transport boxes and mounted on the chassis frame.

153 Für den Transport wird der Dipol mit Lafette vom Anhänger über die Lager aufgenommen, hin zur Zugöse umgekippt und in einer Halterung festgezurrt. Die empfindlichen Dipole sind in eigenen Transportkisten verstaut und am Fahrgestellrahmen montiert.

Deutsches Museum, München, Archiv- FA014/B908/44 T5224 [CD74714]

154 A equipment-vehicle Fz.206 pulls the two coupled antenna trailers Fz.207 on the march to an guide beam position. The vehicles are painted in RAL 7028 dark yellow ›nach Muster‹. With this dark yellow the colors turn out differently, sometimes lighter or darker. Since the assemblies are delivered already painted by the various manufacturers, their paintwork differs. The tracked vehicle (Maultier) is brighter and shinier than the LC box body, and while the superstructure appears darker and duller, the trailers appear lighter.

154 Ein LS-Geräte-Kraftwagen Fz.206 zieht die beiden miteinander gekoppelten LS-Antennen-Anhänger Fz.207 auf dem Marsch zur LS-Stellung. Die Fahrzeuge sind in RAL 7028 dunkelgelb nach Muster gestrichen. Die Farbe fällt in der Helligkeit unterschiedlich aus, denn die Baugruppen wurden bereits lackiert von den verschiedenen Herstellerfirmen angeliefert. Der Anstrich des LC-Koffers erscheint dunkler und matter gegenüber dem Gleisketten-Lastkraftwagen (Maultier). Auch der Anhänger erscheint in einem helleren Farbton.

155

155 The dipoles of the Fz.204 and Fz.207 are nearly identical, differing only in details. If the trailer Fz.204 still had an HF cable reel on the chassis frame, it was omitted from the guide beam-antenna trailer Fz.207. The HF cables of approx. 125 m length each, necessary for operation, were transported on two cable reels in the equipment truck Fz.206.

155 Die Dipole der Fz.204 und Fz.207 sind fast baugleich und unterscheiden sich nur im Detail. Hatte der Anhänger Fz.204 seine HF-Kabeltrommel noch auf dem Fahrgestellrahmen montiert, so entfielen diese an dem LS-Antennen-Anhänger Fz.207. Die zum Betrieb notwendigen HF-Kabel von jeweils ca. 125 m Länge wurden auf zwei Kabeltrommeln im LS-Geräte-Kraft-wagen Fz.206 befördert.

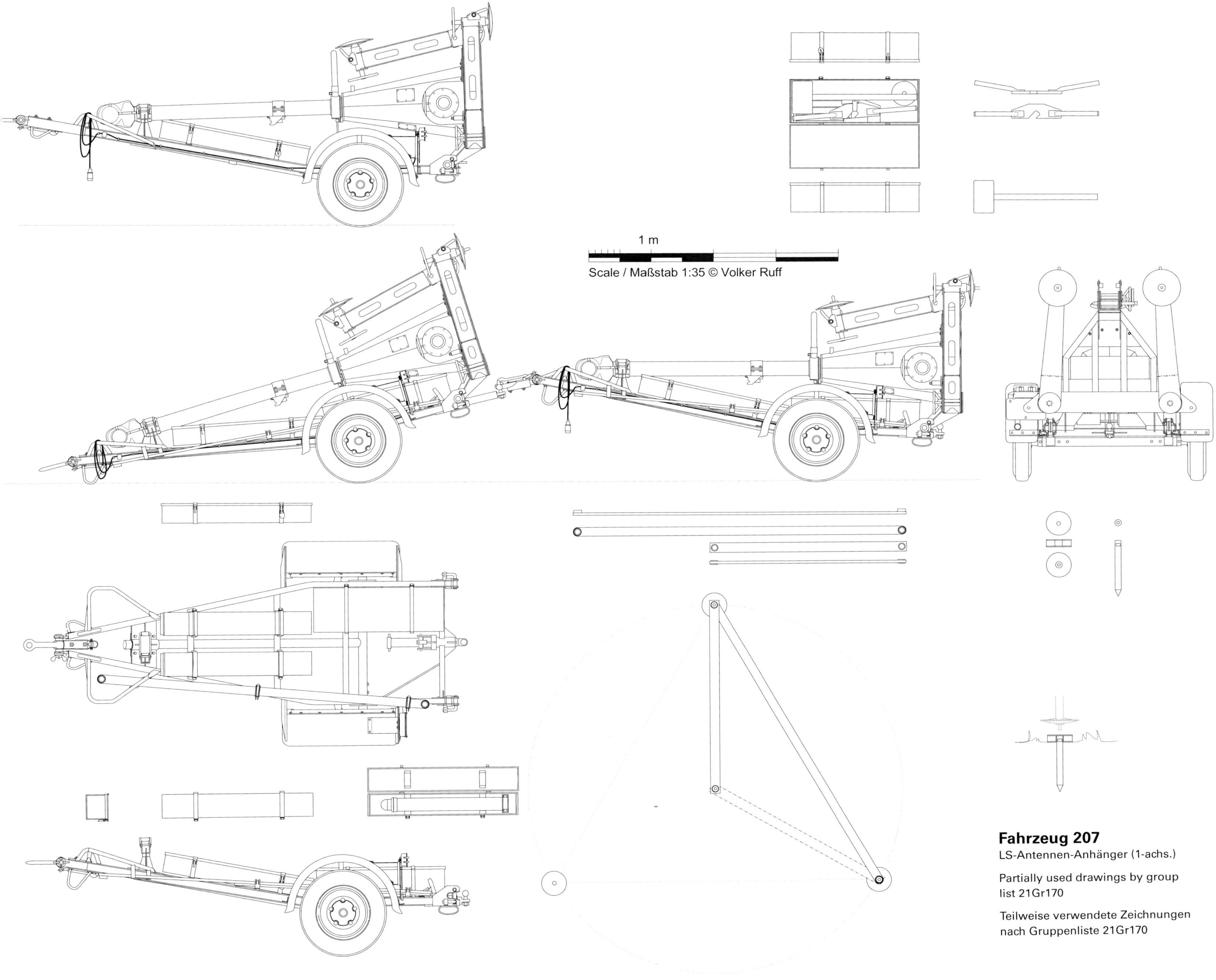

Fahrzeug 207

LS-Antennen-Anhänger (1-achs.)

Partially used drawings by group list 21Gr170

Teilweise verwendete Zeichnungen nach Gruppenliste 21Gr170

Deutsches Museum, München, Archiv- FA014/B381/44 TD5224 [BN39754]

The LS Antenna

The dipole antenna consists of

- antenna mast
- dipole head
- blocking pot
- and two dipolar arms

A support for a directional telescope is attached to the dipole mast of Fz.207. This is needed for setting up and aligning the fixture to the direction of fire. The mast can be rotated conditionally with a directional swivel arm.

The LS-antenna trailer has the following dimensions and weights:

Total length:	3,840/3,830 mm late version w/o dipole
Width:	1,670 mm
Height:	975 mm w/o dipole, 3,200 mm dipole
Track width:	1,400 mm, 3,000 mm dia dipole-lafette
Service weight:	650 kg
Wheels:	5.50-16

Die LS Antenne

Die Dipolantenne besteht aus:

- Antennenmast
- Dipolkopf
- Sperrtopf
- und zwei Dipolarmen.

Am Dipolmast von Fz.207 ist eine Auflage für ein Richtfernrohr angebracht. Diese wird zum Aufstellen und Ausrichten auf die Schussrichtung benötigt. Der Mast kann mit einem Richtschwenkarm bedingt gedreht werden.

Der LS-Antennen-Anhänger hat folgende Maße und Gewichte:

Gesamtlänge:	3.840 mm ohne Dipol
Breite:	1.670 mm
Höhe:	975 mm ohne Dipol, 3.200 mm Dipol
Spurweite:	1.400 mm, 3.000 mm mit Dipol-Lafette
Gefechtsgewicht:	650 kg
Räder:	5,50-16

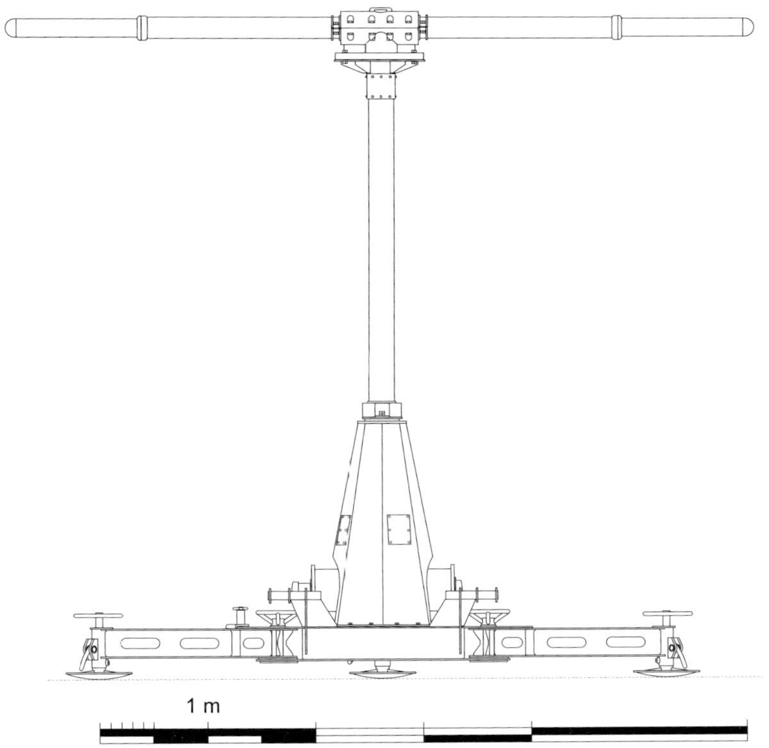

Scale / Maßstab 1:35 © Volker Ruff

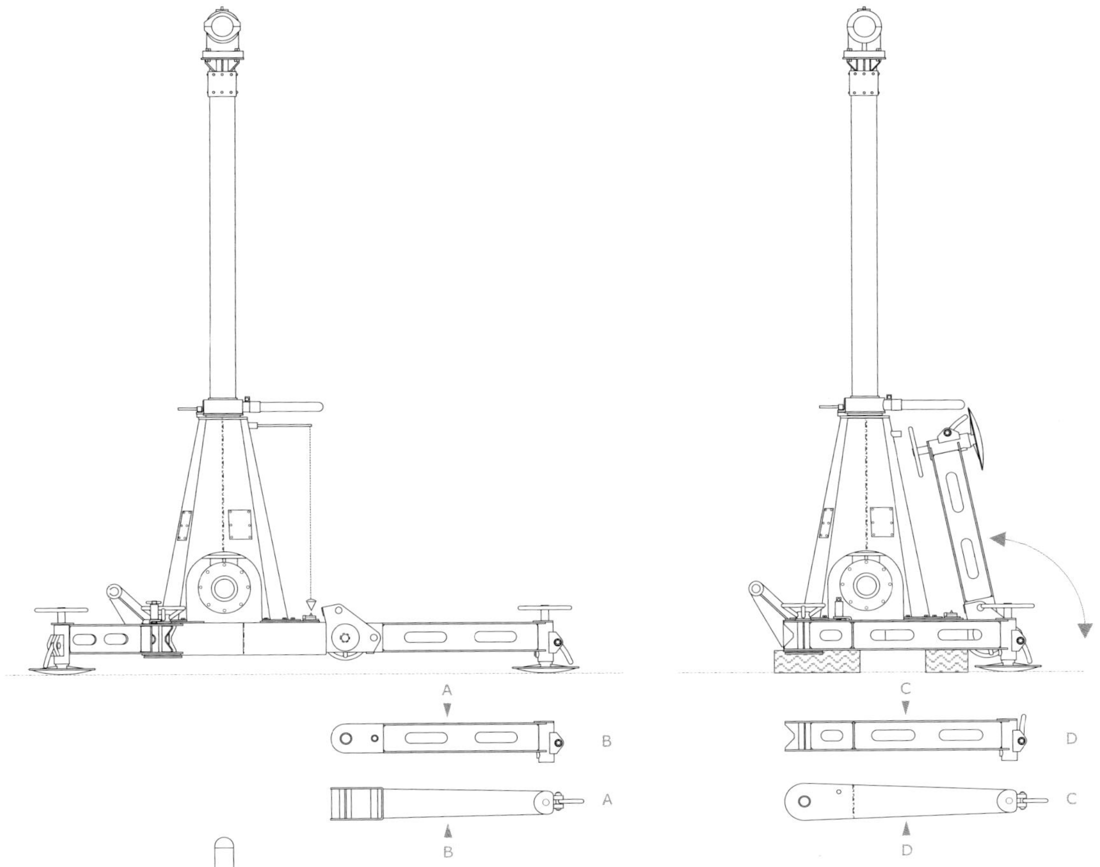

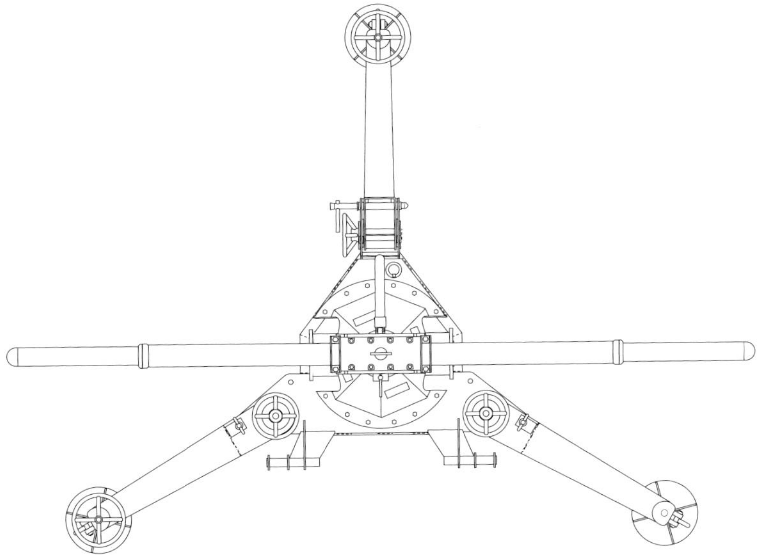

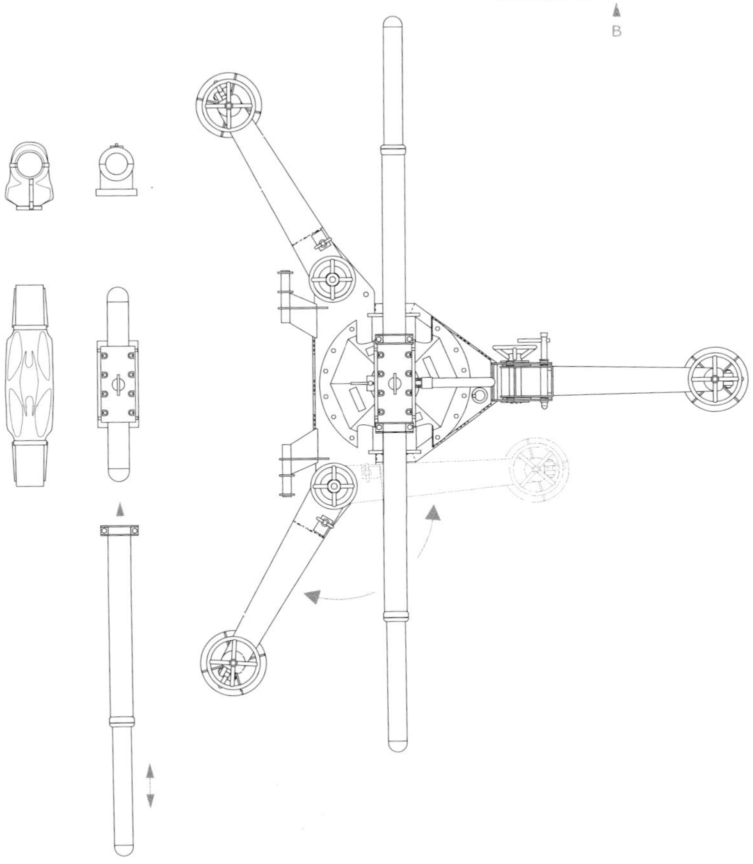

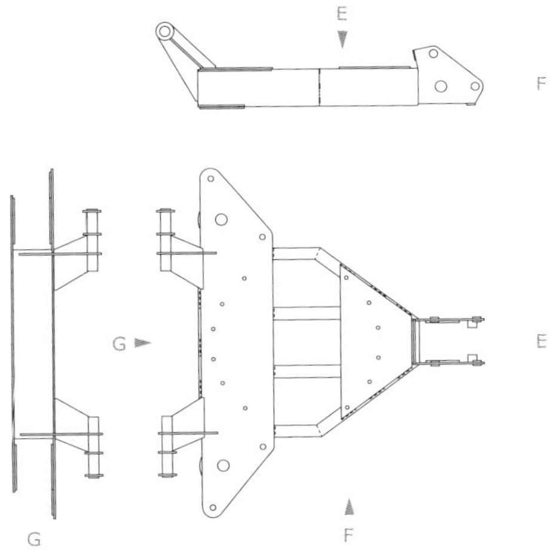

Drawing Dipol Fz.207 is also valid for Dipol for Fz.204.

Zeichnung Dipol Fz.207 gilt auch für Dipol Fz.204.

Vehicle 208

Internal or former designation: Fz.-Pos.-Nr.208
Guide beam controlling trailer (1-axle)
Koblenz

Fahrzeug 208

Interne oder frühere Bezeichnung: Fz.-Pos.-Nr.208
LS-Kontroll-Anhänger (1-achs.)
Koblenz

DMA, GD639.5.18
HAP11/T 507
Karlshagen, 1.6.1944
Bd.Nr. E 3195/44 g

DMA, GD639.5.1
T507
Fz.-Pos.-Nr. 208

DMA, GD630.02
A4 Lehrbuch
der FR-Schule
Mai 1944
Auszug aus Kapitel III
Technische Fahrzeuge
in der LS-Stellung
Ing. Gassmann

Intended use

The guide beam signal apparatus is housed in a single-axle trailer, which contains the crank mast, three 12 V accumulators, a converter, the control receiver, an antenna amplifier and the control panel.

Verwendungszweck

Die LS-Kontrollanlage ist in einem einachsigen Anhänger untergebracht, der den Kurbelmast, drei Sammler 12 V, einen Umformer, den LS-Kontroll-Empfänger, einen Antennenverstärker und die Bedienungstafel enthält.

157

157 A 25 m mast (type K136-23/7) was used for the controlling trailer. The dipole standing pipe is fixed to the antenna mast with clamps for transport. The antenna mast is mounted to rotate on a boom stand connected to the superstructure. Canvas covers are pulled over the support spindles, the antenna head and the dipole standpipe to protect against dirt.

157 Für den Kontroll-Anhänger wurde ein 25 m Kurbelmast (Type 136-23/7) verwendet. Das Dipolstandrohr ist zum Transport auf dem Antennenmast befestigt. Der Antennenmast ist an einem Auslegerbock drehbar montiert, der mit dem Aufbau verbunden ist. Gegen Verschmutzung sind Segeltuchhauben über die Stützspindeln, den Antennenkopf und das Dipolstandrohr gezogen.

158 The controlling trailer Fz.208 is pulled here by a guide beam transmission truck Fz.205. The vehicles are painted in RAL7028 dark yellow as »nach Muster«. The Fz.205 clearly shows the color corrections of subsequently introduced changes to the LC box body. Interesting is the use of different road wheel types on the bogie suspension of the tracked running gear. The road wheel are either in pairs, of one type only or mixed in use. On the roof is a camouflage net/mat made of wire (chicken wire) with riveted snippets of Mika or Astralon in different shapes and colours.

158 Der LS-Kontroll-Anhänger Fz.208 wird hier von einem LS-Sende-Kw Fz.205 gezogen. Die Fahrzeuge sind in RAL7028 dunkelgelb nach Muster gestrichen. Am Fz.205 sind deutlich die Farbausbesserungen von nachträglich eingebrachten Änderungen am LC-Koffer zu erkennen. Interessant ist die Verwendung verschiedener Laufrollentypen am Rollenwagen des Gleiskettenlaufwerks. Die Laufrollen sind entweder paarweise von einem Typ oder gemischt in Verwendung. Auf dem Dach ist ein Tarnnetz oder Tarnmatte aus Draht (Hühnerdraht) mit angenieteten Schnipseln aus Mika (Glimmer) oder Astralon in unterschiedlichsten Formen und Färbung abgelegt.

The chassis

The chassis used for the control trailer was the »Light vehicle trailer, single-axle, demountable« A1 (D.(Luft)T.8204. A cross beam with extendable swivel supports and plate spindles was mounted on the frame, and an additional third support was mounted at the bend of the drawbar. The final version in the picture at right had a longer body integrating equipment boxes for tools and antenna mast accessories.

Das Fahrgestell

Als Fahrgestell für den LS-Kontrollanhänger wurde ein Anhänger »Leichter Kfz.-Anhänger, einachsig, zerlegbar« A1 (D.(Luft)T.8204 verwendet. Dieser wurde um eine Quertraverse mit ausziehbaren Schwenkstützen und Tellerspindeln ergänzt. Im Knickpunkt der Zugdeichsel wurde eine Deichselstütze montiert. Die Abschlussausführung im rechten Bild hatte einen längeren Aufbau mit integrierten Gerätekästen für Werkzeug- und Antennenmastzubehör.

Deutsches Museum, München, Archiv- FA014/B908/44 T5224 [CD74725]

159

BArch, RH8II Bild-B0222-45, o.Ang.

160

The LS-controlling trailer has the following dimensions and weights:

Total length:	~ 3,500 mm w/o mast
Width:	1,635 mm
Height:	~ 1,560 mm
Track width:	~ 1,400 mm
Service weight:	830 kg
Wheels:	5.50-16

Der LS-Kontrollanhänger hat folgende Maße und Gewichte:

Gesamtlänge:	~ 3.500 mm ohne Mast
Breite:	1.635 mm
Höhe:	~ 1.560 mm
Spurweite:	~ 1,400 mm
Gewicht:	830 kg
Räder:	5,50-16

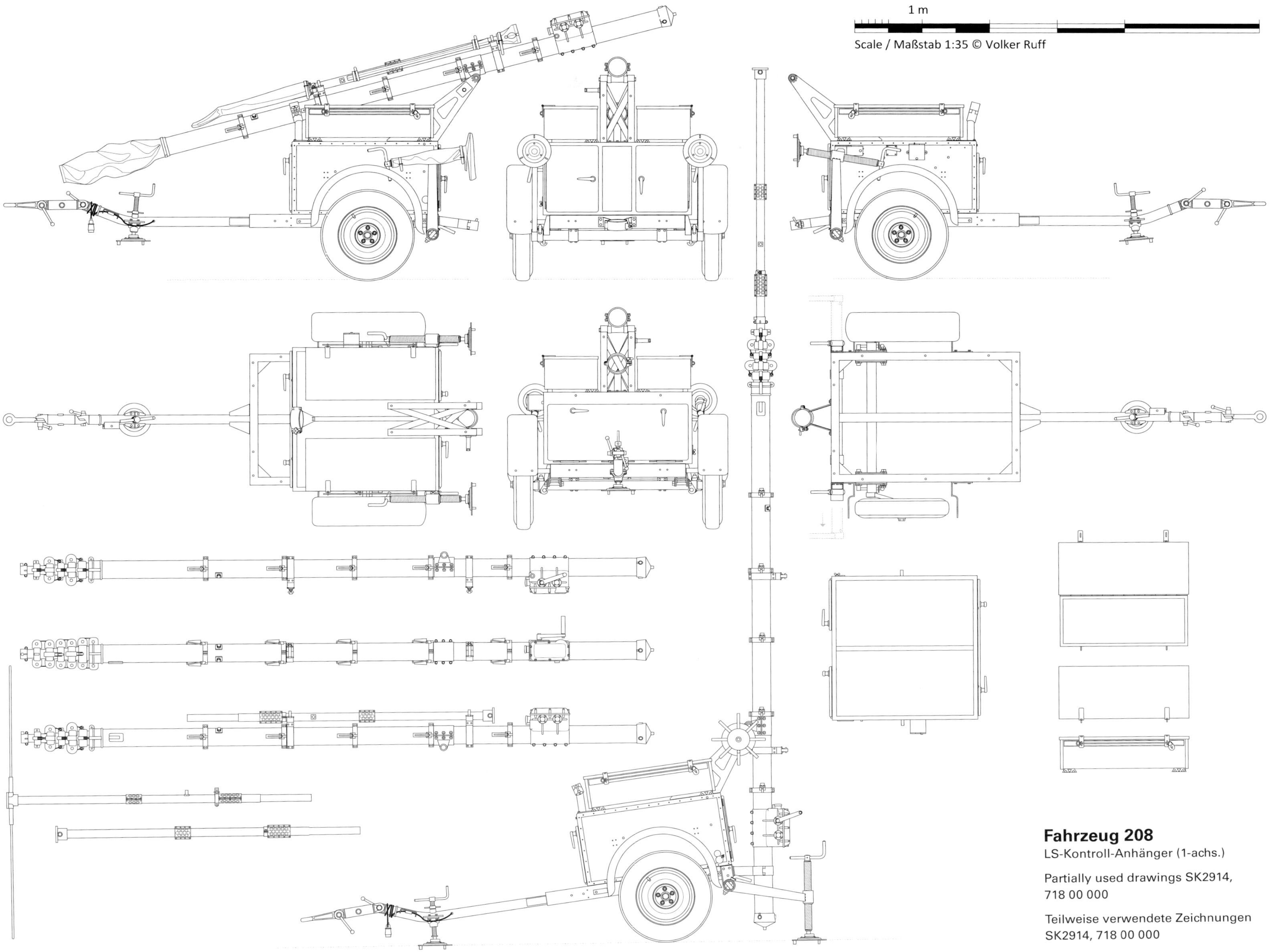

Scale / Maßstab 1:35 © Volker Ruff

Fahrzeug 208

LS-Kontroll-Anhänger (1-achs.)

Partially used drawings SK2914, 718 00 000

Teilweise verwendete Zeichnungen SK2914, 718 00 000

Deutsches Museum, München, Archiv- FA014/B936/44 T5224 (CD74715)

Body – rear view

A framework of angle profiles was placed on the chassis and with sheet metal panels. Two doors were hinged at the rear to seal the two installation spaces airtight. An AF amplifier and a potentiometer for the rectifier were mounted in the installation space on the left, below a HF amplifier was fitted. In the installation space on the right was a control panel with Oemig converter and a battery. A small reel for the antenna cable was mounted on the boom support. The HF cable was plugged into the left side wall and routed along the mast to the HF socket.

Aufbau – Rückansicht

Auf das Fahrgestell wurde ein Gerüst aus Winkelprofilen gesetzt und mit Blechtafeln verkleidet. An der Rückseite wurden zwei Türen angeschlagen, die die beiden Einbauräume luftdicht abschlossen. Im Einbauraum links war ein NF-Verstärker und ein Potentiometer für den Gleichrichter eingebaut, darunter befand sich der HF-Verstärker. Im Einbauraum rechts war eine Schalttafel mit Oemig-Umformer und einer Batterie untergebracht. Der Auslegerbock war für das Antennenkabel mit einer kleine Trommel ausgestattet. Das HF-Kabel wurde in der linken Seitenwand eingesteckt und am Kurbelmast entlang zur HF-Steckdose geführt.

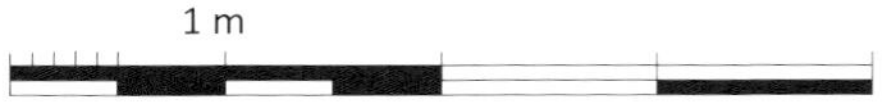

Scale / Maßstab 1:35 © Volker Ruff

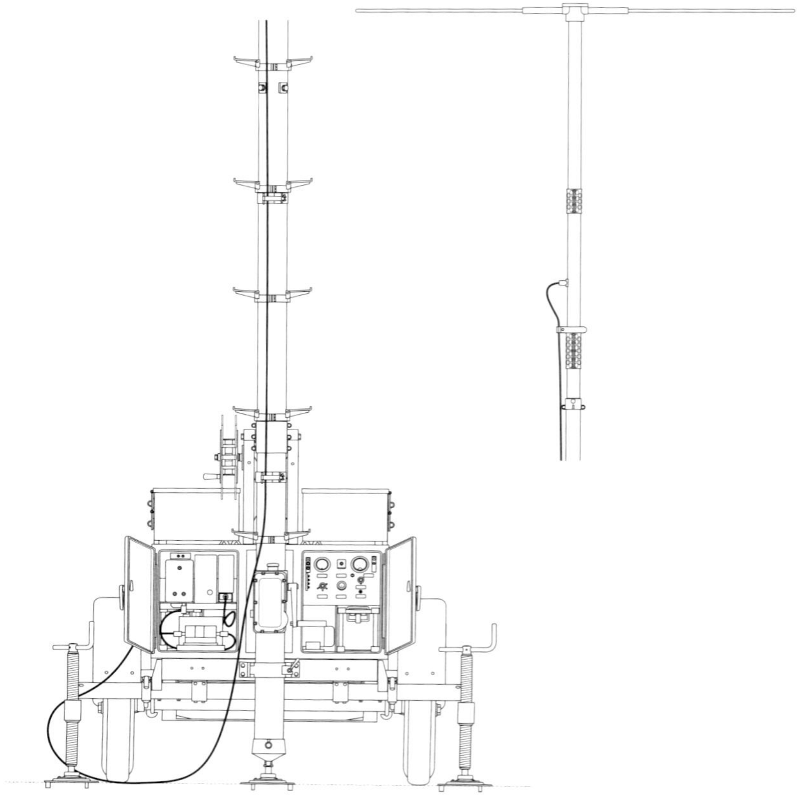

Body – front view

A large hatch was mounted on the front side to seal the installation space airtight. The LS control transmitter was operated exclusively with battery power. A power connection was not provided. The necessary batteries (numbered) were located in holders on the left and right side of the installation space. In the middle was a shaft for tools for antenna bracing. Several reels for the guy ropes to secure the mast were threaded onto two shafts. The shaft bearings were riveted on the inside of the cover. The two hammers for the piles of the guy ropes were not fixed to the cover but stowed in the central shaft.

Aufbau – Vorderansicht

An der vorderen Seite war eine große Klappe montiert, die den Einbauraum ebenfalls luftdicht abschloss Der LS-Kontrollsender wurde ausschließlich mit Batteriestrom betrieben. Ein Stromanschluss war nicht vorgesehen. Die notwendigen Batterien (mit Nummer) waren in Halterungen links und rechts im Einbauraum untergebracht. In der Mitte war ein Schacht für Werkzeug zur Antennenverspannung eingesetzt. Mehrere Trommeln für die Abspannseile zur Mastsicherung waren in zwei Wellen aufgefädelt. Die Wellenlager waren auf der Deckelinnenseite aufgenietet. Die beiden Pfahlhämmer waren nicht auf dem Deckel befestigt, sondern in dem mittigen Schacht verstaut.

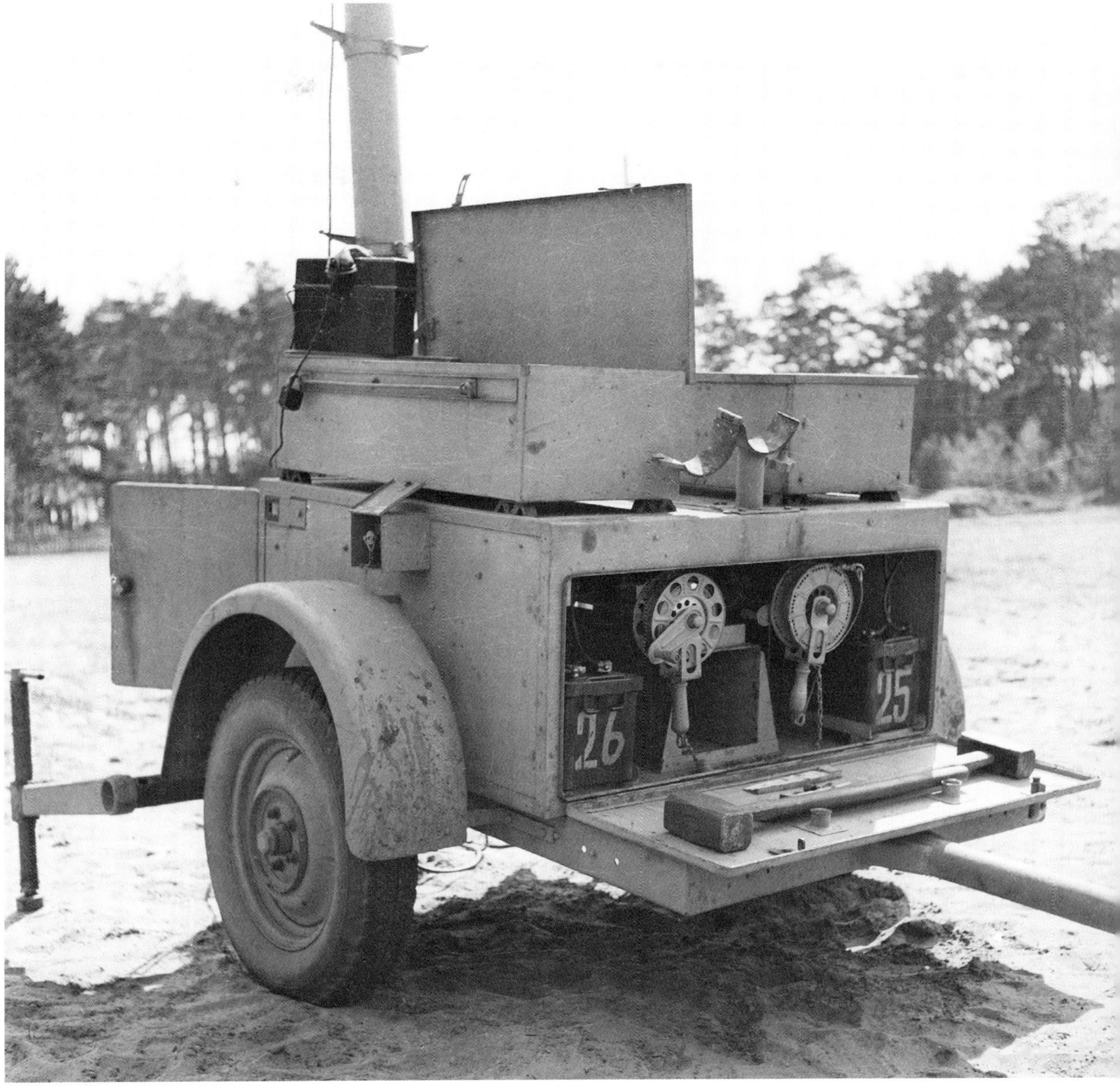

Deutsches Museum, München, Archiv- FA014/B935/44 T5224 (CD79115)

162

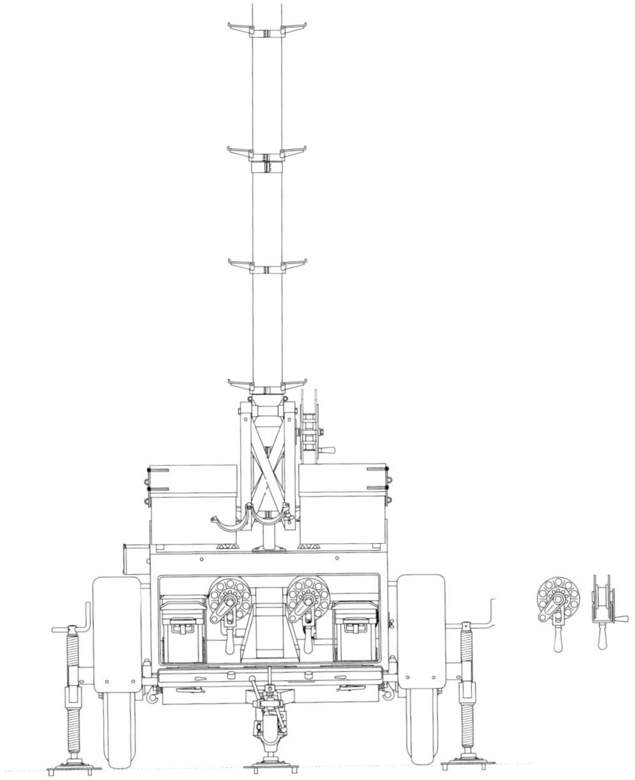

162 A box with a FF cable connection for a field telephone was attached to the side wall. The FF33 was transported in one of the two equipment boxes and removed if needed. The two boxes were mounted on rubber buffers on the body.
Between the two equipment boxes was a clamp (half shell) that secured the mast for transport.

162 An der Seitenwand war ein Kästchen mit einem FF-Kabelanschluss für einen Feldfernsprecher angebracht.
Der FF33 wurde in einem der beiden Kästen transportiert und bei Bedarf entnommen. Die beiden Gerätekästen waren auf Gummipuffern auf dem Aufbau montiert. Zwischen den beiden Gerätekästen befand sich eine Halteschelle (Halbschale) als Transportsicherung für den Mast.

VEHICLE GROUP III
FAHRZEUGGRUPPE III

DMA, GD639.5.12
OKH/Wa Prüf10
Karlshagen, 31.3.1944
Bd.Nr. 774/44 gK
Auszug
Besprechung der
Fahrzeuggruppe 3

Limber position

On call, a Werkstatttrupp, vehicle for spare parts and unneeded trailers, towing and supply vehicles of the firing platoon were parked near the firing position, the Limberposition (Protzenstellung).

Purpose and task
To repair the vehicles of the Division z.V. vehicle fleet, a workshop unit with an weapons specialist was assigned. Besides general maintenance work on the vehicles, maintenance and repair work on the special vehicles also had to be carried out.

Selected spare parts for the FR ground installations were carried by a separate vehicle (an Fz.304).

The equipment of this Werkstattrupps has been increased above the normal amount so that smaller repairs can be carried out on the *Gerät*. As a result it is a KW workshop with a weapons workshop. A cross-country vehicle is therefore not required.

The acquisition of additional Werkstatttrupps of this type is causing exceptional difficulties. Further development and equipment of the units at the Batteries is currently being clarified in conjunction with AHA, the Chief of WKW, and Wa Prüf 6 through B.z.b.V. Heer. The additional 9 units currently being acquired through Wa Prüf 11/III will at first be set up in the old version, and then be adapted to the revised requirements.

Protzenstellung

Ein Werkstattteam auf Abruf, Fahrzeuge für Ersatzteile sowie nicht benötigte Anhänger, Abschlepp- und Versorgungsfahrzeuge des Schiesszuges wurden in der Nähe der Feuerstellung, der Protzenstellung, geparkt.

Zweck und Aufgabe
Zur Instandsetzung der Fahrzeuge im Fuhrpark der Division z.V. wurde ein Werkstatttrupp mit Waffenmeisterei zugewiesen. Neben allgemeinen Wartungsarbeiten an den Fahrzeugen waren auch Instandsetzungs- und Reparaturarbeiten an den Sonderfahrzeugen auszuführen.

In einem weiteren Fahrzeug (Fz.304) wurden ausgewählte Ersatzteile für die FR-Bodenanlagen mitgeführt.

Die Ausrüstung dieser Werkstattrupps ist über die normale Ausrüstung noch so ergänzt worden, dass kleinere Instandsetzungen am Gerät durchgeführt werden können. Es handelt sich also hier um einem KW-Werkstatttrupp mit Waffenmeistereibetrieb. Es ist hier also ein geländegängiger Lkw nicht erforderlich.

Die Beschaffung weiterer Werkstatttrupps dieser Art macht außerordentliche Schwierigkeiten. Die weitere Entwicklung und Ausstattung des Werkstatttrupps bei den Batterien wird in Verbindung mit AHA, Chef WKW und Wa Prüf 6 durch B.z.b.V. Heer z.Zt. geklärt. Die z.Zt. durch Wa Prüf 11/III in Beschaffung befindlichen weiteren neun Trupps werden zunächst in alter Ausführung weiter gefertigt und den geänderten Forderungen dann angepasst.

BArch, RH 8/1310
Spezialfahrzeuge,
Baugruppen und
wesentliche Bauteile
der Bodenanlage FR

Special vehicles in the limber position
Fz.-Pos.-Nr.301 4,5 t truck closed I
Fz.-Pos.-Nr.302 4,5 t truck closed II
Fz.-Pos.-Nr.303 4,5 t truck open III
Fz.-Pos.-Nr.304 Spare items vehicle I on 3t truck closed (305)

Spezialfahrzeuge in der Protzenstellung
Fz.-Pos.-Nr.301 4,5 t Lkw geschl. I
Fz.-Pos.-Nr.302 4,5 t Lkw geschl. II
Fz.-Pos.-Nr.303 4,5 t Lkw offen III
Fz.-Pos.-Nr.304 Vorratsteilwagen I auf 3t Lkw geschl. (305)

Deutsches Museum, München, Archiv- FA014/B552/44 TD5224 [CD74797]

163

163 The Werkstatttrupp of Abt.836 with its vehicles on the army training ground of the z.V. Division in Groß-Born (Borne Sulinowo). Its vehicles are lined up here as follows: An Fz.301 and an Fz.302 with workshop box body on a 4.5 t Mercedes 4x4 truck, behind them an Fz.303 4.5 t open type Büssing 4x2 truck, a closed-body FR-spare parts Fz.304 truck (305), a motorcycle with sidecar, and a Steyr 2 to truck with a platform body.

163 Der Werkstatttrupp der Abt.836 mit seinen Fahrzeugen auf dem Truppenübungsplatz der Division z.V. in Groß-Born. Aufgereiht sind die Fahrzeuge: Fz.301 und Fz.302 mit Werkstattkofferaufbau auf einem Lkw 4,5 t Typ Mercedes 4x4 dahinter folgt ein Fz.303 Lkw 4,5 t offen Typ Büssing 4x2, ein FR-Ersatzteilwagen Fz.304 geschlo. (305), ein Beiwagenkrad und ein Steyr Lkw 2 t mit Pritschenaufbau.

Deutsches Museum, München, Archiv – FA014/B552/44 TD5224 [CD74797]

164

DMA, GD639.5.12
OKH/Wa Prüf10
Karlshagen, 31.3.1944
Bd.Nr. 774/44 gK
Auszug
Besprechung der
Fahrzeuggruppe 3
vom 31.3.44
in Karlshagen I

Production status: Werkstatttrupp

To date, deliveries between July and October 1943 consist of the following: 3 Werkstatttrupps for Battery 836 (without vehicle 304), 3 Werkstatttrupps for Battery 485 (without vehicle 304), 1 Werkstatttrupp for Battery 836 (without vehicle 304). Equipment for 3 additional Werkstatttrupps has remained in Rebstock since November.

Wa Prüf 11/III will immediately arrange the acquisition of vehicles for the three remaining Werkstatttrupps. They consist of: 6 trucks, 4.5-tonnes, with closed superstructure; 3 trucks, 4.5-tonnes, open; and 10 trucks, 3-tonnes, closed superstructure (incl. the missing vehicles of previous deliveries).

Current defects and their remedy

Unit experience is currently only available from Batteries 444 and 836. The repair work on the *Gerät* turned out differently than initially foreseen. No remarks about the vehicles and the equipment.

No change of equipment for the repair of the *Sondergeräte* can be deduced from previous unit experience.

The additional Werkstatttrupps, which first have to be clarified in terms of equipment according to #1, will be given the following suggestions on the occasion of this meeting:

A Werkstattzug (41) complement for its staff and for repair work on the batteries (according to Manual D 621/1) including additional equipment for device repair (according to Manual D 621/2). Due to this simplification it is possible to equip three additional units by using the three workshop equipment sets (LR) which have not been expended.

Suggestions for tool equipment for the repair of the *Sondergerät* at the battery cannot yet be provided by the meeting's participants. A list of spare parts for stocks including the electrical and engine-related sectors is available. The electrical sector alone indicates that the spare parts demanded by B.z.b.V. Heer - taking into account the 9 existing frequency changes - cannot be accommodated on one truck.

Fertigungslage: Werkstatttrupp

Bisher wurde ausgeliefert in der Zeit vom Juli bis Oktober 1943: 3 Werkstatttrupps an die Abteilung 836 (ohne Fahrzeug 304), 3 Werkstatttrupps an die Abteilung 485 (ohne Fahrzeug 304), 1 Werkstatttrupp an die Abteilung 836 (ohne Fahrzeug 304).

Für 3 weitere Werkstatttrupps lag die Ausrüstung seit November in Rebstock.

Die Beschaffung der Fahrzeuge für die drei letzten Werkstatttrupps veranlasst umgehend Wa Prüf 11/III. Es handelt sich dabei um: 6 Lkw 4,5 Tonner mit geschl. Aufbau. 3 Lkw 4,5 Tonner offen und 10 Lkw 3 Tonner geschl. (einschl. der fehlenden Fahrzeuge der bisher gelieferten Trupps).

Derzeitige Mängel und ihre Behebung

Truppenerfahrungen liegen z.Zt. nur von den Batterien 444 und 836 vor. Die Instandsetzungsarbeiten am Gerät waren hierbei jedoch anders, als sie künftig vorgesehen sind. Über die Kraftfahrzeuge und Ausstattung ist nichts zu bemerken.

Eine Änderung der Ausstattung für Instandsetzungen des Sondergerätes ist aus den bisherigen Truppenerfahrungen nicht abzuleiten.

Für die weitere Werkstatttrupps die nach 1 erst ausstattungsmäßig geklärt werden müssen, werden anlässlich dieser Tagung folgende Anregungen gegeben:

Die Abteilung erhält einen Werkstattzug (41) für den Stab und die Instandsetzungsstaffeln bei den Batterien (nach D-Vorschrift 621/1) diese mit zusätzlicher Ausstattung für Geräteinstandsetzung (nach D-Vorschrift 621/2). Durch die Vereinfachung ist die Ausstattung für drei weitere Abteilungen unter Verwendung der drei noch nicht verausgabten Werkstattausrüstungen (LR) möglich.

Anregung für die Werkzeugausstattung zur Sondergerät-Instandsetzung durch die Batterie können von den Tagungsteilnehmern noch nicht vorgebracht werden. Von HAP liegt eine Ersatzteilliste für Vorratsteile sowohl auf dem elektrischen – wie auch Triebwerkseitigen Sektor vor. Schon aus dem elektrischen Sektor ist ersichtlich, dass die vom B.z.b.V. Heer geforderten Ersatzteile unter Berücksichtigung der 9 vorhandenen Frequenzwechsel, nicht annähernd auf einen Lkw untergebracht werden können.

DMA, GD639.5.18
HAP11/T 507
Karlshagen, 1.6.1944
Bd.Nr. E 3195/44 g

Vehicle 301
Vehicle 302

Internal or former designation: Fz.-Pos.-Nr.301, Fz.-Pos.-Nr.302
4,5 t truck closed I, 4,5 t truck closed II
Werkstatttrupp

Intended use

As workshop vehicle the 4.5 t chassis of suppliers Büssing or Mercedes was choosen and a closed standard superstructure was mounted on it.

Installations and equipment corresponded to the manual for a »Kfz.-Werkstattzug der Matra-Werke« and were provided by the Army Ordnance Office according to their intended use. Manufacture of such box bodies was, among others, by the company Fahrzeug und Waggonbau Lindner in Ammendorf/Thuringia.

Fahrzeug 301
Fahrzeug 302

Interne oder frühere Bezeichnung: Fz.-Pos.-Nr.301, Fz.-Pos.-Nr.302
4,5 t Lkw geschl. I, 4,5 t Lkw geschl. II
Werkstatttrupp

Verwendungszweck

Als Werkstattfahrzeug wurden das 4,5 t Fahrgestell der Anbieter Büssing oder Mercedes gewählt und geschlossener Einheitsaufbau darauf montiert. Einbauten und Ausstattung entsprachen dem Handbuch für einen Kfz.-Werkstattzug der Matra-Werke und wurden von den Heereszeugämtern entsprechend ihrer Verwendung angepasst. Hersteller dieser Kofferaufbauten war unter anderem die Firma Fahrzeug und Waggonbau Lindner in Ammendorf/Thüringen.

165

165 The picture shows an issued provided Fz.301 (Mercedes 4.5 t truck chassis) with closed standard superstructure without mounted attachment walls. In rail transport the loading profile (railway) had to be complied with. The attachment walls were removed and carried in the interior, there the installations were adapted to the requirements and equipped with special tools for the FR battery. The vehicle Fz.302 was identical in its design.

165 Das Bild zeigt ein bereitgestelltes Fz.301 (Mercedes 4.5 t Lkw Fahrgestell) mit geschlossenem Einheitsaufbau ohne montierte Aufsatzwände. Im Schienentransport war das Ladeprofil (Eisenbahn) einzuhalten. Die Aufsatzwände wurden entfernt und im Innenraum mitgeführt, dort waren die Einbauten den Erfordernissen angepasst und mit Sonderwerkzeugen für die FR-Batterie ausgerüstet. Das Fahrzeug Fz.302 war baugleich.

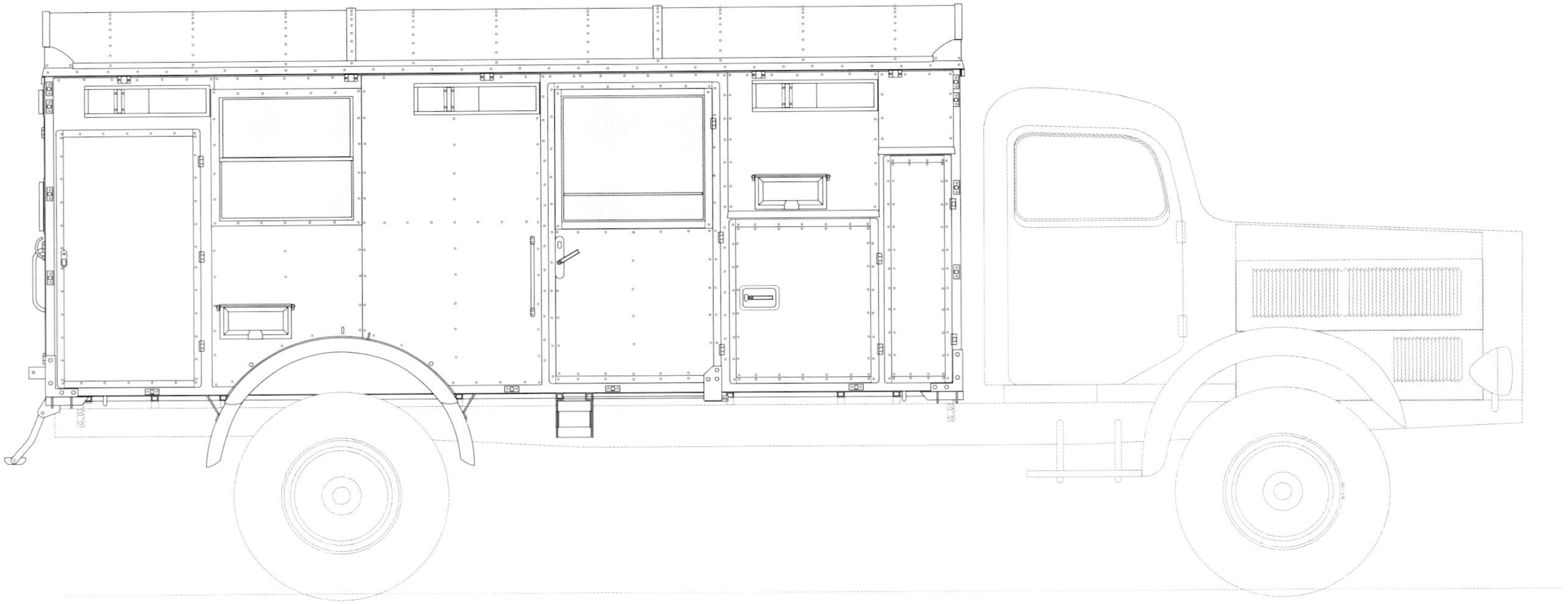

Scale / Maßstab 1:35 © Volker Ruff

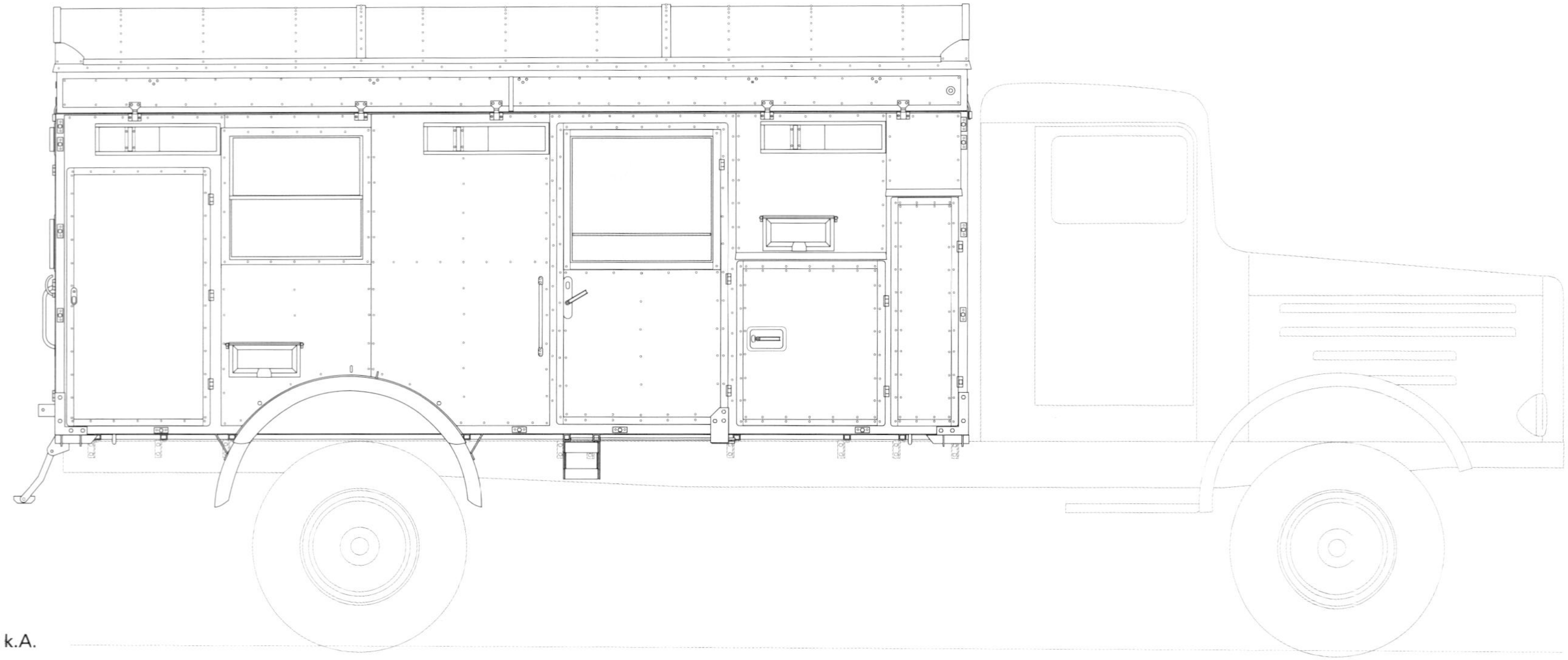

Fahrzeug 301
Fahrzeug 302

4,5 t Lkw geschl. I
4,5 t Lkw geschl. II

Drawing set used n.a.

Verwendeter Zeichnungssatz k.A.

Installations and equipment

The necessary tools and equipment of the workshop cars were set by the specifications of the D 621/1 and D 621/2 tables for a KW Werkstatttrupp with weapon repair and up keeping were provided by the Army Ordnance Office according to their intended use.

Einbauten und Ausstattung

Die notwendigen Werkzeuge und Ausstattung der Werkstattwagen entsprachen den Vorgaben der D 621/1 und D 621/2 für einen KW-Werkstatttrupp mit Waffenmeisterei. Diese wurden von den Heereszeugämtern entsprechend ihrer Verwendung bereitgestellt.

BArch, RH8II Bild-B1990-44, o.Ang.

166

BArch, RH8II Bild-B1989-44, o.Ang.

167

A view into the interior
Blick in den Innenraum

166, 167 In front of the bulkhead were additional, brightly painted boxes holding special tools and devices for the FR vehicles stored. Note on the rear access door the camouflage of the inside of the door. The separation point above the door between the fixed part of the superstructure and the lowerable roof assembly can also be seen. The roof part was locked by quick-release fasteners on its front and side walls. To ensure sufficient headroom the fasteners between the side walls and the roof were opened, the roof itself lifted upward, and narrow attachment walls (height approx. 250 mm) were placed on all side walls and re-fastened. The attachment walls had to be removed again for rail transport in order not to exceed the railroad loading profile.

166, 167 Vor der Stirnwand waren in zusätzlichen, hell lackierten Kisten Spezialwerkzeuge für die FR-Fahrzeuge und Geräte gelagert. An der Rückwand der Zugangstüre, man beachte die Tarnung der Türinnenseite, ist die Trennstelle oberhalb der Tür für den festen Aufbau (Koffer) und der Dachbaugruppe zu erkennen. Die Dachbaugruppe war mit Schnellverschlüssen an den Stirnseiten und Seitenwänden verriegelt. Für eine ausreichende Durchgangshöhe wurden die Verschlüsse an den Seitenwänden zur Dachbaugruppe geöffnet, die Dachbaugruppe nach oben gehoben, schmale Aufsatzwände (Höhe ca. 250 mm) auf alle Seitenwände gesetzt und verriegelt. Für den Bahntransport mussten die Aufsatzwände wieder entfernt werden, da sonst das Ladeprofil (Eisenbahn) überschritten worden wäre.

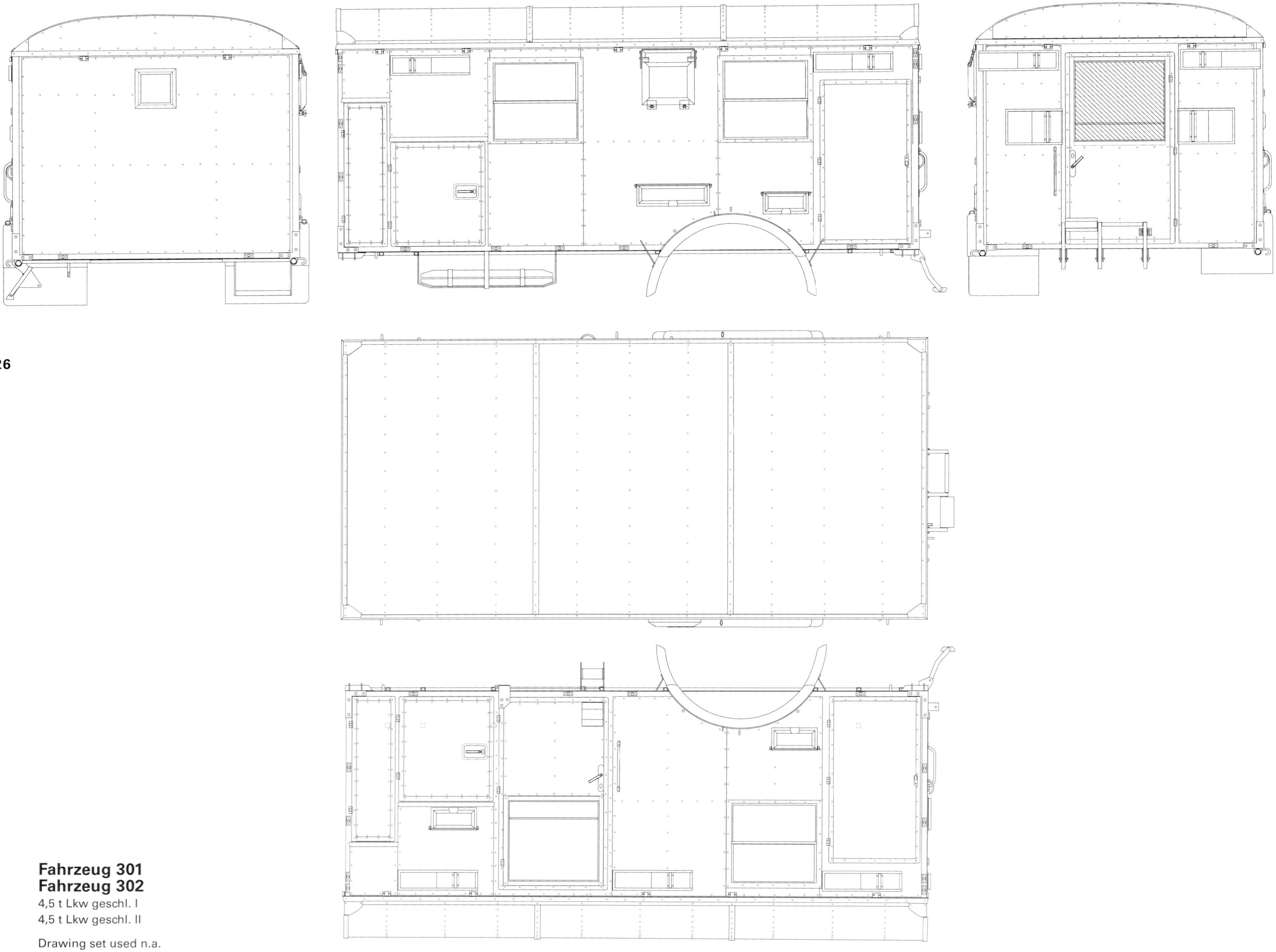

Fahrzeug 301
Fahrzeug 302

4,5 t Lkw geschl. I
4,5 t Lkw geschl. II

Drawing set used n.a.
Verwendeter Zeichnungssatz k.A.

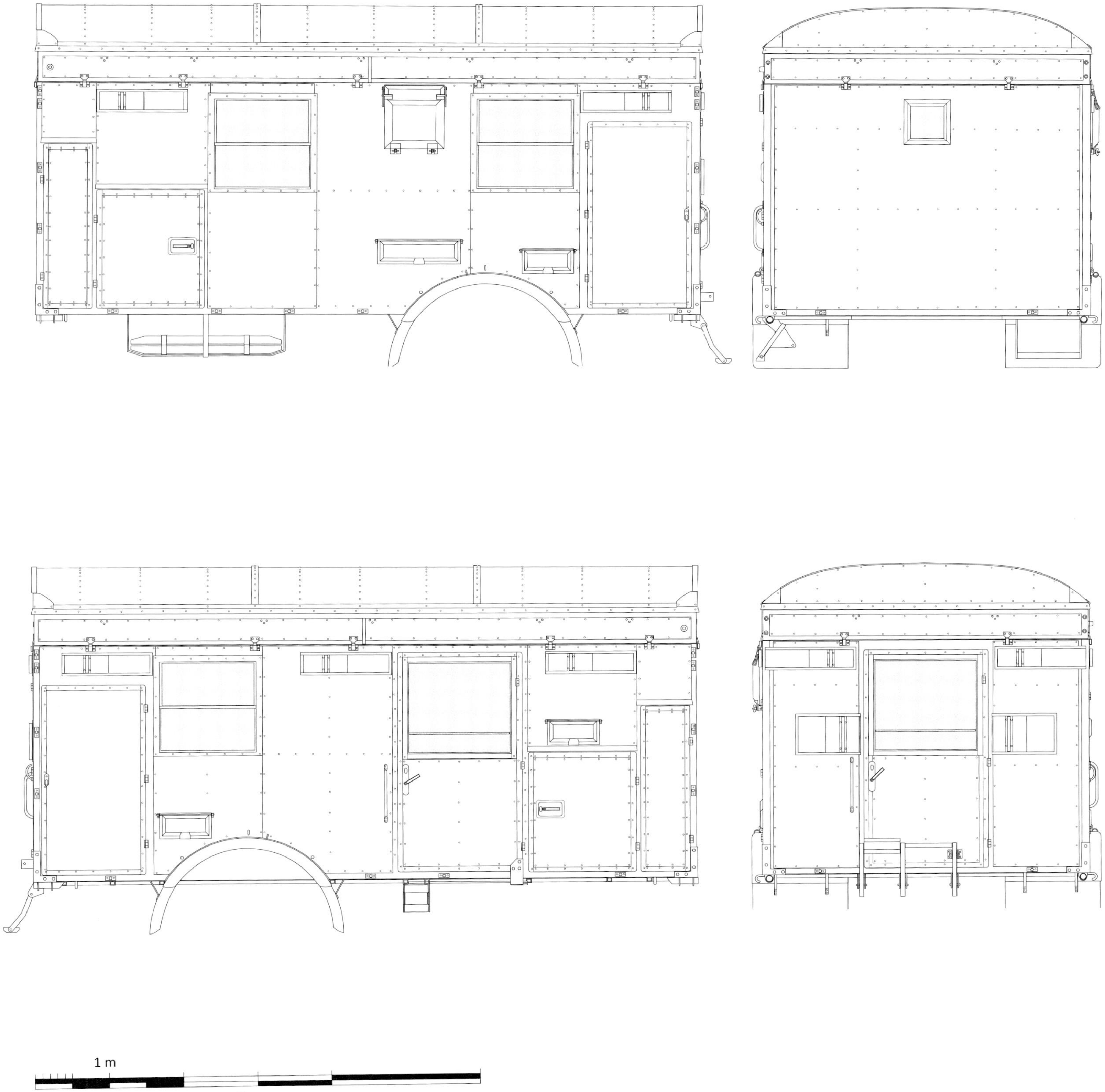

Scale / Maßstab 1:35 © Volker Ruff

DMA, GD639.5.18
HAP11/T 507
Karlshagen, 1.6.1944
Bd.Nr. E 3195/44 g

Vehicle 303

Internal or former designation: Fz.-Pos.-Nr.303
4,5 t truck open III
Werkstatttrupp

Intended use

This spare parts vehicle is a 4.5 t standard truck with platform body equipped with standard workbenches and six spare parts racks. These installations are protected from the weather by a tarpaulin cover. The parts stowed in these installations consist mainly of spare parts for the motorcycles, small trucks and other wheeled vehicles that are used by the FR Abteilungs.

Fahrzeug 303

Interne oder frühere Bezeichnung: Fz.-Pos.-Nr.303
4,5 t Lkw offen III
Werkstatttrupp

Verwendungszweck

Bei diesem Ersatzteilwagen handelt es sich um einen 4,5 t Standard-Lkw mit Pritschenaufbau, ausgerüstet mit Einheitswerkbänken und sechs Ersatzteilgestellen. Die Einbauten sind durch ein Planenverdeck vor Witterungseinflüssen geschützt. Der Vorrat an Teilen, die in den Einbauten im Fahrzeug verstaut sind besteht vor allem aus Kfz-Ersatzteilen für Motorräder, Kleinlastwagen und Räderfahrzeuge, die auf den Fuhrpark der FR-Abteilungen abgestimmt sind.

APG

168

Excerpt from manual D662/6 Kfz.-Werkstattzug (workshop) status 6.8.42

Six small standard workbenches (EWB I) are mounted on the left sidewall and three large standard workbenches (EWB II) on the right sidewall. A rack for storing six steel cylinders for oxygen and acetylene and various sheets and pipes is attached to the front wall. Spare parts racks are mounted on the workbenches on the left and on the right.

On the platform floor a cold sawing machine with electric motor is mounted on the left and a hand lever plate shear on the right.

Auszug aus der D662/6 Kfz.-Werkstattzug Ausf.1941 Stand 6.8.42

Entlang der Bordwände sind links sechs kleine Einheitswerkbänke (EWB I) und rechts drei große Einheitswerkbänke (EWB II) montiert. An der Stirnwand ist ein Gestell zur Lagerung von sechs Stahlflaschen (für Sauerstoff und Azetylen), Blechtafeln und Rohren angebracht. Ersatzteilgestelle sind links und rechts auf den Werkbänken befestigt.

Auf dem Pritschenboden sind auf Bodenplatten links eine Kaltsägemaschine mit Elektromotor und rechts eine Hebelblechschere montiert.

Auszug aus D662/2

167 Gesamtansicht von hinten

170 Gesamtansicht linke Fahrzeugseite

171 Gesamtansicht rechte Fahrzeugseite

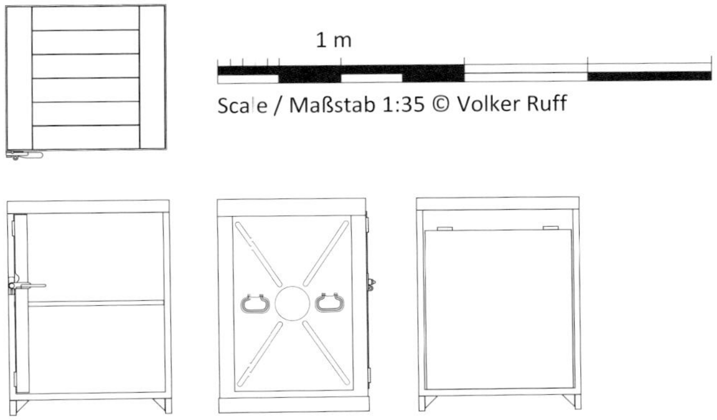

Scale / Maßstab 1:35 © Volker Ruff

Vehicle 304

Internal or former designation: Fz.-Pos.-Nr.304
Spare parts truck I on 3 t truck closed (305)
Spare parts vehicle

DMA, GD639.5.18
HAP11/T 507
Karlshagen, 1.6.1944
Bd.Nr. E 3195/44 g

DMA, GD639.5.3
T 507
Fz.-Pos.-Nr. 304

Intended use

The medium truck (o) closed superstructure (305) LC box body on the 3 t Opel 6700 Type A chassis was chosen as the base vehicle for the FR-Ers.Teile (Werkstatt) Fz.304. Four racks were fitted in its interior to store transport boxes for the spare parts (*Sondergerät*) and ground equipment. On the reinforced roof were mounts for fixed ladders and a rack for spare hoses.

Fahrzeug 304

Interne oder frühere Bezeichnung: Fz.-Pos.-Nr.304
Vorratsteilewagen I auf 3 t Lkw geschl. (305)
Ersatzteilwagen

Verwendungszweck

Der mittlere Lastkraftwagen (o) mit geschlossenem Aufbau (305) LC-Koffer auf Fahrgestell 3 t Opel 6700 Type A wurde als Basisfahrzeug für das FR-Ers.Teile (Werkstatt) Fz.304 gewählt. Im Innenraum wurden vier Gestelle zur Lagerung von Transportkisten für Ersatzteile (Sondergerät) und den Bodenaufwand mitgeführt. Auf dem verstärkten Dach sind Aufnahmen für Steigleitern und eine Galerie für Ersatzschläuche montiert.

Deutsches Museum, München, Archiv- FA014/B993/44 T5224 [CD74712]

172

172 The Luftwaffe's so-called »LC box body« was made in a lightweight board construction and could be dismantled. It had a wall thickness of 30 mm, with internal struts at uniform spacings for the attachment of installations. The arrangements of flaps, windows and doors were also standardized

172 Der sogenannte »LC-Koffer« der Luftwaffe wurde in Leichtplattenbauweise gefertigt und war zerlegbar. Dieser Aufbau hatte eine Wandstärke von 30 mm mit Holzständerwerk und Füllhölzern, die in gleichmäßigen Abständen montiert zur Befestigung von Einbauten vorgesehen waren. Auch die Anordnung der Klappen, Fenster und Türen war genormt.

Deutsches Museum, München, Archiv- FA014/B994/44 T5224 [CD74711]

173

173 The construction (wooden frame design with wooden post with filling and pressed wooden panels) of the LC box bodies also let them be dismantled for easy adaptation of their interior and exterior layouts to the equipment requirements of the Luftwaffe and Heer or of supply office. On the Fz.304 an additional luggage box was mounted on the step of the cabin roof.

173 Die Bauweise (Holz-Rahmenwerk, Holzständer mit Füllhölzern und Holzpresstafeln) der zerlegbaren LC-Kofferaufbauten erlaubte auch eine einfache Anpassung im Innen- und Außenbereich an die Ausrüstungsbedürfnisse von Luftwaffe, Heer und Zeugämter. Am Fz.304 war zusätzlich eine Gepäckkiste auf der Trittfläche des Fahrerhauses montiert.

174

DMA, GD639.5.12
OKH/Wa Prüf10
Karlshagen, 31.3.1944
Bd.Nr. 774/44 gK
Auszug
Besprechung der
Fahrzeuggruppe 3

Production status: Vehicle 304
Total initial requirement 8 as:
3 for 836 immediately,
1 as a replacement for 836,
3 for 485,
1 as a replacement for 485.
As present 9 vehicles are on the way from LR to Abt. 271. These vehicles will be managed at 271 when at the beginning of this week (4.4.) the meeting of the B.z.b.V. Heer has clarified the scope of the spare parts to be carried. Under certain circumstances, 3 vehicles of these 9 are assigned to unit 485 and 836 for their own installation.

Current deficiencies and their elimination

A decision about vehicle 304 cannot be made during the meeting. There is a consensus that the spare parts of the engine and electrical sectors which have to be carried will have to be stowed in individual dust-tight and moisture-proof transport boxes – not in a mounting that is connected to the body. In the event of vehicle failure, only individually transportable device boxes can make possible the transfer of the parts without endangering them.

HAP 11 will carry out a test loadings on both a closed and an open 3-tonne truck by April 4, 1944, and propose a shortened loading plan for vehicle 304 based on volumes and weights. With experience of the two loaded vehicles (closed and open) it is intended to call for a new meeting with representatives of both B.z.b.V. Heer and the units. A decision about the quantity of spare parts to be carried with is still to be made by B.z.b.V. Heer. Reference is made to the respective request of February 28.

Fertigungslage: Fahrzeug 304
Gesamtbedarf zunächst 8 Stück und zwar:
3 für 836 sofort,
1 als Ersatz für 836,
3 für 485,
1 als Ersatz für 485.
Z.Zt. sind 9 Fahrzeuge von LR zur Abt. 271 unterwegs. Diese Fahrzeuge werden bei 271 in Angriff genommen, wenn Anfang dieser Woche (4.4.) durch die Besprechung des B.z.b.V. Heer Klarheit geschaffen ist über den Umfang der mitzuführenden Ersatzteile. Unter Umständen werden je 3 Fahrzeuge von diesen 9 an Abt.485 und 836 zum eigenen Einbau zugewiesen.

Derzeitige Mängel und ihre Behebung

Über das Fahrzeug 304 kann bei der Tagung nicht entschieden werden. Einmütigkeit besteht darüber, dass die mitzuführenden Ersatzteile auf dem Triebwerk- und elektrischen Sektor in einzelnen, staub- und feuchtigkeitsdichten Transportkisten untergebracht werden müssen und nicht in einem mit dem Aufbau verbundenen Einbau. Nur transportable Gerätekisten ermöglichen bei Ausfall eines Fahrzeuges eine ordnungsgemäße Umladung ohne Gefährdung der Teile selbst.

HAP 11 macht bis 10.4.1944 einen Beladeversuch sowohl auf einem geschl. 3 t Lkw, wie auf einem offenen Lkw und schlägt dahin auf Grund der Volumen- und gewichtsmäßigen Möglichkeiten einen verkürzten Beladeplan für das Fahrzeug 304 vor. Es ist beabsichtigt, dass dann an Hand dieser beiden beladenen Fahrzeuge (geschl. und offenen Lkw) eine erneute Besprechung mit Vertretern von B.z.b.V. Heer und Truppe einberufen wird. Eine Entscheidung über Umfang der mitzuführenden Ersatzteile steht von B.z.b.V. Heer noch aus. Es wird an den diesbezüglichen Antrag vom 28.2. verwiesen.

Production and sourcing documents respectively

Production documents for the workshop units delivered to date have not been additionally prepared by HAP 11 or Wa Prüf 11/III. Their subsequent installation is unnecessary, as other equipment must be provided anyway.

An additional list supplementing the normal loading plan for workshop motor vehicles is planned. This list must deal with the parts being dropped or added respectively for the vehicles 301, 302, and 303. b) For the new kind of Werkstattzug or maintenance squad respectively the following manufacturing and acquisition documents have to be clarified after making a decision about the suggestion mentioned before:

1. Mobile workshop vehicles loading plan for the Werkstattzug. To be established by the supplementing or cancelling of 612/2 by B.z.b.V. Heer.
2. Battery loading plan by addition or cancellation respectively of D622/1. A list of existing special tools for the repair and maintenance of the *Sondergeräte*, housed in the test vehicle, will be recorded by HAP.11 by 4.4.1944 and assigned to B.z.b.V. Heer. A decision has to be taken by B.z.b.V. Heer as to which maintenance activities can be done by the battery (J.-Staffel) and which have to be done by the base workshop. The resulting final equipment of the I.-Staffel, including the necessary changes to internal stowage, will be communicated to HAP11 by B.z.b.V. Heer. HAP 11 will produce all drawings for these internal changes and provide the complete document survey to Wa Prüf 11/III.
3. A loading plan for the spare part vehicle has to be decided, after a test load and it is decided if a closed or open truck will be used in the future. Besides the space and weight balancing for this available vehicle it is important to know the maintenance activity taken by the battery (J.-Staffel) and which has to be done by the field storage in the future.

Because spare parts vehicles are still missing from the delivered Werkstatttrupps this decision has to be taken as soon as possible for the final solution which should remain unchanged. It is proposed that this loading plan be done by the units themselves. It will then be used to draw up the necessary documents for production and procurement.

draft signed by Rossmann
F.d.R.

Fertigungsunterlagen bzw. Beschaffungsunterlagen

Fertigungsunterlagen für die bisher ausgelieferten Werkstatttrupps sind seitens HAP 11 bzw. Wa Prüf 11/III nicht zusätzlich aufgestellt. Ihre nachträgliche Aufstellung erübrigt sich, da ohnehin eine andere Ausstattung vorgesehen werden muss. Eine Zusatzliste zu dem normalen Beladeplan für Werksattkraftwagen ist vorgesehen. In dieser Liste müssen aufgeführt werden: fortfallende bzw. hinzukommende Teile für Fahrzeuge 301, 302 und 303.

Für die neue Art des Werkstattzuges bzw. Instandsetzungstrupps sind folgende Fertigungs- bzw. Beschaffungsunterlagen klarzustellen nach Entscheidung des vorher erwähnten Vorschlages:

1. Beladeplan der Werkstattkraftwagen für den Werkstattzug. Aufstellung durch Ergänzung bzw. Streichung der D612/2 durch B.z.b.V. Heer.
2. Beladeplan der Batterien durch Streichung bzw. Ergänzung der D622/1. Eine Liste über die zur Sondergeräte-Instandsetzung erforderlichen Werkzeuge erstellt unter der im Prüfwagen schon untergebrachten Werkzeuge, HAP 11 bis 4.4.1944 an B.z.b.V. Heer. Durch B.v.b.V. Heer ist zu entscheiden, welche Arbeiten zukünftig durch die Batterie (J-Staffel), und welche Arbeiten durch den Feldspeicher bzw. auszuführen sind. Die daraus sich ergebene endgültige Ausstattung der J-Staffel einschl. der notwendigen Änderung der Einbauten wird durch B.z.b.V. Heer an HAP 11 mitgeteilt. HAP 11 stellt erforderlichenfalls Zeichnungen für diese Einbauänderungen auf und gibt die Gesamtbeschaffungsunterlagen an Wa Prüf 11/III.
3. Beladeplan für den Ersatzteilwagen, wobei noch zu entscheiden ist, nach dem oben erwähnten Beladeversuch, ob der Ersatzteilwagen künftig mit geschl. Aufbau oder offen in Frage kommt. Neben den Raum- und Gewichtsverhältnissen dieser zur Verfügung stehenden Fahrzeuge, ist von ausschlaggebender Bedeutung für die Festlegung der Beladung, welche Arbeiten durch die Batterie, bzw. durch Feldspeicher usw. künftig hin vorzunehmen sind.

Da für die bisher ausgelieferten Werkstatttrupps Ersatzteilwagen überhaupt noch fehlen, muss diese Entscheidung baldigst getroffen werden, zweckmäßig für die Endlösung, die auch für die bisherige dann unverändert in Frage kommen muss. Es wird vorgeschlagen, dass dieser Beladeplan durch die Truppe aufgestellt wird. Er dient dann als Unterlage für die Aufstellung der erforderlichen Fertigungs- und Beschaffungsunterlage.

Im Entwurf gezeichnet Roßmann
F.d.R.

Deutsches Museum, München, Archiv- FA014/B995/44 T5224 [CD74710]

DMA, GD639.5.9
Auszug
Beladeplan Fz.304

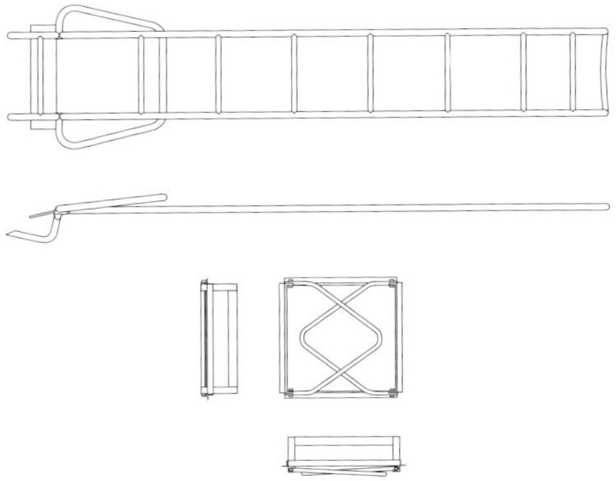

Roof with railing

The roof of the LC box body was reinforced by 30 wooden slats and made walkable. In the middle, the wooden slats were longer and surrounded by a wooden frame used to store spare hoses. The following stowed parts were carried here along:

- lifting basket, complete
- basket Rope
- ladder to heating tank Fz.407
- A-Stoff hose* NW 100 x 1000 mm
- A-Stoff hose* NW 70 x 3500 mm 2 pieces
- A-Stoff hose* NW 70 x 1200 mm 2 pieces
- B-Stoffl hose NW 70 x 2500 mm 2 pieces
- B-Ztoff hose NW 70 x 4000 mm
- B-Stoffl hose NW 60 x 3500 mm
- T-Stoff hose NW 32 x 3000 mm 2 pieces
- collapsible ladder with 6 m total length

* So-called Tombak hose – hose reinforced with a brass mesh.

Dachgalerie

Das Dach des LC-Koffers wurde mit 30 Holzlatten begehbar verstärkt. In der Mitte sind die Holzlatten länger und von einem Holzrahmen eingefasst. Dieser dient zur Aufbewahrung von Ersatzschläuchen. Auf dem Dach werden folgende Vorratsteile mitgeführt:

- Aufzugskorb, vollständig
- Aufzugseil
- Leiter zum Heizbehälter Fz. 407
- A-Stoff-Schlauch* NW 100 x 1000 mm
- A-Stoff-Schlauch* NW 70 x 3500 mm 2 Stück
- A-Stoff-Schlauch* NW 70 x 1200 mm 2 Stück
- B-Stoff-Schlauch NW 70 x 2500 mm 2 Stück
- B-Stoff-Schlauch NW 70 x 4000 mm
- B-Stoff-Schlauch NW 60 x 3500 mm
- T-Stoff-Schlauch NW 32 x 3000 mm 2 Stück
- Leiter geteilt mit 6 m Gesamtlänge

* Sog. Tombak-Schlauch (mit einem Tombak-Gewebenetz verstärkter Schlauch).

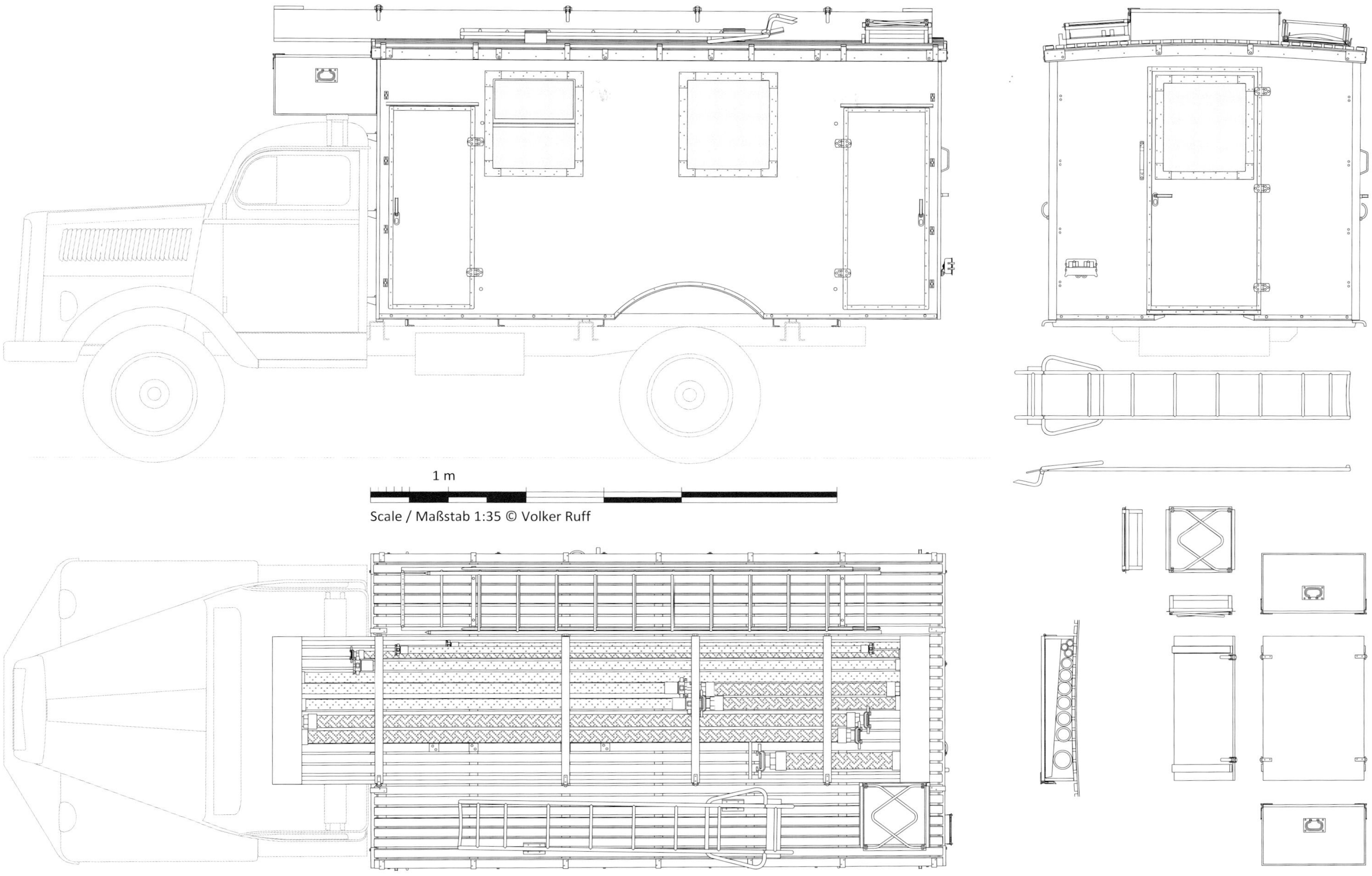

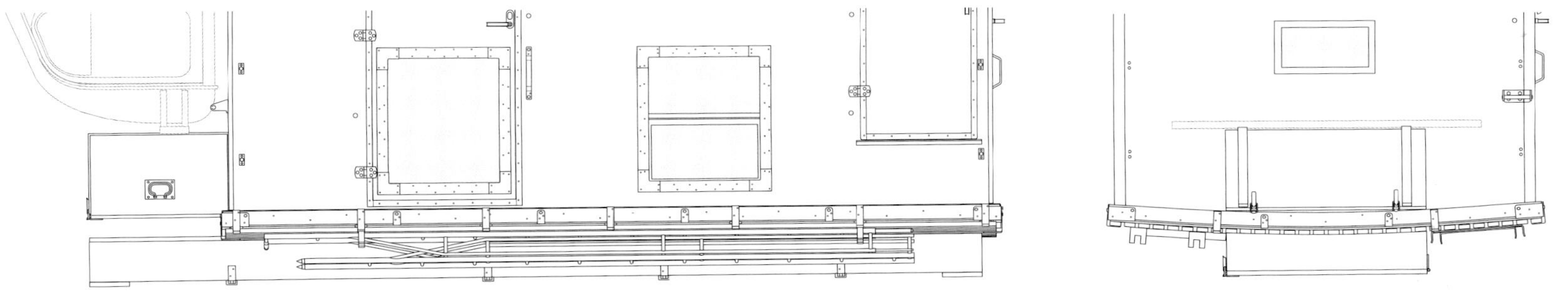

Fahrzeug 304

Vorratsteilewagen I auf 3 t Lkw
geschl. (305)

Partially used drawings by group
list 21Gr360

Teilweise verwendete Zeichnungen
nach Gruppenliste 21Gr360

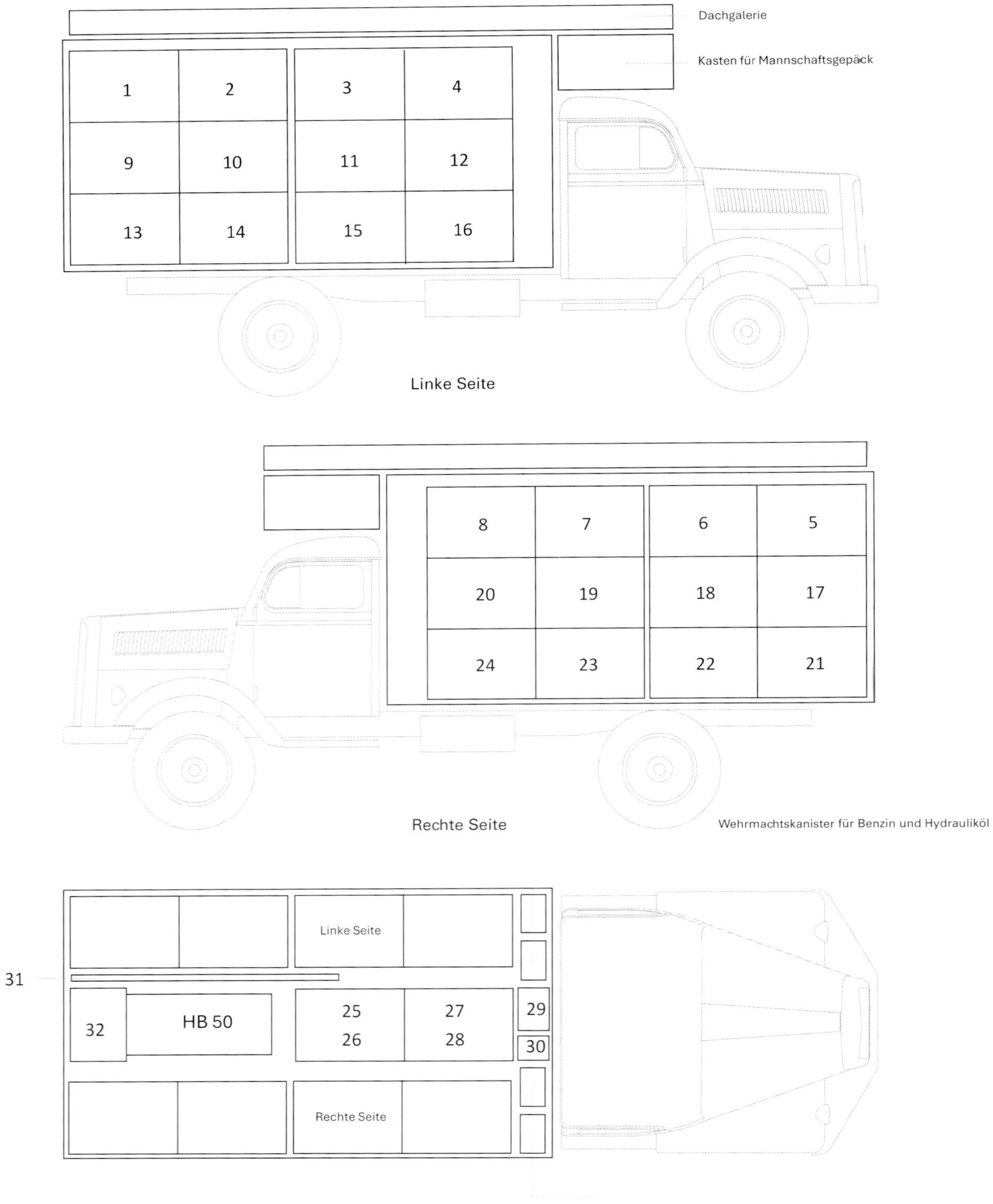

Unterbringung der Vorratsteilekisten 1-32 in den Einheitsgestellen auf der linken und rechten Fahrzeugseite und im Mittelgang des Fahrzeuges (304)

Auszug GD639.5.9 – Abbildung 3

DMA, GD639.5.9
Auszug
Beladeplan Fz.304

Loading plan

Transport in the motor vehicle is on four wooden frames (two each at right and left in the direction of travel), used to hold 24 transport boxes. The wooden racks are interchangeable and can be loaded onto other vehicles (similar group) if required. Six more transport boxes are carried on the center floor between them. Over 350 different stock items are stored in these transport boxes.

Sondergerät

- Engine exterior parts
- Engine Internal parts equipment compartment
- engine rear internal parts
- engine Internal parts T-unit
- standard parts and screw connections
- electrical internals
- frequency changer

Ground equipment

- test vehiclea I and III
- FR Trailer
- A-Stoff trailer
- B-Stoff truck
- T-Stoff truck
- T-Stoff heating equipment
- pump trailer

Beladeplan

Als Transportmittel befinden sich im Kraftfahrzeug vier Holzgestelle (je zwei Stück rechts und links in Fahrtrichtung). Diese dienen zur Aufnahme von 24 Transportkästen. Die Holzgestelle sind austauschbar und können bei Bedarf auf andere Kraftfahrzeuge (ähnlicher Gruppe) verlastet werden. Sechs weitere Transportkästen sind im Mittelgang untergebracht. Über 350 verschiedene Vorratsartikel sind in den Transportkästen verstaut.

Sondergeräte

- Triebwerk Außenteile
- Triebwerk Innenteile Geräteraum
- Triebwerk Innenteile Heck
- Triebwerk Innenteile T-Anlage
- Normteile und Verschraubungen
- Elektrische Einbauten
- Frequenzwechselmittel

Bodenaufwand

- Prüfwagen I und III
- FR-Anhänger
- A-Stoff-Anhänger
- B-Stoff-Lkw
- T-Stoff-Lkw
- T-Stoff-Anwärmgeräte
- Pumpenanhänger

A view into the interior
Blick in den Innenraum

176 Wooden frames were mounted at right and left and loaded with transport boxes 1 to 24. Wooden bars were inserted in front of the boxes for security during transport and are easy to remove. The transport boxes followed a loading plan and were marked with the box number and vehicle number. Four Wehrmacht canisters for petrol and hydraulic oil are stowed on the rear wall. On the floor center are boxes 29, 30 (Z-Stoff preheating container) and box 31 on the floor. Boxes 25 to 28 and a hot air blower HB/50 are also located on the floor center (not in the picture).

176 Holzgestelle waren rechts und links montiert und mit den Transportkästen 1 bis 24 belegt. Einfach zu entfernende Holzriegel vor den Holzkästen dienten als Transportsicherung. Die Transportkisten folgten einem Beladeplan und waren mit Kasten- und Fahrzeugnummer markiert. An der Rückwand waren vier Wehrmachtskanister für Benzin und Hydrauliköl verstaut. Auf diesem Bild stehen im Mittelgang die Kästen 29, 30 (Z-Stoff-Vorwärmbehälter) und Kasten 31 auf dem Boden. Die Kästen 25 bis 28 und ein Heissluftbläser HB/50 sind ebenfalls im Mittelgang (nicht im Bild) untergebracht.

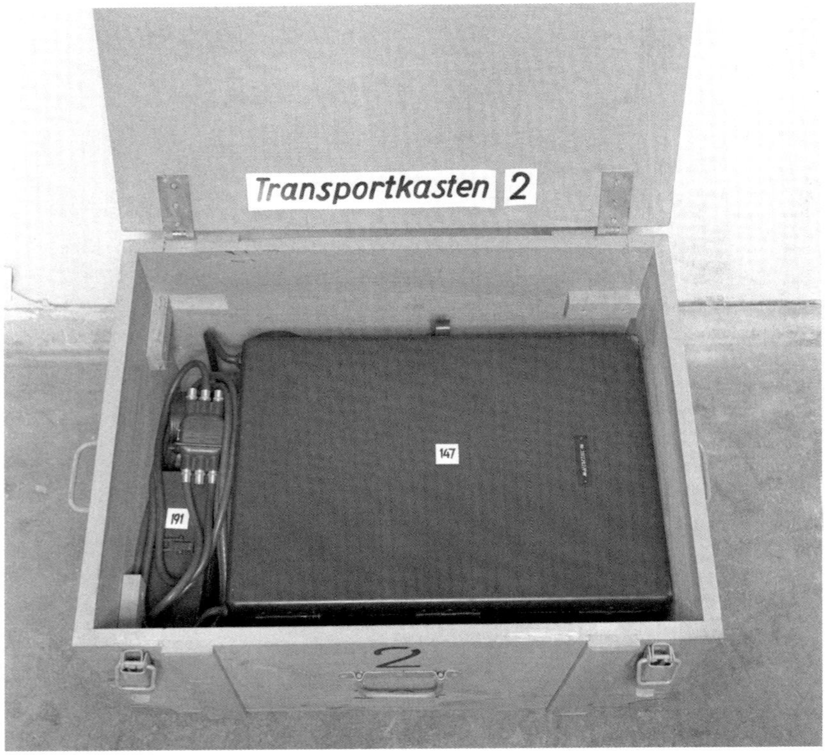

177

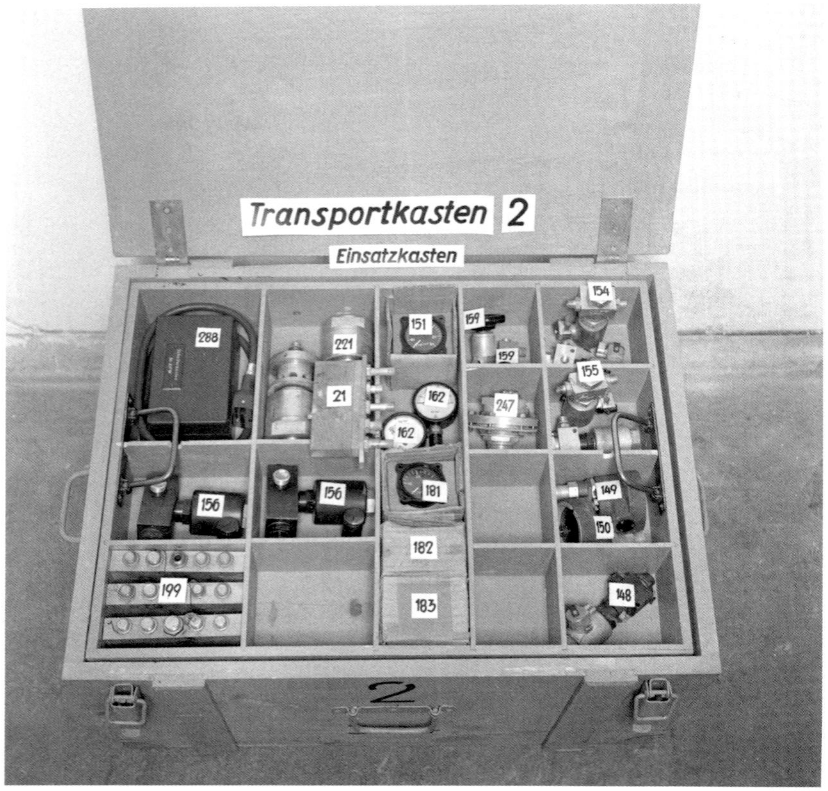

178

Transport boxes

Example of the equipment of the transport boxes

Transportbehälter

Beispiel für die Ausstattung der Transportkästen

177 Transport box 2

Transport box 2 contains the valve control (147), an attachment for the start table Fz.103 and an electrical switch box (191). Both components are also contained in transport box 3.

In an insert box various smaller components such as pressure gauges (162, 181,182), shut off valves (159), pressure valves (148), electrical switch box (288), five-way couplings (21, 199) are stored. The insert box with the same components is again stored in the transport box 6.

The attachments to the launch table Fz.103 are particularly exposed to enormous heat stress and the pressure of the exhaust gas jet, and must be checked thoroughly after firing. In order not to delay the firing operations unnecessarily sufficient spare parts must be kept in stock. The number of parts stocked was determined according to statistical values and experience from the test firings in Heidelager.

177 Transportkasten 2

Im Transportkasten 2 ist die Ventilsteuerung (147), ein Anbauteil für den Starttisch Fz.103 und ein elektrischer Schaltkasten (191) bevorratet. Beide Bauteile sind nochmals im Transportkasten 3 enthalten.

In einem Einsatzkasten sind verschiedene kleinere Bauteile wie Manometer (162, 181,182), Schließventile (159), Druckventile (148), elektrische Schaltkästen (288), elektrische Schliessventile (156), Fünffachkupplungen (21, 199), enthalten.

Gerade die Anbauteile am Starttisch Fz.103 sind enormen Belastungen durch Hitze und dem Druck des Abgasstrahles ausgesetzt und müssen nach dem Abschuss gründlich überprüft werden. Um den Schiessbetrieb nicht unnötig zu verzögern müssen ausreichend Ersatzteile bevorratet werden. Die Anzahl der Bevorratungsteile ist nach statistischen Werten und Erfahrungen aus den Versuchsschießen in Heidelager ermittelt worden.

178 Transport box 6

In transport box 6 the five-way coupling (198) is disassembled into components and fixed in holders. Two A-Stoff oxygen filling nozzles (243) are stored above. Also disassembled into individual parts is the oxygen topping up device (220). It is stowed in the box holders.

178 Transportkasten 6

Im Transportkasten 6 ist die Fünffachkupplung (198) in Bauteile zerlegt und in den Halterungen fixiert. Zwei A-Stoff-Betankungsstutzen (243) lagern darüber. Ebenfalls in Einzelteile zerlegt ist die Sauerstoff-Nachtankvorrichtung (220). Sie ist in den Kastenhalterungen verstaut.

Transportkasten 6

Adjustable
five-way coupling

Justierbare
Fünffachkupplung

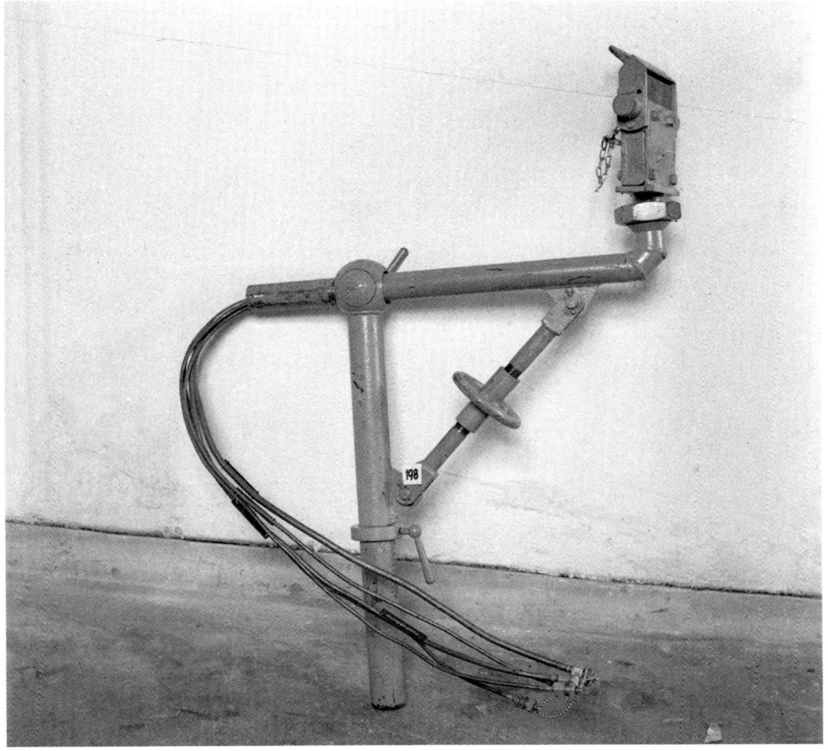

BArch, RH8II Bild-B1101-44, o. Ang.

179

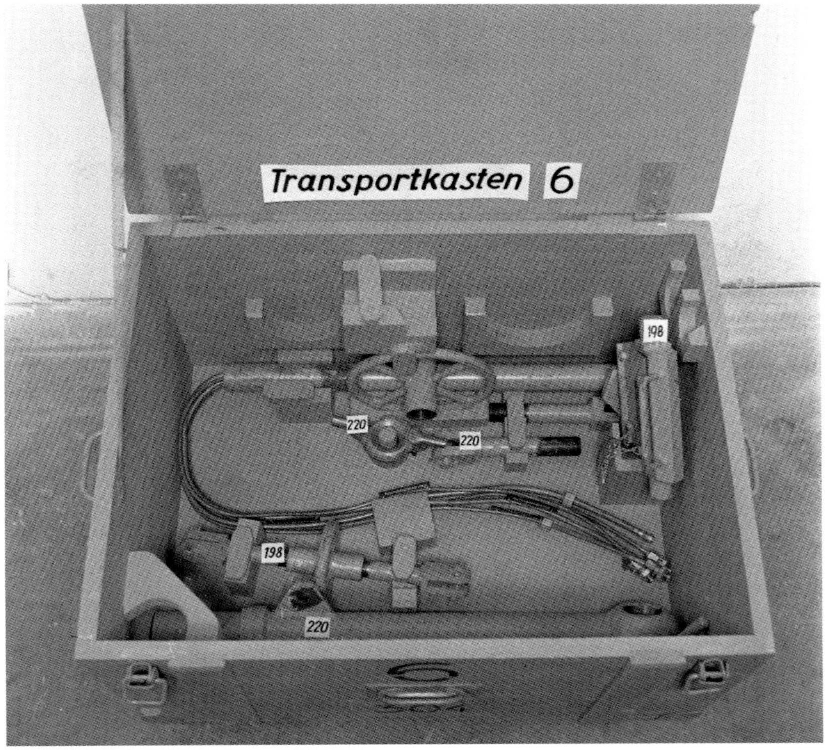

BArch, RH8II Bild-B1098-44, o. Ang.

181

Oxygen topping up
device

Sauerstoff-Nachfüll-
vorrichtung

BArch, RH8II Bild-B1100-44, o. Ang.

180

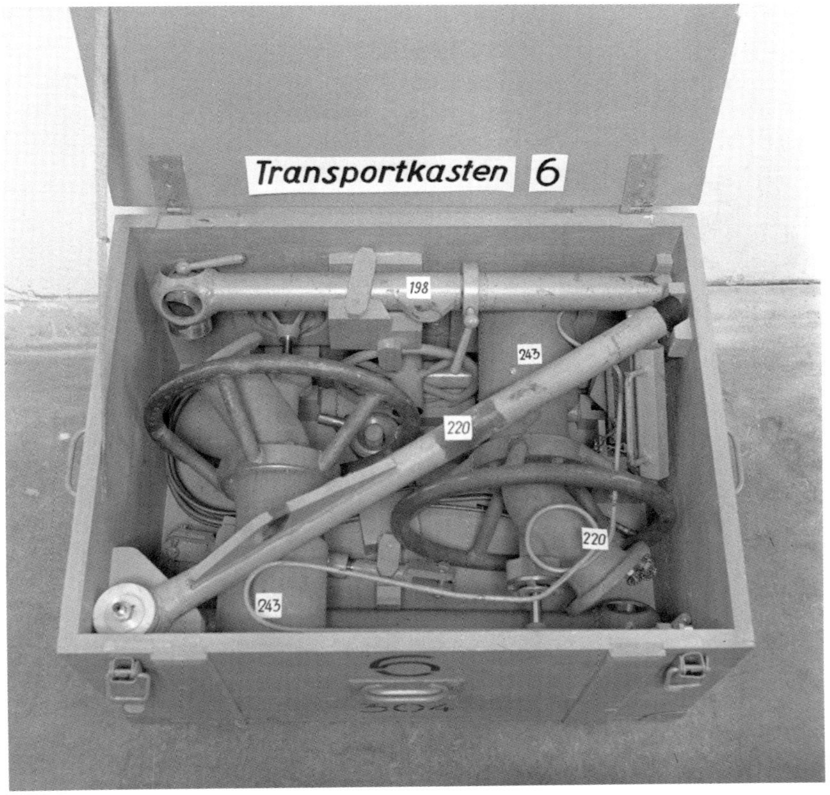

BArch, RH8II Bild-B1099-44, o. Ang.

182

VEHICLE GROUP IV
FAHRZEUGGRUPPE IV

DMA, GD630.02
A4 Lehrbuch
der FR-Schule
Mai 1944
technische Fahrzeuge
Kapitel IV Betankungsfahrzeuge

Fuel column

Purpose and task

Taking over the fuels A-, B- and T-Stoff from the filling station, if necessary storing them temporarily, and fueling the A4 rocket in the firing position. After the tests carried out on the upright and free-standing *Gerät* have been completed, the 'ready for fuelling' message is passed. At an order from the battery officer, the tankers for A, B and T-Stoff move into the firing position. This is followed by the clearing of the pumps and equipment, the laying of the refuelling hoses, and the test runs of the drive motors for the pumps. The actual refuelling is carried out jointly by the three groups of the fuel convoy, the wagon crew and the engine crew and takes about 20 minutes in total.

After flushing the hoses, pipelines and the measuring container for T-Stoff, dismantling and re-locating the hoses and closing the connecting branches with blind caps, the tankers leave the firing position again.

Tankkolonne

Zweck und Aufgabe

Übernahme der Treibstoffe A-, B- und T-Stoff von der Tankstelle, ggf. Zwischenlagern und Betanken der A4-Rakete mit Treibstoffen in der Feuerstellung. Nachdem die am aufgerichteten und frei stehenden Gerät durchgeführten Prüfungen beendet worden sind, erfolgt die Klarmeldung zum Betanken. Auf Befehl des Batterieoffiziers fahren die Tankfahrzeuge für A-, B- und T-Stoff in die Feuerstellung. Es schließen sich das Klarmachen der Pumpen und Geräte, das Auslegen der Betankungsschläuche sowie der Probelauf der Antriebsmotoren für die Pumpen an.
Das eigentliche Betanken wird in gemeinsamer Arbeit von den drei Gruppen der Treibstoffkolonnen, dem Wagentrupp und dem Triebwerkstrupp durchgeführt und nimmt insgesamt etwa 20 Minuten in Anspruch.

Nach Spülen der Schläuche, Rohrleitungen und des Messbehälters für T-Stoff, dem Abbauen und Anortbringen der Schläuche sowie Verschließen der Anschlussstutzen mit Blindkappen verlassen die Tankfahrzeuge wieder die Feuerstellung.

BArch, RH 8/1310
Spezialfahrzeuge
Baugruppen und
wesentliche Bauteile
der Bodenanlage FR

Special vehicles of the fuel column

Fz.-Pos.-Nr.401 A-Stoff trailer 5 (2-axle) *
Fz.-Pos.-Nr.402 A-Stoff trailer 6 (2-axle)
Fz.-Pos.-Nr.403 Tanker truck 2900 l B-Stoff
Fz.-Pos.-Nr.404 Tanker truck 3500 l B-Stoff *
Fz.-Pos.-Nr.405 Tanker truck 2100 l T-Stoff
Fz.-Pos.-Nr.406 Pump trailer Type A (1-axle)
Fz.-Pos.-Nr.407 T-Stoff preheater trailer (1-axle)
Fz.-Pos.-Nr.408 Accumulator truck on 4,5 t truck closed (heated)
Fz.-Pos.-Nr.409 Extinguishing trailer (1-axle) **

* Production stopped, Fz.401 replaced by Fz.402, Fz.403 replaced by Fz.404.
** Production stopped, replaced by LW extinguishing trailer.

Spezialfahrzeuge der Tankkolonne

Fz.-Pos.-Nr.401 A-Stoff-Anhänger 5 (2-achs.) *
Fz.-Pos.-Nr.402 A-Stoff-Anhänger 6 (2-achs.)
Fz.-Pos.-Nr.403 Kessel-KW 2900 l B-Stoff
Fz.-Pos.-Nr.404 Kessel-KW 3500 l B-Stoff *
Fz.-Pos.-Nr.405 Kessel-KW 2100 l T-Stoff
Fz.-Pos.-Nr.406 Pumpen-Anhänger Type A (1-achs.)
Fz.-Pos.-Nr.407 T-Stoff-Vorwärmer-Anhänger (1-achs.)
Fz.-Pos.-Nr.408 Sammler-Lkw auf Lkw 4,5 t geschl. (heizbar)
Fz.-Pos.-Nr.409 Lösch-Anhänger (1-achs.) **

* Fertigung eingestellt, Fz.401 ersetzt durch Fz.402, Fz.403 ersetzt durch Fz.404.
** Fertigung eingestellt, ersetzt durch LW Lösch-Anhänger.

183

V29 preparations test stand X
V29 Vorbereitungen am Prüfstand X

183 Mid-June 1943 – An A4 *Gerät* (V29) is ready for refueling. Two B-Stoff and one T-Stoff vehicle have driven up, and a Type A pump trailer is positioned between them. The A-Stoff trailer is still expected.

183 Mitte Juni 1943 – Ein A4-Gerät (V29) ist bereit zum Betanken. Zwei B- und ein T-Stoff-KW sind aufgefahren. Zwischen den Fahrzeugen ist ein Pumpen-Anhänger Typ A in Stellung gebracht. Der A-Stoff-Anhänger wird noch erwartet.

DMA, GD639.5.18
HAP11/T 507
Karlshagen, 1.6.1944
Bd.Nr. E 3195/44 g

DMA, GD639.5.2
OKH
Chef H.Rüst und B.d.E
Bz.b.V (Heer) Abt.Ib
Bd.Nr.3429/44 g.Kdos.
Stand März 1944

Auszug aus
Gerätebeschreibung
Bedienungsanweisung
Wartungsanweisung

Vehicle 401

Internal or former designation: Fz.-Pos.-Nr.401
A-Stoff trailer 5 (2-axle)
A-Stoff trailer

Intended use
The A-Stoff trailer 5 is used to transport A-Stoff between the filling station and fire position, and to refuel the A4 *Gerät* with A-Stoff.

The trailer is a two-axle road vehicle towed by a tractor. It has air brakes operated from the tractor.

The A-Stoff trailer has the following dimensions and weights

Total length:	8,550 mm		
Width:	2,400 mm		
Height:	2,750 mm		
Wheel track:	5,400 mm		
Spurweite:	1,915 mm		
Tare weight:	4,500 kg		
Gross weight:	10,400 kg		
Wheels:	190-20		
Pump			
Drive motor:	Type	Stamo 10:	Stamo 14:
	Power N	4	6,5 HP
	Speed n	3,000	3,000 rpm
Pump:	Speed n	= 1,670 rpm	
	Delivery rate	approxiamtely 8-10 kg/sec.	
Hose connections:	Filling nozzle		NW 70
	Corner valve		NW 100
Pump	suction connection		NW 100
	discharge connection		NW 70

The main components of the A-Stoff trailer are:
- Chassis
- Tank
- Valve and fittings
- Accessories

The A-Stoff is carried in an inner container* with a capacity of 5,000 l. This is isolated from the chassis so that no harmful thermal stresses can arise due to the large temperature difference. The inner tank is insulated from a steel outer tank by *Iporka*** synthetic resin foam.

* Inner container = inner tank made of brass with 2.5 mm wall thickness, working pressure 1.5 atu
** *Iporka* (trade name) a plastic foam made from formaldehyde resin and urea.

A-Stoff = Oxygen – green – RAL 6010* (LOX liquid oxygen -183 degrees temperatur)
* Technical colour – Stealth term for operating materials

DM, 8001D
Tarnschlüssel für
Betriebsstoffe, mpe

Fahrzeug 401

Interne oder frühere Bezeichnung: Fz.-Pos.-Nr.401
A-Stoff-Anhänger 5 (2-achs.)
A-Stoff-Anhg.

Verwendungszweck
Der A-Stoff-Anhänger 5 dient zum Transport des A-Stoffes zwischen Tankstelle und Feuerstellung, sowie zum Betanken des A4-Gerätes mit A-Stoff.

Der Anhänger ist ein zweiachsiges Straßenfahrzeug, das durch eine Zugmaschine fortbewegt wird. Es wird von der Zugmaschine aus mit einer Druckluftbremse gebremst.

Der A-Stoff-Anhänger hat folgende Maße und Gewichte:

Gesamtlänge:	8.550 mm		
Breite:	2.400 mm		
Höhe:	2.750 mm		
Radstand:	5.400 mm		
Spurweite:	1.915 mm		
Leergewicht:	4.500 kg		
Gefechtsgewicht:	10.400 kg		
Räder:	190-20		
Pumpe			
Antriebsmotor:	Bauart	Stamo 10:	Stamo 14:
	Leistung N	4	6,5 PS
	Drehzahl n	3.000	3.000 Uml./Min.
Pumpe:	Drehzahl n	= 1.670 Uml./Min	
	Liefermenge	ca. 8-10 kg/sek.	
Schlauchanschlüsse:	Füllstutzen		NW 70
	Eckventil		NW 100
Pumpe	Saugstutzen		NW 100
	Druckstutzen NW 70		

Die Hauptteile des A-Stoff-Anhängers sind:
- Fahrgestell
- Behälter
- Armaturen
- Zubehör

Der A-Stoff wird in einem 5.000 l fassenden Innenbehälter* untergebracht. Er ist auf dem Fahrgestell beweglich gelagert, damit durch den großen Temperaturunterschied keine schädlichen Wärmespannungen entstehen können. Der Innenbehälter ist mit einem Kunstharzschaum aus *Iporka*** isoliert und von einem Außenbehälter aus Stahl umkleidet.

* Innenbehälter = Innentank gefertigt aus Messing mit 2,5 mm Wandstärke, Betriebsdruck 1,5 atü
** *Iporka* (Handelsname) ist ein Schaumstoff, der aus Formaldehydharz und Harnstoff hergestellt wurde.

A-Stoff = Sauerstoff – grün – RAL 6010*
* Technische Farbe – Tarnbezeichnung für Betriebsstoffe

Deutsches Museum, München, Archiv- FA014/B541/44 TD5224 [CD74787]

184 A Hanomag SS100 LS is used here as the tractor for the Fz.401 A-Stoff trailer. Both vehicles are painted in dark yellow RAL 7028 and without camouflage paint, they belong to Abt.836.

184 Als Zugmaschine für den A-Stoff-Anhänger Fz.401 ist hier ein Hanomag SS100 LS in Verwendung. Beide Fahrzeuge sind in dunkelgelb RAL 7028 gestrichen und ohne Tarnanstrich, sie gehören zur Abt.836.

Deutsches Museum, München, Archiv- FA014/B1058/43 TD5224 [CD55634]

185

DM
GD639.5.13
OKH/Wa Prüf10
Karlshagen, den 1.4.1944
Bd.Nr. 788/44 gK
Auszug
Besprechung
Fahrzeuggruppe 4
vom 31.3.44
in Karlshagen I

Production status: Fahrzeug 401
Delivered to date: 32 (up to the end of 1943) of which: 22 at 836 (of which 2 still in HL)6 at 444 (of which 3 at HL, 3 sent by HAP11 to the Mako company for repairs) 4 at HAP 11 (of which 2 sent to 836 on 28.3).
No more vehicles 401 will be manufactured. Replaced by Fz.402.

Current defects and their remedy
It proved to be necessary that the oxygen discharging from the pump is deflected from the vehicle, so that it does not come into contact with the tires. The units used wooden gratings for this, which have the disadvantage that they sink into soft ground and become ineffective.

HAP 11 produces drain channels. Production made from 2 ½ m lengths of roof gutters and stored in the flexible hose box. Responsible engineer Ing. Nettesheim.

To be manufactured:
24 for 836 urgently before 11.4.44, 4 for 444 after completion (change case A).

With the angle valve a corrective action was taken; to avoid the risk of freezing the spindle threat was moved inward and an ice stripper was mounted outside. This change is already effected on current production. 50 valves with the modified jack bolt will be delivered during April for exchange. Distribution to the unit for exchange will be organized by TD 41 – Ing. Nettesheim, via B.z.b.V. Heer in the course of this month (change case A).

In place of the hand wheel to operate the angle valve a crossbeam handle was developed by TD 41. This change has already been made by the Mako company for current production. An immediate purchase for replacement by the units has been ordered from the Mako company.

Delivery to the units during April will be ensured. Responsible: TD 41 – Ing. Nettesheim (change case A).

Fertigungslage: Fahrzeug 401
Bisher ausgeliefert: 32 Stück (bis Ende 1943)
Davon: 22 Stück bei 836 (davon 2 noch in HL) 6 Stück bei 444 (davon 3 in HL, 3 über HAP11 zu der Fa. Mako in Reparatur) 4 Stück bei HAP 11 (davon 2 am 28.3 zu 836).
Weitere Fahrzeuge 401 werden nicht gefertigt. Ersetzt durch Fz.402.

Derzeitige Mängel und ihre Behebung
Es hat sich als notwendig erwiesen, dass der allenfalls aus der Pumpe austretende Sauerstoff vom Fahrzeug abgeleitet wird, sodass er mit Sicherheit nicht mit den Reifen in Berührung kommt. Die Truppe hat hierfür Holzroste vorgesehen, die den Nachteil haben, dass sie bei weichem Boden einsinken und wirkungslos werden.

HAP 11 fertigt Ablaufrinnen. Herstellung aus Dachrinnen, Länge 2 ½ m. Unterzubringen im Schlauchkasten. Verantwortlich Ing. Nettersheim.

Es sind herzustellen:
24 Stück für 836 vordringlich bis 11.4.44, 4 Stück für 444 nach Fertigstellung (Änderungsfall A).

Beim Eck-Ventil wurde zur Behebung der Einfriergefahr das Spindelgewinde nach innen verlegt und ein Eisabstreifer außen angebracht. Diese Änderung ist bereits bei der laufenden Fertigung berücksichtigt. Für den Austausch bei der Truppe werden im Laufe April 50 Ventile mit geänderter Spindel angeliefert. Verteilung an die Truppe mit Austauschanleitung veranlasst TD 41 – Ing. Nettersheim, über B.z.b.V. Heer im Laufe dieses Monats (Änderungsfall A).

An Stelle des bisher angebrachten Handrades zur Bedienung des Eckventiles ist ein Stellkreuz entwickelt worden (durch TD 41). Für die laufende Fertigung ist diese Änderung bereits bei der Fa. Mako veranlasst. Für den Austausch bei der Truppe erfolgt sofort entsprechende Bestellung bei der Fa. Mako.

Auslieferung an die Truppe im Laufe des April sichergestellt. Verantwortlich: TD 41 – Ing. Nettersheim (Änderungsfall A).

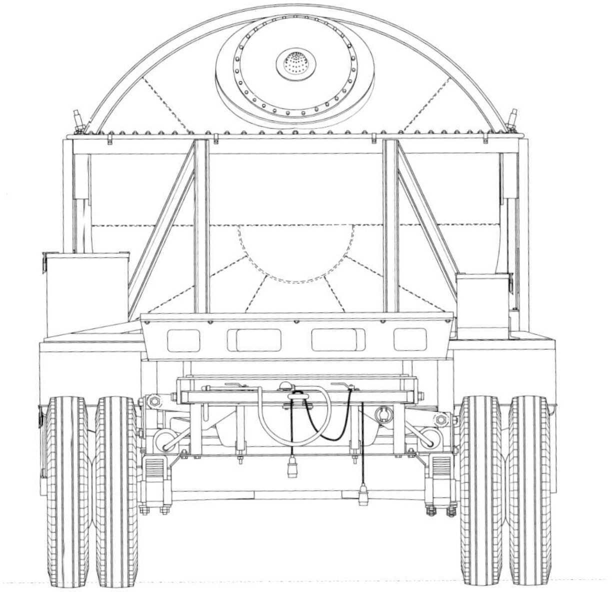

Chassis
Fahrgestell

186 The frame of the twin-axle carriage was made from welded profiles and raised above the steering rack. The front axle was spring mounted on the steering rack and could be turned axle 90 degrees to each side. The towbar was attached to the steering rack. The hoses for the compressed air brakes and cable for the rear lights were also attached to it. The front and rear axles were equipped with 190-20 cross-country dual tires. The early version runs on 7,25-20 dual tires.

186 Der Rahmen des zweiachsigen Fahrgestells war aus Profilleisten geschweißt und über dem Lenkgestell hochgezogen. Die Vorderachse war mit Federn im Lenkgestell gelagert und nach jeder Seite um 90 Grad schwenkbar. An dem Lenkgestell war die Zuggabel befestigt. An dieser waren die Schlauchleitungen für die Druckluftbremse und Schlusslichter angebracht. Vorder- und Hinterachse waren mit Geländereifen 190-20 zwillingsbereift. Die frühe Variante war bestückt mit Reifen 7,25-20.

Deutsches Museum, München, Archiv- FA014/B1067/43TD5224 [CD74706]

186

Deutsches Museum, München, Archiv- FA014/B169/44 TD5224 [CD74958]

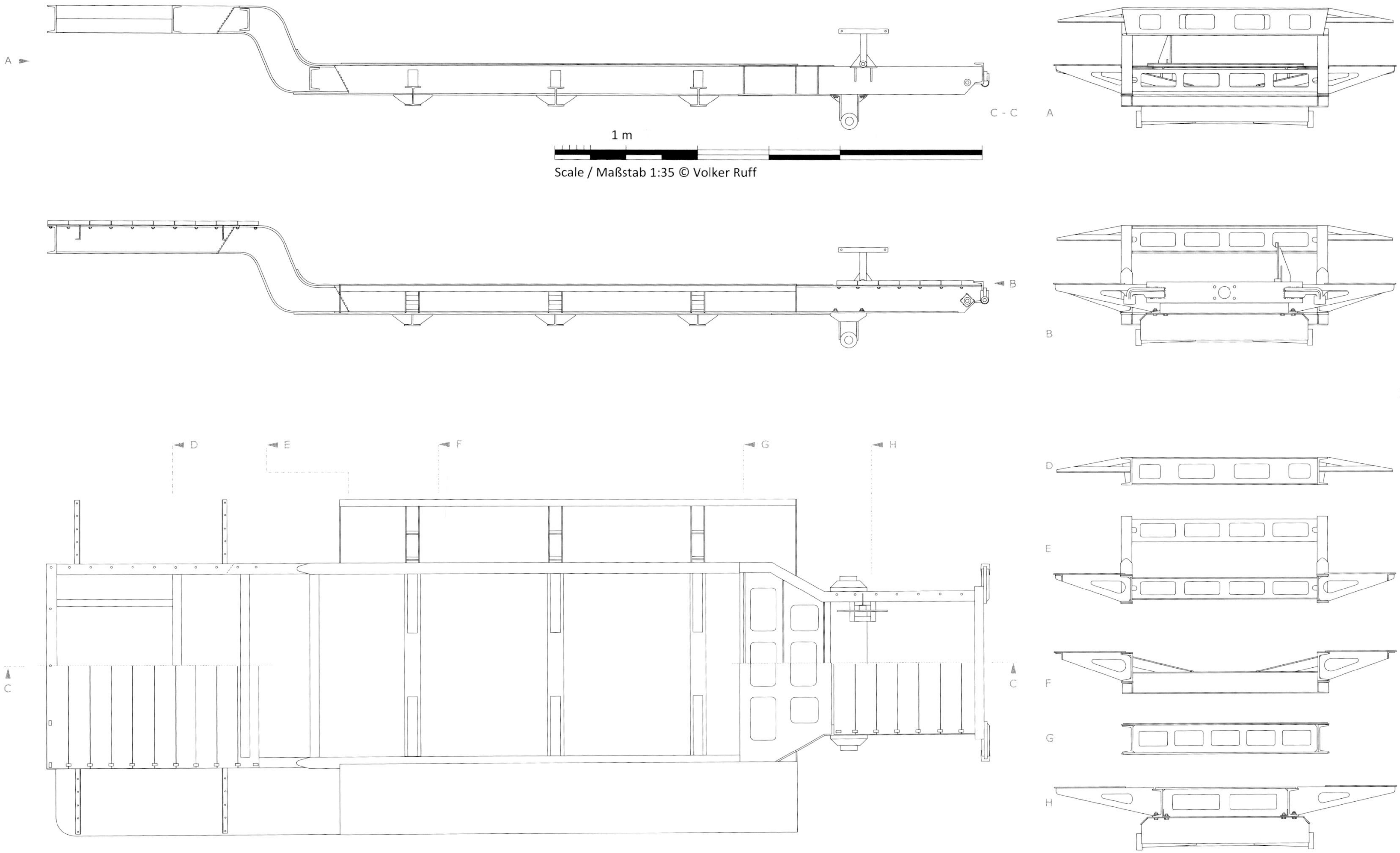

Fahrzeug 401

A-Stoff-Anhänger 5 (2-achs.)

Partially used drawings E1528A, E1944B, E2278A

Teilweise verwendete Zeichnungen E1528A, E1944B, E2278A

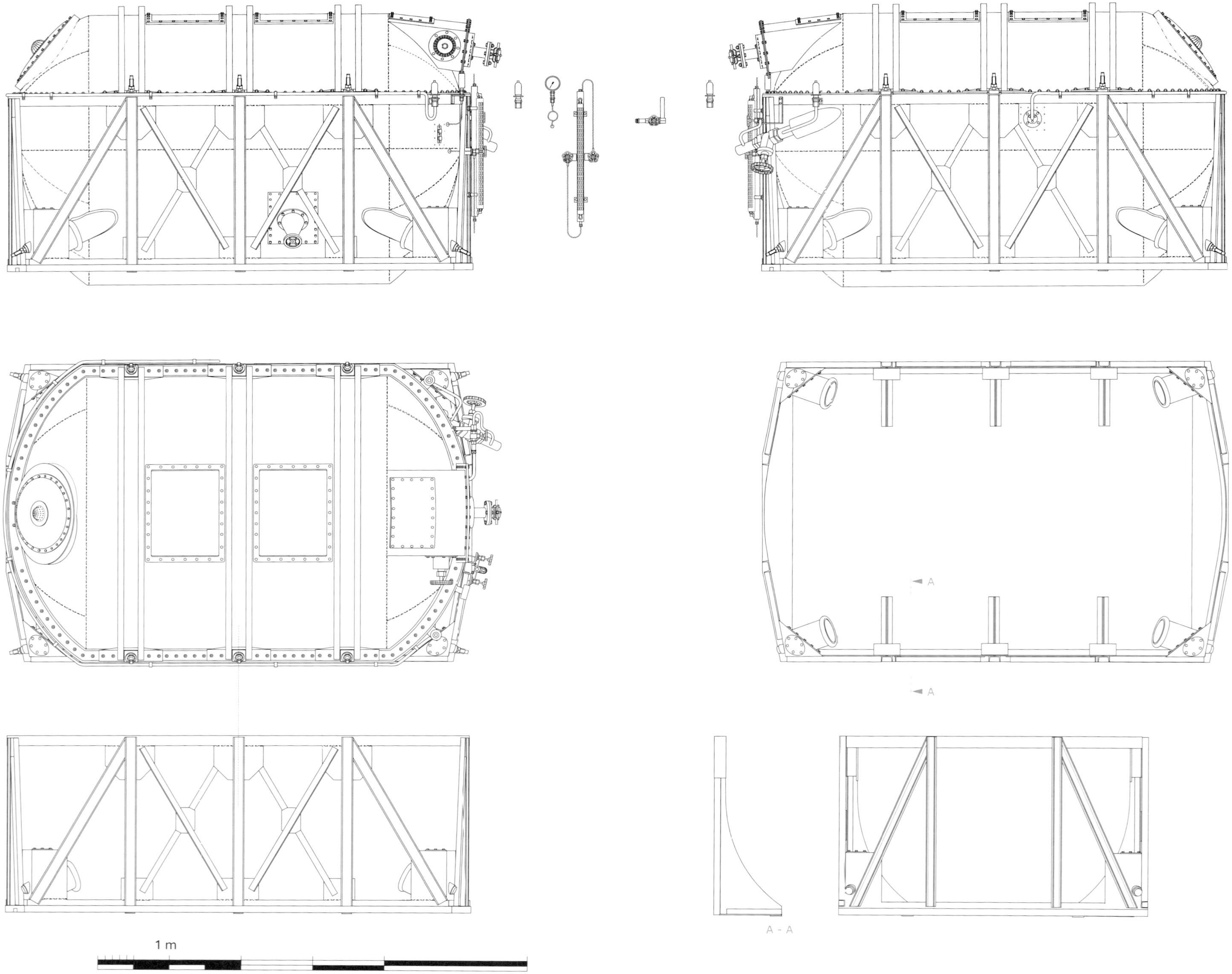

Scale / Maßstab 1:35 © Volker Ruff

Deutsches Museum, München, Archiv- FA014/B1058/43 TD5224 [CD55634]

188

Container
Behälter

188 The container consisted of an inner tank divided by two baffle plates* and the outer container, which was mounted on a frame made of profile strips welded to the chassis. The upper part of the outer container was removable. The inner tank was held by chains on the framework of the outer container, to avoid thermal stress from the temperature differences. The space between the outer container and inner tank was filled with *Iporka* for insulation.

* A baffle is a partition, usually perforated, in liquid containers which prevent the liquid inside the tank from being swelled up too much during transport due to starting or braking maneuvers.

188 Der Behälter bestand aus dem durch zwei Schwallblech* unterteilten Innentank und dem Außenbehälter, der in einem auf dem Fahrgestell angeschweißten Gerüst aus Profilleisten gelagert war. Das Oberteil des Außenbehälters war abnehmbar. Damit durch die großen Temperaturschwankungen keine schädlichen Wärmespannungen entstehen konnten wurde der Innentank mit Ketten am Gerüst des Außen-behälters beweglich gelagert. Der Raum zwischen Außen- und Innenbehälter (Innentank) wurde zur Isolation mit *Iporka* ausgefüllt.

* Als Schwallblech wird eine meist durchbrochene Trennwand in Flüssigkeitsbehältern bezeichnet, die verhindern soll, dass Flüssigkeit im Inneren des Tanks beim Transport nicht zu stark durch Anfahr- oder Bremsmanöver aufgeschaukelt wird.

Deutsches Museum, München, Archiv- FA014/B1055/43 TD5224 [CD55635]

189

Accessories
Zubehör

189 The outer container had a manhole cover in which was a safety membrane which prevented a possible overpressure in the insulated space.

The manhole cover was used to reach into the space between the inner tank and the outer shell in order to refill or re-stuff the insulation if necessary. The required A-Stoff refueling hoses with NW40 and NW70 were carried in the hose box on the chassis frame. Filling and refueling equipment and copper seals were stored in the boxes on the chassis. Tools, wheel nut wrench, grease gun, and ramp planks are stowed in another box, and the spare wheel was stored between the boxes.

189 Der Außenbehälter besaß einen Mannloch-Deckel mit innen eingebauter Sicherheitsmembran. Ein etwaiger Überdruck sollte hierdurch im Isolierraum verhindert werden.

Über den Mannlochdeckel gelangte man in den Zwischenraum, um im Bedarfsfall die Isolierung nachzufüllen bzw. nachzustopfen. Die benötigten A-Stoff-Betankungsschläuche mit NW40 und NW70 wurden im Schlauchkasten am Fahrgestellrahmen mitgeführt. Füll- und Betankungsgeräte sowie Kupferdichtungen wurden in den Kästen auf dem Fahrgestell untergebracht. Werkzeuge, Radmutterschlüssel, Fettpresse und Auflaufbohlen waren in einem weiteren Kasten verstaut, dazwischen lagerte das Reserverad.

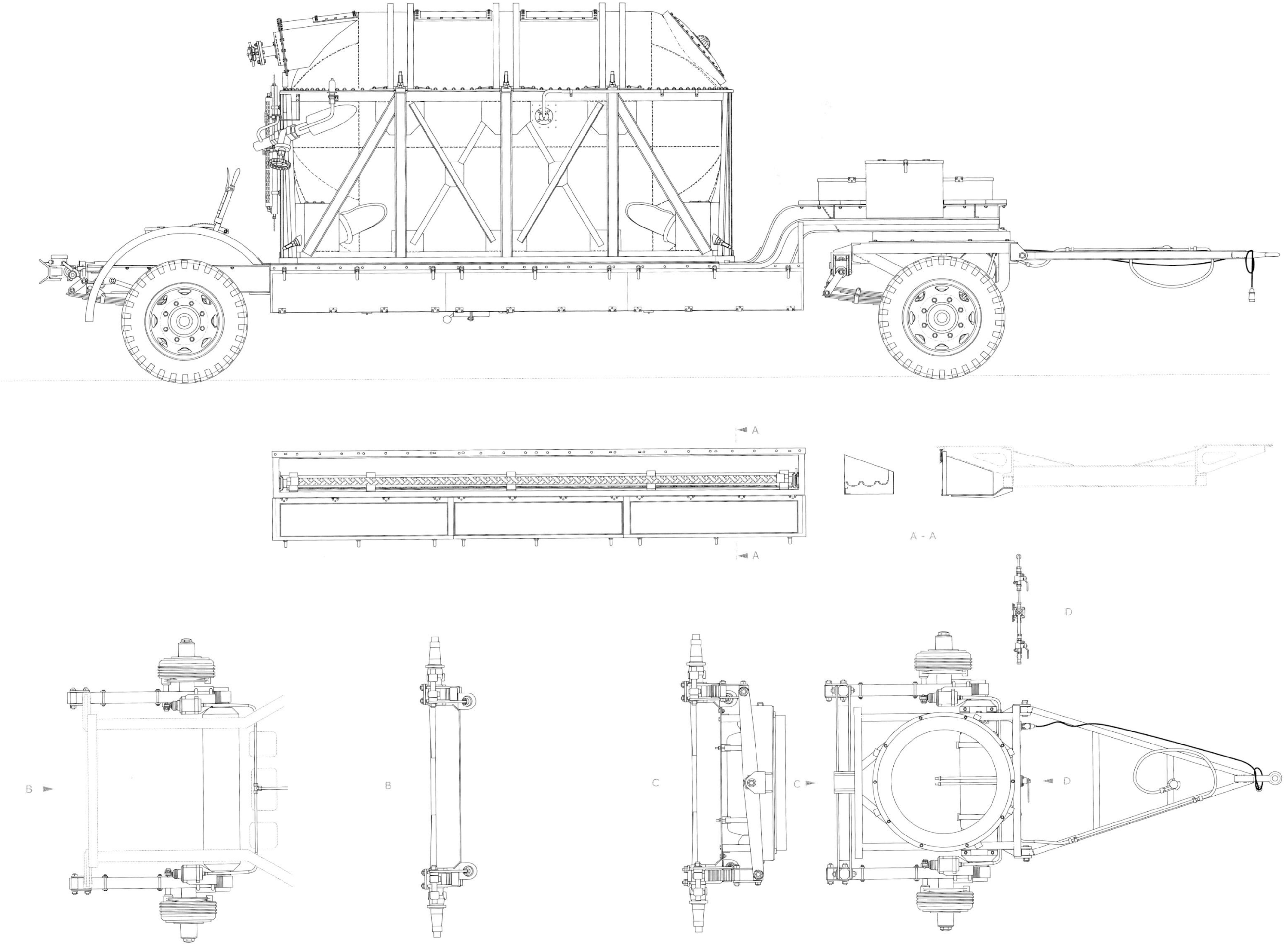

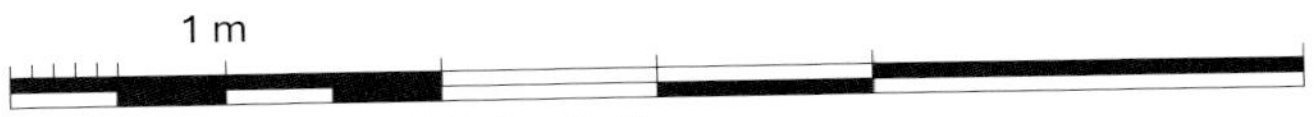

Scale / Maßstab 1:35 © Volker Ruff

Final version, manufacture year 1943,
deployed to Art. Abt. 836

Abschlussausführung Baujahr 1943,
verwendet von der Art Abt.836

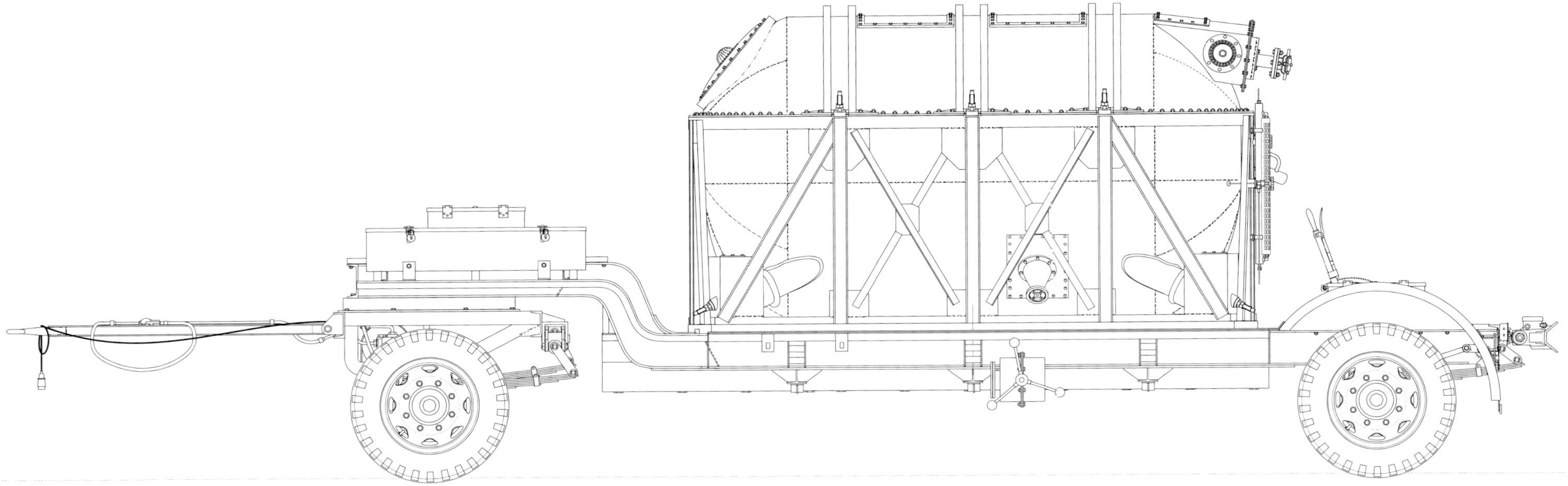

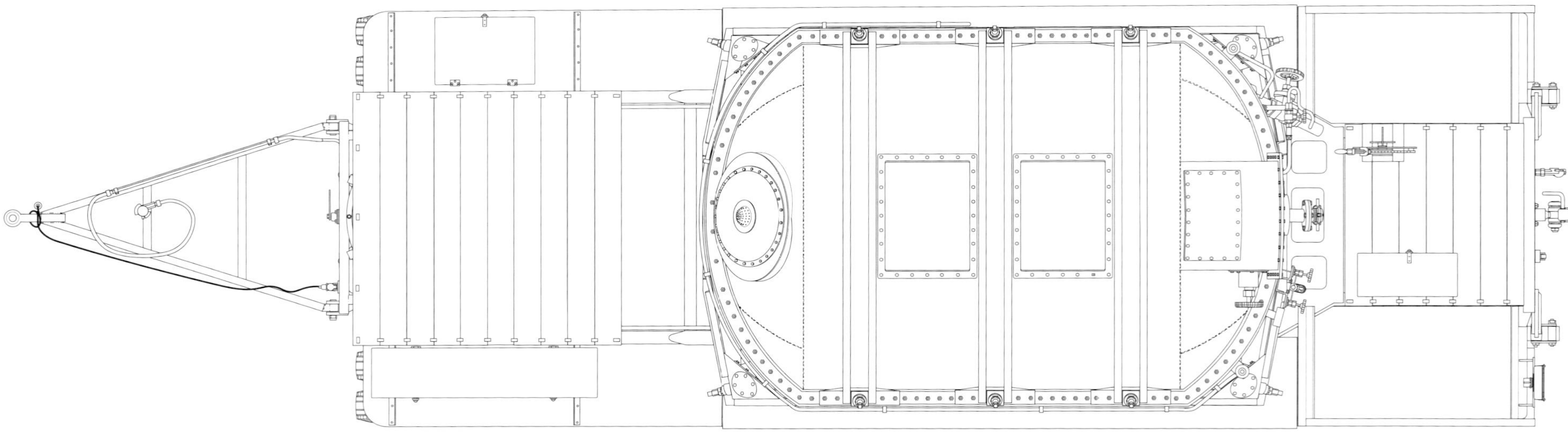

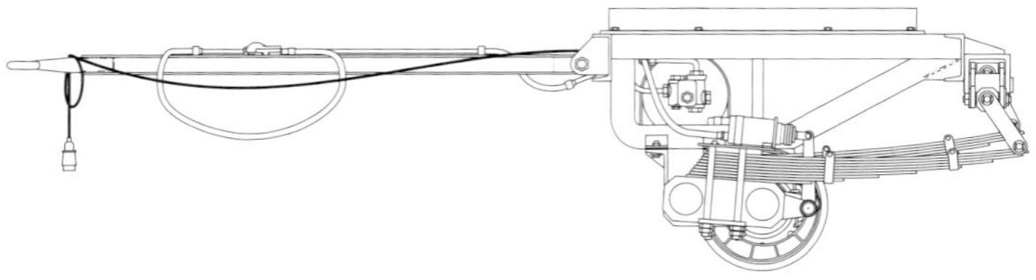

1 m

Scale / Maßstab 1:35 © Volker Ruff

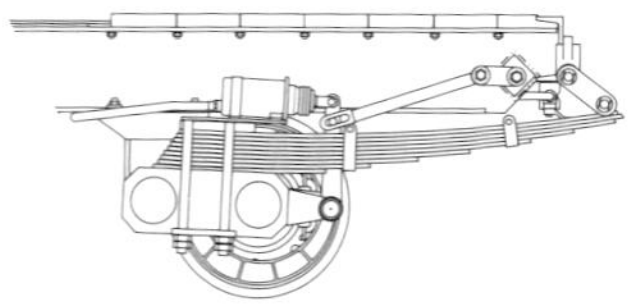

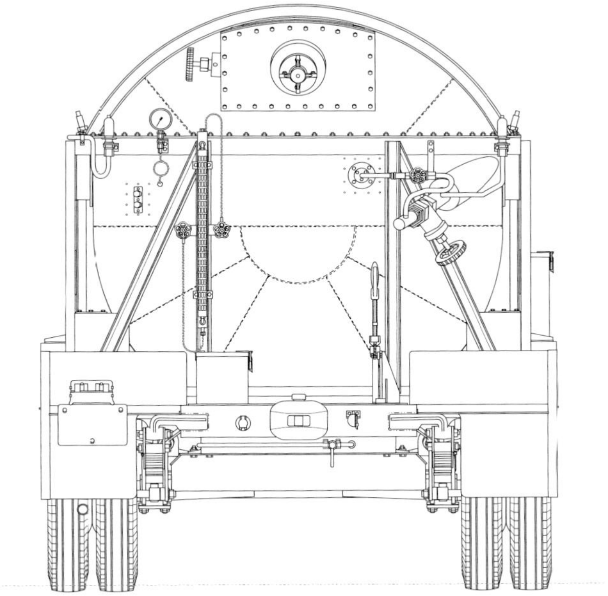

Valve and fittings
Armaturen

190 The container was filled using a filler nozzle located at the top of the rear end, to which a hose coupling for the 70 mm filling hose was flanged. The filling pipe was extended to the bottom of the inner tank and was closed by a hand wheel valve mounted in an insulated metal box. On the right side of the front wall were the outlet vents. In them the main ventilation pipe and the snake ventilation pipe ran through.

A number is painted on the fairing boxes of the chain suspension for the inner container, here a white 68 and a large white X on a black background. The A-Stoff trailer belongs to group c of the training battery. See also the chapter about the »division of duties«.

190 Zum Befüllen des Behälters diente ein an der hinteren Stirnseite oben befindlicher Füllstutzen, an welchem eine Schlauchkupplung für den 70er-Füllschlauch angeflanscht war. Das Füllrohr reichte bis auf den Boden des Innentanks, geschlossen wurde das Füllrohr durch das in einem Blechkasten isoliert angebrachte Füllventil mit Handrad. An der rechten Seite der Stirnwand befanden sich die Entlüftungsstutzen. In ihnen verlief die Hauptentlüftungsleitung und die Schlangenentlüftungsleitung.

An den Verkleidungskästen der Kettenaufhängung für den Innenbehälter war eine Zählnummer, hier eine weiße 68 und ein großes weißes X auf schwarzem Grund aufgemalt. Der A-Stoff-Anhänger gehörte zur Gruppe c Lehrbatterie. Siehe auch Kapitel »Arbeitsteilung«.

Deutsches Museum, München, Archiv FA011/B1088/43 TD5224 [CD74707]

190

Deutsches Museum, München, Archiv- FA014/B1087/43 TD5224 [CD79117]

191

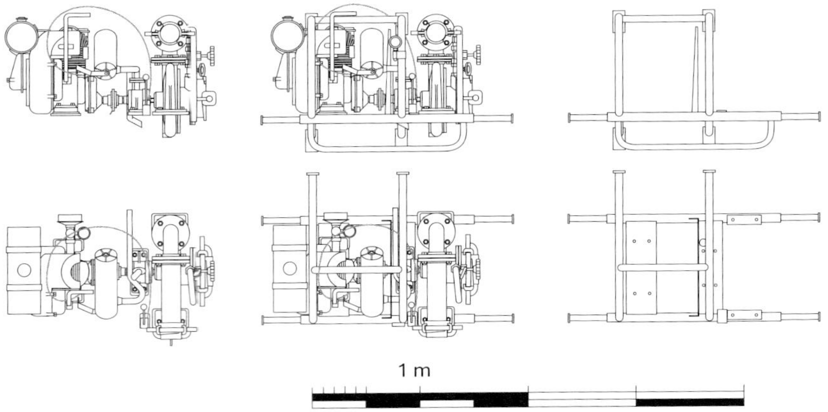

Scale / Maßstab 1:35 © Volker Ruff

DMA, GD630.02
A4 Lehrbuch
der FR-Schule
Auszug aus
technische Fahrzeuge
Kapitel IV Betankungs-
fahrzeuge

DMA, GD639.5.13
OKH/Wa Prüf10
Karlshagen, 1.4.1944
Bd.Nr. 788/44 gK
Auszug
Besprechung der
Fahrzeuggruppe 4
vom 31.3.44
in Karlshagen I

A-Stoff pump set

The A-Stoff pump set is transported in the B-Stoff trailer Fz.406. The set is attached to a bracket on the left longitudinal beam of the A-Stoff trailer for refueling. It consists of a 5 hp petrol engine* coupled to a centrifugal pump with flying impeller. The suction and discharge connections, here still fitted with blind caps, are each fitted with a vent valve to discharge the A-Stoff fumes when the pump has not cooled down yet. A pressure gauge connected to the discharge port indicates the pressure in the discharge line.

The requirement that the same A-Stoff pump must be used for both railway and motor vehicles needs the following immediate action:

Uses of the A-Stoff pump in the form provided for railroad vehicles (31 have been) and creation of a suitable mechanical stand for vehicles 401 and 402 as follows:

The tube rack of the pump is pushed onto 2 pins fastened to the vehicle. The tubes are secured to the vehicle frame in a weight-relieving manner. The depth gauge must also be used for the hoses. This change is already in force for current production. It is not possible for the units to change the pumps of vehicles already delivered so this must await the delivery of an appropriate number of new pumps. (Revision letter A).

HAP11/TD 41 checks whether the mountings on the A-stoff trailers, described above, can be arranged moveably. Completion of testing in April (Revision letter B).

Mechanical stand for the heater is already developed and advised as a modification to B.z.b.V. Heer. This mechanical bracket is still to be mounted on all vehicles with the units. Ing. Nettesheim, TD41, is responsible for installation in current production. (Revision letter B)

* STAMO: stationary engine from Fichtel & Sachs

A-Stoff-Pumpensatz

Der A-Stoff-Pumpensatz wird im B-Stoff-Anhänger Fz.406 befördert. Er ist zum Betanken an einer Halterung am linken Längsträger des A-Stoff-Anhängers angebracht und besteht aus einem 5 PS Otto-Motor*, der mit einer Kreiselpumpe mit fliegend angeordnetem Laufrad gekuppelt wird. Am Saug- und Druckstutzen, hier noch mit Blindkappen versehen, befindet sich je ein Entlüftungsventil zum Abführen der A-Stoff-Dämpfe bei nicht abgekühlter Pumpe. Ein an den Druckstutzen angeschlossenes Manometer zeigt den Druck in der Druckleitung an.

Die Forderung, für Eisenbahn und mot. Fahrzeuge die gleiche A-Stoff-Pumpe zu verwenden, zwingen zu folgender Sofortmaßnahme:

Verwendung der A-Stoff-Pumpe in der für die Eisenbahnfahrzeuge bisher vorgesehenen Form (31 Stück hierfür geliefert) und Schaffung einer geeigneten Halterung für die Fahrzeuge 401 bzw. 402 in folgender Weise:

Das Rohrgestell der Pumpe wird auf zwei am Fahrzeug befestigte Dornen aufgeschoben. Die Rohre werden am Fahrzeugrahmen gewichtsentlastend verriegelt. Stichmaß muss auch für die Schläuche eingehalten werden. Für die laufende Fertigung ist diese Ausführung bereits sichergestellt. Die Änderungen der bereits ausgelieferten Pumpen und Fahrzeuge ist bei der Truppe nicht möglich und muss bis zur Auslieferung einer entsprechenden Anzahl neuer Pumpen zurückgestellt werden (Änderungszustand A).

HAP11/TD 41 prüft, ob die oben geschilderten Aufnahmen an den A-Stoff-Anhängern verschiebbar angeordnet werden kann. Abschluss der Prüfung bis April (Änderungszustand B).

Halterung für das Heizgerät ist entwickelt und bereits als Änderungsmitteilung an B.z.b.V. Heer ergangen. Diese Halterung ist an sämtlichen Fahrzeugen bei der Truppe noch anzubringen. Für die Anbringung in der laufenden Fertigung ist verantwortlich Ing. Nettersheim, TD 41 (Änderungszustand B).

* STAMO: Stationär-Motor von Fichtel & Sachs

192

192 The A-Stoff pump set is ready for use. An NW 100 suction line with fabric reinforcement is connected to the angle valve of the discharge line and the suction connection on the A-pump set. Operating pressure is monitored by a pressure gauge. A layer of ice forms very quickly around the handwheel on the angle valve. The ice can only be removed mechanically by careful hammer blows.

192 Der A-Stoff-Pumpensatz ist einsatzbereit. Eine Saugleitung NW 100 mit Gewebeverstärkung ist mit dem Eckventil der Entleerungsleitung und dem Saugstutzen am A-Pumpensatz verbunden. Der Betriebsdruck wird mit einem Manometer überwacht. Sehr schnell bildet sich ein Eispanzer um das Handrad am Eckventil. Das Eis kann nur mechanisch durch vorsichtige Hammerschläge entfernt werden.

DMA, GD639.5.18
HAP11/T 507
Karlshagen, 1.6.1944
Bd.Nr. E 3195/44 g

DMA, GD639.5.12
EW 4141
Karlshagen, 31.8.1944
EW23/331/44g
VTL (Entwurf)
Fz. 402

DMA, GD639.5.2
T13
Stand 1.4.1944
Auszug
Gerätebeschreibung

DMA, GD639.5.1
EW234
Fz.-Pos.-Nr.402

Vehicle 402

Internal or former designation: Fz.-Pos.-Nr.402
A-Stoff-Trailer 6 (2-axle)
A-Stoff-Trailer

Intended use
The A-Stoff-Trailer 6 is the successor of the A-Stoff-Trailer 5. Its use and construction is similar to the trailer 5, differing only in a larger 6,000 l inner tank for the A-Stoff.

The Fz.402 trailer was, after the Meillerwagen and the Vidalwagen, the largest vehicle of the FR ground system.

Excert from the preliminary technical delivery terms (VTL)
for the Fz.402 vehicle.

Technical requirements for the tank superstructure
Dimensions, weights and specifications for the complete vehicle:

Total length:	8,600 mm
Width:	2,400 mm
Height:	3,000 mm
Wheelbase:	5,500 mm
Track width:	1,800 mm center distance between dual tires on axle
Tare weight:	6,030 kg
Weight loaded:	12,560 kg
Wheels:	9.75-20
A-Stoff container:	6,000 l

The main parts of the A-Stoff trailer 6 are:
- Chassis
- Inner tank for A-Stoff with carrying- and tension chains
- Outer container with support frame and insulation
- Fittings
- Tools and accessories

The chassis is delivered in an already adapted condition together with the support frame.

The dimensions of the inner tank shall be adapted to the profile of the vehicle, taking into account the required insulation and the outer container, and shall be manufactured in rolled sheet metal* with a capacity of 6,000 l. The insulation in the space between it and the outer container made of 2 mm sheet steel is filled with dry flakes of *Iporka*** and compressed to a density of 27 kg/m³.

* Made of Al Mg 3 in wall thickness of 5 + 7 mm, operating pressure 1.5 atu
** *Iporka* (trade name) is foam made of formaldehyde resin and urea.

Fahrzeug 402

Interne oder frühere Bezeichnung: Fz.-Pos.-Nr.402
A-Stoff-Anhänger 6 (2-achs.)
A-Stoff-Anhänger

Verwendungszweck
Der A-Stoff-Anhänger 6 ist das Nachfolgemodell von A-Stoff-Anhänger 5. Die Verwendung und der Aufbau ist analog zum Anhänger 5 ausgeführt, unterscheidet sich im Wesentlichen nur durch einen größeren Innentank mit 6000 l Fassungsvolumen für den A-Stoff.

Der Anhänger Fz.402 ist nach dem Meillerwagen und Vidalwagen das größte Fahrzeug der FR-Bodenanlage.

Auszug aus der vorläufigen technischen Lieferbedingung (VTL)
für das Fahrzeug Fz.402.

Technische Forderungen für den Tankaufbau
Abmessungen, Gewichte und Angaben für das komplette Fahrzeug:

Gesamtlänge:	8.600 mm
Breite:	2.400 mm
Höhe:	3.000 mm
Radstand:	5.500 mm
Spurweite:	1.800 mm Mittenabstand zwischen Doppelreifen
Leergewicht:	6.030 kg
Gewicht beladen:	12.560 kg
Räder:	9,75-20
A-Stoffbehälter:	6.000 l

Die Hauptteile des A.-Stoff-Anhängers 6 sind:
- Fahrgestell
- Innentank für A-Stoff mit Trage- und Spannketten
- Außenbehälter mit Tragegerüst und Isolierung
- Armaturen
- Werkzeug und Zubehör

Das Fahrgestell wird in bereits abgenommenem Zustand zusammen mit dem Tragegerüst angeliefert.

Der Innentank ist in seinen Abmessungen unter Berücksichtigung der erforderlichen Isolierung und des Außenbehälters dem Wagenprofil anzupassen und in Walzenform aus Blechen* mit einem Fassungsvermögen von 6.000 l zu fertigen. Die Isolierung im Zwischenraum des aus 2 mm Stahlblech gefertigten Außenbehälters ist zum Innenbehälter mit trockenen Flocken aus *Iporka** befüllt und auf ein Dichte von 27 kg/m³ gestopft.

* Gefertigt aus Al Mg 3 in der Wandstärke von 5 + 7 mm, Betriebsdruck 1,5 atü
** *Iporka* (Handelsname) ist ein aus Formaldehydharz und Harnstoff hergestellter Schaumstoff.

BArch, RH8II Bild-B1388-44, o.Ang.

193

193 Tombak* hoses were stowed in the opened hose box. On the vehicle itself some changes had already been carried out. For example, a bracket for a drain channel was mounted on the upper tank cover. The drain channel was attached to the drain valve and routed over the right rear fender, to prevent cold damage caused by liquid oxygen being blown directly onto the tires. Later, a drain pipe was connected directly to the drain valve, attached to the chassis frame and routed past the right-hand mudguard. The brackets on the support frame and manhole cover were designed to hold camouflage mats.

* Hose reinforced with a tombac fabric net. Tombak = high copper brass alloy.

193 Im geöffneten Schlauchkasten wurden Tombakschläuche* verstaut. Am Fahrzeug selbst sind bereits einige Änderungen ausgeführt worden. Zum Beispiel wurde eine Halterung für eine Abflußrinne auf dem oberen Tankdeckel montiert. Die Abflussrinne wurde am Ablassventil angesetzt und über den Kotflügel rechts hinten geführt, um Kälteschäden durch direktes Ausblasen von flüssigem Sauerstoff auf die Reifen zu verhindern. Später wurde ein Abflussrohr direkt an das Ablassventil angeschlossen, am Fahrgestellrahmen befestigt und am rechten Kotflügel vorbeigeführt. Die Halterung am Gerüst und Mannlochdeckel war zur Aufnahme von Tarnmatten vorgesehen.

* Schlauch mit einem Gewebenetz aus Tombak verstärkt. Tombak = eine hoch kupferhaltige Messinglegierung.

BArch, RH8II Bild-B1389-44, o.Ang.

194

DMA, GD639.5.12
EW4141
Karlshagen,1.9.1944
EW4141
Karlshagen den 1.9.1944
Auszug
Versuchsbericht
EW23/333/44g

194 Incorrect handling of the valves, longer waiting times or intermediate storage of Fz.402 filled with A-Stoff repeatedly resulted in considerable evaporation losses. On behalf of the developement plant (Entwicklungswerk EW) a series of tests was started to determine the evaporation losses and to take countermeasures. The losses with open valves were 29 kg/h and with closed valves 16 kg/h. Losses during driving on the transport route depended on the speed and were between
20 to 24 kg/h at driving speeds of 20 and 30 km/h.

194 Bei unsachgemäßer Handhabung an den Ventilen, längeren Wartezeiten oder dem Zwischenlagern mit A-Stoff befüllter Fz.402 kam es immer wieder zu empfindlichen Verdampfungsverlusten. Im Auftrag des Entwicklungswerks (EW) wurde eine Versuchsreihe gestartet, um die Verdampfungsverluste zu ermitteln und Gegenmaßnahmen zu ergreifen. Die Verluste bei offenen Ventilen betrug 29 kg/h und bei geschlossenen Ventilen 16 kg/h. Der Fahrverlust auf dem Transportweg war von der Fahrgeschwindigkeit abhängig. Diese lagen zwischen 20 bis 24 kg/h bei Fahrgeschwindigkeiten von 20 und 30 km/h.

DMA, GD639.5.11
EW236/Wö
Karlshagen, 1.8.1944
EW236/Wö
Karlshagen, den 1.8.1944
Bd.Nr.2/44 gK
Auszug
Teileliste Fz.402

Vehicles on order | Fahrzeugbestellungen

	Item Lfd. Nr.	Qty. ordered Best. Stückzahl	Qty. per vehicle Stück je Fahrzeug	Subject drawing number Gegenstand Zeichnungs-Nr.	Ordered by order number Besteller Bestell-Nr.	Supplier Lieferfirma
Fahrzeugbestellung	1	44	1	Sonder-Tieflader-Anhänger ohne Außenbehälter	WaPrüf 11/III 1433-SS-361-0006/43	Ackermann Fahrzeugbau Wuppertal-Vohwinkel
	2	300	1	dito, mit Außenbehälter	WaPrüf 11/III 1433-SS-361-0028/43	

Customizing/outfitting | Ausrüstung

	Item Lfd. Nr.	Qty. ordered Best. Stückzahl	Qty. per vehicle Stück je Fahrzeug	Subject drawing number Gegenstand Zeichnungs-Nr.	Ordered by order number Besteller Bestell-Nr.	Supplier Lieferfirma
Ausrüstung	1	44	1	Transportbehälter 6.000 l (Außenbehälter und Innenbehälter) für A-Stoff, mit Montage aller Ausrüstungsteile	WaPrüf 11/III 1433-SS-361-0005/43	Maschinenbau u. Co Mako Erfurt
	2	100	1	Montage und Ausrüstung an den angelieferten Fahrz. Mit Aussenbehälter (von Ackermann) und Innenbehälter (von Canzler) gemäß nachstehenden Baugruppen	WaPrüf 11/III 1433-SS-361-0041/43	dito
	3	100	1	dito	WaPrüf 11/III 1433-SS-361-0005/43	dito
	4	100	1	Innenbehälter nach ZG C9255a und Montage aller Baugruppen (Fahrzeug v. Ackermann)	WaPrüf 11/III 4925-SS-1433-0017/44	Wilhelm Schmidding, Hannover-Linden

Additional informations on the customizing items 1-4

1
44 vehicles were completed by Mako-Erfurt after delivery of the flatbed trailers. The company received an order for the delivery of of the outer and inner containers and all equipment parts assembled. Orders for hoses and coupling elements were placed separately. The opening valves were provided as government-furnished equipment (GFE).

2 + 3
An additional 200* vehicles were completed by Mako-Erfurt after receiving flatbed trailers with outer tanks. To do so the supplier received an order for completion of the whole equipment. Orders for vendor parts are shown separately in the list.

4
100 vehicles were completed by Schmidding-Hannover, the angle valves as usual provided as GFE parts.

* Davon 100 Stück bei Fa. Ziemann (Ausweichfirma) beauftragt.

Zusatzinformationen zur Ausrüstung der lfd.Nr. 1-4

1
von Mako-Erfurt wurden 44 Fahrzeuge nach Auslieferung der Anhänger-Fahrgestelle ausgerüstet. Dazu erhielt die Firma Auftrag auf Lieferung der Außen- und Innenbehälter mit Montage dieser selbigen und aller Ausrüstteile. Gesondert wurden auf Schläuche und Kupplungsteile Aufträge erteilt. Die Öffnungsventile sind amtsseitig zugestellt worden.

2 + 3
200* weitere Fahrzeuge rüstete Mako-Erfurt nach Anlieferung der Anhänger-Fahrgestelle mit Außenbehälter aus. Dazu erhielt die Firma Auftrag auf die Montage der gesamten Ausrüstung. In der Liste sind gesondert die Aufträge auf Zulieferteile genannt.

4
100 Fahrzeuge rüstete Schmidding-Hannover komplett aus, wobei nur die Eckventile wie bei sämtlichen Ausrüstungsaufträgen gestellt wurden.

* Davon 100 Stück bei Fa. Ziemann (Ausweichfirma) beauftragt.

IWM BU10274

195

DMA, GD639.5.13
OKH/Wa Prüf10
Karlshagen, den 1.4.1944
Bd.Nr. 788/44 gK
Auszug
Besprechung
Fahrzeuggruppe 4
vom 31.3.44
in Karlshagen I

Production status: Vehicle 402
Delivered to date: 25 (14 January to 11 February).
These 25 vehicle are available now at LR (the missing 9 angle valves are on the way from HAP 11 by courier), 17 are available at the Mako Company. These will be refitted with nickel steel chains (2 already equipped), 2 under construction at the Mako Company.
Also ordered: 200 from the Mako Company (from May onward monthly delivery of 24). 100 from the Ziemann Company (second source) with monthly delivery of 2 from June onwards.

Current defects and their remedy

Support of the inner tank with tensioning chains has led to fractures in the rosette suspension of vehicle 402. As a remedy 3 links of the chain are made of nickel steel. This solution still requires a check to be made of stress on the tank in the supercooled state. For this purpose extensive driving trials will be made of fueled vehicles. Two of these vehicles will arrive at HAP11 in the next few days and be fueled by B.z.b.V. Heer for unit testing. Action TD 41 Ing. Nettesheim.

This driving trial is for testing this suspension point of the internal tank and will at the same time be the driving test requested by General d.Art. Heidemann with 2 trailers towed one behind the other. After this driving test they will return to the company for thorough investigation in the presence of representatives of Wa Prüf11/III. Conversion of further vehicles already delivered to LR will be on hold until the results are known.

In order to compensate for the shrinkage process that occurs in the tensioning chains and to ensure that the inner container is held without play even when warm, disk springs are installed on the tensioning chain holders on a trial basis.

The sheet metal boxes on the front platform of the vehicle must have water drain holes in their bottoms. Action by Wa Prüf 11/III for current production. Change by the units themselves.

Fertigungslage: Fahrzeug 402
Bisher ausgeliefert: 25 Stück (14. Januar bis 11. Februar)
Diese 25 Fahrzeug sind z.Zt. LR (die fehlenden 9 Eckventile sind durch Kurier vom HAP 11 dorthin unterwegs), 17 Stück befinden sich z.Zt. bei Fa. Mako. Diese werden umgebaut auf Nickelstahlglieder (2 davon bereits damit ausgerüstet). 2 weitere Fahrzeuge sind bei der Fa. Mako im Bau.
Ferner sind bestellt: 200 Stück bei Fa. Mako (ab Mai monatliche Lieferung von 24 Stück). 100 Stück bei Fa. Ziemann (Ausweichfirma) zunächst ab Juni monatlich 2 Stück.

Derzeitige Mängel und ihre Behebung

Halterungen des Innenbehälters mit Spannketten haben in der Rosettenaufhängung des Fahrzeuges 402 zu Brüchen geführt. Als Abhilfe werden 3 Glieder der Kette aus Nickelstahl gefertigt. Diese Lösung bedarf noch der Überprüfung hinsichtlich Beanspruchung des Kessels selbst im unterkühlten Zustand. Hierzu sind gründliche Fahrversuche mit betankten derartigen Fahrzeugen durchzuführen. Zwei dieser Fahrzeuge treffen in den nächsten Tagen beim HAP11 ein und werden B.z.b.V. Heer betankt zum Truppenversuch gemeldet. Veranlassung TD 41 Ing. Nettersheim.
Dieser Fahrversuch bezweckt die Erprobung dieser Aufhängung des inneren Behälters und wird gleichzeitig als Fahrversuch gemäß Antrag General d. Art. Heidemann ausgenützt, wozu 2 Anhänger hintereinander gehängt werden. Nach diesem Fahrversuch erfolgt Rückführung an die Firma und genaue Untersuchung in Gegenwart von Vertretern von Wa Prüf11/III. Umbau weiterer bereits an LR ausgelieferter Fahrzeug wird bis zum Vorliegen des Ergebnisses zurückgestellt.

Um den bei der Unterkühlung auftretenden Schrumpfvorgang in den Spannketten auszugleichen und auch bei warmem Zustand eine spielfreie Halterung des inneren Behälters zu gewährleisten, werden versuchsweise an den Spannkettenhalterungen Tellerfedern eingebaut.

Die Blechbehälter auf der vorderen Plattform des Fahrzeuges müssen im Boden Wasserablauflöcher erhalten. Veranlassung Wa Prüf 11/III für die laufende Fertigung. Änderung durch die Truppe selbst.

Bracing of the inner tank on the A-Stoff trailer 6 (2-axle)

Verspannung des Innenbehälters am A-Stoff-Anhänger 6 (2-achs.)

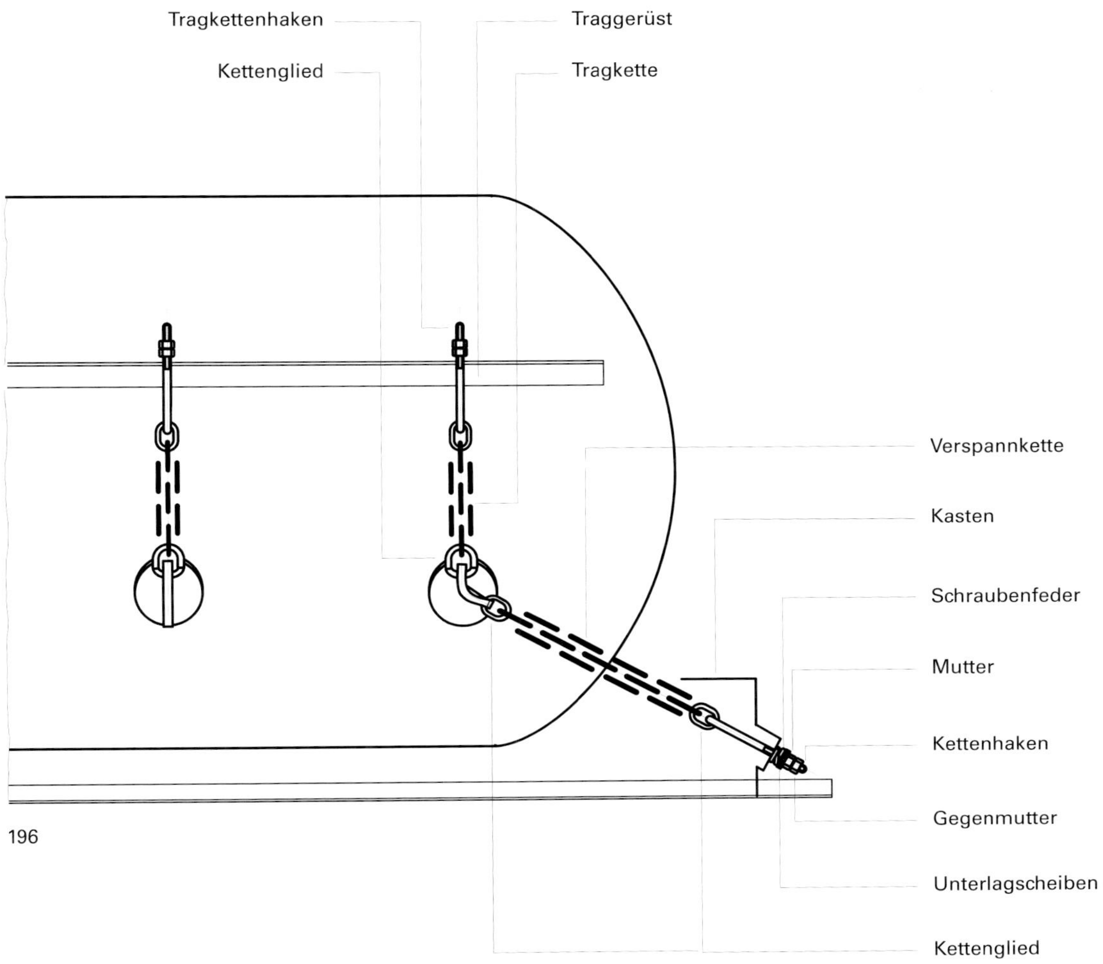

196

197

BArch, RH8II Bild-B1388-44, o.Ang (Ausschnitt)

Auszug
VTL (Entwurf)
Fz.402

In order to compensate for the shrinkage process that occurs in the tensioning chains and to ensure that the inner tank is held without play even when warm, disk springs are installed on the tensioning chain holders on a trial basis. The first links (calculated from the boiler) of the carrying chains, as well as the first 3 links of the tension chains are to be made of *Remanit**. The remaining chain links and the chain hooks are to be made of St 34.12 Din1612. The tensioning chains are to be fastened with springs in the support frame.

* *Remanit*: brand name of Witten-Krefeld stainless steel, DRP 360467 (1926)

Um schädliche Wärmespannungen durch den großen Temperaturunterschied zu vermeiden, wurde der Innencontainer beweglich gelagert. Zur Lagerung auf dem Tragegerüst sind auf jeder Seite 3 Trage- und 2 Spannketten vorgesehen. Die ersten Glieder (vom Kessel ausgerechnet) der Trageketten, sowie die ersten 3 Glieder der Spannketten sind aus *Remanit** zu fertigen. Die übrigen Kettenglieder, sowie die Kettenhaken sind aus St 34.12 Din1612 herzustellen. Die Spannketten sind mit Federn im Tragegerüst zu befestigen.

* *Remanit*: Markenname der Edelstahl Witten-Krefeld, DRP 360467 (1926)

Emptying the 402 vehicle with compressed air

Test report dated 25.7.44

DMA, GD639.5.12
EW23
o.O., den 30.11.44
EW23/1627/44 g
Auszug

On the front platform of the 402 vehicle is mounted a set of 6 high-pressure cylinders holding 60 l at 250 atü operating pressure (the same cylinders as used in the Fz.102). The ventilation line leads from this set via a pressure reducer to the tank's ventilation line. Emptying is done with the valve NW 70, which is now only used for filling and has a riser pipe connection.

Overloading of the chassis due to the additional load of the P-battery does not occur, as was determined by weighing. A permissible spring load of 6.250 kg at the front and 7.260 kg at the rear is taken as a basis. The measured axle pressures are 6.110 kg at the front and 6.720 kg at the rear with a full container and P-battery installed.

Compared to refueling with an A-Stoff pump the following advantages and disadvantages result:

Entleerung des Fahrzeuges 402 mit Pressluft

Versuchsbericht vom 25.7.44

Auf der vorderen Plattform des Fahrzeugs 402 ist eine Batterie aus 6 Hochdruckflaschen von 60 l bei 250 atü Betriebsdruck (gleiche Flaschen wie beim Fz.102 verwendet) aufgebaut. Die Belüftungsleitung führt von dieser Batterie über einen Druckminderer in die Entlüfterleitung des Tanks. Die Entleerung erfolgt über das jetzt nur zur Füllung dienende Ventil NW 70, das Steigrohranschluss hat.

Eine Überlastung des Fahrgestelles tritt durch die zusätzliche Beladung mit der P-Batterie nicht auf, wie durch Wiegen festgestellt wurde. Es ist hierbei eine zulässige Federbelastung von 6.250 kg vorne und 7.260 kg hinten zu Grunde gelegt. Die gemessenen Achsdrücke betragen bei gefüllten Behälter und aufgebauter P-Batterie vorne 6.110 kg und hinten 6.720 kg.

Gegenüber dem Tanken mit A-Stoffpumpe ergeben sich folgende Vor- und Nachteile:

BArch, RH8II_Bild-B1149-44, o.Ang.

198

Advantages

Increased ground clearance

by omission of the suction pipe including accessories; 200 mm more ground clearance is gained.

Reduction of refuelling losses

By eliminating the suction line including the degassing plug and angle valve as well as the degassing valve with piping, a major cause of the still high evaporation losses is removed. Reduction of mist clouds by omission of the pump also eliminates the previous pre-cooling and thus a larger part of the mist clouds. Simplified operation and greater operational safety. Pressure refuelling does not require any pre-cooling time. Similarly no delivery problems are be expected. Operational maintenance is much easier than for the Otto engine with pump, and the number of malfunctions is lower.

Saving of bottlenecked parts.

The petrol engine and the NW 100 tombak hose including the associated coupling parts can be eliminated.

Disadvantages

Reduction of delivery pressure and the flow rate/ time unit. Delivery pressure is sufficient for normal cases and would only be of significance if the tank lines were extended or the delivery head increased, as in the »Regenwurm« project, for example. The lower flow rate (increasing defueling time from 10 to 12 minutes) is compensated by the omission of the pre-cooling time, so that overall a saving of 3 minutes is still made.

Need for compressed air

For the complete emptying of the tank with a single airing of 1.5 atü pressure, three bottles with a 60 l capacity each, a pressure of 200 atü is required. This bottle filling must either take place at the refilling point of the A-Stoff-EKW, or the vehicle must drive to the technical battery for this purpose. If the filling is done, as expected, by a compressor Fz.506 with 100 cbm suction capacity, a filling time of approx. 25 min is necessary.

Procurement of high pressure bottles.

At present there is already a bottleneck, albeit a temporary one, in the procurement of high-pressure cylinders for the vehicle Fz.102.

Introduction of a change to a series.

We respectfully ask you to check these considerations and, if necessary, to place an order with us for further investigations and development of a model ready for reproduction.

Gez. Electromechanical Works Department – Ground Services

Vorteile

Erhöhung der Bodenfreiheit

durch Wegfall der Saugleitung. Einschl. Zubehör werden 200 mm an Bodenfreiheit gewonnen.

Erniedrigung der Betankungsverluste

Durch den Fortfall der Saugleitung einschl. Entgasungstopf und Eckventil sowie des Entgasungsventiles mit Rohrleitungen wird eine wesentliche Ursache der immer noch hohen Verdampfungsverluste weggenommen. Durch den Wegfall der Pumpe entfällt auch die vorherige Vorkühlung und damit ein großer Teil der Nebelwolken. Vereinfachte Bedienung und größere Betriebssicherheit. Die Druckbetankung erfordert keine Vorkühlungszeit. Ebenfalls sind keine Förderschwierigkeiten zu erwarten. Die betriebliche Wartung ist weitaus einfacher als die des Otto-Motors mit Pumpe, dementsprechend ist die Zahl der Störungen geringer.

Einsparung von Engpassteilen

Der Otto-Motor und der Tombak-Schlauch NW 100 einschl. der dazugehörigen Kupplungsteile kann eingespart werden.

Nachteile

Herabsetzung des Förderdrucks und der Fördermenge/Zeiteinheit. Der Förderdruck reicht für den Normalfall aus und würde nur bei Verlängerung der Tankleitungen oder Vergrößerung der Förderhöhe wie beispielsweise im Regenwurmprojekt ins Gewicht fallen. Die geringere Fördermenge (vergrößerte Enttankungszeit von 10 auf 12 min) wird durch Wegfall der Vorkühlzeit wettgemacht, sodass im Ganzen gesehen noch eine Einsparung von 3 min auftritt.

Notwendigkeit der Druckluft

Für die restlose Entleerung des Tanks bei einmaliger Belüftung mit 1,5 atü Druck, sind 3 Flaschen von je 60 l Inhalt bei 200 atü Druck notwendig. Diese Flaschenfüllung muss entweder an der Umfüllstelle vom A-Stoff-EKW erfolgen, oder aber das Fahrzeug muss zu diesem Zweck an der techn. Batterie vorbeifahren. Erfolgt die Füllung, wie zu erwarten, durch einen Kompressor Fz.506 mit 100 cbm Ansaugleistung, so ist eine Füllzeit von ca. 25 min notwendig.

Beschaffung von Hochdruckflaschen

Zurzeit besteht schon in der Beschaffung von Hochdruckflaschen für Fahrzeug Fz.102 ein, wenn auch vorübergehender, Engpass.

Einführung einer Änderung in eine Serie

Wir bitten diese Überlegung zu prüfen und uns ggf. einen Auftrag über weitere Untersuchungen und Entwicklungen eines nachbaureifen Musters zu erteilen.

Gez. Elektromechanische Werke Abteilung – Bodendienst

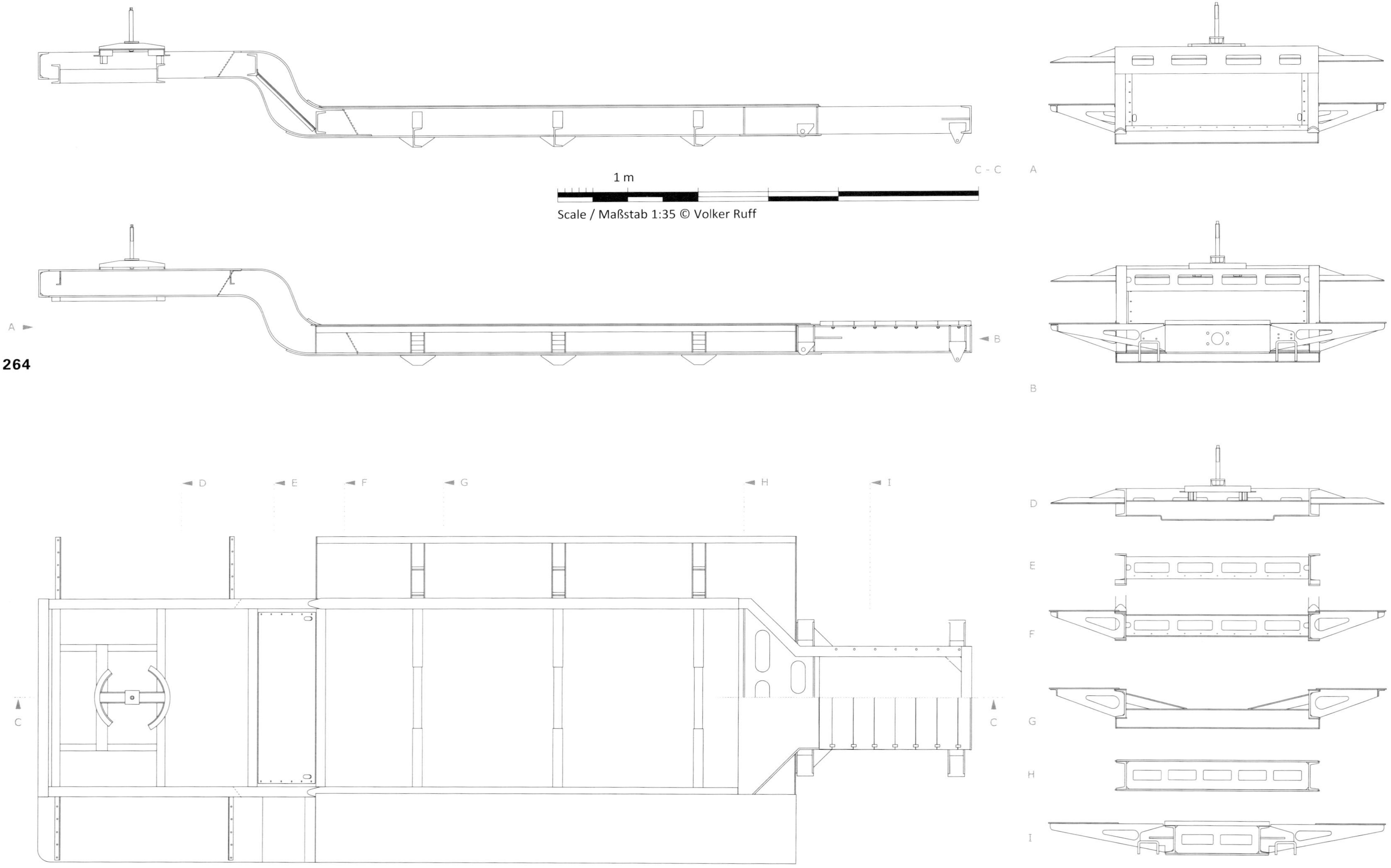

Fahrzeug 402

A-Stoff-Anhänger 6 (2-achs.)

Partially used drawings acc. to
group list 21Gr159

Teilweise verwendete Zeichnungen
nach Gruppenliste 21Gr159

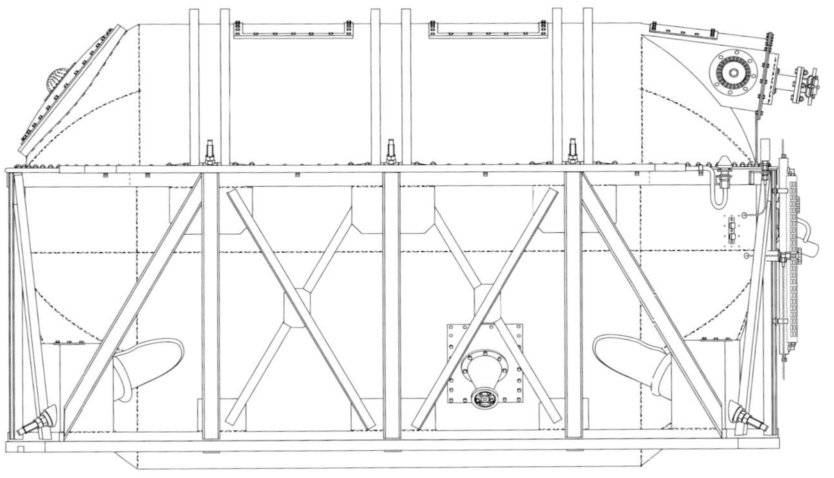

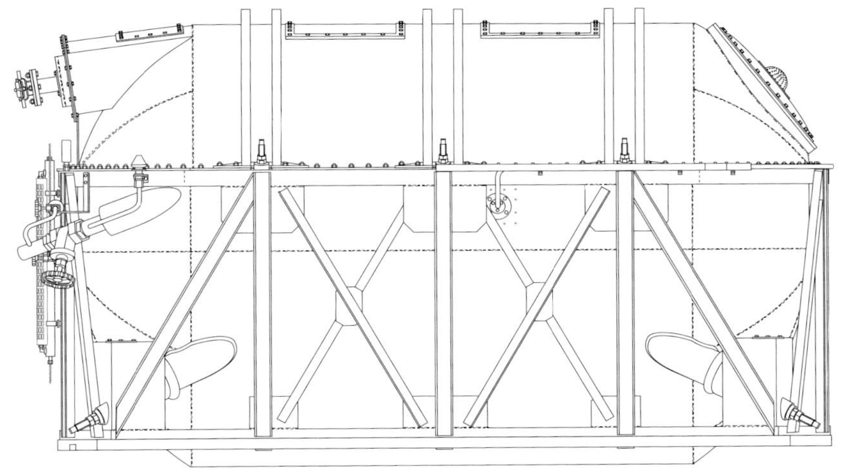

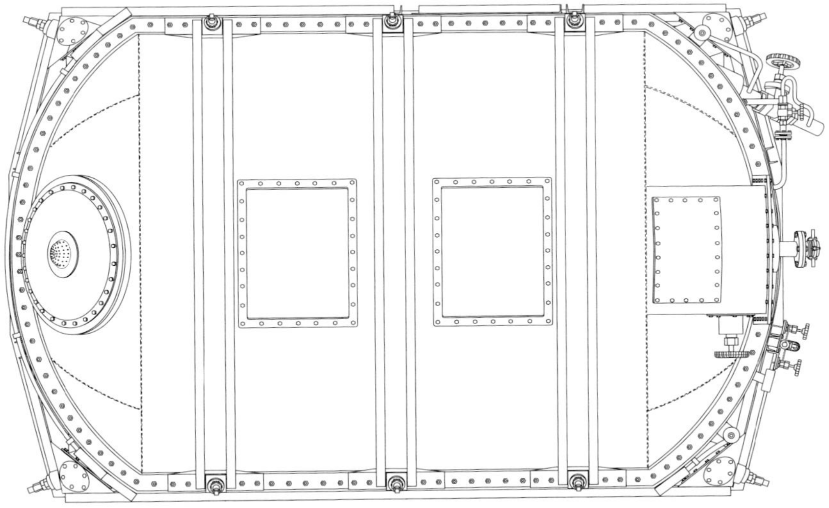

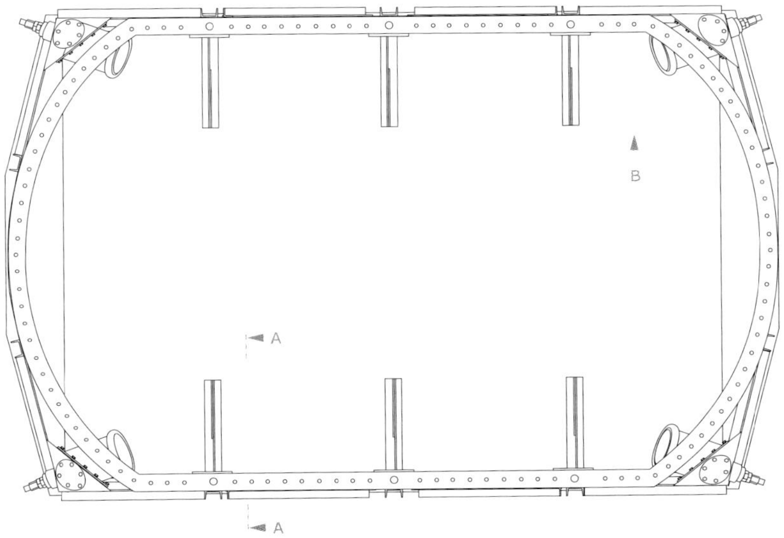

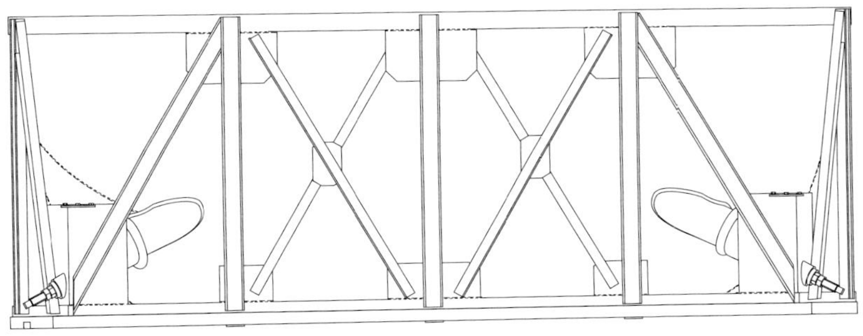

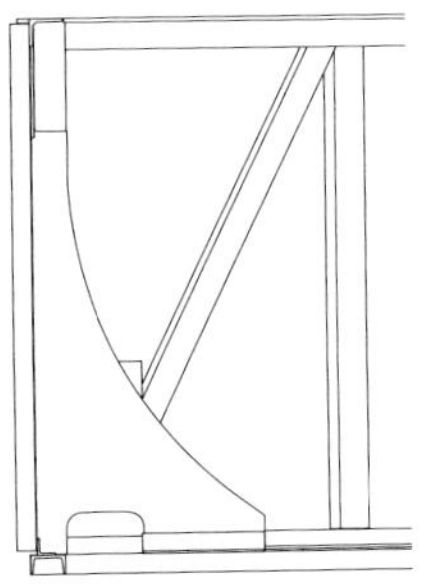

A - A

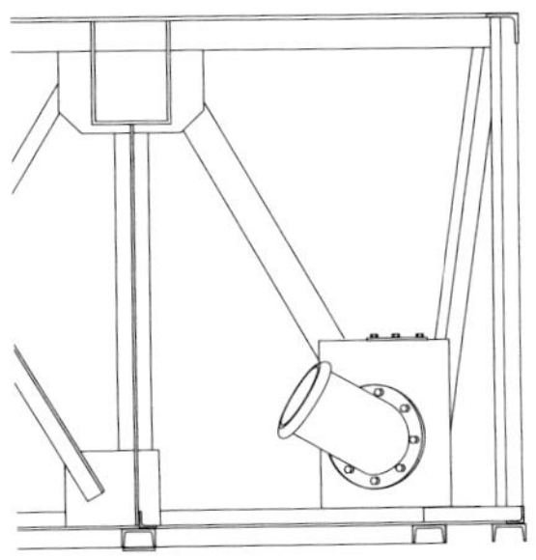

B

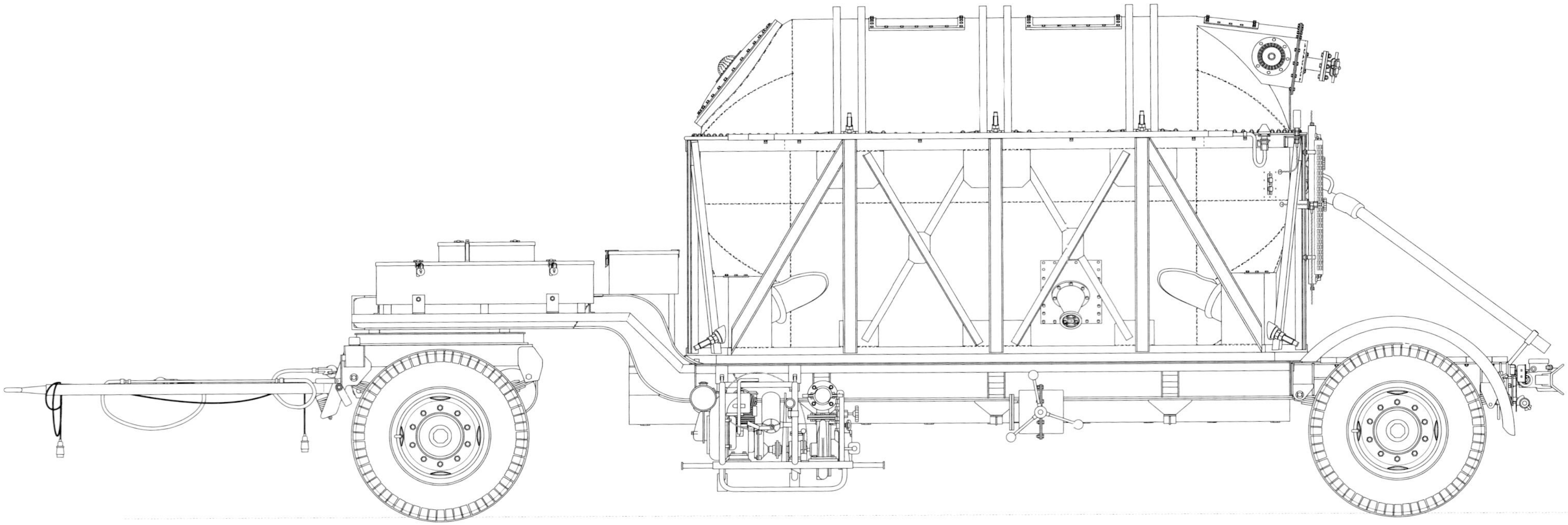

1 m

Scale / Maßstab 1:35 © Volker Ruff

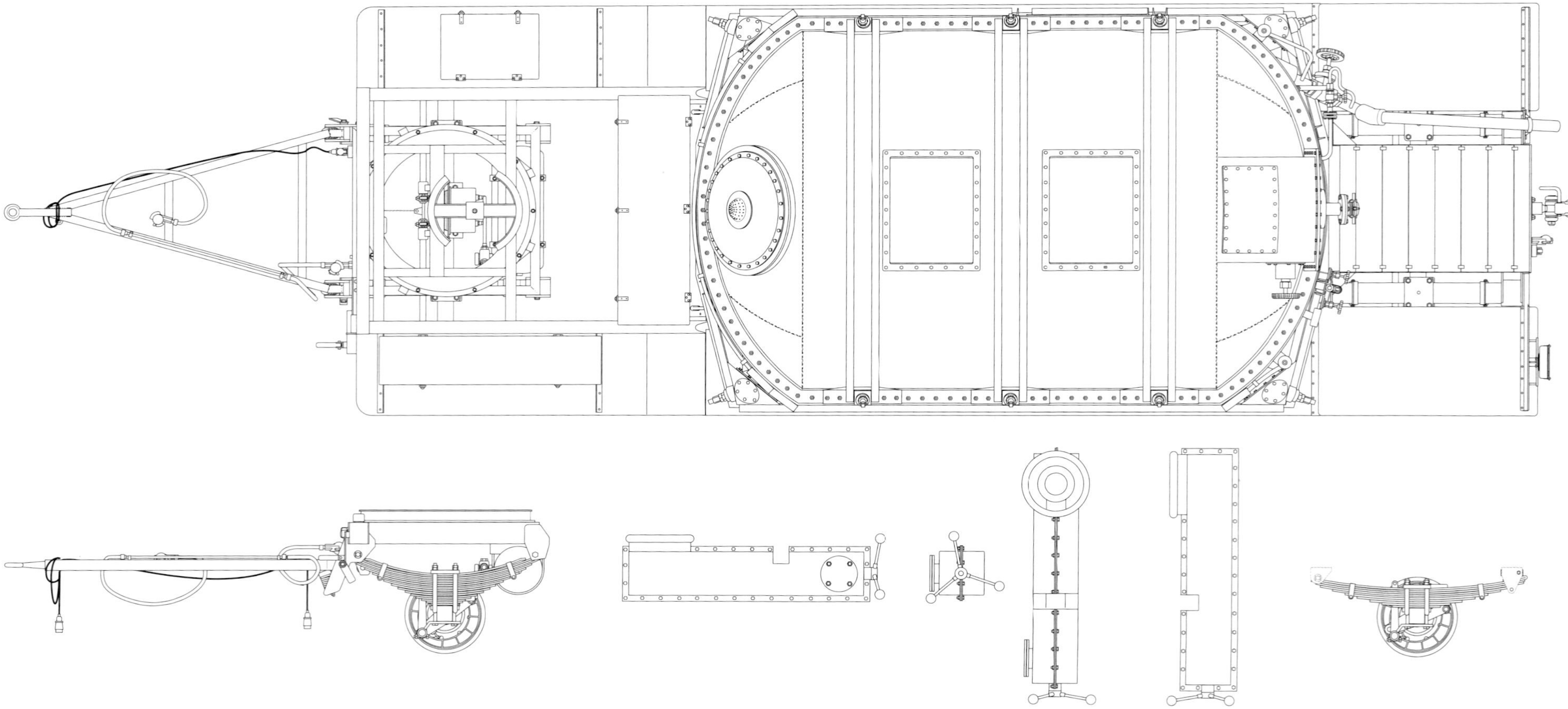

BArch, RH8II Bild-B1150-44, o.Ang.

199

199 The A-Stoff pump is connected and in operation. A sheet of ice very quickly forms around the valves, Tombak hoses and supply line. Refueling losses depend on the hose length and refueling times. Their values were determined by tests on nine vehicles. With a hose length of 7 m there was an average refueling loss of 290 kg at ~25 minutes refuelling time, and with a hose length of 50 m a refuelling loss of an average of ~750 kg ~35 minutes refueling time was determined. The sum of all evaporation losses had to be considered when calculating the filling quantity.

199 Die A-Stoff-Pumpe ist angeschlossen und in Betrieb. Sehr schnell bildet sich ein Eispanzer um die Ventile, Tombakschläuche und Zulaufleitung. Die Betankungsverluste sind abhängig von der Schlauchlänge und den Betankungszeiten. Im Versuch wurden an neun Fahrzeugen die Werte ermittelt. Bei einer Schlauchlänge von 7 m wurde ein Betankungsverlust von durchschnittlich 290 kg bei ~25 Minuten Betankungszeit und bei einer Schlauchlänge von 50 m ein Betankungsverlust von durchschnittlich ~750 kg bei ~35 Minuten Betankungszeit ermittelt. Die Summe aller Verdampfungsverluste musste bei der Berechnung der Füllmenge berücksichtigt werden.

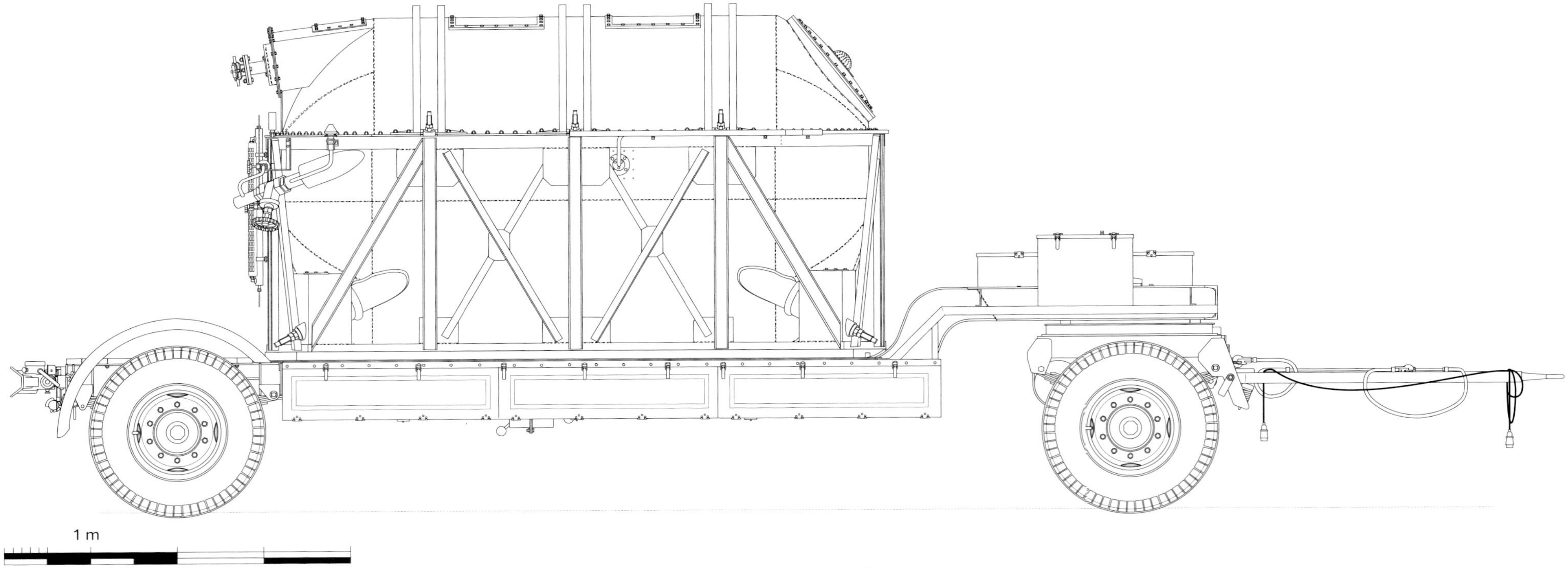

Scale / Maßstab 1:35 © Volker Ruff

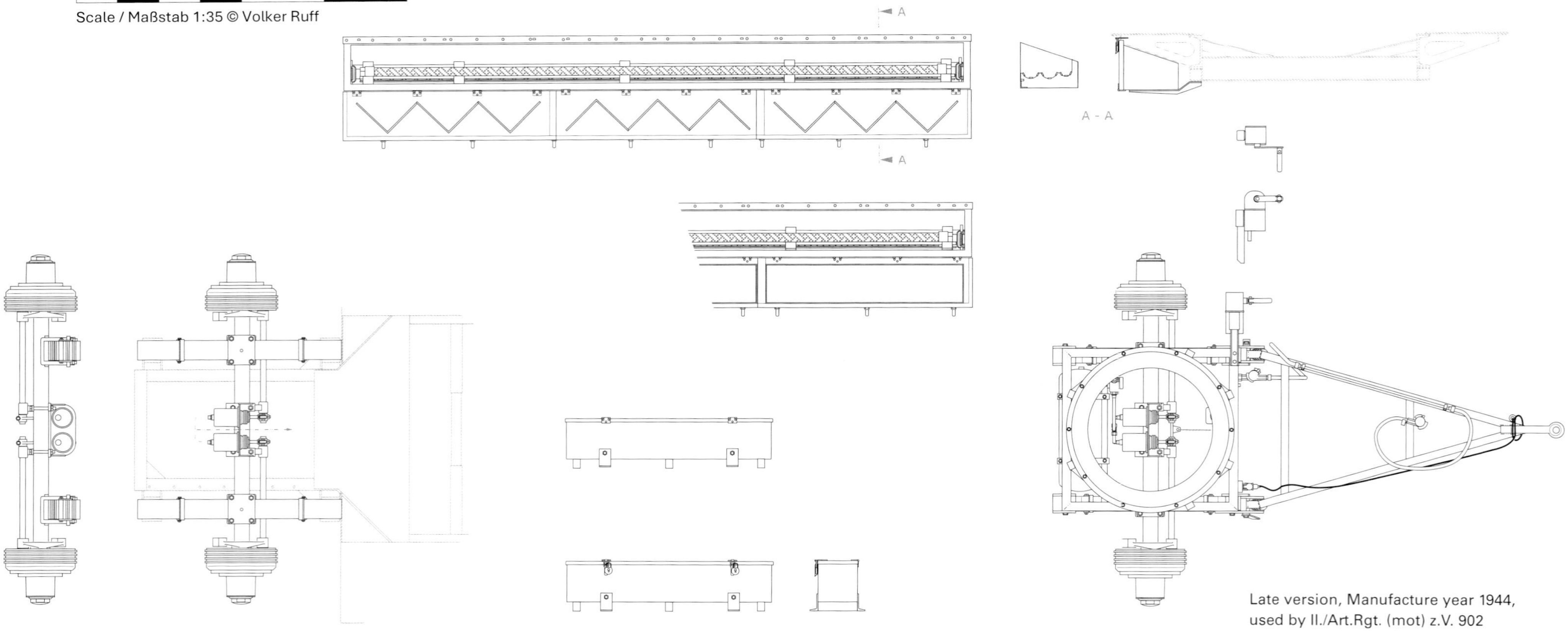

Late version, Manufacture year 1944,
used by II./Art.Rgt. (mot) z.V. 902

Abschlussausführung Baujahr 1944,
verwendet von der II./Art.Rgt. (mot) z.V. 902

BArch, RH8II Bild-B2000-44, o.Ang.

200

BArch, RH8II Bild-B2001-44, o.Ang.

201

View of the fittings
Blick auf die Armaturen

200 The following fittings are mounted on the support frame on the left side in the direction of travel: safety valve for the inner tank, pressure gauge, and on the support a liquid level indicator with shut-off valves at L+R. The two hexagon closures on the outer container belong to the defrosting line. A nameplate is attached to the support frame and stamped with the manufacturer code »hnj«.

201 On the cover of the outer cladding is the mounted in the middle a sheet metal box for the liquid separator with filling supports and and quick-release coupling for the filling hose.
The 1" pipe of the flue gas pipe coil leads from the outer cladding to the degassing connection and can be closed with a valve.
A further safety valve leads to the shut-off valve at the degassing connection.

200 Am Tragegerüst sind in Fahrrichtung links folgende Armaturen montiert: Sicherheitsventil für den Innenbehälter, Manometer und an der Stütze ein Flüssigkeitsanzeiger mit Absperrventilen L+R. Die beiden Sechskantverschlüsse am Außenbehälter gehören zur Auftauleitung. Am Traggerüstrahmen ist ein Typenschild angebracht und mit Herstellercode »hnj« gestempelt.

201 Auf dem Deckel der Außenverkleidung ist mittig der Blechkasten für den Flüssigkeitsabscheider mit Füllstützen und Schnellverschlusskupplung für den Füllschlauch montiert.
Aus dem Außenbehälter führt ein 1-Zoll-Rohr der Abgasrohrschlange zum Entgasungsstutzen. Dieser kann mit einem Ventil verschlossen werden. Ein weiteres Sicherheitsventil führt zum Absperrventil am Entgasungsstutzen.

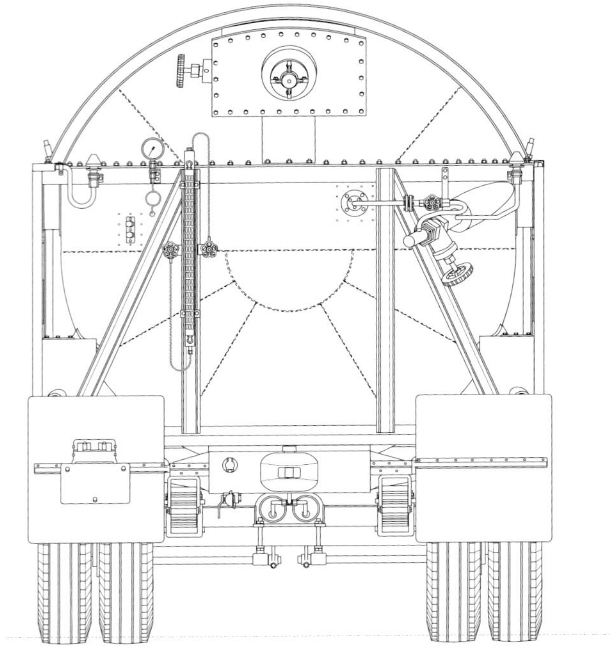

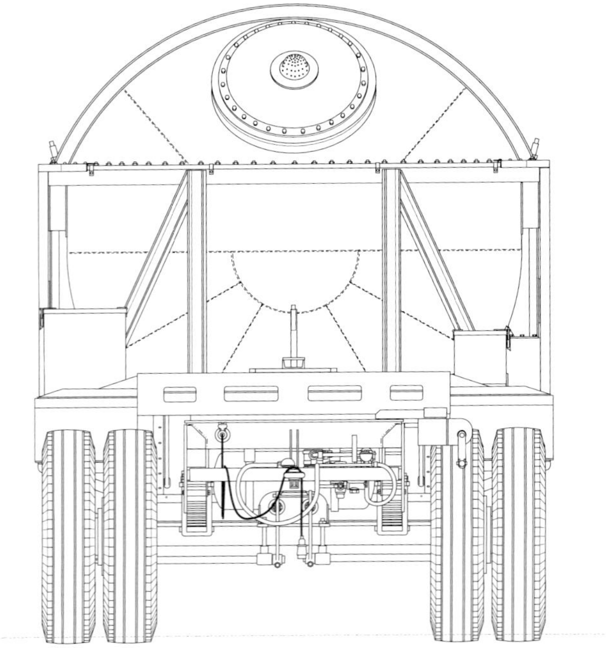

1 m

Scale / Maßstab 1:35 © Volker Ruff

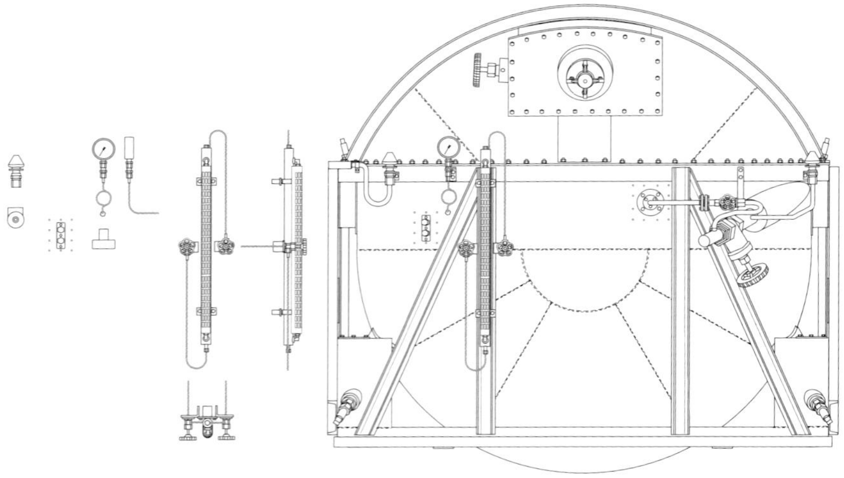

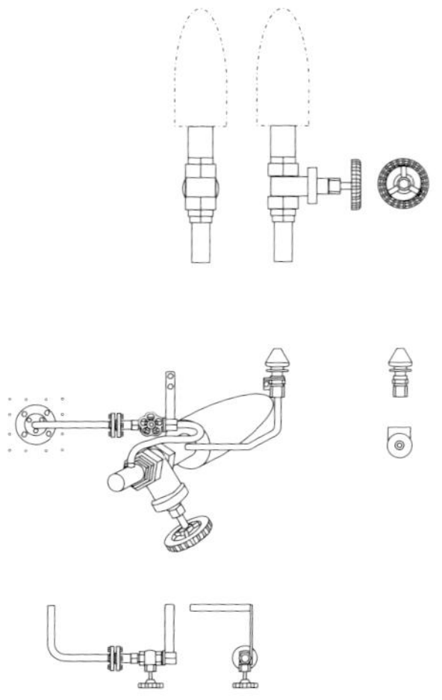

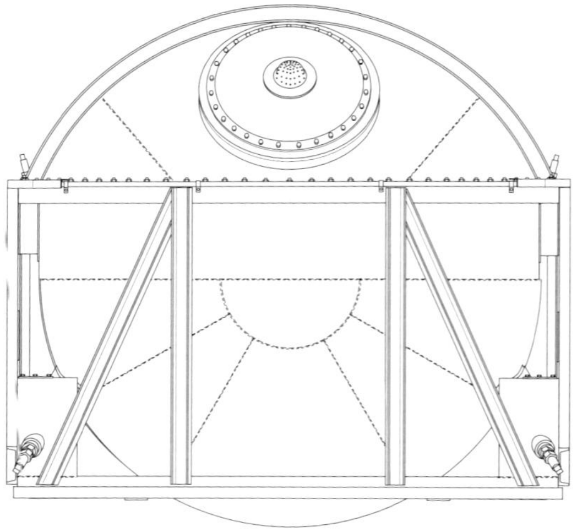

IWM: BU3320

202

IWM: BU10973

203

202 This A-Stoff trailer 402, left by II./Art.Rgt. (mot) z.V.902, was found by British units in April 1945 at the railroad station of Steyerberg (district of Nienburg/Weser). After the destruction of the railroad bridges over the Weser in March 1945, the transport of A- and B-Stoff was transferred to the roads. Almost 4,500 km were driven in the last two weeks of operations until the end of March 45 between the collecting stations in Leese and Wunstorf respectively to the fuel depot of Fst.117 (Schöppingen). The operation has left its scars. Spare tires for dual wheels were hardly available and longer marching distances now had to be covered with single tires.

203 The best-preserved vehicles of those found, here an Fz.402, were repaired and used for Operation Backfire in October 1945. This top view of the cover of the outer panel shows many details. In the front is the manhole cover with protective screen, two access covers between the frames for suspension chain hook stowage, and another smaller one on the sheet metal housing for the liquid separator.

* Various vehicles were found in the aeria around Steyerberg, Leese, Niendorf and Fallingbostel by British units.

202 Dieser von der II./Art.Rgt. (mot) z.V.902 zurückgelassene A-Stoff-Anhänger 402 wurde von britischen Truppen im April 1945 am Bahnhof von Steyerberg (Landkreis Nienburg/Weser) aufgefunden. Nach Ausfall der Eisenbahnbrücken über die Weser im März 1945 wurde der Nachschub für A- und B-Stoff auf die Straße verlegt. Fast 4.500 km wurden in den letzten zwei Einsatzwochen bis Ende März 45 zwischen den Abholbahnhöfen in Leese bzw. Wunstorf zum Treibstofflager der Fst.117 (Schöppingen) zurückgelegt. Der Einsatz hat seine Spuren hinterlassen. Ersatzreifen für eine Zwillingsbereifung waren kaum vorhanden und längere Marschstrecken mussten jetzt einfachbereift zurückgelegt werden.

203 Aus dem Fundus der aufgefundenen Fahrzeuge* wurden die am besten erhaltenen Fahrzeuge, hier ein Fz.402, instandgesetzt und im Oktober 1945 für die Operation Backfire verwendet. Die Draufsicht auf den Deckel der Außenverkleidung zeigt viele Details. Vorne der Mannlochdeckel mit Schutzsieb, zwischen den Spanten für die Tragkettenhakenlagerung sind zwei Zugangsdeckel und ein weiterer auf dem Blechgehäuse für den Flüssigkeitsabscheider zu sehen.

* Verschiedenste Fahrzeuge wurden im Raum Steyerberg, Leese, Nienburg und Fallingbostel von britischen Truppen aufgefunden.

DMA, GD639.5.18
HAP11/T 507
Karlshagen, 1.6.1944
Bd.Nr. E 3195/44 g

DMA, GD639.5.3
OKH
Chef H.Rüst und
B.d.E
B.z.b.V (heer) Abt.Ib
Bd.Nr.1843/44
g.Kdos.
Januar 1944

Auszug aus
Gerätebeschreibung
Bedienungs- und
Wartungsanweisung

DMA, GD639.5.3
EW234
Fz.-Pos.-Nr. 403

DM, 8001D
Tarnschlüssel für
Betriebsstoffe, mpe

Vehicle 403

Internal or former designation: Fz.-Pos.-Nr.403
Tanker truck 2900 l B-Stoff
B-Stoff truck

Intended use

The Kw 2900 l B-Stoff tanker is used to transport the B-Stoff between filling stations and the firing position.

The Luftwaffe's Kfz.385 aircraft fuel tank truck, an Opel Blitz 3 t S with open unit cab, was selected as the base vehicle and converted for B-Stoff operations with pipe connections and a pump.

The K.Kw. 2900 l B-Stoff has the following dimensions and weights:

Total length:	6,105 mm
Total width:	2,100 mm
Overall height:	2,400 mm
Wheel base:	3,600 mm
Track width:	1,540/1,620 mm front/rear
Tank capacity:	2,900 l
Tare weight:	3,180 kg
Payload:	2,870 kg
Tires:	190-20

hand pump capacity 80 l/minute
discharge nozzle NW60
connection thread M 78x2
filler nozzle NW 60, connection thread M78x2

The main parts of the K.Kw. 2900 l are:

- Chassis
- Tank
- Equipment compartment
- Fittings compartmentroom
- Valves and fittings
- Accessories

Chassis

The chassis is the normal chassis of the Opel 3 t truck type 3,6/36 S (see technical manual D 669/19B)

B-Stoff = Ethyl alcohol – red – RAL 3000*
* Technical colour – Stealth term for operating materials

Fahrzeug 403

Interne oder frühere Bezeichnung: Fz.-Pos.-Nr.403
Kessel-Kw 2900 l B-Stoff
B-Stoff Lkw

Verwendungszweck

Der Kessel Kw 2900 l B-Stoff dient zum Transport des B-Stoffes zwischen Tankstellen und Feuerstellung.

Der Flugbetriebsstoff-Kesselkraftwagen (Kfz.385) der Luftwaffe auf Opel Blitz 3 t S mit offenen Einheitsfahrerhaus wurde als Basisfahrzeug ausgewählt und mit Rohranschlüssen und Pumpe für den B-Stoffbetrieb umgerüstet.

Der K.Kw. 2900 l B-Stoff hat folgende Maße und Gewichte:

Gesamtlänge:	6.105 mm
Gesamtbreite:	2.100 mm
Gesamthöhe:	2.400 mm
Radstand:	3.600 mm
Spurweite:	1.540/1620 mm vorne/hinten
Behälterinhalt:	2.900 l
Leergewicht:	3.180 kg
Nutzlast:	2.870 kg
Räder:	190-20

Förderleistung der Handpumpe 80 l/Minute
Entleerungsstutzen NW 60
Anschlussgewinde M 78x2
Füllstutzen NW 60 Anschlussgewinde M78x2

Die Hauptteile des K.Kw 2900 l sind:

- Fahrgestell
- Behälter
- Geräteraum
- Armaturenraum
- Armaturen
- Zubehör

Fahrgestell

Als Fahrgestell dient das normale Fahrgestell des Opel 3 t Lastkraftwagens Typ 3,6/36 S (Gerätebeschreibung siehe D 669/19B).

B-Stoff = Spititus – Aethylalkohol – rot – RAL 3000*
* Technische Farbe – Tarnbezeichnung für Betriebsstoffe

Tank

The B-Stoff tank is welded from steel sheet and has the cross-section of a flattened ellipse. The tank holds 2,900 l. The tank is fastened to the chassis support frames with four tensioning screws accessible through four lockable doors in the sides of the vehicle.

In order to avoid expansion stresses caused by temperature differences in the sheet metal construction of the vehicle body, the B-Stoff tank is separated by a small air gap from the sheet metal body of the equipment- and of the fitting compartment. The gaps are sealed by fabric strips to prevent moisture penetration.

Behälter

Der B-Stoffbehälter ist aus Stahlblech geschweißt und hat den Querschnitt einer abgeflachten Ellipse. Der Behälter fasst 2.900 l. Der Behälter ist an den Fahrgestellrahmen mit vier Spannschrauben befestigt, die durch vier verschließbare Türen an den Längsseiten des Fahrzeuges zugänglich sind.

Um durch Temperaturunterschiede hervorgerufene Dehnungsspannungen in der Blechkonstruktion des Fahrzeugaufbaues zu vermeiden ist der B-Stoffbehälter von der Blechverkleidung des Geräteraumes und des Armaturenraumes durch einen Luftspalt getrennt. Je ein die beiden Spalten überbrückender Stoffstreifen verhindert das Eindringen von Feuchtigkeit.

Deutsches Museum, München, Archiv- FA014/B1053/43 TD5224 [CD74759]

204

Deutsches Museum, München, Archiv- FA014/B1276/44 T5224 [CD80827]

205

DMA, GD639.5.13
OKH/Wa Prüf10
Karlshagen den
1.4.1944
Bd.Nr. 788/44 gK

Auszug
Besprechung der
Fahrzeuggruppe 4
vom 31.3.44
in Karlshagen I

Status of Production

Vehicle 403

delivered at date: 48
of which 36 at 836
4 at 444
8 at HAP 11

No further production will take place. Replaced by vehicle 404.

Fertigungslage

Fahrzeug 403

bisher ausgeliefert: 48 Stück
davon 36 bei 836
4 bei 444
8 bei HAP 11

Weitere Fertigung erfolgt nicht. Ersetzt durch Fahrzeug 404.

Equipment compartment

Between the B-Stoff tank and the driver's cab is located the equipment compartment. It holds the fill and refueling euipment. The equipment compartment occupies the whole width of the vehicle. It is accessible from both sides by two lockable hatches. The floor and walls of the equipment compartment are fitted with devices for attaching the accessories. In the equipment compartment, on the lower front of the B-Stoff tank a thermometer is installed which shows the temperature of the B-Stoff in the tank.

Accessories

Motor vehicle equipment is as per vehicle handbook: filling and refueling equipment, railroad filling connector, B-Stoff flexible hoses and connectors, various wrenches, bucket, funnel, fire-extinguishers, pump handle and Dominit lamps*.

* Mining hand lamp operated with rechargeable battery (safety hand lamp)

Geräteraum

Zwischen B-Stoffbehälter und Führerhaus befindet sich der Geräteraum. Er dient zur Aufnahme des Füll und Betankungsgerätes. Der Geräteraum nimmt die ganze Fahrzeugbreite ein. Er ist von beiden Längsseiten aus durch je zwei verschließbare Klappen zugänglich. Am Boden und an den Wänden des Geräteraumes sind Vorrichtungen zur Befestigung des Zubehörs angebracht. Im Geräteraum befinden sich am vorderen unteren Teil des B-Stoffbehälters ein Thermometer, welches die Temperatur des B-Stoffes im Behälter anzeigt.

Zubehör

Kraftfahrzeuggerät gemäß Fahrzeugbegleitheft Füll- und Betankungsgerät, Eisenbahnabfüllkupplung, B-Stoffschläuche und Kupplungsstücke, verschiedene Schlüssel, Eimer, Trichter, Feuerlöscher, Pumpenschwengel und Dominitlampen*.

* Akku-betriebene Grubenhandleuchte (Sicherheitshandleuchte)

Deutsches Museum, München, Archiv- FA014/B1052/43 TD5224 [CD74758]

206

Fitting compartment
The fitting compartment is enclosed by a sheet metal housing which adjoins the rear wall of the B-Stoff tank.

Valve and fittings
The following fittings are located in the fittings compartment located behind the right-hand outside door: from bottom to top, 1 drain connection, 1 four-way cock, and a bypass line with a quick-acting tap. On the inside of the door is located a box with a Dominit lamp*. In the cent right is the hand pump with attached pump handle. The resting bracket for the pump swivel is located on the base plate in front of the container. On the left is the B-Stoff level indicator with supply line and two shut-off valves.

Armaturenraum
Der Armaturenraum wird von einem Blechgehäuse umschlossen, das sich an die Rückwand des B-Stoffbehälters anschließt.

Armaturen
Im Armaturenraum befinden sich hinter der rechten Außentür folgende Armaturen. Von unten nach oben: ein Entleerungsstutzen, ein Vierwegehahn und eine Umführungsleitung mit Schnellschlusszapfventil. Auf der Türinnenseite befindet sich ein Kasten mit Dominitlampe. Im Mittelteil ist rechts die Handflügelpumpe mit angesetztem Pumpenschwengel angebracht. Die Ruhehalterung für den Pumpenschwengel befindet sich auf dem Bodenblech vor dem Behälter. Links ist der B-Stoff-Standanzeiger mit Zuleitung und zwei Absperrventilen montiert.

Deutsches Museum, München, Archiv- FA014/B1278/43 TD5224 [CD80834]

207

Deutsches Museum, München, Archiv- FA014/B1277/43 TD5224 [CD80833]

208

DMA, GD639.5.18
HAP11/T 507
Karlshagen, 1.6.1944
Bd.Nr. E 3195/44 g

DMA, GD639.5.1
EW234
Fz-Pos.-Nr.404

Vehicle 404

Internal or former designation: Fz.-Pos.-Nr.404
Tanker truck 3500 l B-Stoff
B-Stoff truck

Intended use

The successor model for the Fz.403 had a larger container and was put into service as a tanker truck 3500 l of B-Stoff to carry it from the filling station to the firing position. The basic vehicle was developed for the Luftwaffe as an aviation fuel tanker, and for use as an Fz.404 equipped with pipe connections and a hand pump for B-Stoff operations.

The K.Kw. 3500 l B-Stoff had the following dimensions and weights:

Total length:	6,040 mm
Total width:	2,100 mm
Total height:	2,300 mm
Wheelbase	3,600 mm
Track width:	1,540/1,620 mm front/rear
Tank capacity:	3,500 l
Tare weight:	3,115 kg
Payload:	3,180 kg
Maximum weight	6,100 kg
Loaded weight:	6,740 kg
Wheels:	190-20

Delivery rate of the hand pump 80 l-minute
- Discharge nozzle NW 60
- Connection thread M 78x2
- Filling nozzle NW 60 Connection thread M78x2

The main parts of the K.Kw 3500 l are:
- Chassis
- Tank frame
- Tank
- Fitting compartment
- Fittings
- Accessories

Chassis

The normal chassis of the Opel 3 t truck type 3,6/36 S (D 669/19B), or as an alternative the chassis of the Opel 3 t truck type A 6700 (D 666/5), are used.

B-Stoff = Ethyl alcohol - red - RAL3000*
* technical colour - Stealth term for operating materials

Fahrzeug 404

Interne oder frühere Bezeichnung: Fz.-Pos.-Nr.404
Kessel-Kw 3500 l B-Stoff
B-Stoff Lkw

Verwendungszweck

Das Nachfolgemodell für das Fz.403 hatte einen größeren Behälter und wurde als Kessel-Kw 3500 l B-Stoff für den Transport von B-Stoff von der Tankstelle zur Feuerstellung in Dienst genommen. Das Basisfahrzeug wurde für die Luftwaffe entwickelt und als Flugbetriebsstoff-Kesselkraftwagen eingesetzt und für die Verwendung als Fz.404 mit Rohranschlüssen und einer Handpumpe für den B-Stoffbetrieb ausgerüstet.

Der K.Kw. 3500 l B-Stoff hat folgende Maße und Gewichte:

Gesamtlänge:	6.040 mm
Gesamtbreite:	2.100 mm
Gesamthöhe:	2.300 mm
Radstand:	3.600 mm
Spurweite:	1.540/1.620 mm vorne/hinten
Behälterinhalt:	3.500 l
Eigengewicht:	3.115 kg
Nutzlast:	3.180 kg
Zul. Höchstgew.	6.100 kg
Gefechtsgewicht:	6.740 kg gewogen
Räder:	190-20

Förderleistung der Handpumpe 80 l-Minute
- Entleerungsstutzen NW 60
- Anschlussgewinde M 78x2
- Füllstutzen NW 60 Anschlussgewinde M78x2

Die Hauptteile des K.Kw 3500 l sind:
- Fahrgestell
- Tankrahmen
- Behälter
- Armaturenraum
- Armaturen
- Zubehör

Fahrgestell

Als Fahrgestell dient das normale Fahrgestell des Opel 3 t Lastkraftwagens Typ 3,6/36 S (Gerätebeschreibung D 669/19B) aber auch Fahrgestelle des Lastkraftwagen 3 t Opel 6700 Typ A (Gerätebeschreibung D 666/5) wurden verbaut.

DM, 8001D
Tarnschlüssel für Betriebsstoffe, mpe

B-Stoff = Spititus – Aethylalkohol – rot – RAL3000*
* Technische Farbe – Tarnbezeichnung für Betriebsstoffe

BArch, RH8II Bild-B1390-44, o. Ang.

209

209 On this Fz.404 the tank body is mounted on the chassis of the Opel 3 t truck 3,6-36. A frame with three brackets for tools and a tubular ladder is welded to the tank body. The tank frame is slightly tapered to mount the tank horizontally. The wheels have only single tires, not the normal double-tired rear axle of these Opel vehicles. In accordance with its TL (technical delivery condition – January 45) the vehicle is painted in olive green RAL 6003.

209 Bei diesem Fz.404 ist der Tankaufbau auf dem Fahrgestell des Opel 3 t Lastkraftwagen 3,6-36 aufgebaut. Ein Gestell mit drei Einschüben für Werkzeug und einer Rohrleiter ist am Tankaufbau angeschweißt. Der Tankrahmen ist leicht konisch ausgeführt, um den Tank in die Waagerechte zu bringen.
Das Fahrzeug besitzt nur eine Einfachbereifung anstelle der normalerweise doppelbereiften Hinterachse der Opel-Fahrzeuge. Gemäß TL (technische Lieferbedingung – Januar 45) ist es in RAL 6003 (olivgrün) gestrichen.

BArch, RH8II Bild-B1391-44, o. Ang.

210

210 The vehicle does not have a segregated equipment compartment, its accessories and hoses are stowed in the equipment area or in the drawers of a frame mounted on the right. Simple brackets for camouflage nets / mats are welded onto the tank. A bracket for ramps is attached to the side of the tank frame. These ramps are used to level the terrain in order to park the vehicle as horizontally as possible.

210 Am Fz.404 gibt es keinen abgetrennten Geräteraum. Die Zubehörteile und Schläuche sind im Armaturenraum verstaut oder in den Schubladen des rechts angebauten Gestelles untergebracht. Einfache Halterungen für Tarnnetze oder Tarnmatten sind auf dem Tank angeschweißt. Am Tankrahmen ist seitlich eine Halterung für Auffahrrampen angebracht. Die Auflaufbohlen dienen zum Geländeausgleich, um das Fahrzeug möglichst waagerecht abstellen zu können.

DMA, GD639.5.11
T6627
Stand Juli 1944
Bd.Nr.E2790/44 gK
Auszug
Teileliste Fz.404

Vehicles on order

Fahrzeugbestellungen

	Item Lfd. Nr.	Qty. ordered Best. Stückzahl	Qty. per vehicle Stück je Fahrzeug	Subject drawing number Gegenstand Zeichnungs-Nr.	Ordered by order number Besteller Bestell-Nr.	Supplier Lieferfirma
Fahrzeugbestellung	1	284	1	Fahrgestell	WaPrüf 11/III	Opel

Customizing/outfitting

Ausrüstung

	Item Lfd. Nr.	Qty. ordered Best. Stückzahl	Qty. per vehicle Stück je Fahrzeug	Subject drawing number Gegenstand Zeichnungs-Nr.	Ordered by order number Besteller Bestell-Nr.	Supplier Lieferfirma
Aufbau- und Ausrüstungsfirmen	1	72	1	kompletter Behälteraufbau n. Z. 269 T	WaPrüf 11/III 361-0014/42	Schwelmer Eisenwerk, Schwelm i.W.
		90	1	mit kompletter Ausrüstung	WaPrüf 11/III 361-0032/43	
		122	1	bestehend aus	WaPrüf 11/III 361-0051/43	
		284	1	Tanklagerung	dito	dito
		284	1	Stahltank	dito	dito
		284	1	Armaturenkasten usw.	dito	dito

DMA, GD639.5.13
OKH/Wa Prüf10
Karlshagen, 1.4.1944
Bd.Nr. 788/44 gK

Auszug
Besprechung der
Fahrzeuggruppe 4
vom 31.3.44
in Karlshagen I

Production status: Vehicle 404
Delivered to date: 72
of which 36 at 485 (Febr. / March)
4 at 444 (Febr.).
The remainder will be distributed by B.z.b.V. Heer.
Delivery of a further 356 will be starting from May at 50 per month.

Current defects and their remedy
Vehicle 404/406:
The 70 mm hoses are regarded as part of the pump trailer and therefore located on it. The roof is installed accordingly for this. Reason is already given by Wa Prüf 11/III. Replacements are provided for the pump trailers already delivered.

Fertigungslage: Fahrzeug 404
Bisher ausgeliefert: 72 Stück
davon 36 bei 485 (Febr. März)
4 bei 444 (Febr.).
Die restlichen verteilt B.z.b.V. Heer.
Weitere Lieferung von 356 Stück – ab Mai monatlich 50 Stück.

Derzeitige Mängel und ihre Behebung
Fahrzeug 404/406:
Die 70 mm-Schläuche werden zum Pumpenanhänger gehörig betrachtet und deshalb auf ihm untergebracht. Hierzu wird das Dach entsprechend verlegt. Veranlassung ist bereits durch Wa Prüf 11/III erfolgt. Für die bereits ausgelieferten Pumpen-Anhänger wird Austausch vorgesehen.

211

211 This photo shows an Fz.404 on the Opel truck 3 t 3.6-6700 chassis with four-wheel drive. The vehicle is higher overall and looks more compact than the vehicle without cross-country ability Opel 3 t Lastkraftwagen 3,6-36. The tank frame here is designed with stiffening holes. The camouflage mats were placed between the cab and the tank on a support on the tank frame and secured with chains. The vehicle was painted in dark yellow RAL 7028 (according to sample), camouflage paint applied by the unit. Using a stencil, the lettering »Flammable« is dabbed on both sides of the tank and the »Loading class« on the drivers door.

211 Das Bild zeigt ein Fz.404 auf dem Opel Lastkraftwagen 3 t 3,6-6700 mit Allradantrieb. Das Fahrzeug ist insgesamt höher und wirkt kompakter im Vergleich zum Fahrzeug ohne Allradantrieb auf Opel 3 t Lastkraftwagen 3,6-36. Der Tankrahmen ist hier mit Versteifungslöchern ausgeführt. Die Tarnmatten wurden auf eine Lagerung am Tankrahmen zwischen Fahrerhaus und Tank abgelegt und mit Ketten gesichert. Das Fahrzeug ist in dunkelgelb RAL 7028 (nach Muster) angestrichen. Der Tarnanstrich wurde durch die Truppe selbst aufgebracht. Mittels einer Schablone wurde der Schriftzug »Feuergefährlich« beidseitig am Tank und die »Verladeklasse« an der Fahrertür aufgetupft.

BArch, RH8II Bild-B1476-44, o. Ang.

212

Deutsches Museum, München, Archiv- FA014/B742/44 TD5224 [CD80831]

213

212, 213 In this side view from right, the attachment of the frame with three toolboxes to the tank container is clearly visible. The tubular ladder is bolted to the tank at two angles and to the front strut of the frame. It is not directly mounted on the tank put away from it. The tank frame is slightly tapered and has three brackets on both the left and right, welded to the tank and bolted to the chassis. See detail at top left.

212, 213 Die Anbringung des Gestelles mit drei Werkzeugkästen am Tankbehälter ist in der Seitenansicht gut zu erkennen. Die Rohrleiter ist an zwei Winkeln auf dem Tank und an der Strebe des Gestelles verschraubt. Zum Tank hin ist sie auf Abstand geführt. Der Tankrahmen ist leicht konisch zulaufend ausgeführt. Daran sind je drei Halterungen links und rechts angeschweißt und mit den Aufnahmen am Fahrgestell verschraubt. Siehe Detail oben links.

BArch, RH8II Bild-B1392-44, o. Ang.

214

BArch, RH8II Bild-B1472-44, o. Ang.

215

Container
Behälter

214, 215 The B-Stoff tank made of sheet steel in the shape of a flattened ellipse and manufacured out of several parts and welded to the frame. The tank holds 3,500 l. The equipment compartment with a double wing door is made of sheet metal and riveted to the tank on a connecting frame. A manhole cover with attachments is mounted on the front of the tank. In the middle from top to bottom: Filler cap, direction finding device and filling level indicator. On the left and right of it are the safety plug and ventilation cap. The locking caps belong to the fittings.

214, 215 Der B-Stoffbehälter aus Stahlblech ist in Form einer abgeflachten Ellipse aus mehreren Teilen gefertigt, auf einen Rahmen gesetzt und verschweißt. Der Behälter fasst 3.500 l. Der Geräteraum mit einer Doppelflügeltür ist aus Blech gefertigt, am Tank angesetzt und vernietet. Ein Mannlochdeckel mit Einhängearmaturen ist am Tank vorne montiert. In der Mitte von oben nach unten: Einfüllarmatur, Peilvorrichtung und Schwimmerstandanzeiger. Links und rechts davon die Sicherungsstopfen und Belüftungskappe. Die Verschlusskappen gehören zu den Armaturen.

BArch, RH8II Bild-B1473-44, o. Ang.

216

BArch, RH8II Bild-B1393-44, o. Ang.

217

View into the equipment bay
Blick in den Armaturenraum

216, 217 View into the equipment bay
A B-Stoff level indicator is mounted centrally on the rear wall of the tank, a CO_2 fire extinguisher is mounted on the left, and the filling hoses in tubular containers are mounted underneath. On the bottom floor are a funnel, the filler neck, and the railway refuelling coupling. The guide line, hand wing pump with attached pump handle, four-way cock with drain connection, and overpressure gauge are mounted on the right. On the left side of the bay are various accessories, a bucket secured with a tension spring and a holder with a Dominit lamp.*

* Dominit lamp: Mining hand lamp. A safety hand lamp operated with rechargeable battery.

216, 217 Blick in den Armaturenraum
An der Tankrückwand ist mittig ein B-Stoff-Standanzeiger montiert, links ein CO_2-Feuerlöscher, darunter die Füllschläuche in Rohrbehältern. Auf dem Bodenblech sind Trichter, Füllstutzen und eine Eisenbahnbetankungskupplung gelagert. Führungsleitung, Handflügelpumpe mit angesetzten Pumpenschwengel, Vierwegehahn mit Entleerungsstutzen und Überdruckmesser sind im rechten Bereich montiert. Auf der linken Seite der Nische befinden sich verschiedene Zubehörteile, ein mit einer Zugfeder gesicherter Eimer und ein Halter mit einer Dominitlampe.*

* Dominitlampe: Grubenhandleuchte. Eine mit einem Akku betriebene Sicherheitshandleuchte.

Deutsches Museum, München, Archiv- FA014/B761/44 TD5224 [CD80832]

218

Deutsches Museum, München, Archiv- FA014/B737/44 TD5224 [CD80826]

219

218, 219 The free space between the tank and the rear wall of the driver's cab was used to stow four camouflage mats made of astralon. For this purpose a floor plate was mounted on the tank frame and braced to the vehicle frame. Different versions of the brackets were delivered. Brackets with chains were bolted to the rear wall of the vehicle. On the left side of the tank a metal strip with hooks was welded onto the manhole cover.

218, 219 Der Freiraum zwischen Tank und Rückwand der Fahrerkabine wurde zum Verstauen von vier Tarnmatten aus Astralonschnüren verwendet. Dazu wurde ein Bodenblech auf dem Tankrahmen montiert und zum Fahrzeugrahmen verstrebt. Verschiedene Ausführung der Halterungen kamen zur Auslieferung. Sechs Halterungen mit Ketten wurden an der Fahrzeugrückwand angeschraubt. Links am Tank befindet sich bis zum Mannlochdeckel ein angeschweißter Blechstreifen mit Haken.

Deutsches Museum, München, Archiv- FA014/B762/44 TD5224 [CD80828]

220

Deutsches Museum, München, Archiv- FA014/B738/44 TD5224 [CD80829]

221

220, 221 Hooks were attached to the manhole cover and the ladder to attach the chains. Two camouflage mats were placed in the space between tank and cabin, one camouflage mat each on the left and right side held by these chains. Astralon* camouflage mats were used, width 1 m, length 20 m.

* Astralon = Polyvinylchlorid

220, 221 Zum Einhängen der Ketten wurden Haken am Mannlochdeckel und an der Leiter angebracht. Zwei Tarnmatten wurden in den Zwischenraum von Tank und Führerhaus abgelegt, links und rechts je eine Tarnmatte eingestellt und mit den Sicherungsketten gehalten. Als Tarnmaterial wurden Astralon-Tarnmatten in 1 m Breite und 20 m Länge verwandt.*

* Astralon = Polyvinylchlorid

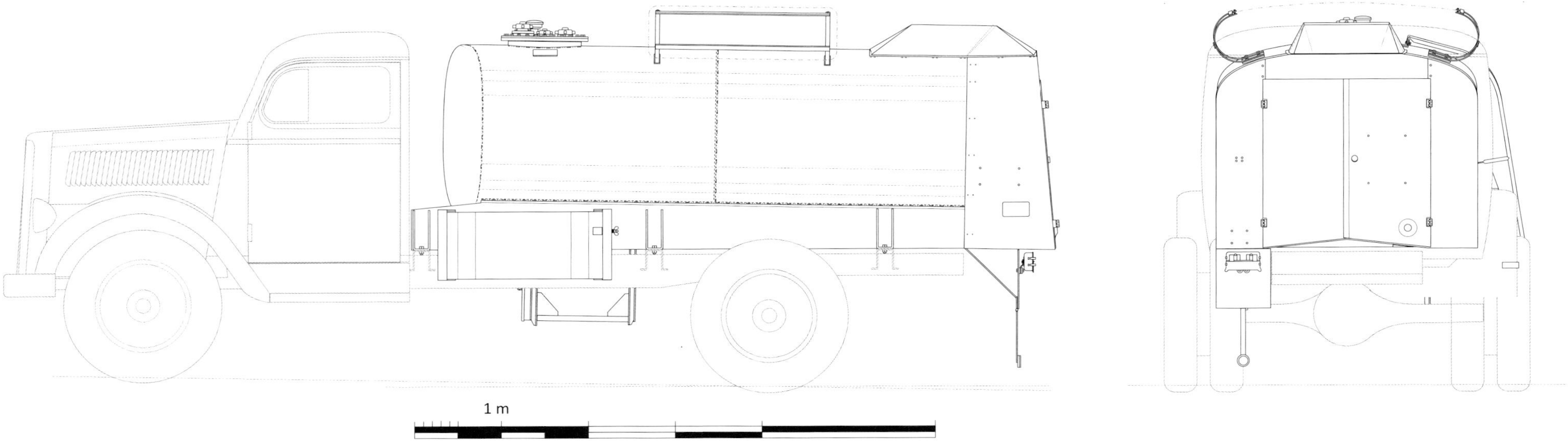

Scale / Maßstab 1:35 © Volker Ruff

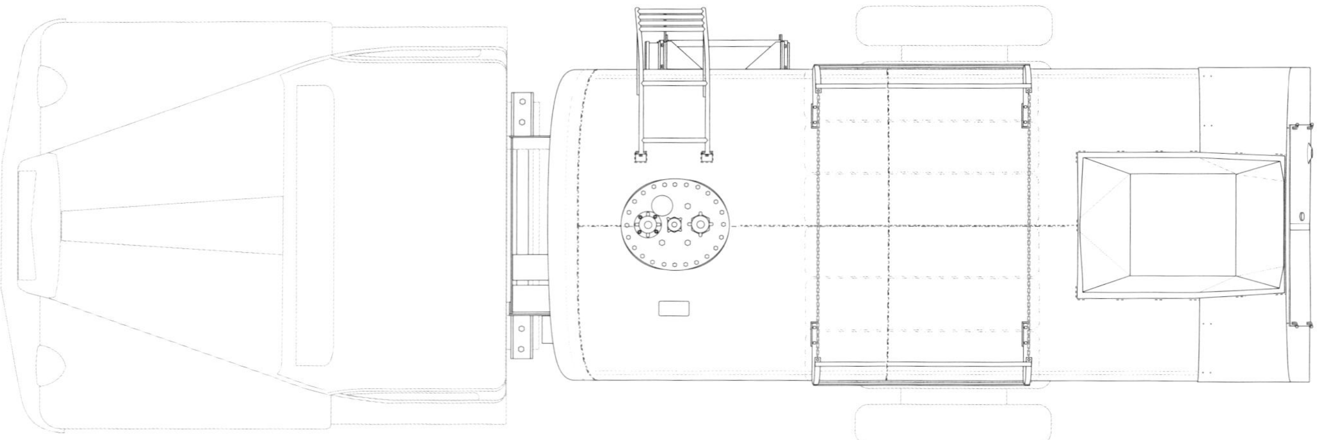

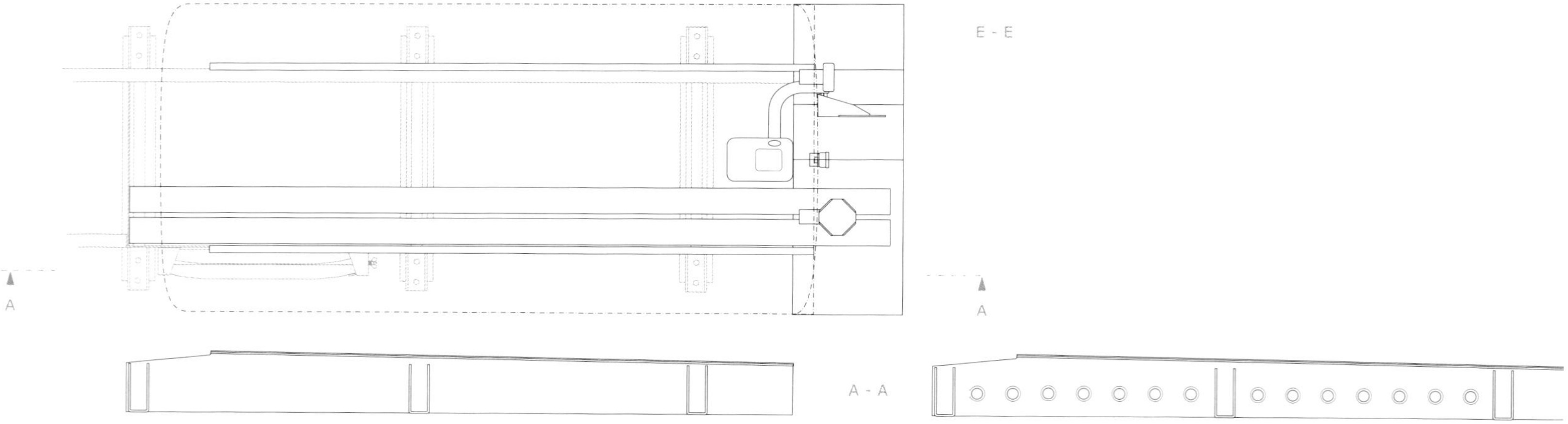

Fahrzeug 404

Kessel-Kw 3500 I B-Stoff

Drawing set used n.a.
Verwendeter Zeichnungssatz k.A.

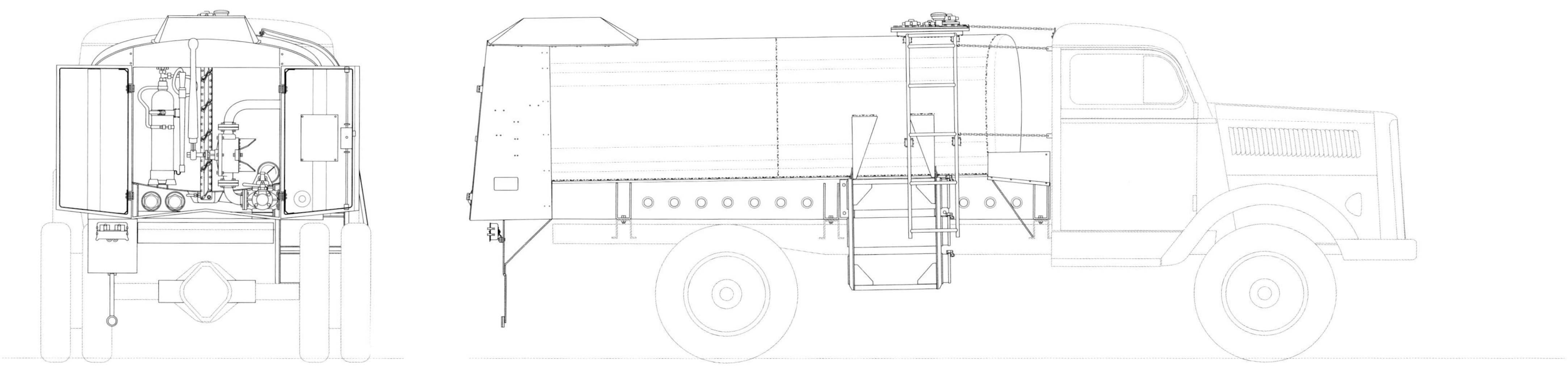

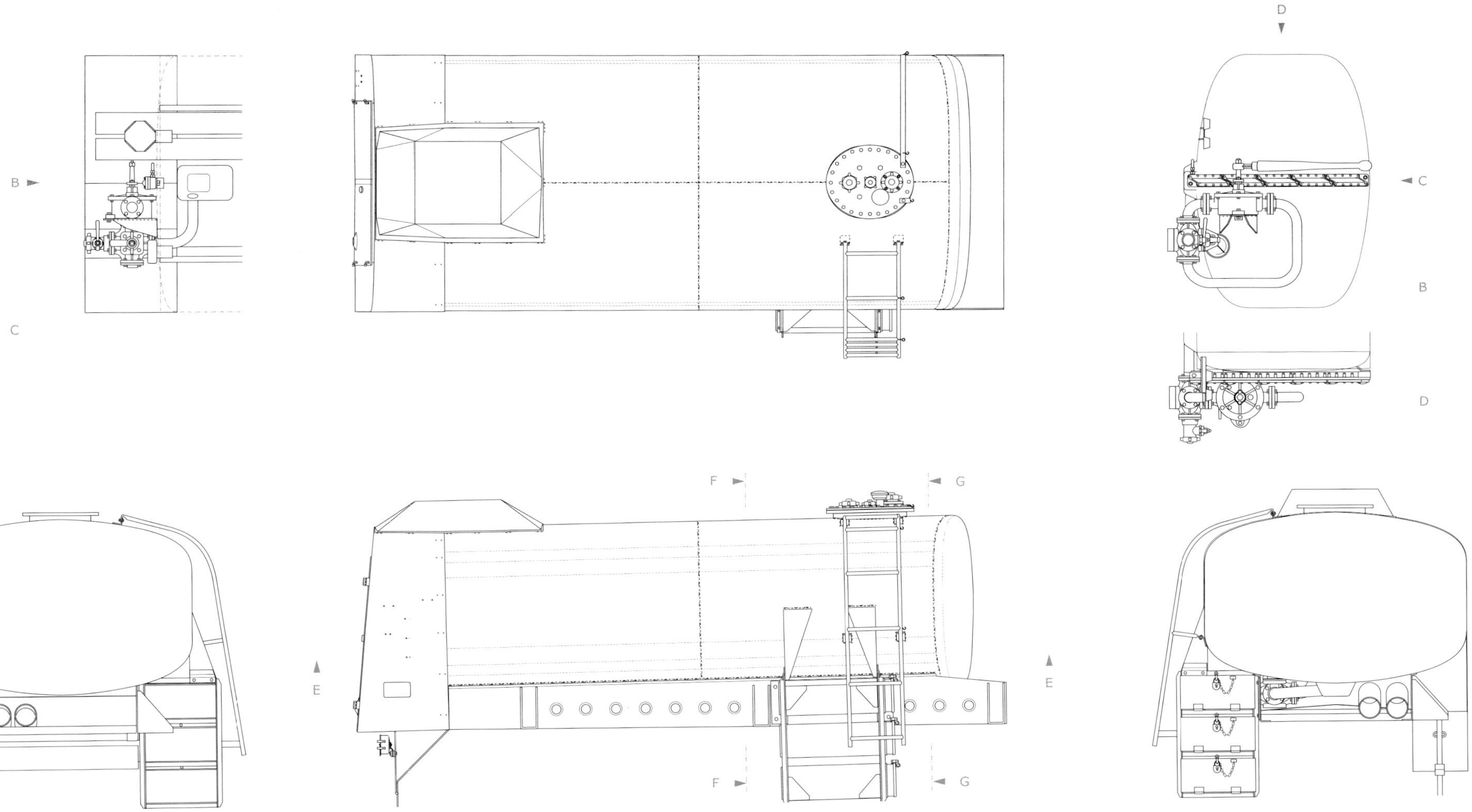
D
B
C
C
B
D
F
G
E
E
F
G
F-F
G-G

DMA, GD639.5.18
HAP11/T 507
Karlshagen, 1.6.1944
Bd.Nr. E 3195/44 g

DM, GD639.5.15
OKH
Chef H.Rüst und B.d.E
B.z.b.V (Heer) Abt.Ib
Januar 1944

Auszug aus
Gerätebeschreibung
Bedienungsanweisung
Wartungsanweisung

DMA, GD639.5.1
EW234
Fz.-Pos.-Nr. 405

TL 21/1019
Januar 1945

Vehicle 405

Internal or former designation: Fz.-Pos.-Nr.405
Tank truck 2100 l T-Stoff
T-Stoff truck

Intended use

The K.Kw. 2100 l was used to transport the T-Stoff, the T-Stoff pump and 300 l of water. The liquid (T-Stoff) was pumped from a rail tank car into the tank of the truck, transported to the launch site and pumped into the rocket with the help of a motor pump installed in the rear of the vehicle.

The base vehicle was developed by the Luftwaffe and was used, among other things, to refuel the Me 163. For the use by the units of the Division z.V. was the tank vehicle retrofitted with pipe connections, a motor pump for T-Stoff and a hand pump.

The K.Kw. 2100 l T-Stoff had the following dimensions and weights:

Total length:	5,900 mm
Total width:	2,120 mm
Total height:	2,240 mm
Wheelbase:	3,600 mm
Track width:	1,540/1,620 mm front/rear
Tank capacity:	2,100 l
Water tank capacity:	300 l
Tara weight:	3,290 kg
Payload:	3,310 kg
Max. permissible weight:	6,100 kg
Service weight:	6,500 kg
Wheels:	190-20
Connections:	Suction nozzle, pressure nozzle and filling nozzle NW 50, round thread 85x1/6"

The main parts of the K.Kw 2100 l are:

- Chassis
- T-Stoff tank
- Equipment compartment
- Water tank with wing pump
- Fittings compartment
- Pump set
- Fittings
- Accessories

Chassis

The normal chassis of the Opel 3 t truck type 3.6/36 S (equipment description in D 669/19 B) is used, but chassis of the truck 3 t Opel 6700 type A (equipment description in D 666/5) were also used.

Fahrzeug 405

Interne oder frühere Bezeichnung: Fz.-Pos.-Nr.405
Kessel-Kw 2100 l T-Stoff
T-Stoff Lkw

Verwendungszweck

Der K.Kw. 2100 l diente zum Transport des T-Stoffes, der T-Stoffpumpe und 300 l Wasser. Die Flüssigkeit (T-Stoff) wurde aus einem Eisenbahnkesselwagen in den Tank des Lastwagens gepumpt, zum Abschussort transportiert und mit Hilfe einer im Heck des Fahrzeugs installierten Motorpumpe in die Rakete gefördert.

Das Basisfahrzeug wurde von der Luftwaffe entwickelt und unter anderem zur Betankung der Me 163 eingesetzt. Für die Verwendung in den Abteilungen der Division z.V. wurde das Tankfahrzeug mit Rohranschlüssen, Motorpumpe für T-Stoff und einer Handpumpe für Wasser umgerüstet.

Der K.Kw. 2100 l T-Stoff hat folgende Maße und Gewichte:

Gesamtlänge:	5.900 mm
Gesamtbreite:	2.120 mm
Gesamthöhe:	2.240 mm
Radstand	3.600 mm
Spurweite:	1.540/1.620 mm vorne/hinten
Behälterinhalt:	2.100 l
Wassertank:	300 l
Leergewicht:	3.290 kg
Nutzlast:	3.310 kg
Zul. Höchstgew.	6.100 kg
Gefechtsgewicht:	6.500 kg gewogen
Räder:	190-20
Anschlüsse:	Saugstutzen, Druckstutzen und Füllstutzen NW 50, Rundgewinde 85x1/6"

Die Hauptteile des K.Kw. 2100 l sind:

- Fahrgestell
- T-Stoff-Behälter
- Geräteraum
- Wasserbehälter mit Flügelpumpe
- Armaturenraum
- Pumpensatz
- Armaturen
- Zubehör

Fahrgestell

Als Fahrgestell dient das normale Fahrgestell des Opel 3 t Lastkraftwagens Typ 3,6/36 S (Gerätebeschreibung siehe D 669/19B). Aber auch Fahrgestelle des Lastkraftwagen 3 t Opel 6700 Typ A (Gerätebeschreibung siehe D 666/5) wurden verbaut.

BArch, RH8II Bild-B0048-45, o.Ang.

222 The T-Stoff tank of Fz.405 with with equipment- and fittings department was mounted on a frame and mounted on the chassis of the Opel 3 t truck 3.6-36 or type 3.6-6700 (all-wheel drive). This T-Stoff tanker is seen with a freshly painted camouflage pattern of large and small sprayed spots in RAL 6003 olive green and RAL 8017 red-brown on a base coat in RAL 7028 dark yellow. The paint was applied to all added parts. Note that the wheel rims were painted after assembly, so that a mist of paint has settled on the tires.

222 Der T-Stoff-Tank des Fz.405 wurde mit Geräte- und Armaturenraum auf einen Rahmen montiert und anschließend auf das Fahrgestell des Opel 3 t Lastkraftwagen 3,6-36 oder Typ 3,6-6700 (Allrad) gesetzt. Der T-Stoff-Tanker zeigt auf dem Foto ein frisch lackiertes Tarnmuster aus großen und kleineren aufgespritzten Flecken in RAL 6003 (olivgrün) und RAL 8017 (rotbraun) auf einem Grundanstrich in RAL 7028 (dunkelgelb). Der Anstrich wurde über alle Anbauteile gesetzt. Man beachte, dass die Felgen nach der Montage lackiert wurden, so dass sich ein Farbnebel auf den Reifen niedergelegt hat.

DMA, GD639.5.11
EW236/Wö
Karlshagen, den 1.8.1944
Bd.Nr.2/44 gK
Auszug
Teileliste Fz. 405

Vehicles on order

Fahrzeugbestellungen

	Item Lfd. Nr.	Qty. ordered Best. Stückzahl	Qty. per vehicle Stück je Fahrzeug	Subject drawing number Gegenstand Zeichnungs-Nr.	Ordered by order number Besteller Bestell-Nr.	Supplier Lieferfirma
Fahrzeugbestellung	1	117	1	Fahrgestell 3 t mit 3,6 m Achsabstand mit Führerhaus ohne Pritsche	WaPrüf 11/III	Opel, Brandenburg

Customizing/outfitting

Ausrüstung

	Item Lfd. Nr.	Qty. ordered Best. Stückzahl	Qty. per vehicle Stück je Fahrzeug	Subject drawing number Gegenstand Zeichnungs-Nr.	Ordered by order number Besteller Bestell-Nr.	Supplier Lieferfirma
Aufbau- und Ausrüstungsfirmen	1	2 4	1 1	T-Stoff-Kesselwagen Type Kkw 2100 n. Nr. 21B229 mit nachfolgenden Baugruppen	WaPrüf 11/III SS 048-5165/41 WaPrüf 11/III SS 011-5310/42	Volkmar Häring & Co. Heidenau-Süd/bez. Dresden
		40	1	Behälter Lagerung	WaPrüf 11/III 1433 SS-361-0008/42	
		20	1	Behälter 21 St.22903	WaPrüf 11/III 1436 SS-361-0036/43	

DMA, GD639.5.13
OKH/Wa Prüf10
Karlshagen, den 1.4.1944
Bd.Nr. 788/44 gK
Auszug
Besprechung der Fahrzeuggruppe 4
vom 31.3.44
in Karlshagen I

Production status: Vehicle 405
Delivered to date: 43 (to the end of February)
of which: 4 at 444,12 at 836,
4 at 485,
3 at HAP11.
The whereabouts of the remaining vehicles are not clear at the moment. The vehicles delivered to the units have anodized pumps. The 20 questionable vehicles have pumps that are not yet anodized. Further deliveries of 120 from March onward with anodized pumps at 12 per month.

Current defects and their remedy

The emptying port of the water tank is led outward by an extension tube so the water does not run into the tool box. TD 41 – Ing. Edler will send the drawings to Wa Prüf 11/III by 12.4.44 for further production and to B.z.b.V. Heer for the units to do the modification (Revision letter B).

The discharge tube in the water tank must be led closely over the floor. Instructions given by Wa Prüf 11/III to the production company. TD41 – Ing. Edler, giving form change notification via B.z.b.V. Heer to the units (Revision letter A).

The floor of the pump chamber must slope downward to the drain gutter. Instructions by Wa Prüf 11/III. Delivered vehicles do not meet this change (revision letter B).

The foundation bolts for pumps with motors are currently inappropriately arranged, so that the screw bolts turn when tightened. Wa Prüf 11/III has instructed an appropriate change by the company (revision letter B).

Recently the seals of the glass tubes of the liquid content indicators have been leaking. The company is prompted by Wa Prüf 11/III to remedy the situation (revision letter B).

Fertigungslage: Fahrzeug 405
Bisher ausgeliefert: 43 Stück (bis Ende Februar)
Davon: 4 bei 444, 12 bei 836,
4 bei 485,
3 bei HAP11.
Der Verbleib der restlichen Fahrzeuge ist z.Zt. unklar. Die an die Truppe ausgelieferten Fahrzeuge haben eloxierte Pumpen. Bei den 20 fraglichen Fahrzeugen sind die Pumpen noch nicht eloxiert. Weitere Lieferung von 120 Stück – ab März monatlich 12 Stück und zwar mit eloxierten Pumpen.

Derzeitige Mängel und ihre Behebung

Der Entleerungsstutzen am Wasserbehälter wird durch ein Verlängerungsrohr nach außen geführt, damit das Wasser nicht in den Werkzeugkasten läuft. TD 41 – Ing. Edler gibt die Zeichnungsunterlagen an Wa Prüf 11/III bis 12.4.44 für die weitere Fertigung und an B.z.b.V. Heer zur Selbständerung bei der Truppe (Änderungszustand B).

Das Stichrohr im Wassertank muss dicht über den Boden geführt werden. Entsprechendes wird durch Wa Prüf 11/III bei der Fertigungsfirma veranlasst. TD41 – Ing. Edler gibt Formänderungsmitteilung über B.z.b.V. Heer an die Truppe (Änderungszustand A)

Der Boden des Pumpenraumes muss ein Gefälle zur Ablaufrinne bekommen. Veranlassung durch Wa Prüf 11/III. Bei den ausgelieferten Fahrzeugen tritt diese Änderung nicht ein (Änderungszustand B).

Die Fundamentschrauben für Pumpen mit Motor sind z.Zt. unzweckmäßig angeordnet, so dass der Schraubenbolzen sich beim Anziehen mitdreht. Eine entsprechende Änderung bei der Firma veranlasst Wa Prüf 11/III (Änderungszustand B).

Neuerdings sind bei den Standanzeigern die Abdichtungen der Glasröhren undicht. Die Firma wird durch Wa Prüf 11/III veranlasst, Abhilfe zu schaffen (Änderungszustand B).

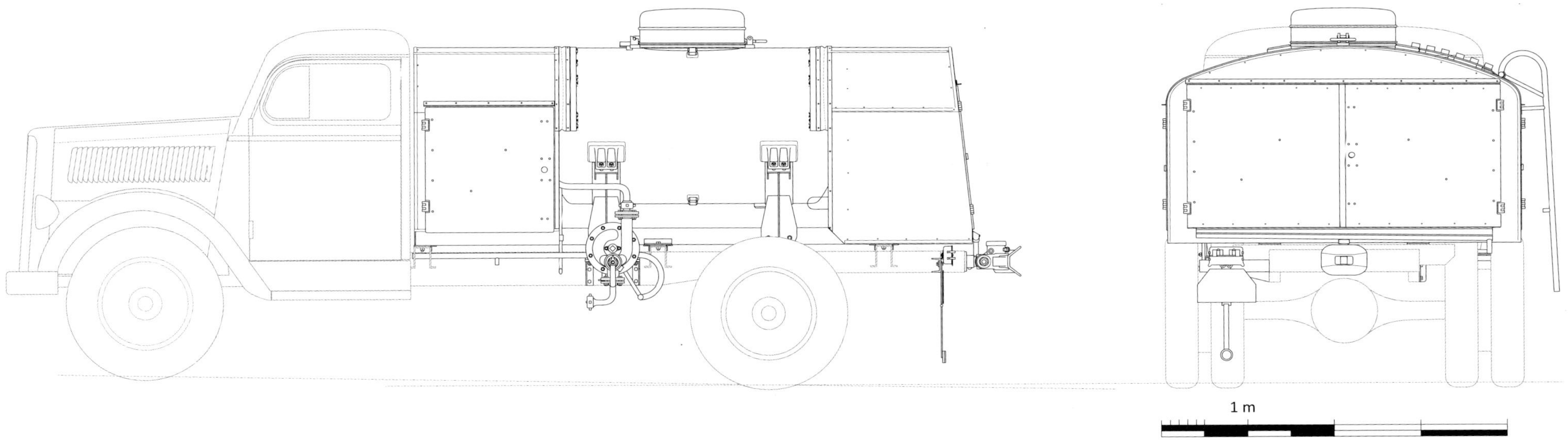

Scale / Maßstab 1:35 © Volker Ruff

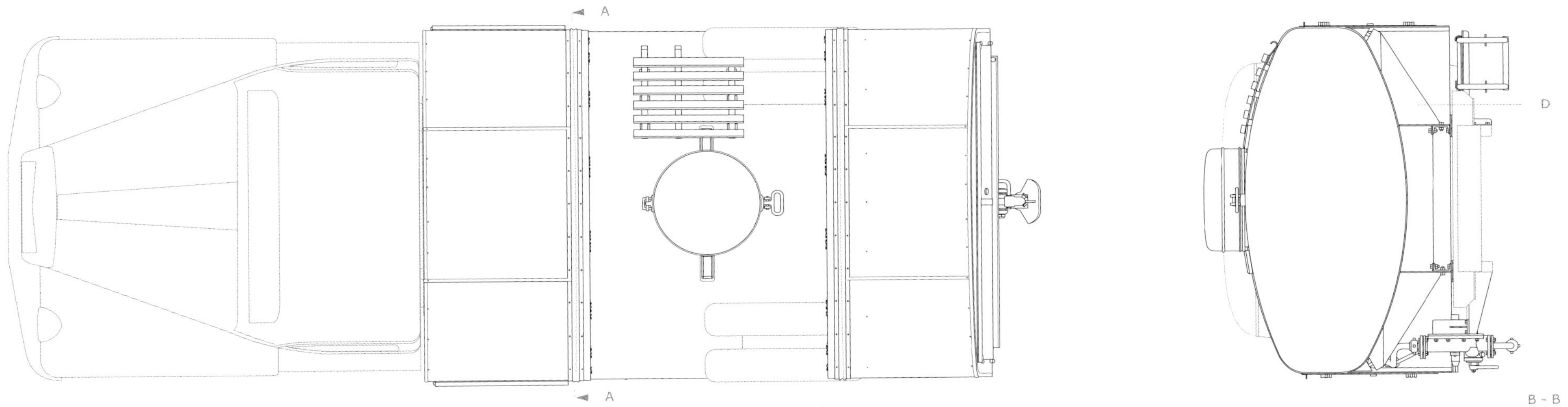

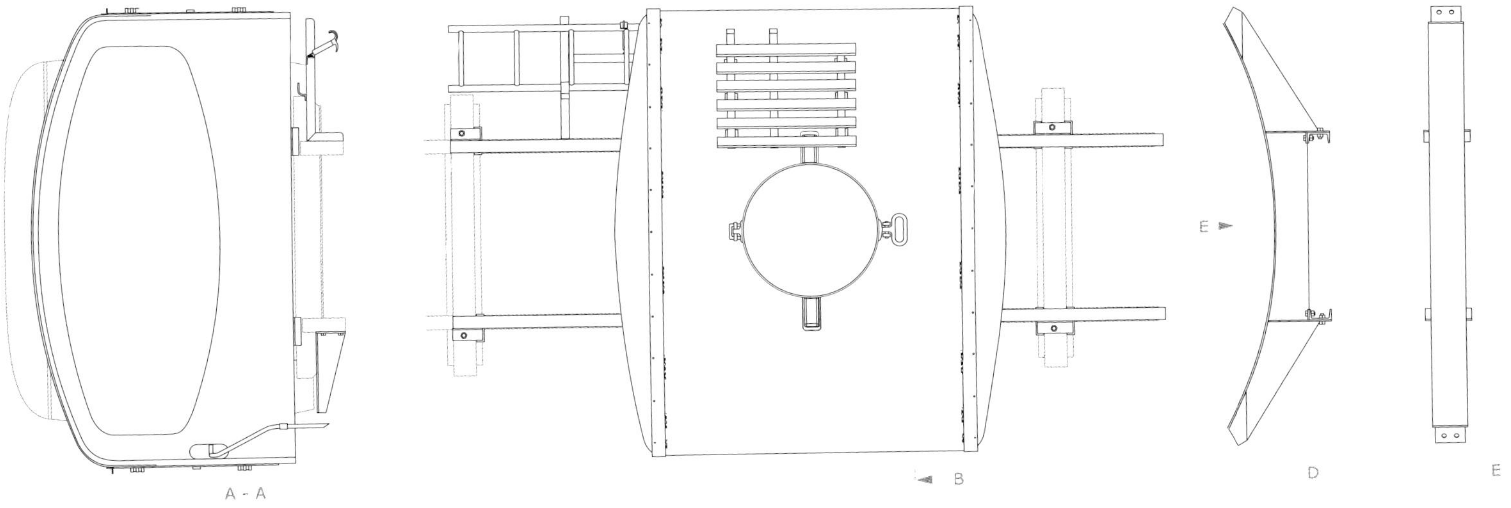

Fahrzeug 405

Kessel-Kw 2100 l T-Stoff

Partially used drawings acc. to group list 21Gr229, 57-438

Teilweise verwendete Zeichnungen nach Gruppenliste 21Gr229, 57-438

T-Stoff tank

The suitcase-shaped T-Stoff tank had a wall thickness of 6 mm and a capacity of 2,100 liters. As the T-Stoff chemically attacks iron, all parts that came into contact with T-Stoff, such as sheet metal pipes, nuts, etc. were made of aluminum. For corrosion protection the tank was anodized on the inside. The tank was mounted with two brackets riveted to the chassis, to which it was bolted on both sides with two retaining claws welded to the container. As it was not connected to the other superstructures apart from this mounting, the tank was free from additional stresses and could expand freely in any direction according to temperature fluctuations.

T-Stoff-Behälter

Der T-Stoff-Behälter in Kofferform besaß eine Wandstärke von 6 mm und ein Fassungsvermögen von 2.100 l. Da der T-Stoff Eisen chemisch angriff, waren alle mit T-Stoff in Berührung kommenden Teile wie Bleche, Rohre, Muttern usw. aus Aluminium hergestellt. Zum Korrosionsschutz war der Behälter innen eloxiert. Er war auf zwei mit dem Fahrgestell vernieteten Haltern gelagert, an denen er beiderseits mit je zwei am Behälter angeschweißten Haltepratzen angeschraubt war. Da er außer dieser Befestigung mit den übrigen Aufbauten nicht verbunden war, blieb der Behälter von zusätzlichen Spannungen frei und konnte gemäß den Temperaturschwankungen frei nach jeder Richtung arbeiten.

BArch, RH8II Bild-B2040-44, o.Ang.

223

223 In the two sheet metal housings; the front housing held a water tank, hoses, fittings and tools, and the rear one held the fittings compartment with valves, connections and pump. The housings were mounted on the chassis frame, not connected to the T-stoff tank. The spaces in front of and behind the tank were covered by canvas to protect against the weather.

223 In den beiden Blechgehäusen, den Geräteräumen, waren vorne ein Wassertank, Schläuche, Armaturen und Werkzeug und hinten im Armaturenraum, Ventile, Anschlüsse und Pumpe untergebracht. Die Gehäuse waren auf dem Fahrgestellrahmen montiert und nicht mit dem T-Stoff-Behälter verbunden. Die Zwischenräume vor und hinter dem Behälter wurden zum Schutz gegen Witterungseinflüsse durch Segeltuch abgedeckt.

Equipment compartment – water tank with wing pump

Between the driver's cab and the T-Stoff tank was the equipment compartment. It was accessible from both sides each with a lockable door.

In the upper part of the equipment compartment was a water tank made of 2 mm sheet metal with a capacity of 300 l. This water was needed to flush the pipes and hoses after filling or refueling. The water tank was attached to the toolbox frame in the equipment compartment by 2 retaining straps. On top of the tank there was a refueling nozzle and an overflow pipe. At the lower right rear corner there was a plug for emptying the tank. A pipe led from the refueling nozzle to the hand pump mounted on the left side of the chassis behind the equipment compartment. The hand pump served to fill and empty the water tank. The hand wing pump had a delivery rate of 80l/min. The water tank could be filled and emptied with it in about 4 minutes.

Geräteraum – Wasserbehälter mit Flügelpumpe

Zwischen dem Führerhaus und dem T-Stoff-Behälter befand sich der Geräteraum. Er war von beiden Seiten zugänglich und mit je einer verschließbaren Tür versehen.

Im oberen Teil des Geräteraumes befand sich ein Wasserbehälter aus 2 mm starkem Eisenblech und einem Inhalt von 300 l. Das Wasser wurde zum Durchspülen der Rohrleitungen und Schläuche nach erfolgtem Füllen oder Tanken benötigt. Der Wasserbehälter war durch zwei Haltebänder auf dem Werkzeugkastengestell im Geräteraum befestigt. Am Behälter befand sich oben ein Betankungsstutzen und ein Überlaufrohr. An der rechten hinteren unteren Ecke befand sich ein Stopfen zur Restentleerung des Behälters. Vom Betankungsstutzen führte ein Rohr zur Handflügelpumpe, welche an der linken Seite des Fahrgestelles hinter dem Geräteraum angebracht war. Die Handflügelpumpe diente zum Füllen und Entleeren des Wasserbehälters. Sie hatte eine Förderleistung von 80 l/min. Der Wasserbehälter konnte mit ihr in etwa vier Minuten gefüllt und entleert werden.

BArch, RH8II Bild-B1925-44, o.Ang.

224

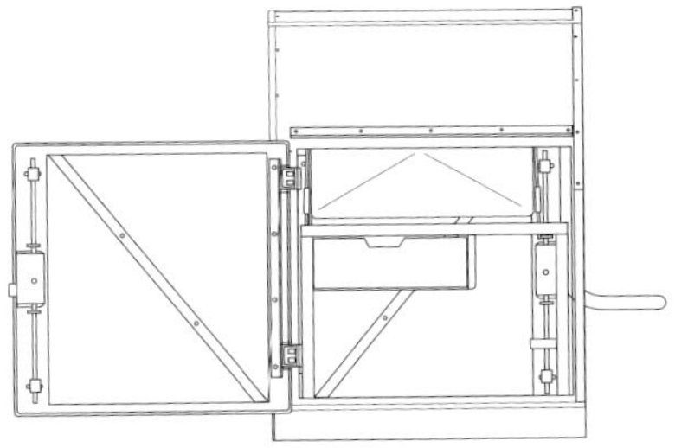

224 With the door open you can see on this picture the equipment compartment with its tool rack consisting of drawers and storage compartments. Hoses, tank accessories and two ramps were stowed here. The tank was covered with a camouflage net and pine branches.

224 Das Bild zeigt bei geöffneter Tür den Geräteraum mit dem Werkzeuggestell aus Schubladen und Ablagefächern. Es wurden darin Schläuche, Tankzubehör und zwei Auffahrrampen verstaut. Der Behälter war mit einem Tarnnetz und Kiefernzweigen bedeckt.

Equipment compartment – Accessories

The equipment compartment was used to store the standard tools, spare parts and accessories for a 3 t Opel Blitz truck, and also the hoses, fittings, tools and accessories for T-Stoff refueling.

Geräteraum – Zubehör

Der Geräteraum diente zur Aufnahme des Werkzeuges, der Ersatzteile und des sonstigen Zubehörs eines 3 t Opel-Blitz Lkw. Zusätzlich waren die Schläuche, Armaturen, Werkzeuge und das Zubehör für die T-Stoff-Betankung untergebracht.

Deutsches Museum, München, Archiv- FA014/B657/44 TD5224 [CD74961]

225

Deutsches Museum, München, Archiv- FA014/B655/44 TD5224 [CD74960]

226

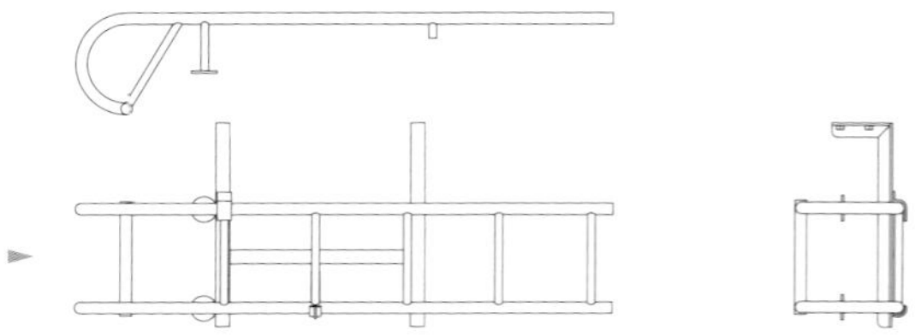

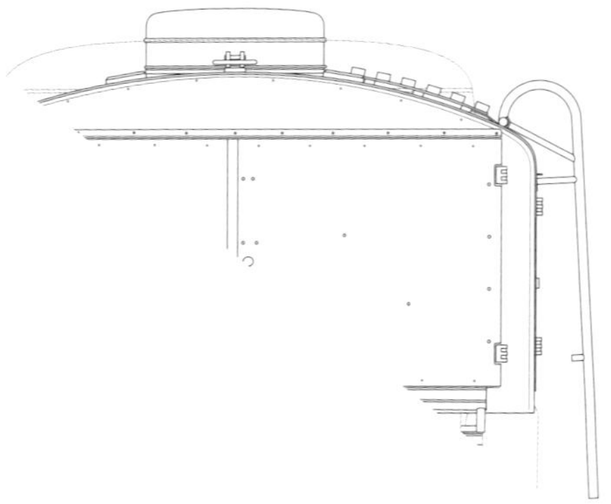

225, 226 On the closed access door on the right, the inscription »Fittings and tools« indicated the contents. The left door was labelled »Hoses and tools«. Below the equipment compartment a bracket made of angle sections was mounted on the vehicle frame for stowing the tubular steel ladder. With the door open you could look into the equipment compartment with its tool rack consisting of drawers and storage compartments. A wooden grate was mounted on the tank lid. Hooks were attached to it, to which the tubular steel ladder was attached.

225, 226 Die Beschriftung der geschlossenen Zugangstür des Geräteraums beschrieb dessen Inhalt mit »Armaturen und Werkzeug«. Die Beschriftung der linken Tür lautete »Schläuche und Werkzeug«. Unterhalb des Geräteraumes war eine Lagerung aus Winkelprofilen am Fahrzeugrahmen montiert. Dort wurde die Stahlrohrleiter verstaut. Bei geöffneter Tür blickte man in den Geräteraum mit dem Werkzeuggestell aus Schubladen und Ablagefächern. Am Kesseldeckel war ein Holzrost montiert. Daran waren Haken befestigt, an der die Stahlrohrleiter eingehängt war.

Fittings room – pump set and fittings

This compartment was connected to the T-Stoff tank and was welded together from 1 mm thick sheet metal. The two sides had ventilation slots. The rear wall was closed by a double-leaf lockable door through which the fittings inside were accessible. Inside was an aluminum drip tray, on which the T-Stoff pump set is sitting. The pipelines, valves and connection pipes for filling and refueling with T-Stoff were also located here. The pump set consisted of a base plate with a petrol engine (Fichtel & Sachs Stamo 10 with 4 PS performance) bolted onto it, and the self-priming centrifugal pump.

Armaturenraum – Pumpensatz und Armaturen

Der Armaturenraum schloß sich an den T-Stoff-Behälter an und war aus 1 mm starkem Eisenblech zusammengeschweißt. Die beiden Seiten hatten Entlüftungsschlitze. Die hintere Stirnwand wurde durch eine zweiflügelige verschließbare Tür abgeschlossen, durch welche die im Inneren untergebrachten Armaturen zugänglich waren. Im Inneren befand sich eine Auffangwanne aus Aluminium, auf der sich der T-Stoff-Pumpensatz befand. Außerdem befanden sich dort die Rohrleitungen, Ventile und Anschlussstutzen zum Befüllen und Betanken von T-Stoff. Der Pumpensatz bestand aus dem auf einer Grundplatte aufgeschraubten Benzinmotor (Fichtel & Sachs Stamo 10 mit 4 PS Leistung) und der selbstsaugenden Kreiselpumpe.

Deutsches Museum, München, Archiv- FA014/B1272/43 TD5224 [CD79116]

227

BArch, RH8II Bild-B1924-44, o.Ang.

228

227, 228 From mid-1944 these vehicles were delivered without the pump set due to delivery problems; a hand pump was mounted in its place. The venting slots in the side walls were also omitted, and an overflow gutter was added to the drip pan.

227, 228 Ab Mitte 1944 wurden die Fahrzeuge wegen Lieferschwierigkeiten ohne den Pumpensatz ausgeliefert. An deren Stelle wurde eine Handpumpe montiert. Auch waren die Entlüftungsschlitze in den Seitenwänden entfallen. Die Auffangwanne wurde um eine Überlaufrinne ergänzt.

DM, 8001D
Tarnschlüssel für Betriebsstoffe, mpe

T-Stoff = Thymiol – Hydrogen peroxide – yellow – RAL 1012*
* Technical colour – Stealth term for operating materials

T-Stoff = Thymiol – Wasserstoffperoyxd – gelb – RAL 1012*
* Technische Farbe – Tarnbezeichnung für Betriebsstoffe

DMA, GD639.5.18
HAP11/T 507
Karlshagen, 1.6.1944
Bd.Nr. E 3195/44 g

DMA, GD639.5.4
OKH
Chef H.Rüst und B.d.E
Bz.b.V (Heer) Abt.Ib
Bd.Nr.1844/44 g.Kdos.
Januar 1944

Auszug aus
Gerätebeschreibung
Bedienungsanweisung
Wartungsanweisung

DMA, GD639.5.1
EW234
Fz.-Pos.-Nr. 406

Vehicle 406

Internal or former designation: Fz.-Pos.-Nr.406
Pump trailer type A (1-axle)
Pump trailer

Intended use

The pump trailer type A (1-axle) is primarily intended for refueling the A4 *Gerät* with B-Stoff. However, it can also be used to pump B-Stoff into other containers. The pump trailer is towed by the B-Stoff motor vehicle (Fz.404).

The pump trailer is a pneumatic-tired single-axle road trailer in which the motor pump set, the fittings and pipelines and the disk meter with quantity presetting and shut-off valve are installed.

The A-Stoff pump set for the A-Stoff trailer (Fz.402) is also transported in the vehicle, but removed before the B-Stoff pump set is put into operation. The trailer is fixed by supports.

The pump trailer has the following dimensions and weights:

Total length:	3,330 mm
Width:	1,740 mm
Height:	1,795 mm (1,905 mm with raised structure)
Distance:	2,265 mm from axle centre to eyelet centre
Wheelbase:	1,560 mm
Track width:	1,915 mm
Tare weight:	450 kg
Payload:	900 kg
Service weight:	1,310 kg
Wheels:	6.00-20
Connection:	Suction and pressure nozzle NW60 connecting thread M78x2

Motor pump set:

power of the drive motor:	STAMO* 10 with 4 HP
Pump speed:	1,430 rpm
Delivery rate of the pump:	approx. 250 l/min

(Leaflet 5475 of Fichtel and Sachs A.G.)

The pump machine set consists of the following main components:

- Chassis
- Base frame with housing and A-Stoff pump mounting
- Motor pump set with pressure contact
- Fittings and pipelines
- Disc meter with flow presetting and shut-off valve

* STAMO: stationary engine from Fichtel & Sachs

Fahrzeug 406

Interne oder frühere Bezeichnung: Fz.-Pos.-Nr.406
Pumpen-Anhänger Type A (1-achs.)
Pumpen-Anhänger

Zweck und Aufgabe

Der Pumpenanhänger Typ A (1-achs.) ist in erster Linie zum Betanken des Gerätes A4 mit B-Stoff bestimmt. Er kann aber auch zum Pumpen von B-Stoff in andere Behälter verwendet werden. Gezogen wird der Pumpenanhänger vom B-Stoff Kraftwagen (Fz.404).

Der Pumpenanhänger ist ein luftbereifter Einachs-Straßenanhänger, in den der Motor-Pumpensatz, die Armaturen und Rohrleitungen und der Scheibenzähler mit Mengenvoreinstellung und Absperrventil eingebaut sind.

In dem Fahrzeug wird außerdem der A-Stoff-Pumpensatz für den A-Stoff-Anhänger (Fz.402) befördert, der vor Inbetriebnahme des B-Stoff-Pumpensatzes herausgenommen wird. Durch Stützen wird der Anhänger fest gestellt.

Der Pumpen-Anhänger hat folgende Maße und Gewichte:

Gesamtlänge:	3.330 mm
Breite:	1.740 mm
Höhe:	1.795 mm (1.905 mm mit hohem Aufbau)
Abstand:	2.265 mm von Achsmitte bis Ösenmitte
Radstand:	1.560 mm
Spurweite:	1.915 mm
Leergewicht:	450 kg
Nutzlast:	900 kg
Gefechtsgewicht:	1.310 kg
Räder:	6,00-20
Anschlüsse :	Saug- und Druckstutzen NW60 Anschlussgewinde M78x2

Motor-Pumpensatz:

Leistung des Antriebmotors:	STAMO* 10 mit 4 PS
Pumpendrehzahl:	1.430 U/min
Förderleistung der Pumpe:	ca. 250 l/min

(Druckschrift 5475 der Fa. Fichtel und Sachs A.G.)

Der Pumpenmaschinensatz besteht aus folgenden Hauptbauteilen:

- Fahrgestell
- Grundrahmen mit Verkleidung und A-Stoff-Pumpenhalterung
- Motor-Pumpensatz mit Druckkontakt
- Armaturen und Rohrleitungen
- Scheibenzähler mit Mengenvoreinstellung und Absperrventil

* STAMO: Stationär-Motor von Fichtel & Sachs

Chassis

The vehicle for the pump trailer is the single-axle trailer with pneumatic tires Type »A« for 900 kg load capacity. The equipment description is in manual D691/21 B.

Fahrgestell

Das Fahrzeug für den Pumpenanhänger ist der luftbereifter Einachsanhänger Typ »A« für 900 kg Tragfähigkeit. Gerätebeschreibung siehe D691/21 B.

Base frame, housing and A-Stoff pump bracket

The base frame carries the housing, the motor pump set, some of the fittings, the shut-off valve for the disk meter and the A-Stoff pump mounting. The base frame is welded together from angles and U-irons and fixed to the chassis frame.

The housing consists of a sheet metal box with a slightly curved roof which is attached to the base frame. In each of the side walls is a double-leaf lockable door. The entire rear wall is designed as a lockable sheet metal door with two leafs that can be locked in the open position.

Astralon camouflage mats were used, width 1 m, length 20 m.
Astralon = Polyvinylchlorid

Grundrahmen, Verkleidung und A-Stoff-Pumpenhalterung

Der Grundrahmen trägt die Verkleidung, den Motor-Pumpensatz, einen Teil der Armaturen, das Absperrventil für den Scheibenzähler und die A-Stoff-Pumpen-Halterung. Der Grundrahmen ist aus Winkeln und U-Eisen zusammengeschweißt und auf dem Fahrgestellrahmen befestigt.

Die Verkleidung besteht aus einem Blechkasten mit schwach gewölbten Dach, der am Grundrahmen befestigt ist. In den Seitenwänden ist je eine zweiflügelige verschließbare Türe angeordnet. Die ganze Rückwand ist als verschließbare Blechtüre mit zwei in geöffneter Stellung feststellbaren Flügeln ausgebildet.

Als Tarnmatten wurden verwandt: Astralon-Tarnmatten, Breite 1 m, Länge 20 m.
Astralon = Polyvinylchlorid

Deutsches Museum, München, Archiv- FA014/B1059/43 BSM [CD74732]

229

Deutsches Museum, München, Archiv- FA014/B189/44 TD5224 [CD74761]

230

229, 230 At the front of the housing was a simple holder for stowing two camouflage mats.

229, 230 An der Stirnseite der Verkleidung war eine einfache Halterung zum Verstauen von zwei Tarnmatten angebracht.

DM, GD639.5.11
EW236/Wö
Karlshagen, den 1.8.1944
Bd.Nr.2/44 gK
Auszug
Teileliste Fz. 406

Vehicles on order — Fahrzeugbestellungen

	Item Lfd. Nr.	Qty. ordered Best. Stückzahl	Qty. per vehicle Stück je Fahrzeug	Subject drawing number Gegenstand Zeichnungs-Nr.	Orderec by order number Besteller Bestell-Nr.	Supplier Lieferfirma
Fahrzeugbestellung	1	138	1	Anhänger Type A (1-achs.)	WaPrüf 11/III	unbekannt (u.a. Lindner, Ammendorf)

Customizing/outfitting — Ausrüstung

	Item Lfd. Nr.	Qty. ordered Best. Stückzahl	Qty. per vehicle Stück je Fahrzeug	Subject drawing number Gegenstand Zeichnungs-Nr.	Ordered by order number Besteller Bestell-Nr.	Supplier Lieferfirma
Aufbau- und Ausrüstungsfirmen	1	48	1	kompl. Ausrüstung f. Pumpen-anhänger Type A (1-achs.) nach Zeichnung Z. 261 T	WaPrüf 11/III 361-0015/42	Schwelmer Eisenwerk, Schwelm i.W.
		90	1		WaPrüf 11/III 361-0033/43	

DMA, GD639.5.13
OKH/Wa Prüf10
Karlshagen, den 1.4.1944
Bd.Nr. 788/44 gK
Auszug
Besprechung der Fahrzeuggruppe 4
vom 31.3.44
in Karlshagen I

Production status: Vehicle 406
Delivered to date: 47
(14 in October, 17 in November and 16 in December)
of which 1 at 444, 9 at 836, 3 at 485.

Those delivered to the units have A-Stoff-Pumps. The remainder lack the A-Stoff pumps which are short of supply. Further deliveries of pump trailers are delayed because of the B-Stoff-pumps (bomb damage at Siemens which supplies the gauges) und still depends on the timely supply by KSP of wheels. The delay occurs by deliveries of defective wheels on one hand, on the other hand by renewed bomb damage at the Fichtel and Sachs Company.

For the A-Stoff-pump purchase order it can be said:
A total of 450 are on order.
Of these 264 are from the Dickow Company. The remainder from the Ziemann Company. The purchase order covers the requirements for mot.- and rail-mounted vehicles. 31 are specifically for railroad vehicles. At present 17 motors are available, so that in April a further 17 A-Stoff-pumps deliveries by the Ziemann company can be expected at best.
Planned delivery was
by Ziemann Company: March 50, from April 30 monthly.
By the Dickow Company from May 30 per month.

Current defects and their remedy
The round iron bar hooks suggested to hold the A-Stoff-pump in the pump trailer are long enough to be detached from the pump frame without removing the wing nut. Complete unscrewing of the wing nut is prevented by a cotter pin.

Fertigungslage: Fahrzeug 406
Bisher ausgeliefert: 47 Stück
und zwar: (14 Oktober, 17 November und 16 Dezember)
Davon 1 Stück bei 444, 9 Stück bei 836, 3 Stück bei 485

Diese an die Truppe ausgelieferten Fahrzeuge haben A-Stoff-Pumpen.Für die übrigen Fahrzeuge fehlen z.Zt. die A-Stoff-Pumpen und sind Engpass. Die weitere Auslieferung von Pumpenanhängern ist ferner verzögert wegen der B-Stoff-Pumpen (Bombenschaden bei Fa. Siemens als Zulieferant der Messuhren) und von der rechtzeitigen Zulieferung von Laufrädern seitens KSP abhängig. Die Verzögerung ist eingetreten einerseits durch Lieferungen fehlerhafter Laufräder, andererseits durch erneuten Bombenschaden bei Fa. Fichtel und Sachs (Motorpumpe).

Zur A-Stoff-Pumpen Bestellung ist zu sagen:
Insgesamt sind 450 Stück A-Stoff-Pumpen bestellt.
Davon sind 264 Stück bei Fa. Dickow. Die restlichen bei Fa. Ziemann.
Diese Bestellung umfasst den Bedarf für mot.- und Schienenfahrzeuge. 31 Stück sind hiervon ausschließlich für Eisenbahnfahrzeuge ausgeliefert. Zurzeit sind 17 Stück Motoren vorhanden, so dass im April günstigstenfalls mit der Lieferung von 17 weiteren A-Stoff-Pumpen durch die Fa. Ziemann zu rechnen ist.
Vorgesehene Lieferung war
von Fa. Ziemann: März 50 Stück ab, April monatlich 30 Stück
Von Fa. Dickow ab Mai monatlich 30 Stück.

Derzeitige Mängel und ihre Behebung
Die zur Halterung der A-Stoff-Pumpe im Pumpenanhänger vorgeschlagenen Rundeisenhaken werden soweit verlängert, dass sie ohne Abnehmen der Flügelschraube vom Pumpenrahmen gelöst werden können. Das gänzliche Abschrauben der Flügelmutter wird mit Splint verhindert.

231

231 It can be observed that there are always minor deviations in the design and equipment of the special vehicles and trailers. Here, for example, the rims of the wheels are designed with a different hole cutout, than those used in the previous pictures. This also applies to the shorter exhaust pipe.

231 Immer wieder sind kleinere Abweichungen in der Aus-führung und Ausstattung der Spezialfahrzeuge und Anhänger feststellbar. Hier wurden zum Beispiel die Felgen der Räder im Vergleich zu den auf den vorherigen Seiten abgebildeten Fahrzeugen mit einem anderen Lochausschnitt ausgeführt. Das gilt auch für das kürzere Auspuffrohr.

Motor pump set with pressure contact, fittings and piping
On the rear side, the motor pump set was accessible through the opened double leaf sheet metal doors. A bulkhead was mounted on a base frame in the middle. On the left was a STAMO 8 and on the right the B-Stoff pump (SiHi type). The assembly was mounted with the housing on the chassis frame.

Motor-Pumpensatz mit Druckkontakt, Armaturen und Rohrleitungen
Auf der Rückseite war der Motorpumpensatz durch die geöffnete zweiflügelige Blechtür zugänglich. Auf einem Grundgestell war in der Mitte ein Schottblech montiert, links davon der STAMO 8 und rechts die B-Stoff-Pumpe (Bauart SiHi). Die Baugruppe war mit dem Gehäuse auf dem Fahrgestellrahmen montiert.

Deutsches Museum, München, Archiv- FA014/B1083/44 T5224 [CD74709]

232

IWM BU-10281

233

232 The exhaust pipe was led on the left side wall upward through the roof. The pump had two outlets, the left of which leads via the pump suction nozzle to the B-Stoff filter which was mounted in the middle behind the bulkhead. A pipe led from the right outlet via the pump pressure nozzle to the pressure contact under the roof and to the fitting there, a pressure connection 4 with three-way cock, on the late model the fittings were reorganized.

233 The first 48 pump trailers delivered did not yet have a raised roof with an intermediate floor and storage space for the B-Stoff hoses. The superstructure was retrofitted. The roof was separated above the doors, a floor was installed, and a partition for the storage space was added and closed by the revised roof. The doors were extended upwards and adapted to the new contours, the exhaust pipe was led the side wall.

232 Das Auspuffrohr führte an der linken Seitenwand nach oben durch das Dach. Die Pumpe hatte zwei Ausgänge, davon führte der linke Ausgang über den Pumpensaugstutzen zum B-Stoff-Filter, der in der Mitte hinter dem Schott montiert war. Vom rechten Ausgang führte eine Rohrleitung über den Pumpendruckstutzen zum Druckkontakt unter dem Dach und zu den Armaturen, dem Druckstutzen 4 mit Dreiwegehahn, bei der Abschlussausführung wurden die Armaturen neu positioniert.

233 Die ersten 48 auslieferten Pumpenanhänger hatten noch kein erhöhtes Dach mit Zwischenboden und Stauraum für die B-Stoff-Schläuche. Der Aufbau wurde nachgerüstet. Das Dach wurde oberhalb der Türen getrennt, ein Boden eingebaut und eine Zwischenwand für den Stauraum aufgesetzt und wieder mit dem Dach verschlossen. Die Türen wurden nach oben ver- längert und der Kontur angepasst, das Auspuffrohr durch die Seitenwand geführt.

A-Stoff pump support

The mounting for the A-Stoff pump set was welded together from iron profiles and was attached to the base frame above the axle to balance its weight. The A-Stoff pump set was fastened to the mounting with wing nuts and hook bolts which hook into the pump frame.

A-Stoff-Pumpenhalterung

Die Halterung für den A-Stoff-Pumpensatz war aus Profileisen zusammengeschweißt und wegen des Gewichtsausgleiches über der Achse am Grundrahmen befestigt. Der A-Stoff Pumpensatz wurde auf der Halterung mit Flügelmuttern und Hakenschrauben, die in den Pumpenrahmen eingriffen, befestigt.

BArch, RH8II Bild-B01806-44, o.Ang.

234

Deutsches Museum, München, Archiv- FA014/B1082/43 TD5224 [CD79118]

235

234 The B-Stoff pump is seen in the A-Stoff pump mounting, secured with hooked bolts and wingnuts. The B-Stoff suction nozzles 1+2 with quick-acting stop valve are located on the final model on the left.

235 View into the interior through the left door: The pump set now removed. Behind the door frame on the left is the gas separator with overflow pipe, and behind it is the disc meter with pointer mechanism and volume presetting. In the opposite opening is the gas indicator with sight glass, pipe and discharge connection 3.

234 Auf dieser Fotografie ist die B-Stoffpumpe noch in der A-Stoff-Pumpenhalterung eingeschoben und mit Hakenschrauben und Flügelmuttern gesichert. Die Saugstutzen 1+2 mit Schnellschlusszapfventil sind in der Abschlussausführung links angeordnet.

235 Blick in den Innenraum durch die linke Tür: Der Pumpensatz ist herausgenommen. Hinter dem Türrahmen links befindet sich der Gasabscheider mit Überlaufrohr. Davor ist der Scheibenzähler mit Zeigerwerk und Mengenvoreinstellung und in der gegenüberliegenden Öffnung der Gasanzeiger mit Schauglas, Rohrleitung und Druckstutzen 3 zu sehen.

B-Stoff pump trailer (final model)

Mid 1944 a false floor was installed to accommodate four B-Stoff hoses NW70 x 2000 mm, the side walls and door were raised by approx. 110 mm, and a flat gable roof was installed. Previously delivered pump trailers were retrieved, converted and handed back to the unit troops. The previous flat vaulted roof was reused.

B-Stoff-Pumpenanhänger (Abschluss Ausführung)

Mitte 1944 wurde für die Unterbringung von vier B-Stoffschläuchen NW70 x 2000 mm ein Zwischenboden eingezogen, die Seitenwände und Tür um ca. 110 mm erhöht und ein flaches Satteldach aufgesetzt. Bisher ausgelieferte Pumpenanhänger wurden zurückgeholt, umgebaut und wieder an die Truppe übergeben. Das bisherige flachgewölbte Dach kam zur Wiederverwendung.

BArch, RH8II Bild-B01807-44, o.Ang.

236

BArch, RH8II Bild-B01805-44, o.Ang.- Ausschnitt

237

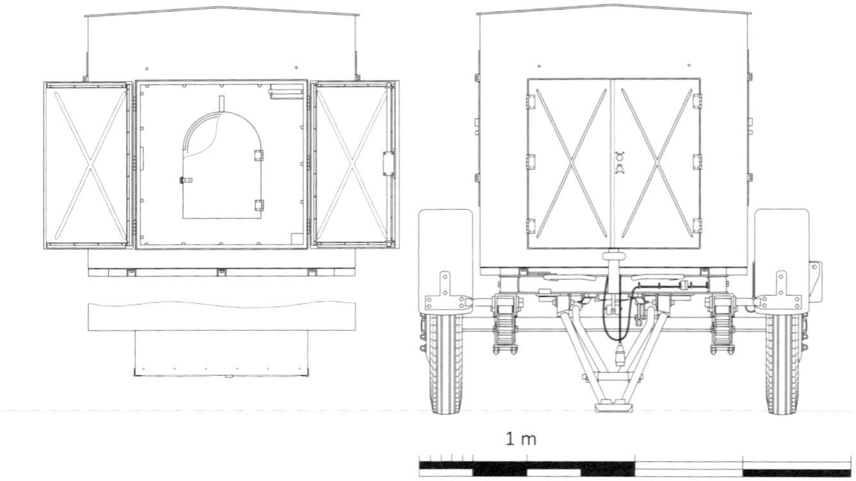

Scale / Maßstab 1:35 © Volker Ruff

236 Another box for hoses, accessories, tools and protective clothing was attached at the front of the raised superstructure.

236 An der Stirnseite des erhöhten Aufbaus wurde noch ein weiterer Kasten für Schläuche, Zubehörteile, Werkzeug und Schutzbekleidung angebracht.

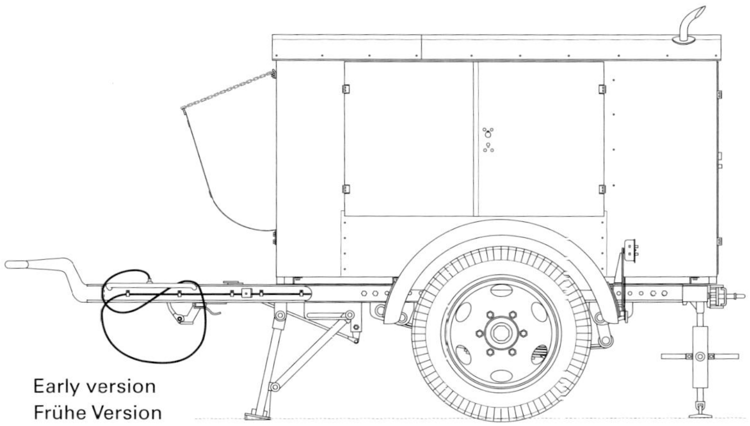

Early version
Frühe Version

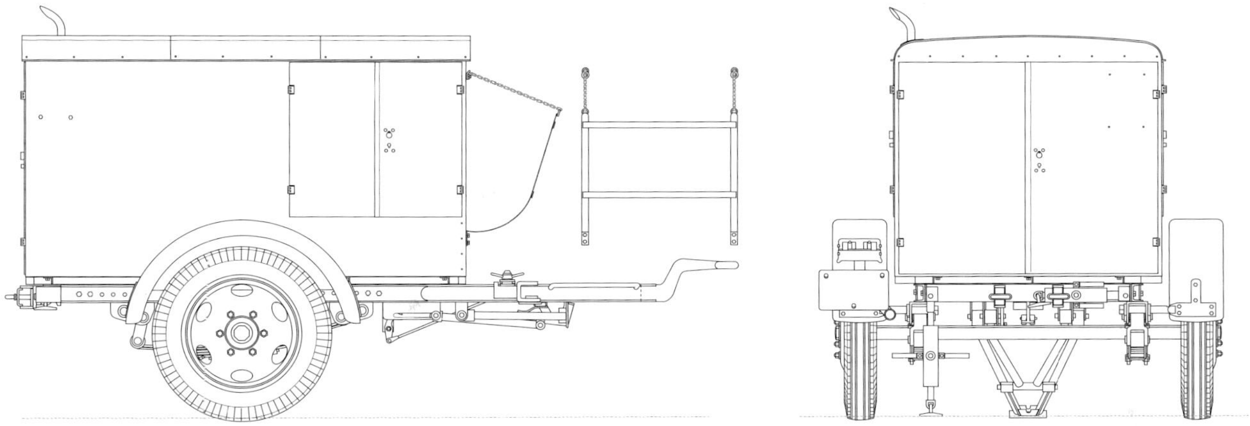

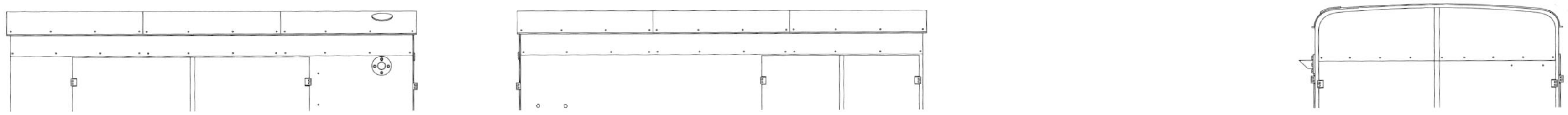

Raised roof
Erhöhtes Dach

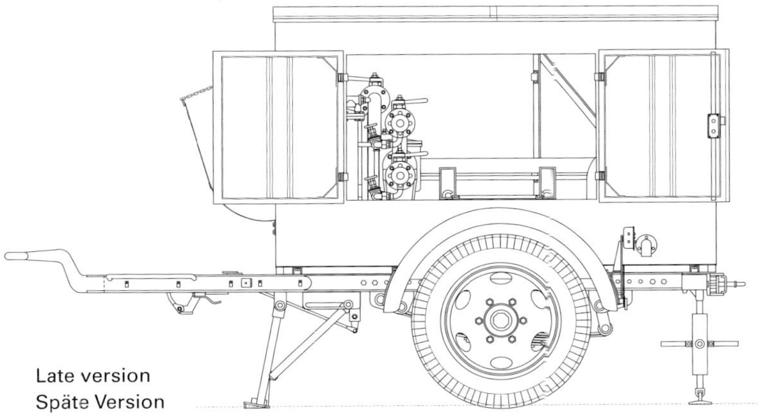

Late version
Späte Version

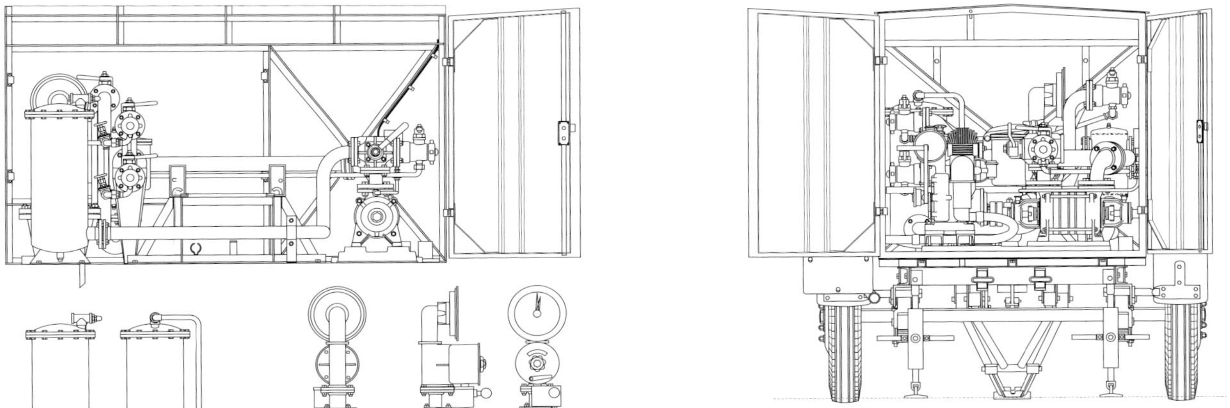

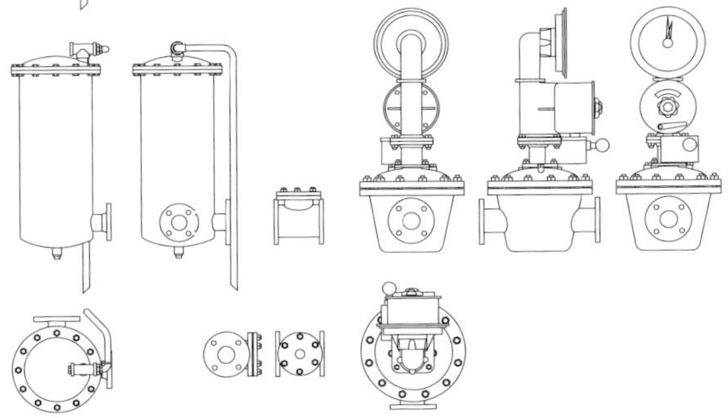

1 m

Scale / Maßstab 1:35 © Volker Ruff

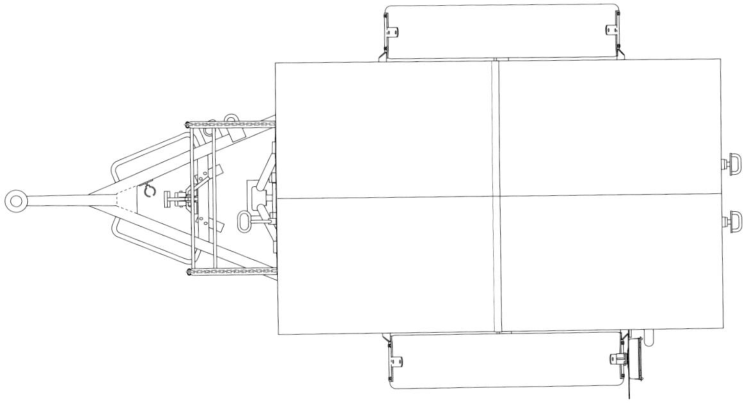

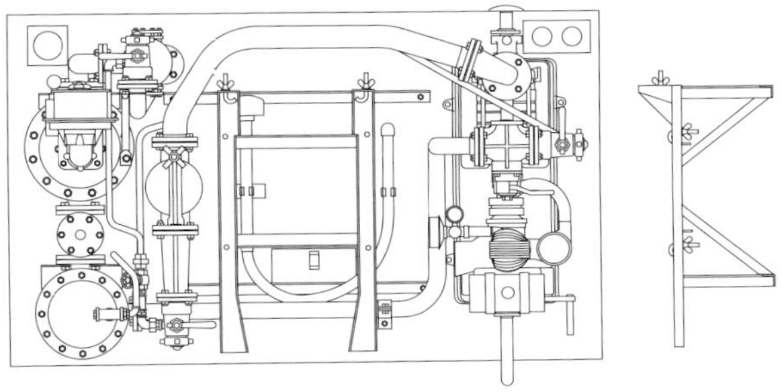

Fahrzeug 406

Pumpen-Anhänger Type A (1-achs.)

Partially used drawings acc. to group list 21F17305, 21B159-1, TW1.05

Teilweise verwendete Zeichnungen nach Gruppenliste 21F17305, 21B159-1, TW1.05

Vehicle 407

Internal or former designation: Fz.-Pos.-Nr.407
T-Stoff preheater trailer (1-axle)
T-Stoff preheater

DMA, GD639.5.18
HAP11/T 507
Karlshagen, 1.6.1944
Bd.Nr. E 3195/44 g

DMA, GD639.5.1
EW234
Fz.-Pos.-Nr. 407

DMA, GD630.02
A4 Lehrbuch
der FR-Schule
Mai 1944
Auszug aus
Kapitel
Betanken & Enttanken
des A4-Gerätes

Intended use

In winter, do not fill up with T-Stoff* directly from the vehicle (405), but fill the measuring tank (on vehicle 102) with pre-heated T-Stoff.

Heating to 60-80 °C is carried out with a heating device which is built on an single-axle trailer. The T-Stoff preheater consists of an inner container with a capacity of 150 l, surrounded by a water jacket. The water is heated by an electric blower using a spirit flame and the T-Stoff temperature is monitored by a plug-in thermometer. A signaling device sounds a horn at a set temperature.

In Summer the T-Stoff measuring container on the Fz.102 is filled directly with the vehicle (Fz.405), heating is not necessary.

Protective clothing and goggles must be worn when working with T-Stoff. T-Stoff causes etching on the skin. The damage caused by T-Stoff is alleviated if the affected parts are rinsed off immediately with plenty of water.

The T-Stoff preheater trailer has the following dimensions and weights:

Total length:	3,400 mm
Width:	1,800 mm
Height:	1,800 mm
Distance:	2,300 mm from axle centre to eyelet centre
Track width:	1,560 mm
Tare weight:	646 kg
Payload:	900 kg
Service weight:	933.5 kg
Wheels:	6.00-20

The T-Stoff preheater consists of the following main components:

- Chassis
- Base frame with T-Stoff preheater and burner unit

Chassis

The vehicle for the T-Stoff preheater is the psingle-axle trailer with pneumatic tires Type »A« for 900 kg load capacity. The equipment description is in manual D691/21 B.

* T-Stoff = Thymiol – hydrogen peroxide – yellow – RAL 1012
(Technical colour – stealth term for operating materials)
T-Stoff starts to freeze at -20 °C.

DM, 8001D
Tarnschlüssel für
Betriebsstoffe, mpe

Fahrzeug 407

Interne oder frühere Bezeichnung: Fz.-Pos.-Nr.407
T-Stoff-Vorwärmer-Anhänger (1-achs.)
T-Stoff Vor-Gerät

Zweck und Aufgabe

Im Winter tankt man T-Stoff* nicht direkt aus dem Fahrzeug (405), sondern man füllt den Messbehälter (am Fz.102) mit erwärmtem T-Stoff.

Mit einem Anwärmgerät, das auf einen Einachsanhänger aufgebaut ist, wird der T-Stoff auf 60-80 °C erwärmt. Der T-Stoff-Vorwärmer besteht aus einem 150 l fassenden Innenbehälter, der von einem Wassermantel umgeben ist. Das Wasser wird von einem Elektrogebläse mittels Spiritusflamme angewärmt und die T-Stoff-Temperatur an einem Einsteckthermometer überwacht. Eine Signalvorrichtung lässt bei einer bestimmten Temperatur eine Hupe ertönen.

Im Sommer wird der T-Stoff Messbehälter (am Fz.102) direkt mit dem Fahrzeug (Fz.405) befüllt, ein Anwärmen entfällt.

Bei Arbeiten mit T-Stoff ist Schutzbekleidung und Schutzbrille zu tragen. Auf der Haut erzeugt T-Stoff Ätzungen. Die Schädigungen durch T-Stoff werden gemildert, wenn die betroffenen Teile sofort mit reichlichen Wassermengen abgespült werden.

Der T-Stoff-Vorwärmer Anhänger hat folgende Maße und Gewichte:

Gesamtlänge:	3.400 mm
Breite:	1.800 mm
Höhe:	1.800 mm
Abstand:	2.300 mm von Achsmitte bis Ösenmitte
Spurweite:	1.560 mm
Leergewicht:	646 kg
Nutzlast:	900 kg
Gefechtsgewicht:	933,5 kg
Räder:	6,00-20

Der T-Stoff-Vorwärmer besteht aus folgenden Hauptbauteilen:

- Fahrgestell
- Grundrahmen mit T-Stoff-Vorwärmer und Brennereinheit

Fahrgestell

Das Fahrzeug für den T-Stoff-Vorwärmer ist der luftbereifte Einachsanhänger Typ »A« für 900 kg Tragfähigkeit. Gerätebeschreibung siehe D691/21 B.

* T-Stoff = Thymiol – Wasserstoffsuperoyxd – gelb – RAL 1012
(Technische Farbe – Tarnbezeichnung für Betriebsstoffe)
T-Stoff gefriert ab -20 C°.

Frame with T-Stoff preheater

A base plate was bolted to the chassis frame and a double-walled boiler was mounted. An inner tank made of aluminum, capacity 150 l, was used in the T-Stoff preheater. For heating, water was poured into the double-walled preheater and heated by a heating unit. The inner container was equipped with a lid, nozzle cover, alarm bell for ventilation, and filler neck for the T-Stoff. Heating temperature was monitored by a thermometer. At the front of the T-Stoff preheater there were a flue gas outlet (grill) and a cleaning cap.

Grundrahmen mit T-Stoff-Vorwärmer

Auf den Grundrahmen des Fahrgestells wurde eine Bodenplatte geschraubt und darauf ein doppelwandiger Kessel montiert. Im T-Stoff-Vorwärmer war ein aus Aluminium gefertigter Innenbehälter mit 150 l Fassungsvermögen eingesetzt. Zum Erwärmen wurde Wasser in den doppelwandigen Vorwärmer eingefüllt und durch eine Anwärmeinheit beheizt. Der Innenbehälter war mit einem Deckel, einem Stutzendeckel und einer Schutzglocke zur Entlüftung sowie einem Einfüllstutzen für den T-Stoff versehen. Über ein Thermometer wurde die Anwärmtemperatur überwacht. An der Vorderseite des T-Stoff-Vorwärmers war ein Rauchgasabzug (Gitter) und eine Reinigungskappe angebracht.

238

238 The paint on the boiler is noticeably darker around the cleaning flap. This discoloration was caused by the heat of the burner unit.

238 Um die Reinigungsklappe ist der Anstrich am Kessel deutlich dunkler. Diese Verfärbung ist durch die Hitze der Brennereinheit verursacht worden.

DMA, GD639.5.11
EW236/Wö
Karlshagen, 1.8.1944
Bd.Nr.86/44 gK
Auszug
Teileliste Fz.407

Vehicles on order — Fahrzeugbestellungen

	Item Lfd. Nr.	Qty. ordered Best. Stückzahl	Qty. per vehicle Stück je Fahrzeug	Subject drawing number Gegenstand Zeichnungs-Nr.	Ordered by order number Besteller Bestell-Nr.	Supplier Lieferfirma
Fahrzeugbestellung	1	150	1	Anhänger für T-Stoff-Vorwärmer Type A (1-achs.)	WaPrüf11/III	unbekannt (u.a. Lindner, Ammendorf)

Customizing/outfitting — Ausrüstung

	Item Lfd. Nr.	Qty. ordered Best. Stückzahl	Qty. per vehicle Stück je Fahrzeug	Subject drawing number Gegenstand Zeichnungs-Nr.	Ordered by order number Besteller Bestell-Nr.	Supplier Lieferfirma
Ausrüstungsfirma	1	60	1	T-Stoff-Anwärmer mit Montage auf dem Anhänger	WaPrüf 10/H 0112-3330/43 WaPrüf 11/III SS 4949 0030-4563/42 WaPrüf 11/III SS 4949 0112-3821/43	Julius Pinsch KG, Berlin

* In August 1944 an additional new batch of 30 T-Stoff preheaters was ordered.
Im August 1944 wurde eine Neufertigung von 30 Stück weiterer T-Stoff-Anwärmer Fz.407 beauftragt.

DMA, GD639.5.13
OKH/Wa Prüf10
Karlshagen, 1.4.1944
Bd.Nr. 788/44 gK
Auszug
Besprechung der Fahrzeuggruppe 4 vom 31.3.44
in Karlshagen I

Production status: Vehicle 407
Delivered to date: 13
of which: 1 at 444
9 at 836
3 at HAP 11
Further deliveries of 10 monthly starting in March
Total ordered: 180
(150 + 30 as new batch with hand pump)

Current defects and their remedy

The support assemblies including the bolts of the T-Stoff preheater equipment on the chassis are too weak. TD41, Ing. Kempini, has arranged an immediate change at the factory. Modification notice given to the units by B.z.b.V. Heer. (Revision letter A).

Grounding of the electrical equipment is not yet done, although 220 V are provided for the operation of the engine. Cables and connectors are already provided for grounding to earth so that the grounding of the electrical parts can easily be accomplished.

Action by Ing. Kempini, TD 41. Action by the units via modification notice from B.z.b.V. Heer by 12.4.44.

Fertigungslage: Fahrzeug 407
Bisher ausgeliefert: 13 Stück
Davon: 1 Stück bei 444
9 Stück bei 836
3 Stück bei HAP 11
Weitere Lieferungen von monatlich 10 Stück ab März
Insgesamt bestellt: 180
(150 Stück + 30 Stück Neufertigung mit Handpumpe)

Derzeitige Mängel und ihre Behebung

Die Halterungen einschließlich Bolzen des T-Stoff-Anwärmgeräts am Fahrgestell sind zu schwach ausgeführt. Änderung bei der Firma veranlasst TD41 – Ing. Kempini, für die weiteren Fahrzeuge sofort. Änderungsmitteilung an die Truppe erfolgt durch B.z.b.V. Heer. (Änderungszustand A).

Bisher ist die Geräteerdung der elektrischen Teile nicht erfolgt, obwohl 220 Volt für den Betrieb des Motors vorgesehen sind. Kabel und Stecker sind bereits für die Erdung vorgesehen, so dass die Erdung der elektrischen Teile ohne weiteres durchgeführt werden kann.

Veranlassung durch Ing. Kempini, TD 41. Veranlassung bei der Truppe durch Änderungsmitteilung durch B.z.b.V. Heer bis 12.4.44.

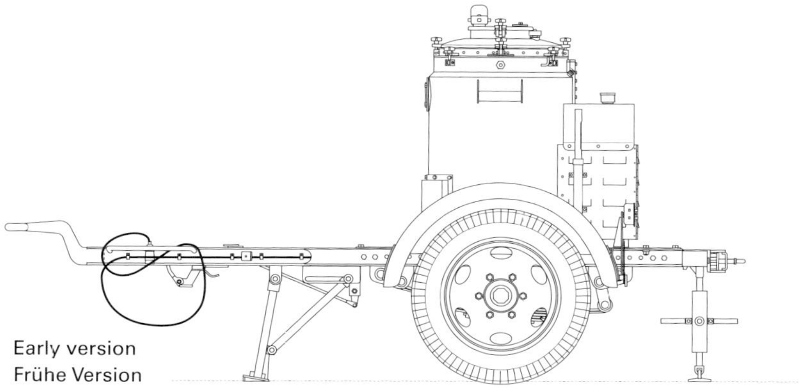

Early version
Frühe Version

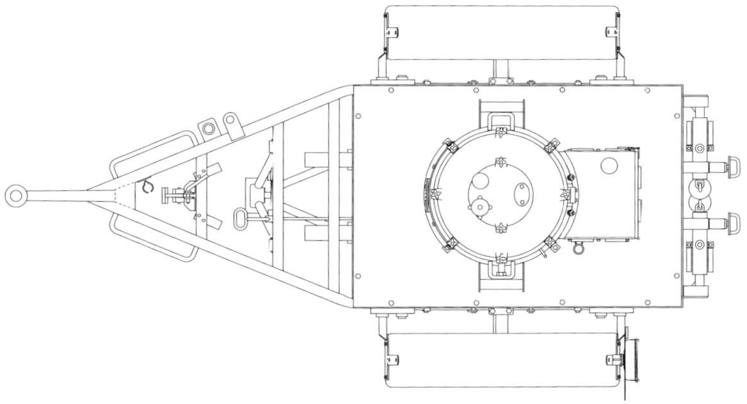

1 m

Scale / Maßstab 1:35 © Volker Ruff

239 The protective box for the heating unit was attached to the rear of the preheater and a fuel container with filler nozzle was fitted on it. A water level glass was mounted on the rear wall of the preheater and a protection tube for the thermometer was fixed between the protection box to the housing of the preheater. The plug-in thermometer was stored in this during travel.

239 An der Rückseite des Vorwärmers war der Schutzkasten für die Anwärmeinheit angebracht und ein Brennstoffbehälter mit Einfüllstutzen aufgesetzt. Ein Wasserstandglas war an der Rückwand des Vorwärmers montiert und im Übergang vom Schutzkasten zum Gehäuse des Vorwärmers ist der Thermometerschutz befestigt. An dieser Stelle wurde das Einsteckthermometer während der Fahrt aufbewahrt.

Deutsches Museum, München, Archiv- FA014/B501/44 TD5224 [CD74757]

239

308

Deutsches Museum, München, Archiv- FA014/B502/44 TD5224 [CD66674]

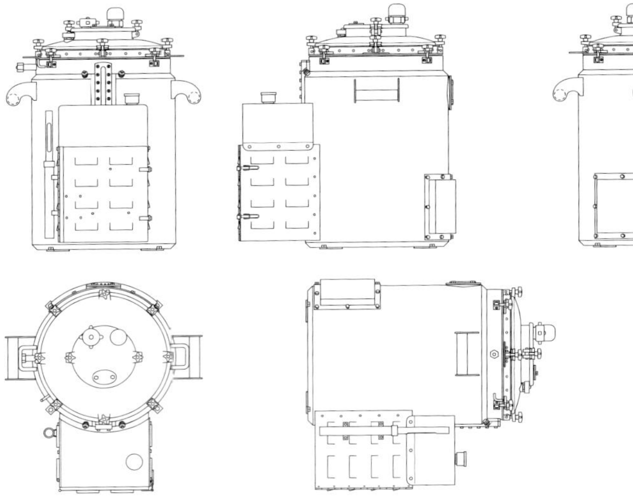

Early version
Frühe Version

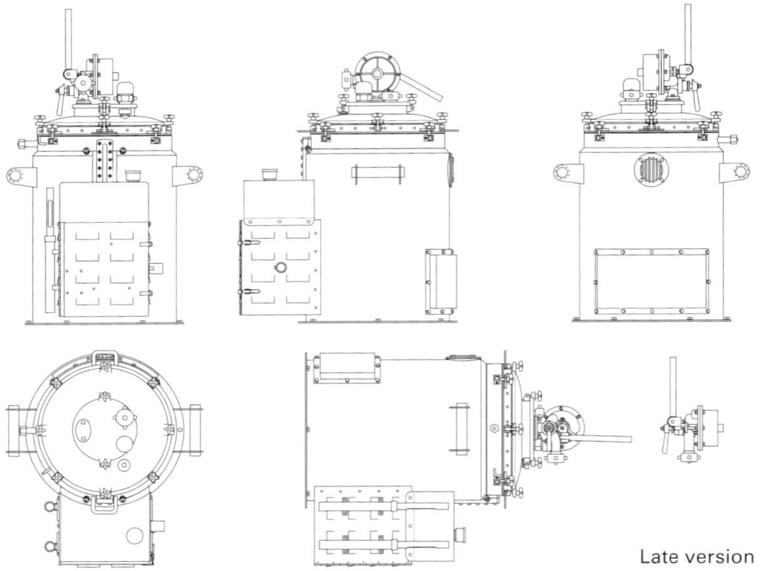

Late version
Späte Version

Burner unit

The burner unit with electric blower and a signal unit (horn) were located in the protective box on the rear of the container. A hose line led from the fuel container to the burner unit. The burner was fired with spirit and ignited by a fuse, which was stored in a tube on the inside of the lid.

Brennereinheit

Im Schutzkasten auf der Behälterrückseite war eine Brennereinheit mit elektrischem Gebläse und eine Signaleinheit (Hupe) untergebracht. Vom Brennstoffbehälter führte eine Schlauchleitung zur Brennereinheit. Der Brenner wurde mit Spiritus betrieben und mit einer Lunte gezündet. Diese wurde auf der Deckel-Innenseite in einem Röhrchen aufbewahrt.

T-Stoff preheater trailer (final model)

The T-Stoff preheater of the final version was equipped with a T-Stoff hand pump on the boiler cover. A cable reel with a cover was mounted on the chassis to supply power to the electric blower of the burner unit. Camouflage mats were stowed on the floor plate to the left and right of the heating container, secured by chains and attached to the container and fenders.

T-Stoff-Vorwärmer (Abschluss-Ausführung)

Die T-Stoff-Vorwärmer der Neubaufertigung waren auf dem Behälterdeckel mit einer T-Stoff-Handpumpe ausgerüstet. Zur Stromversorgung des Elektrogebläses an der Brennereinheit diente eine an der Fahrgestelldeichsel montierte Kabeltrommel mit Abdeckhaube. Tarnmatten wurden links und rechts vom Anwärmbehälter auf dem Bodenblech verstaut und mit Ketten am Behälter und Kotflügel gesichert.

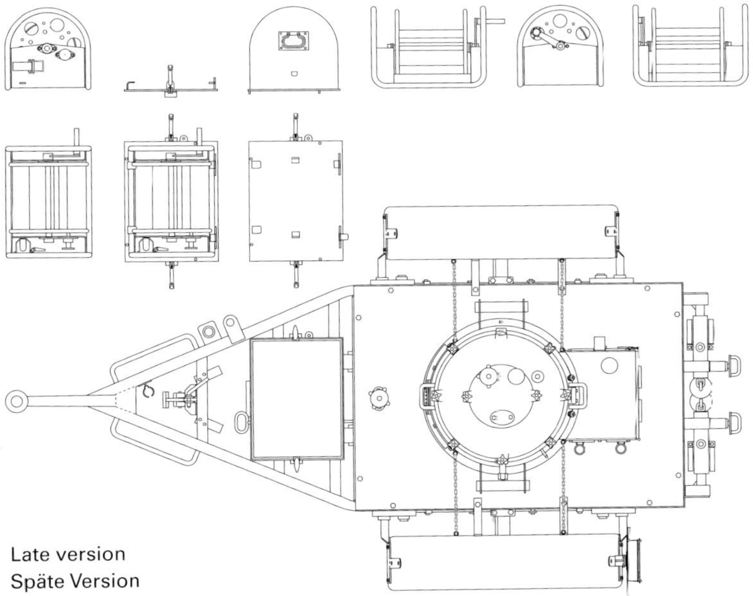

Late version
Späte Version

BArch, RH8II Bild-004, o.Ang.

1 m

Scale / Maßstab 1:35 © Volker Ruff

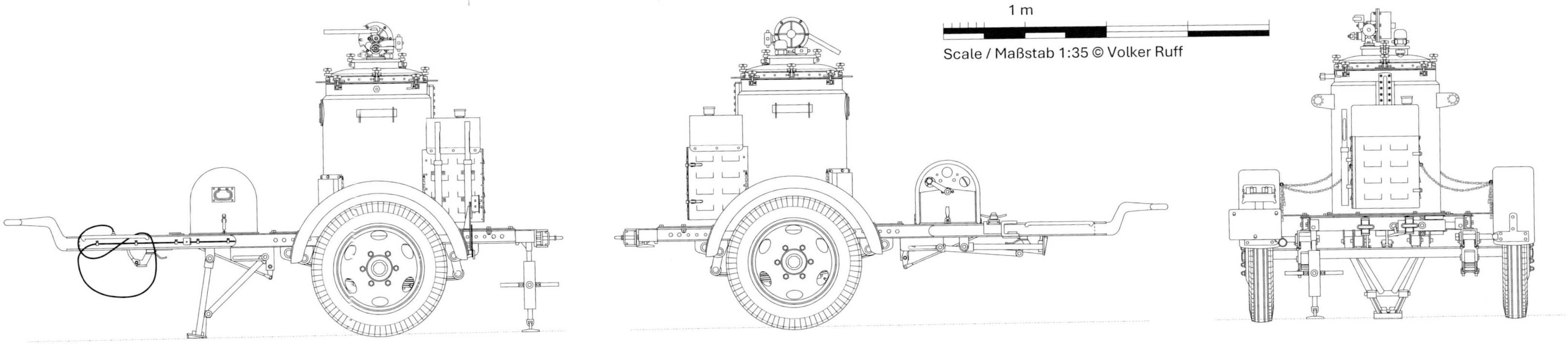

DMA, GD639.5.18
HAP11/T 507
Karlshagen, 1.6.1944
Bd.Nr. E 3195/44 g
Interne oder frühere Bezeichnung

DMA, GD639.5.1
EW234
Fz.-Pos.-Nr. 408/508

DMA, GD630.02
A4 Lehrbuch
der FR-Schule
Mai 1944
Auszug aus
Technische Fahrzeuge
des A4 Gerätes

Vehicle 408

Internal or former designation: Fz.-Pos.-Nr.408
Acummulator truck auf 4,5 t truck closed (heated)
Acummulator truck

Intended use
The accumulator truck is a heated vehicle. The 27-volt on-board battery and 50-volt nickel-cadmium battery are transported in it and kept at a temperature of 30 °C all the time. A charging system in the vehicle enables the batteries to be recharged continuously. The Z-Stoff canisters required for the shooting operation are also carried in the accumulator-truck at all times.

The accumulator truck has the following dimensions and weights:

Total length:	7,570 mm
Width:	2,400 mm
Height:	3,005 mm
Wheelbase:	4,700 mm
Track width:	1,934/1,714 mm rear center distance between dual tires
Tare weight:	6,080 kg
Payload:	4,500 kg
Service weight:	9,250 kg
Wheels:	270-20

The accumulator truck (thermal truck) consists of the following main components:

- Chassis
- Closed standard superstructure with racks and oven with circulation heating

Chassis
The chassis is the chassis of the Büssing-NAG Typ 4500 A 4,5 t truck Type serie A 1 (see technical manual D 667/9)

DMA, GD539.5.13
OKH/Wa Prüf10
Karlshagen, 1.4.1944
Bd.Nr. 788/44 gK
Auszug
Besprechung der Fahrzeuggruppe 4 vom 31.3.44 in Karlshagen I

Production status: Vehicle 408
Delivered to date including the requirement for 508:
27 vehicles with installed heating, but without accumulator rack.
Of which issued with accumulator rack (as vehicle 408):
3 for 836,
1 for 444,
1 for 485,
3 for Kdo. Basse (as vehicle 508).
Remaining vehicles under construction at LR.

DM, 8001D
Tarnschlüssel für Betriebsstoffe, mpe

Z-Stoff = Rubid – Sodium permanganate – weiß – RAL 9002*
* Technical colour – Stealth term for operating materials

Fahrzeug 408

Interne oder frühere Bezeichnung: Fz.-Pos.-Nr.408
Sammler-Lkw auf Lkw 4,5 t geschl. (heizbar)
Sammler-Kw

Zweck und Aufgabe
Der Sammler-Kw ist ein beheizbares Fahrzeug. In ihm werden die 27 V-Bordbatterie und 50 V-Nickel-Kadmium-Batterie befördert und dauernd auf einer Temperatur von 30 °C gehalten. Eine im Fahrzeug befindliche Ladeanlage ermöglicht eine laufende Nachladung der Batterien. Gleichzeitig werden in den Sammler-KW die für den Schiessbetrieb erforderlichen Z-Stoff-Kanister mitgeführt.

Der Sammler-Kw hat folgende Maße und Gewichte:

Gesamtlänge:	7.570 mm
Breite:	2.400 mm
Höhe:	3.005 mm
Radstand:	4.700 mm
Spurweite:	1.934/1.714 mm hinten Mittenabstand zwischen Doppelreifen
Leergewicht:	6.080 kg
Nutzlast:	4.500 kg
Gefechtsgewicht:	9.250 kg
Räder:	270-20

Der Sammler-Kw (Thermowagen) besteht aus folgenden Hauptbauteilen:

- Fahrgestell
- Geschlossener Einheitsaufbau mit Einbauten und Ofen mit Umlaufheizung

Fahrgestell
Als Fahrgestell dient das Fahrgestell des Büssing-NAG Typ 4500 A 4,5 t Lastkraftwagens Baureihe A 1 (Gerätebeschreibung siehe D 667/9).

Fertigungslage: Fahrzeug 408
Bisher ausgeliefert einschließlich des Bedarfs für 508:
27 Fahrzeuge mit eingebauter Heizung, aber ohne Sammlergestell.
Davon mit Sammlergestell ausgegeben (als Fahrzeug 408):
3 Stück für 836
1 Stück für 444
1 Stück für 485
3 Stück für Kdo. Basse (als Fahrzeug 508)
Restliche Fahrzeuge in LR in Arbeit.

Z-Stoff = Rubid – Natriumpermanganat – weiß – RAL 9002*
* Technische Farbe – Tarnbezeichnung für Betriebsstoffe

BArch, RHBII Bild-B1440-44, o.Ang.

242

242 The closed standard superstructure of Fz.408 was derived from the large workshop box body (see Fz.301 and Fz.302) and mounted on a Büssing NAG 4500 A (all-wheel drive) in a late version. Normally the roof was extendable upwards, but this possibility was blocked on the Fz.408 and the corner fittings bolted to the roof. It was otherwise identical to the Fz.508, so for further details see Fz.508.

242 Der geschl. Einheitsaufbau des Fz.408 war vom großen Werkstattkoffer abgeleitet (siehe Fz.301, Fz.302) und auf einem Büssing NAG 4500 A (Allrad) in einer späten Ausführung aufgesetzt. Normalerweise war das Dach nach oben erweiterbar, jedoch wurde diese Möglichkeit am Fz.408 blockiert und die Eckbeschläge am Dach fest verschraubt. Das Fz.408 war identisch mit dem Fz.508, weitere Details siehe Fz.508.

VEHICLE GROUP V
FAHRZEUGGRUPPE V

Technical battery

DMA, GD630.02
A4 Lehrbuch
der FR-Schule
Mai 1944
Auszug aus Kapitel
Nachschub, Umschlag
und Lagerung

Purpose and task

The delivered A4 *Geräte* are loaded onto the Vidal transportation trailer (Fz.601) and driven to the technical battery in a field storage position for testing. After a visual inspection for external damage a detailed technical inspection is carried out. Usable equipment are provided with the warhead and as necessary temporarily stored or moved to the firing position. The built-in equipment that failed this inspection (x) of the *Gerät* are exchanged, minor repairs carried out, provided that this is within the scope of inspection and repair at the field storage position. If short-term failures of tested *Geräte* are detected in the firing position they are returned to the technical battery. There the defect [1] is repaired or a decision is made to return the unit to the field workshop or Army Muna. This needs explanation) for repair. This also applies to *Geräte* tested in inspection (x).

BAMA, RH8-1310
Spezialfahrzeuge
Baugruppen und
wesentliche Bauteile
der Bodenanlage FR

Special vehicles of the technical battery

Fz.-Pos.-Nr.501 Test vehicle II on 3 t truck closed (305)
Fz.-Pos.-Nr.502 Cable truck II on 3 t truck closed (305)
Fz.-Pos.-Nr.503 Spare parts truck II on 3 t truck closed (305)
Fz.-Pos.-Nr.504 Heavy machinery set I (acc. attachment N 3670)
Fz.-Pos.-Nr.505 Converter trailer 220/380 V to 27 V current
Fz.-Pos.-Nr.506 Compressed air generator trailer (2-axle)
Fz.-Pos.-Nr.507 Tool and spare euipment on 3 t truck open*
Fz.-Pos.-Nr.508 Accumulator truck on 4,5 t truck geschl. (heated)
Fz.-Pos.-Nr.509 Charging Station trailer Type B (2-axle)
Fz.-Pos.-Nr.510 Spare parts truck III on 3 t truck closed (305)
Fz.-Pos.-Nr.511 two-axle trailer for Kfz. workshop

* Replaced by 3 t truck open

Technische Batterie

Zweck und Aufgabe

Die angelieferten A4-Geräte werden nach dem Umschlag von der Schiene auf den Vidal-Transportanhänger (Fz.601) verladen und der technischen Batterie im Feldspeicher zur Prüfung vorgefahren. Nach einer Sichtprüfung auf äußerliche Beschädigungen erfolgt eine eingehende technische Prüfung. Brauchbare Geräte werden mit der Nutzlastspitze versehen, ggf. zwischengelagert oder dem Schiessbetrieb zugeführt. Die bei der technischen Prüfung (x) des Geräts ausgefallenen Einbaugeräte werden getauscht, auch kleinere Reparaturen ausgeführt, sofern das im Prüf- und Instandsetzungsumfang der Feldspeicher liegt. Werden kurzfristige Ausfälle an den geprüften Geräten in der Feuerstellung festgestellt, gehen diese wieder zur technischen Batterie zurück. Dort wird der Defekt [1] behoben oder über die Rücksendung an die Feldwerkstatt oder Heimat Muna zur Instandsetzung entschieden, das gilt auch für unter (x) geprüfte Geräte.

Spezialfahrzeuge der technischen Batterie

Fz.-Pos.-Nr.501 Prüfwagen II auf Lkw 3 t geschl.(305)
Fz.-Pos.-Nr.502 Kabelwagen II auf Lkw 3 t geschl.(305)
Fz.-Pos.-Nr.503 Vorratsteilewagen II auf Lkw 3 t geschl.(305)
Fz.-Pos.-Nr.504 s. Maschinensatz I (nach Anlage N3670)
Fz.-Pos.-Nr.505 Umformer-Anhanger 220/380 V auf 27 V
Fz.-Pos.-Nr.506 Luftverdichter-Anhänger (2-achs.)
Fz.-Pos.-Nr.507 Werkzeug- u. Ersatzgeräte auf Lkw 3 t offen*
Fz.-Pos.-Nr.508 Sammler-Lkw auf Lkw 4,5 t geschl. (heizbar)
Fz.-Pos.-Nr.509 Ladestation-Anhänger Type B (2-achs.)
Fz.-Pos.-Nr.510 Vorratsteilewagen III auf Lkw 3 t geschl.(305)
Fz.-Pos.-Nr.511 Zwei-Achs-Anhänger Kfz.-Werkstatt

* Ersetzt durch Lkw 3 t offen

DMA, GD639.5.12
EW22-760/44 gK
Bedienungsanweisung zum Prüfen von Sondergeräten im liegenden Zustand mit dem Prüfwagen 501

Testing *Sondergeräte* in horizontal position (field buffer)

Prüfen von Sondergeräten im liegenden Zustand (Feldspeicher)

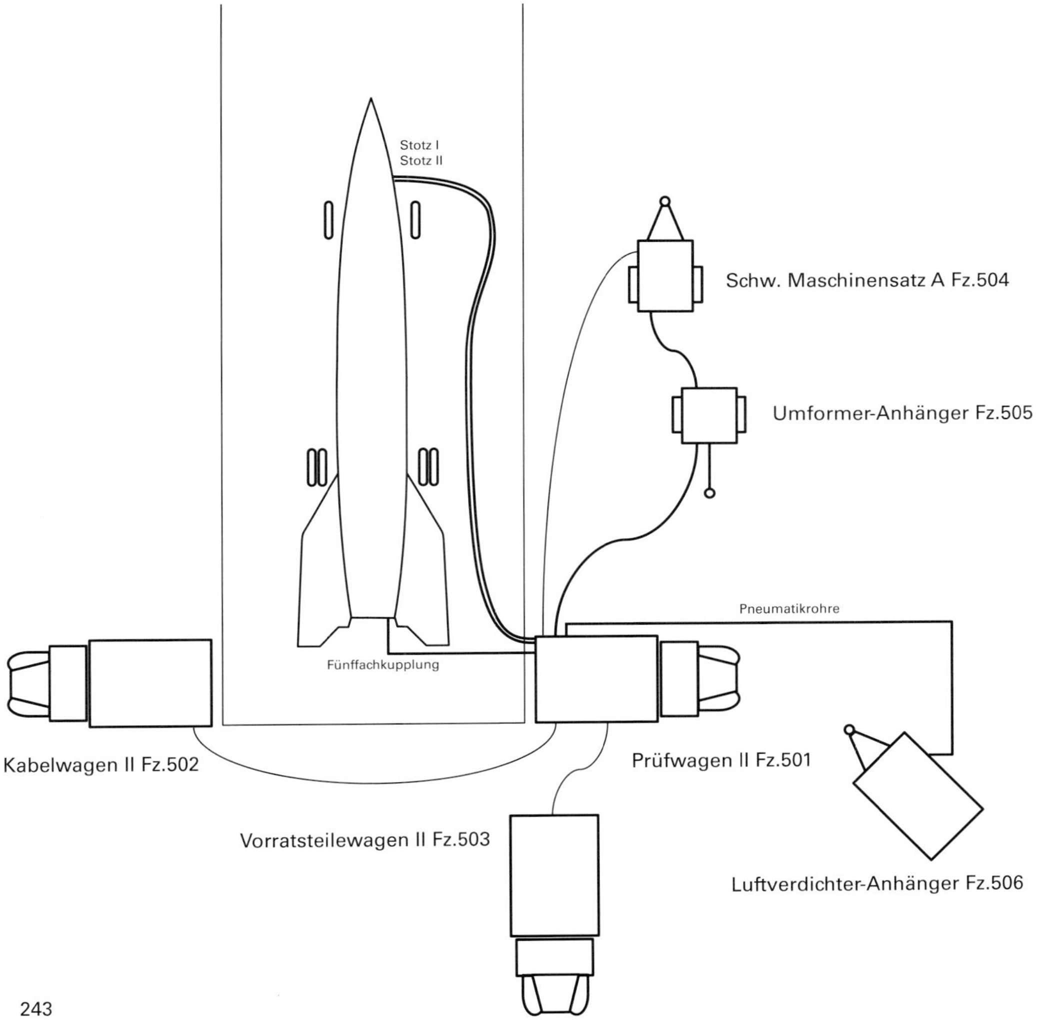

243

* *Mischgerät* = a ampflifier of five independent panels mounted on a common base plate. It tooks signals from the gyroscope and then translated into a signal that directed the servo steering mechanism and steering vans to correct for attitude deviations in powered flight.

Assembly and finishing

The test vehicle II (Fz.501) is to be positioned as horizontally as possible so that its left side faces a long side of the *Gerät*. Only in exceptional cases, such as when there is a lack of space, should it be set up with its left side facing the rear of the *Gerät*. The following are still needed to connect the unit to test vehicle II:

cable car II	(Fz.502)
electric machinery set	(Fz.504)
three-phase converter	(Fz.505)
air compressor	(Fz.506)

The air compressor, electric machinery set and three-phase converter are to be placed as far away from test vehicle II as the connecting cables permit and prepared for operation. The location of the vehicle, its connection to each other and to the *Gerät* is shown in the sketch »KB132«.

The following tests are to be carried out:

- Measuring insulation values
- Steering gear, *Mischgerät* (steering control)*, rudder operation
- Radio, antenna measurement, Doppler, LS guide beam
- Propulsion unit, general switching test and combustion cut-off

Aufbau und Fertigmachen

Der Prüfwagen II (Fz.501) ist möglichst waagerecht so aufzustellen, dass seine linke Seite einer Längsseite des Gerätes zugewandt ist. Nur im Ausnahmefallen, wie bei Platzmangel, ist die Aufstellung mit der linken Seite zum Heck des Gerätes einzunehmen. Zur Prüfung des Gerätes werden in Verbindung mit dem Prüfwagen II noch benötigt:

der Kabelwagen II	(Fz.502)
der Elektro-Maschinensatz	(Fz.504)
der Drehstrom-Umformer	(Fz.505)
der Luftverdichter	(Fz.506)

Luftverdichter, Elektro-Maschinensatz und Drehstrom-Umformer sind möglichst so weit entfernt vom Prüfwagen II aufzustellen, wie es die Verbindungsleitungen zulassen und sind zum Betrieb fertigzumachen. Die Aufstellung der Fahrzeuge, ihre Verbindung untereinander und mit dem Gerät, ist aus der Skizze »KB132« zu ersehen.

Folgende Prüfung werden durchgeprüft:

- Messen der Isolationswerte
- Steuerung, Mischgerät*, Rudermaschinen
- Funk, Antennenmessung, Verdoppler, Leitstrahl
- Triebwerk, Generaldurchschaltversuch und Brennschluß

* Mischgerät = ein Verstärker aus fünf getrennten Schalttafeln, die auf einer gemeinsamen Grundplatte montiert sind. Er nimmt die Signale des Gyroskops auf und setzt sie in ein Steuersignal um, das den Servolenkungsmechanismus der Ruder (Segel) zur Korrektur der Fluglage bei Abweichungen ansteuert.

Vehicle 501

Internal or former designation: Fz.-Pos.-Nr.501
Test vehicle II on 3 t truck closed (305)
Test vehicle II

DMA, GD639.5.18
HAP11/T 507
Karlshagen, 1.6.1944
Bd.Nr. E 3195/44 g

DMA, GD639.5.1
EW234
Fz.-Pos.-Nr. 501

DMA,GD639.55.5
EW234
Stand 01.11.44

Intended use

Test vehicle II is used to carry out various tests of the delivered *Gerät* in field storage. A test desk, B-rack and bench seat box with table are accommodated in an closed superstructure mounted on the chassis of a 3 t Opel 6700 Type A truck. With the test desk the functions for engine, control and pneumatics are tested and the switching and control functions are simulated by means of current pulses. The transmitter and receiver functions are tested with the test transmitter and, among others, with the command transmitter in the B-rack.

The test vehicle II has the following dimensions and weights:

Total length:	6,130 mm
Total width:	2,100 mm
Total height:	2,970 mm
Wheelbase:	3,450 mm
Tare weight:	3,275 kg
Payload:	2,825 kg
Maximum weight:	6,100 kg
Service weight:	4,285 kg
Wheels:	190-20

The main parts of the test truck II are:

- Chassis
- Closed superstructure (305) LC box body
- Test desk
- B-rack
- Bench seat box and storage table

Chassis*

The chassis used is that of the 3 t Opel 6700 Type A truck (equipment description according to D 666/5) but the normal chassis of the Opel 3 t Type 3.6/36 S truck (equipment description according to D 669/19B) was also used.

* Note on Fz.501, Fz.502, Fz.503 and Fz.510 (Fz.304).
None of these vehicles need to be off-road capability, and only Fz. 501 requires a closed box body. Therefore, if there is a bottleneck of such off-road vehicles and superstructures of this type, they must be replaced as soon as possible with commercially available 3 t open trucks in the case of Fz.502, 503 and 510 (or 304). It must be checked beforehand whether it is necessary to remain with the closed body due to the bottleneck in tarpaulins and bows. In any case, an off-road chassis. Is not required.

Fahrzeug 501

Interne oder frühere Bezeichnung: Fz.-Pos.-Nr.501
Prüfwagen II auf Lkw 3 t geschl. (305)
Prüfwagen II

Verwendungszweck

Mit dem Prüfwagen II wird das angelieferte Gerät im Feldspeicher verschiedenen Prüfungen unterzogen. Ein Prüfpult, B-Gestell und Sitzbankkiste mit Tisch sind in einem geschl. Aufbau untergebracht. Der Aufbau ist auf dem Fahrgestell des Lastkraftwagen 3 t Opel 6700 Typ A montiert. Mit dem Prüfpult werden die Funktionen für Triebwerk, Steuerung und Pneumatik überprüft und mittels Stromimpulsen die Schalt- und Steuerungsfunktionen simuliert. Die Sende- und Empfängerfunktionen werden mit dem Prüfsender und u.a. mit dem Kommandogeber im B-Gestell geprüft.

Der Prüfwagen II hat folgende Maße und Gewichte:

Gesamtlänge:	6.130 mm
Gesamtbreite:	2.100 mm
Gesamthöhe:	2.970 mm
Radstand	3.450 mm
Leergewicht:	3.275 kg
Nutzlast:	2.825 kg
Zul.Höchstgew.	6.100 kg
Gefechtsgewicht:	4.285 kg gewogen
Räder:	190-20

Die Hauptteile des Prüfwagen II sind:

- Fahrgestell
- Geschlossener Aufbau (305) LC-Koffer
- Prüfpult
- B-Gestell
- Sitzbankkiste und Ablagetisch

Fahrgestell*

Als Fahrgestell dient das Fahrgestell des Lastkraftwagen 3 t Opel 6700 Typ A (Gerätebeschreibung nach D 666/5) aber auch das normale Fahrgestell des Opel 3 t Lastkraftwagens Typ 3,6/36 S (Gerätebeschreibung nach D 669/19B) wurde verbaut.

* Anmerkung zu Fz.501, Fz.502, Fz.503 und Fz.510 (Fz.304).
Keines dieser Fahrzeuge muss geländegängig sein. Nur das Fahrzeug 501 benötigt einen geschlossenen Aufbau. Es muss also bei einem Engpass derartiger gleicher Fahrzeuge und Aufbauten unter allen Umständen baldigst der Austausch gegen handelsübliche Lkw 3 t offen bei den Fahrzeugen 502, 503 und 510 (bzw. 304) erfolgen. Es ist dabei vorher zu überprüfen, ob durch den Engpass bei Planen und Spiegeln notgedrungenermaßen beim geschlossenen Aufbau verblieben werden muss. Auf alle Fälle entfällt das gl. Fahrgestell.

244 The test vehicle II on the chassis of the 3 t Opel 6700 type A truck did not differ externally from the Kfz.305 with the closed superstructure. A heating hose conducted the waste heat from the muffler into the interior via a connection on the rear wall.

244 Der Prüfwagen II auf dem Fahrgestell des Lastkraftwagen 3 t Opel 6700 Typ A unterschied sich äußerlich nicht vom Kfz.305 mit dem geschlossenen Aufbau. Ein Heizungsschlauch leitete die Abwärme des Schalldämpfers der Auspuffanlage über einen Anschluss an der Rückwand in den Innenraum.

DMA, GD639.5.1
OKH/Wa Prüf10
Karlshagen, den 1.4.1944
Bd.Nr. 789/44 gK
Auszug
Besprechung der
Fahrzeuggruppe 5

Production status: Vehicle 501
Initial requirement 7 of which 1 sample is with Kdo. Basse and 6 vehicles at LR, three of them are worked on.

Requirement for 1st to 3rd technical Batteries of Art Abt. 91, 3; for technical training, 1 vehicle; 1 spare vehicle for Technical Training Abt.

Total quantity of 5 vehicles requested from B.z.b.V. Heer to 15.4.44 and 4 delivered to the units. The remaining 3 will follow as single deliveries (no fixed schedule).

Current defects and their remedy

B-racks
5 B-racks for Technical Battr. are available. In the built-in equipment sets for this, however, the command transmitter (general with inspection racks) is missing, alternating frequencies, controller for the doubler, radio command and directional beam tester so different that individual listing is omitted. Adjustments have to be made by changing the crystals. On 2 of these B-racks the redundant switching assemblies on the upper face of the rack are not yet removed. They must be covered with protection caps. Action Kriegs-Ing. Doerken by 12.4.1944.

Desktop telephone
In lieu of the currently installed Type 37 desktop telephone a field telephone set is to be used and, if possible, fitted with a throat-microphone. Change to field telephone set to be arranged by Kdo. Basse by 12.4.1944.

Pneumatic panel
Before delivery to the firm of Gollnow, the pneumatic consoles must be by HAP 11 checked for correct compression, thereby avoiding later damage (pressure reducing gauge plan). Action done by TD 4, Ing. Poppel.

The relays in the engine control panel are not dust-protected. The case is insufficiently sealed and collects dust. Sealing of the relays and better sealing of the cabinet has to be ensured . Responsible to ensure TD 5, execution by Kdo. Basse. Cabinet sealing Kdo.Basse. To be done by 15.4.44 for all vehicles present or arriving at Kdo. Basse.

All vehicles still at LR are repaired by the firm of Gollnow to improve the relay sealing and cabinet sealing. Delivery delays are not permitted.

Fertigungslage: Fahrzeug 501
Zunächst Gesamtbedarf 7 davon 1 Muster bei Kdo. Basse und 6 Fahrzeug in LR, davon drei in Arbeit.

Bedarf für 1te bis 3te technische Batterie der techn. Art. Abt. 91, 3 Fahrzeuge, für technische Ausb. Abt., 1 Fahrzeug, 1 Ersatzfahrzeug für techn. Art. Abt.

Gesamtmenge von 5 Fahrzeugen bis 15.4.44 von B.z.b.V. Heer gefordert und zwar bei der Truppe ausgeliefert. Davon bis dahin verfügbar 4. Die restlichen 3 werden mit Einzellieferungen folgen (keine Terminforderung).

Derzeitige Mängel und ihre Behebung

B-Gestelle
5 B-Gestelle für techn. Battr. sind vorhanden. An Einbausätzen hierfür fehlen jedoch Kommandogeber (allgemein bei Prüfgestellen), verschiedene Wechselfrequenzen, Einsatz für Verdoppler, Funkkommando- und Leitstrahlprüfung (so verschieden, dass Einzelaufstellung unterbleibt). Die Ergänzung erfolgt stetig durch entsprechende Anlieferungen der fehlenden Quarze. Bei 2 dieser B-Gestelle sind die überflüssigen Stecker auf der Oberseite des Gestelles noch nicht entfernt. Sie müssen mit Schutzkappen abgedeckt werden.
Veranlassung Kriegs-Ing. Doerken bis 12.4.1944.

Tischfernsprecher
An Stelle des z.Zt. eingebauten Tischfernsprechers 37 wird ein Feldfernsprecher verwendet und möglichst mit Kehlkopfmikrophon-Ausstattung ergänzt. Änderung betreffs Feldfernsprecher veranlasst Kdo. Basse bis 12.4.1944.

Pneumatisches Pult
Vor Auslieferung an die Firma Gollnow müssen die pneumatischen Pulte durch HAP 11 vorschriftsmäßig abgedrückt worden sein, damit nachträgliche Schäden (Fall Druckminderer) vermieden werden. Veranlassung ist erfolgt durch TD 4, Ing. Poppel.

Die Relais im Triebwerkspult sind nicht staubgeschützt. Der Kasten selbst ist ungenügend abgedichtet und ein Staubfänger. Abdichtung der Relais und bessere Abdichtung der Kästen ist sicherzustellen und zwar Relaisschutz verantwortlich durch TD 5, Ausführung Kdo. Basse. Kastenabdichtung Kdo. Basse. Termin hierfür für alle bei Kdo. Basse befindlichen und eingehenden Fahrzeuge bis 15.4.44.

Alle noch in LR befindlichen Fahrzeuge werden bezüglich Relaisabdichtung und Kastenabdichtung durch Firma Gollnow in Ordnung gebracht. Lieferverzug dafür nicht gestattet.

Pneumatic hoses

Pneumatic hoses (from the test vehicle to the *Gerät*) by required type and quantity will be checked and fulfilled by the Gollnow Company for all new deliveries of vehicles 501. The missing pneumatic hoses not already delivered with vehicles to Kdo. Basse will be sent with the next vehicles. Arrangement has already been made by Gollnow.

Cold-soldering joints

Bad and incorrect manufacture of engine and control panels (currently made in the Russian camp). Above all, cold-soldering joints are to be feared. And the rest is bad as well.

Remedy:

1. In vehicles at Kdo. Basse resoldering by Kdo.Basse itself is needed.
2. All vehicles at LR are to be checked and resoldering done by the Gollnow Company. Delivery delays not permitted.
3. For deliveries starting with the 8th vehicle the Gollnow Company takes over the production of the panels.
   Action TD 531, Ing. Rosinski, both for stopping assembly in the Russian camp, and for timely transmission of the production documents to the Gollnow Company.

Connector covers

Connector covers for all relay cables concerned are to be ensured immediately.

Action TD 5. Report about present possibilities by 10.4.44 to BfE with practical demonstration of the cables in question.

Test transmitter

The test transmitter installed in the 501 test vehicle II is equipped only with a part of the application/mission frequencies and in addition is susceptible to interference. Therefore a newly developed test transmitter for LS is being completed, forecast for delivery in mid-May.

Druckluftschläuche

Druckluftschläuche (vom Prüfwagen zum Gerät) erforderliche Art und Menge ist klargestellt und wird von Fa. Gollnow bei allen neuen Lieferungen der Fahrzeuge 501 erfüllt und berücksichtigt. Die beiden bereits an das Kdo. Basse ausgelieferten Fahrzeugen fehlenden Druckluftschläuche werden mit den nächsten Fahrzeugen zusätzlich übersandt. Veranlassung ist durch Fa. Gollnow bereits erfolgt.

Kalte Lötstellen

Schlechte und mangelhafte Ausführung der Triebwerks- und Steuerungstafeln (werden z.Zt. im Russenlager hergestellt). Vor allen Dingen sind kalte Lötstellen zu befürchten. Auch die übrige Ausführung ist schlecht.

Abhilfe:

1. In den Fahrzeugen beim Kdo. Basse, nachlöten durch Kdo. Basse selbst.
2. Bei den in LR befindlichen Fahrzeugen überprüfen und nachlöten durch Fa. Gollnow. Lieferverzug nicht gestattet.
3. Für spätere Lieferung ab 8. Fahrzeug übernimmt die Fa. Gollnow selbst die Herstellung der Tafeln.
   Veranlassung TD 531. Ing. Rosinski sowohl in Bezug auf Stopp der Montage im Russenlager, wie auf rechtzeitige Übermittlung der Fertigungsunterlagen an die Fa. Gollnow.

Steckerschutz

Steckerschutz für alle verwandten Relaiskabel ist umgehend sicherzustellen.

Veranlassung TD 5. Meldung über derzeitige Möglichkeiten bis 10.4.44 an BFE unter praktischer Vorführung an den fraglichen Kabelstücken.

Prüfsender

Der im Prüfwagen II (501) eingebaute Prüfsender ist nur mit einem Teil der Einsatzfrequenzen ausgestattet und außerdem zu störungsanfällig. Daher ist z.Zt. ein neu entwickelter Prüfsender für LS in Fertigstellung, der voraussichtlich ab Mitte Mai geliefert werden kann.

BArch, RH8II Bild-B1172-44, o. Ang.

245

Deutsches Museum, München, Archiv – FA014/B326/44 TD5224 [CD88073]

246

Installations
Einbauten

245 A cable junction box and various sockets and connections for the air pipes are located inside the left rear access hatch. The sensitive devices are mounted on rubber buffers.

246 The photo on the right show a modified installation. Here, the pneumatic connections are positioned on the side panel of the pneumatic console. In later versions, the connections were turned for better access to the door.

245 Ein Kabelanschlusskasten mit verschiedenen Steckdosen und Anschlüssen für die Luftleitungen befand sich an der linken hinteren Zugangsklappe. Die empfindlichen Geräte waren auf Gummipuffern gelagert.

246 Das Bild rechts zeigt eine abgeänderte Einbausituation. Die Pneumatikanschlüsse sind hier an der Seitenwand des Pneumatikpultes angebracht. In späteren Ausführungen wurden die Anschlüsse für einen bessseren Zugang zur Klappe gedreht.

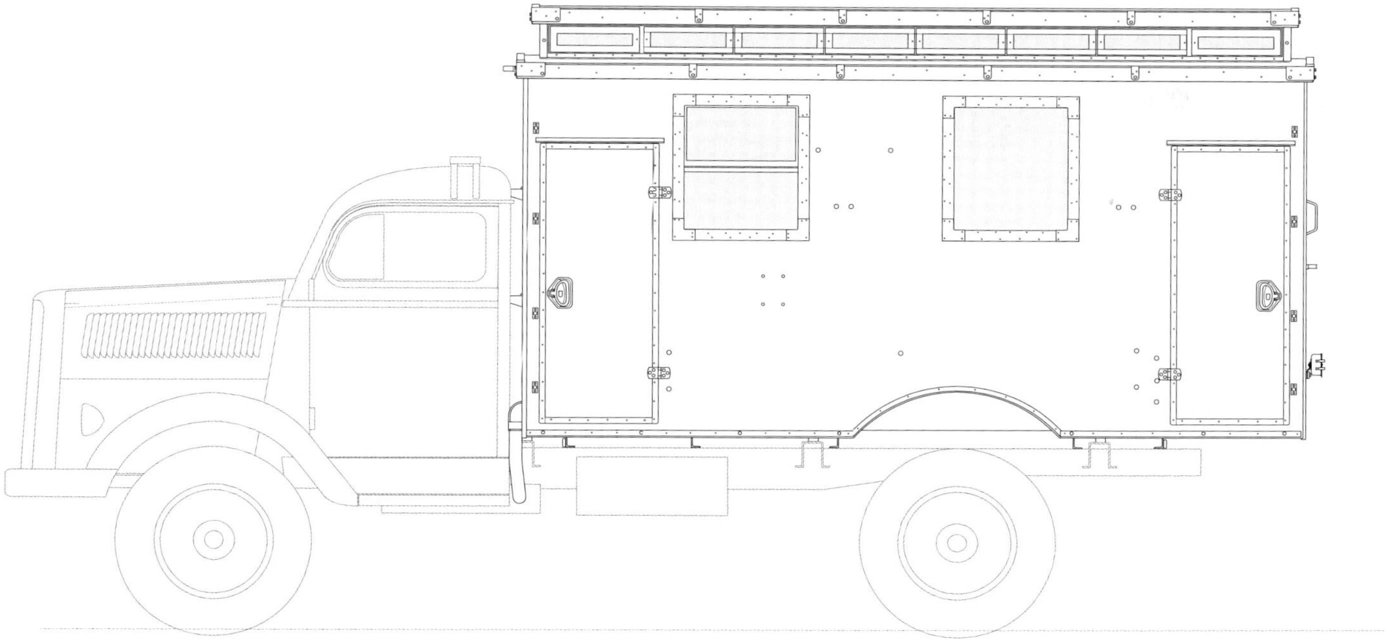

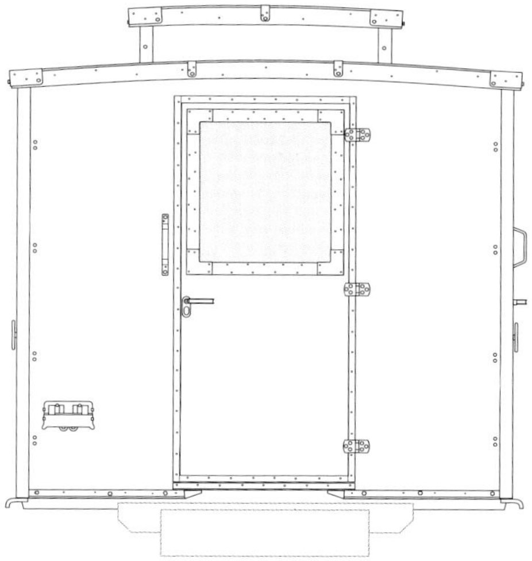

1 m

Scale / Maßstab 1:35 © Volker Ruff

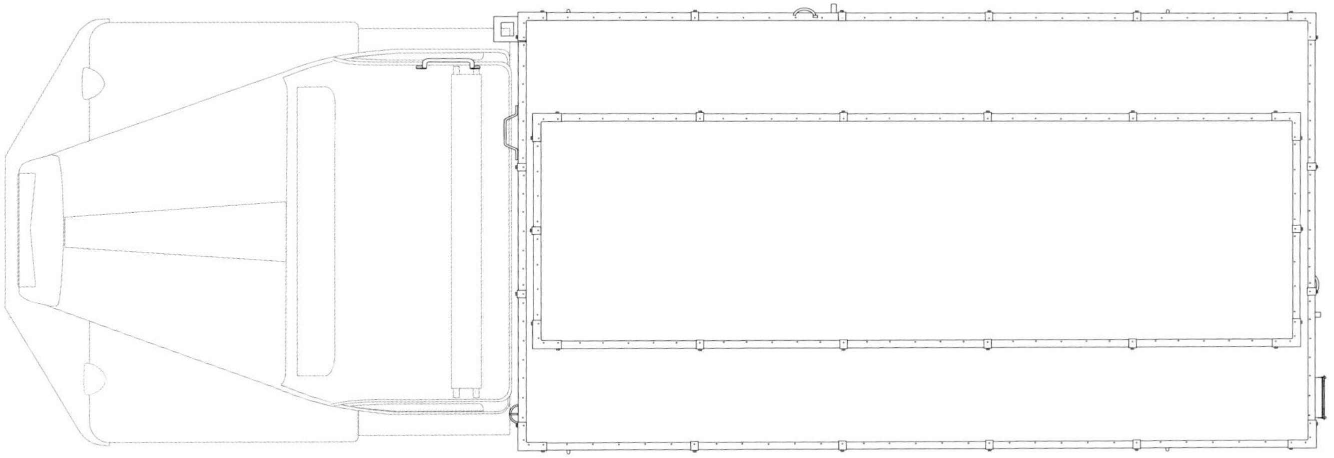

Fahrzeug 501

Prüfwagen II auf Lkw 3 t geschl. (305)

Partially used drawings acc. to group list 21Gr359

Teilweise verwendete Zeichnungen nach Gruppenliste 21Gr359

Deutsches Museum, München, Archiv- FA014/B313/44 TD5224 [CD66673]

247

Deutsches Museum, München, Archiv- FA014/B314/44 TD5224 [CD74739]

248

Test bench
Prüfpult

247 The rear door provides a view of the interior. On the left is the test console, above is the external command unit.
A B-frame is attached to the floor and on the back wall to the compartment, equipped with test transmitter and various control levels.

248 The main components are from left to right:
At the bottom the cable connection box, above it is the pneumatic console, power unit console and control console. On the wall above, the exterior command device and two working lights.
All the windows could be covered with with canvas panels. The interior of all LC box bodies was always painted in a light color.

247 Durch die rückwärtige Tür blickt man in den Innenraum. Links befindet sich das Prüfpult, darüber das Fremdkommandogerät. Ein B-Gestell ist am Boden und an der Rückwand zur Kabine befestigt, ausgerüstet ist es mit Testsender und verschiedenen Steuerstufen.

248 Das Prüfpult bestand aus folgenden Hauptbauteilen (von links nach rechts): der Kabelanschlusskasten unten, darüber das Pneumatische-, Triebwerks- und Steuerungspult. An der Wand darüber befand sich das Fremdkommandogerät und zwei Arbeitsleuchten. Die Fenster und Galeriefenster konnten mit Segeltuchbahnen verdunkelt werden. Der Innenraum bei allen geschlossenen Aufbauten war immer hell ausgestrichen.

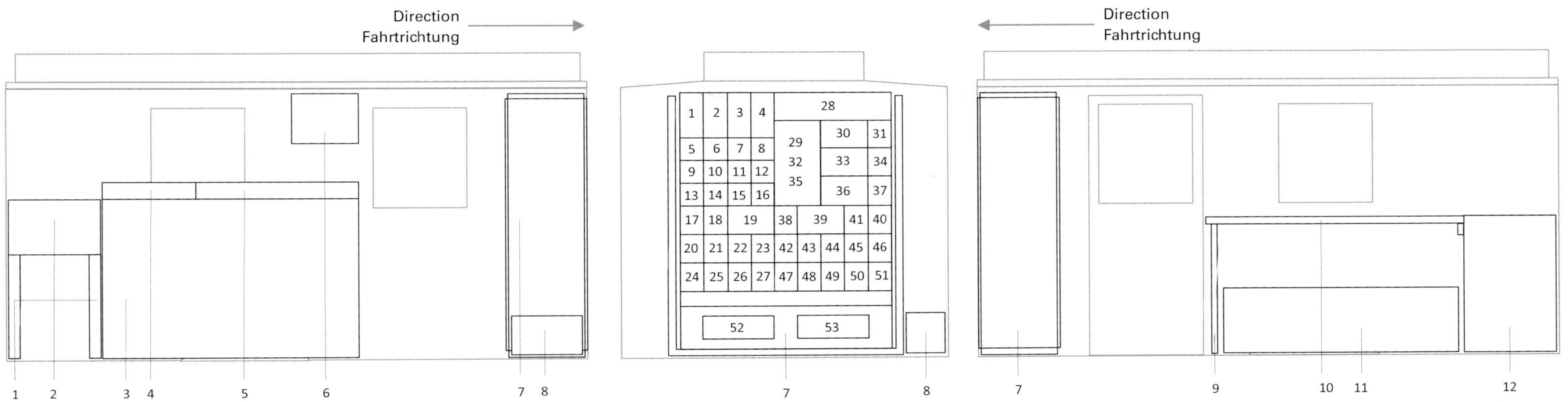

View left side wall
Ansicht linke Seitenwand

View right side wall
Ansicht rechte Seitenwand

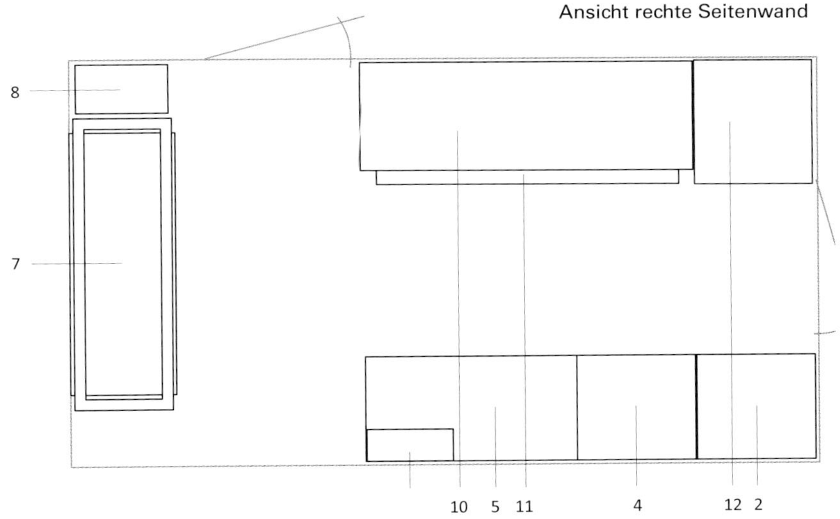

Top view
Ansicht von oben

1 Cable connection box
2 Pneumatic console
3 Control console
4 Propulsion panel
5 Steering control panel
6 External command unit
7 B-rack
8 Command transmitter II
9 Support for storage table
10 Storage table – foldable
11 Bench box for tools and equipment
12 Cabinet for tools and equipment

1 Key stage
2 Control stage
3 Mains stage
4 Control stage for timer
5 AF stage
6/17 Cover plate
7 Test transmitter B-yellow (operating frequency)
8 Preheating and remote sensing stage
9-16 Test transmitter B-yellow (alternating frequency)
18 Test transmitter B-red (operating frequency)
19 stage I
20-27 Test transmitter B-red (alternating frequency)
28 Mains signal field
29/32/35 Cover plate
30 Command feedback field
31/37/41 Cover plate
33 Operating switch
34 Main power switch and fuses
36 Instrument feedback panel
38/42 Power stage f 1-2
39 Tube voltmeter with attenuator
40 Test transmitter B-green (operating frequency)
43-51 Test transmitter B-green (alternating frequency)
47 Cover plate for normal version
52 Power supply unit B I
53 Empty drawer (space for command transmitter II)

1 Kabelanschlusskasten
2 Pneumatisches Pult
3 Bedienkonsole
4 Triebwerkspult
5 Steuerpult
6 Fremdkomandogerät
7 B-Gestell
8 Kommandogeber II
9 Stütze für Ablagetisch
10 Ablagetisch - klappbar
11 Sitzbankkiste für Werkzeug und Gerät
12 Schrank für Werkzeug und Gerät

1 Taststufe
2 Kontrollstufe
3 Netzstufe
4 Steuerstufe für Zeitschaltwerk
5 NF-Stufe
6/17 Abdeckplatte
7 Prüfsender B-gelb (Betriebsfrequenz)
8 Vorheiz- und Ferntaststufe
9-16 Prüfsender B-gelb (Wechselfrequenz)
18 Prüfsender B-rot (Betriebsfequenz)
19 Mischstufe I
20-27 Prüfsender B-rot (Wechselfrequenz)
28 Netzsignalfeld
29/32/35 Abdeckplatte
30 Kommando-Rückmeldefeld
31/37/41 Abdeckplatte
33 Betriebsschalter
34 Netzhauptschalter und Sicherungen
36 Instrumenten-Rückmeldefeld
38/42 Leistungsstufe f 1-2
39 Röhrenspannungsmesser mit Abschwächer
40 Prüfsender B-grün (Betriebsfrequenz)
43-51 Prüfsender B-grün (Wechselfrequenz)
47 Abdeckplatte bei normal Ausführung
52 Netzgerät B I
53 Leerfach (Raum für Kommandogeber II)

Vehicle 502

Internal or former designation: Fz.-Pos.-Nr.502
Cable truck II on 3 t truck closed (305)
Truck for cable and lighting

DMA, GD639.5.18
HAP11/T 507
Karlshagen, 1.6.1944
Bd.Nr. E 3195/44 g

Intended use
The cable truck II is a 3 t Opel 6700 type A truck with closed superstructure. The necessary connecting cables to the Fz.501 and for power supplies to Fz.504 and Fz.505 are accommodated in its interior. There are also four cable reels with NSH-, NSM-cable, a cabinet for low beam spotlights, a workbench, and tool boxes.

DMA, GD639.5.1
EW234
Fz.-Pos.-Nr. 502

DMA, GD639.55.4
EW234
Stand 01.11.44

The cable truck II has the following dimensions and weights:

Total length:	6,130 mm
Total width:	2,100 mm
Total height:	2,990 mm
Wheelbase	3,450 mm
Tare weight:	3,275 kg
Payload:	2,825 kg
Maximum weight	6,100 kg
Service weight:	4,290 kg
Wheels:	190-20

The main parts of the cable truck II are:
- Chassis
- Closed superstructure (305) LC box body
- NSH-, NSM-cable reels and connecting cables
- Cabinet for lighting, workbench and tools

Chassis
The chassis used is that of the 3 t Opel 6700 Type A truck (equipment description according to D 666/5) but the normal chassis of the Opel 3 t Type 3.6/36 S truck (equipment description according to D 669/19B) was also used.

DMA, GD639.5.21OKH/
Wa Prüf 10
Karlshagen, 1.4.1944
Bd.Nr. 789/44 gK
Auszug
Besprechung der
Fahrzeuggruppe 5
vom 1.4.44
in Karlshagen I

Production status: Vehicle 502
Initial requirement 7 vehicles of which 1 sample is with Kdo. Basse and 6 vehicles at LR in different versions and equipped up to 80 %.
Requirements as for 501, 5 pieces by 15.4.44. Delivery ensured until then. 3 vehicles are completed; the 4th and 5th vehicles can only be delivered after completion of their cables (Stotz connectors) and lighting sets.

Fahrzeug 502

Interne oder frühere Bezeichnung: Fz.-Pos.-Nr.502
Kabelwagen II auf Lkw 3 t geschl. (305)
Lkw für Kabel und Beleuchtung

Zweck und Aufgabe
Der Kabelwagen II ist ein Lastkraftwagen 3 t Opel 6700 Typ A mit geschl. Aufbau. Im Innenraum sind die notwendigen Verbindungskabel zum Fz.501 und zur Stromversorgung mit Fz.504 und Fz.505 untergebracht. Ebenfalls sind vier Kabelrollen mit NSH-, NSM-Kabel vorhanden, ein Schrank für Tiefstrahler, eine Werkbank und Werkzeugkästen untergebracht.

Der Kabelwagen II hat folgende Maße und Gewichte:

Gesamtlänge:	6.130 mm
Gesamtbreite:	2.100 mm
Gesamthöhe:	2.990 mm
Radstand	3.450 mm
Leergewicht:	3.275 kg
Nutzlast:	2.825 kg
Zul.Höchstgew.:	6.100 kg
Gefechtsgewicht:	4.290 kg gewogen
Räder:	190-20

Die Hauptteile des Kabelwagen II sind:
- Fahrgestell
- Geschlossener Aufbau (305) LC-Koffer
- NSH-, NSM-Kabelrollen und Verbindungskabel
- Schrank für Beleuchtung, Werkbank und Werkzeug

Fahrgestell
Als Fahrgestell dient das Fahrgestell des Lastkraftwagen 3 t Opel 6700 Typ A (Gerätebeschreibung nach D 666/5) aber auch das normale Fahrgestell des Opel 3 t Lastkraftwagens Typ 3,6/36 S (Gerätebeschreibung nach D 669/19B) wurde verbaut.

Fertigungslage: Fahrzeug 502
Gesamtbedarf zunächst 7 Fahrzeuge, davon 1 Muster bei Kdo. Basse und 6 in LR in anderer Ausführung, ausgerüstet bis zu 80 %.
Gefordert sinngemäß wie bei 501, 5 Stück bis 15.4.44. Lieferung bis dahin sichergestellt. Vollständig 3 Fahrzeuge, 4. und 5. Fahrzeug können erst nach Ergänzung von Kabel (Stotzstecker) und Beleuchtungssatz nachgeliefert werden.

BArch, RH8II Bild-B1176-44, o. Ang.

249

249 To the left of the rear exit was a rack for several toolboxes and a workbench with drawers for special tools. On the right-hand side on the inside wall were two support frames, each with two NSH and NSM cable reels, left and right of a cabinet for spotlights FL56011 (suspended spotlights) mounted.

249 Links neben dem rückwärtigen Ausgang befand sich ein Gestell für mehrere Werkzeugkoffer sowie eine Werkbank mit Schubladen für Sonderwerkzeuge. An der rechten Seite waren an der Innenwand zwei Haltegerüste mit je zwei NSH-, NSM-Kabelrollen links und rechts eines Schrankes für Tiefstrahler FL56011 (Einhängestrahler) montiert.

BArch, RH8II Bild-B1177-44, o. Ang.

250

250 The opening in the rear wall of the driver's cab for the rear window was closed and three brackets for connections and power cables were mounted there. The 104-pin Flakstecker (anti aircraft plug) and connecting plugs rested in holders on the rear and side walls. The side windows were partly darkened with canvas or closed by filling pieces of hardboard.

250 An der Rückwand zum Fahrerhaus war der Durchbruch für das Rückfenster verschlossen. Drei Wandhalter für Verbindungs- und Stromkabel wurden dort angebracht. Die 104-poligen Flak- und Verbindungsstecker ruhten in Halterungen an der Rück- und Seitenwand. Teilweise waren die seitlichen Fenster mit Segeltuch verdunkelt oder durch Füllstücke aus Hartfaserplatten verschlossen.

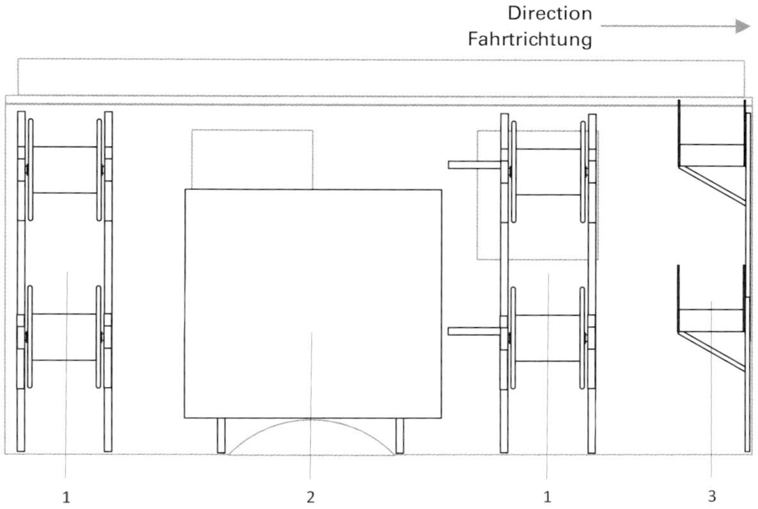

View left side wall
Ansicht linke Seitenwand

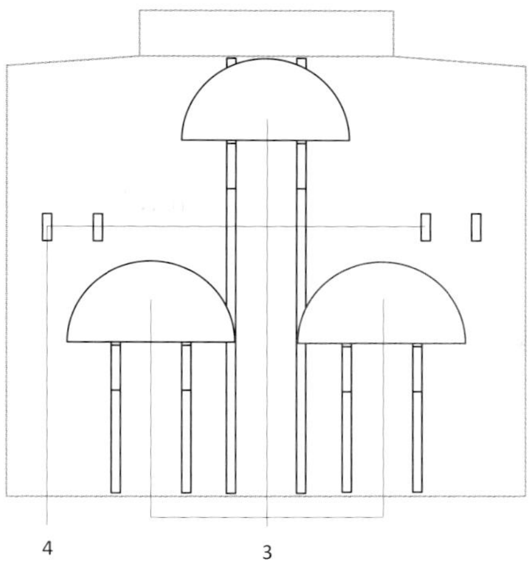

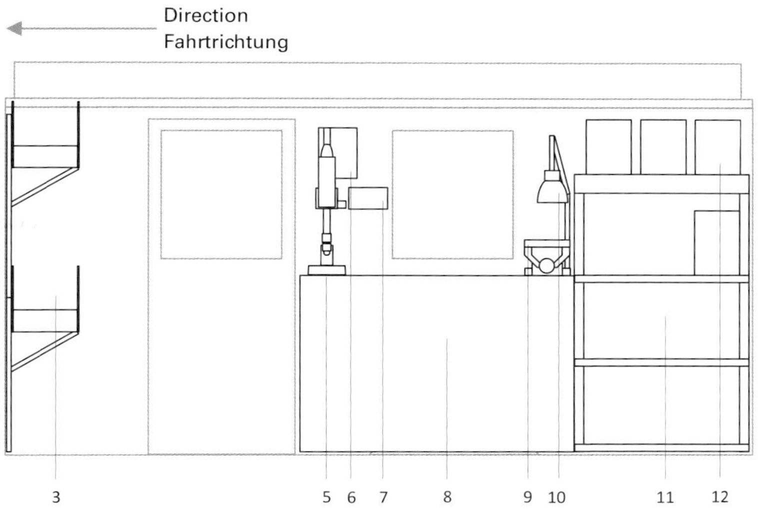

View right side wall
Ansicht rechte Seitenwand

1 cable drums
2 Cabinet for lighting device
3 Cable supports
4 Holder for Flak plug
5 Hand drill with drill stand
6 Resistor of the drill 220 V
7 On-Off switch
8 Standard workbench EWB II
9 Parallel vise
10 Work lamp
11 Compartments for cable drums
12 Tool boxes

1 Kabeltrommeln
2 Schrank für Beleuchtungsgerät
3 Kabelböcke
4 Halterung für Flakstecker
5 Handbohrmaschine mit Bohrständer
6 Vorwiderstand der Bohrmaschine 220 V
7 Ein Aus Schalter
8 Einheitswerkbank EWB II
9 Parallelschraubstock
10 Arbeitsleuchte
11 Fächer für Kabeltrommeln
12 Werkzeugkästen

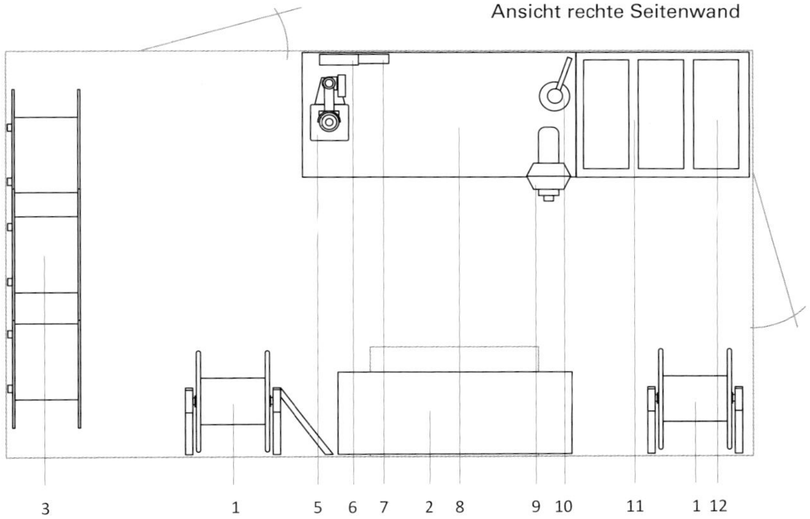

Top view
Ansicht von oben

Fahrzeug 502
Kabelwagen II auf Lkw 3 t geschl.
(305)

Drawing set used n.a.
Verwendeter Zeichnungssatz k.A.

Vehicle 503

Internal or former designation: Fz.-Pos.-Nr.503
Spare parts truck II on 3 t truck closed (305)
Electrical spare parts

DMA, GD639.5.18
HAP11/T 507
Karlshagen, 1.6.1944
Bd.Nr. E 3195/44 g

DMA, GD639.5.1
EW234
Fz.-Pos.-Nr.503

Intended use
The spare parts truck II is intended for transporting electrical supplies. The arrangement of the shelves in the vehicle is the same as in the supply section vehicle I (Fz.304).

In boxes the following stock items were stowed
- 27 boxes code no. 21 with 2 doppler each
- 6 boxes code no. 25 with 9 HF receiver parts each.

The spare parts truck II has the following dimensions and weights:

Total length:	6,130 mm
Total width:	2,100 mm
Total height:	2,940 mm
Wheelbase	3,450 mm
Track width:	1,630/1,642 mm front/rear
Tare weight:	3,275 kg
Payload:	2,825 kg
Maximum weight:	6,100 kg
Service weight:	3,640 kg* weighted
Wheels:	190-20

The main parts of the stock items truck II are:
- Chassis
- Closed superstructure (305) LC box body
- Installations-storage boxes

Chassis
The chassis used is that of the 3 t Opel 6700 Type A truck (equipment description according to D 666/5) but the normal chassis of the Opel 3 t Type 3.6/36 S truck (equipment description according to D 669/19B) was also used.

DMA, GD639.5.21
OKH/Wa Prüf10
Karlshagen, 1.4.1944
Bd.Nr. 789/44 gK
Auszug
Besprechung der
Fahrzeuggruppe 5
vom 1.4.44
in Karlshagen I

Production status: Vehicle 503
Total requirement 7 vehicles, of which
2 samples at Kdo. Basse.
2 from LR (finished) on the way to Kdo. Basse,
3 in LR (not started).
Requested accordingly to 501, 5 pieces by 15.4.44.
Of these, 4 will be available. As in the case of vehicle 502, a decision will be made on the 3 other vehicles after consultation with the special committee for vehicle installations, Mr. Horn.

* Service weight without 24 boxes with contents

Fahrzeug 503

Interne oder frühere Bezeichnung: Fz.-Pos.-Nr.503
Vorratsteilwagen II auf Lkw 3 t geschl. (305)
Elektrische Ersatzteile

Verwendungszweck
Der Vorratsteilwagen II ist für den Transport der elektrischen Vorratsteile vorgesehen. Die Anordnung der Regale im Fahrzeug ist die gleiche wie im Vorratswagen I (Fz.304).

In Kisten wurden die folgenden Lagerartikel verstaut
- 27 Stück Kisten Kenn-Nr. 21 mit je 2 Verdoppler
- 6 Stück Kisten Kenn-Nr. 25 mit je 9 HF-Empfängerteilen.

Der Vorratsteilwagen II hat folgende Maße und Gewichte:

Gesamtlänge:	6.130 mm
Gesamtbreite:	2.100 mm
Gesamthöhe:	2.940 mm
Radstand	3.450 mm
Spurweite:	1.630/1.642 mm vorne/hinten
Leergewicht:	3.275 kg
Nutzlast:	2.825 kg
Zul.Höchstgew.	6.100 kg
Gefechtsgewicht:	3.640 kg* gewogen
Räder:	190-20

Die Hauptteile des Vorratsteilwagen II sind:
- Fahrgestell
- Geschlossener Aufbau (305) LC-Koffer
- Einbauten-Vorratskisten

Fahrgestell
Als Fahrgestell dient das Fahrgestell des Lastkraftwagen 3 t Opel 6700 Typ A (Gerätebeschreibung nach D 666/5), aber auch das normale Fahrgestell des Opel 3 t Lastkraftwagens Typ 3,6/36 S (Gerätebeschreibung nach D 669/19B) wurde verbaut.

Fertigungslage: Fahrzeug 503
Gesamtbedarf 7 Fahrzeug, davon
2 als Muster bei Kdo. Basse.
2 von LR (fertiggestellt) unterwegs zu Kdo. Basse,
3 in LR (nicht in Arbeit).
Gefordert sinngemäß wie bei 501, 5 Stück bis 15.4.44.
Davon werden verfügbar 4. Über die 3 weiteren Fahrzeuge wird wie bei dem Fahrzeug 502 nach Einschaltung des Sonderausschusses für Fahrzeugeinbauten, Herrn Horn, entschieden.

* Gefechtsgewicht ohne 24 Kisten mit Inhalt

BArch, RH8II Bild-B1160-44, o. Ang.

251 Externally, the spare parts truck II was indistinguishable from the Kfz.305. As basic chassis the 3 t Opel 6700 type A truck was choosen. No changes were made to the closed superstructure itself, and no fastenings were made through the outer walls. The storage racks in the interior were fastened with screws to the side wall reinforcements. On this vehicle initial material-saving measures by omission of the mudguards of the superstructure can be seen. It is camouflaged with an almost complete camouflage paint of olive green and brown spots.

251 Der Vorratsteilwagen II war äußerlich vom Kfz.305 nicht zu unterscheiden. Als Basisfahrgestell wurde hier der Lastkraftwagen 3 t Opel 6700 Typ A gewählt. Am geschlossene Aufbau selbst wurden keinerlei Änderungen durchgeführt oder Befestigungen durch die Außenwände geführt. Die Lagergestelle im Innenraum waren mit Schrauben an den Seitenwandverstärkungen befestigt. Erste Materialeinsparmaßnahmen durch Fortfall der Kotflügel am Aufbau sind auf dem abgebildeten Fahrzeug festzustellen. Es ist mit einem fast deckenden Tarnanstrich aus olivgrün und braunen Flecken getarnt.

DMA, GD639.5.18
HAP11/T 507
Karlshagen, 1.6.1944
Bd.Nr. E 3195/44 g
Interne oder frühere
Bezeichnung

DMA, GD639.5.1
EW234
Fz.-Pos.-Nr.504

Vehicle 504

Internal or former designation: Fz.-Pos.-Nr.504
Heavy machinery set I (acc. attachment N 3670)
Machinery set

Intended use
The heavy machinery set I is used to generate electrical energy for the vehicles of the technical battery and also for the combustion cut-off and guide beam position. It contains a petrol engine coupled to a generator providing 230/400 V, 50 Hz 15 kVA. The machine set is mounted on a sledge and can be loaded. The machine set is transported on a trailer chassis (1-axle) 1500 kg. Only a few vehicles were delivered to units before the change to the standard Heer Sd.Ah.24 generator trailer.

The machine set I has the following dimensions and weights:

Total length:	3,350 mm
Total width:	1,750 mm
Total height:	1,900 mm
Distance:	2,225 mm axle centre to eyelet centre
Track width:	1,573 mm
Tare weight:	630 kg
Payload:	1,500 kg
Maximum weight:	2,130 kg
Combat weight:	n.a.
Wheels:	190-20
Engine set:	Phänomen Granit type petrol engine 27-stat. three-phase generator 15 kVA 230/400 Volt, made by Still

The main parts of the heavy machinery set I are:
- Chassis-Ah Fgst (1 axle.)
- Superstructure with electric machinery set I (EMS 1)

Chassis
The trailer chassis Ah Fgst (1-axle) 1500 kg type 41 (equipment description according to D 691/41) serves as chassis.

DMA, GD639.5.21
OKH/Wa Prüf10
Karlshagen, 1.4.1944
Bd.Nr. 789/44 gK
Auszug
Besprechung der
Fahrzeuggruppe 5
vom 1.4.44
in Karlshagen I

Production status: Vehicle 504
Initial requirement 7.
Available from Kdo. Basse 7 vehicles.

Fahrzeug 504

Interne oder frühere Bezeichnung: Fz.-Pos.-Nr.504
S. Maschinensatz I (nach Anlage N 3670)
Maschinensatz

Verwendungszweck
Mit dem schweren Maschinensatz I wird die Stromenergie für die Fahrzeuge der technischen Batterie aber auch die der BS/LS-Stellung erzeugt. Er enthält einen Otto-Motor, der mit einem Generator für 230/400 V, 50 Hz 15 kVA gekuppelt ist. Der Maschinensatz ist auf einen Schlitten aufgebaut und verlastbar. Transportiert wird der Maschinensatz auf einem Anhängerfahrgestell (1-achs.) 1500 kg. Nur wenige Fahrzeuge wurden an die Abteilungen ausgeliefert, bevor auf den heeresüblichen Generator-Anhänger Sd.Ah.24 umgestellt wurde.

Der s. Maschinensatz I hat folgende Maße und Gewichte:

Gesamtlänge:	3.350 mm
Gesamtbreite:	1.750 mm
Gesamthöhe:	1.900 mm
Abstand:	2.225 mm Achsmitte bis Ösenmitte
Spurweite:	1.573 mm
Leergewicht:	630 kg
Nutzlast:	1.500 kg
Zul.Höchstgew.	2.130 kg
Gefechtsgewicht:	o.A.
Räder:	190-20
Maschinensatz:	Benzin-Motor Phänomen Typ Granit 27-stat. Drehstromgenerator 15 kVA 230/400 Volt, Fabrikat Still

Die Hauptteile des schw. Maschinensatz I sind:
- Fahrgestell-Ah Fgst (1-achs.)
- Aufbau mit Elektro-Maschinensatz I (EMS 1)

Fahrgestell
Als Fahrgestell dient das Anhängerfahrgestell Ah Fgst (1-achs.) 1500 kg Baumuster 41 (Gerätebeschreibung nach D 691/41).

Fertigungslage: Fahrzeug 504
Gesamtbedarf 7 Fahrzeuge.
Verfügbar bei Kdo. Basse 7 Fahrzeuge.

Trailer chassis

The frame was made of two NP8 U-shaped longitudinal members and had in its front part the drawbar with overrun brake. The frame was stiffened with inserted cross braces. Spring assemblies, axle and wheels were taken from the E3 unit trailer. The wheels and tires were also used on the 3 t truck. Above the drawbar was a support frame with a toolbox mounted on it, and on the left and right of this were holders for storing one Wehrmacht 20 l fuel canister each.

Anhängerfahrgestell

Der Rahmen war aus zwei U-Längsträgern NP8 gabelförmig hergestellt, mit einsetzbaren Querstreben versteift und bildete im vorderen Teil die Zuggabel mit Auflaufbremse. Federpakete, Achse und Räder wurden vom Einheitsanhänger E3 übernommen. Räder und Bereifung wurden auch an dem Lkw 3 t verwendet. Über der Zuggabel wurde ein Haltegerüst mit Werkzeugkasten montiert. An diesem Gerüst befanden sich links und rechts Halter zur Aufnahme von je einem 20 l-Wehrmachtsbenzinkanister.

252

252 The machinery set could be fixed by supports attached to the frame. The rear supports were ajustable by spindels to the ground.

252 Der Maschinensatz konnte durch am Rahmen angebrachte Stützen fixiert werden. Zum Bodenausgleich waren die hinteren Stützen verstellbar ausgeführt.

Deutsches Museum, München, Archiv- FA014/B921/44 TD5224 [CD74720]

253

Machinery set and superstructure

The electric machinery set consisted of the Still-type three-phase generator 15 kVA 230/400 Volt for power generation and a Phänomen Granit-type gasoline engine 27 V stationary. The machinery set was mounted in a frame work on a tub with rollers. The superstructure, a sheet metal housing was bolted to the hull. Access was ensured by large hatches on the left and right, and an air filter was attached only on the left between them. A switch panel was installed in the front, with the power sockets mounted opposite it. The gasoline engine was accessible by the rear flaps. A double muffler was mounted on the roof in case of stationary use in a tent or hall, the exhaust gases could be discharged by extension tubes. A guide way was mounted on the trailer frame. The superstructure could be pushed off the trailer by the tray with rollers. To secure the load there were pins at the front and claws at the rear.

Maschinensatz und Aufbau

Der Elektro-Maschinensatz EMS1 bestand zur Stromerzeugung aus dem Drehstromgenerator 15 kVA 230/400 Volt, Fabrikat Stil, und einem stationären Benzin-Motor Typ Phänomen Granit 27 V. Der Maschinensatz war in ein Gerüst montiert und auf eine Wanne mit Rollen gesetzt. Der Aufbau, ein Blechgehäuse, wurde mit der Wanne verschraubt. Um den Zugang zu ermöglichen befanden sich auf beiden Seiten zwei große Klappen. Dazwischen befand sich auf der linken Seite ein Luftfilter. Im vorderen Bereich war eine Schalttafel angeordnet. Dem gegenüber waren die Kraftsteckdosen montiert. Der Benzinmotor war über die beiden hinteren Klappen zugänglich. Auf dem Dach war ein Doppelschalldämpfer angebracht. Bei stationärer Verwendung in einem Zelt oder einer Halle konnten die Abgase über Verlängerungsrohre abgeleitet werden. Auf dem Anhängerrahmen war auf der linken und rechten Seite eine Führungsbahn montiert. Der Aufbau konnte über die Wanne mit Rollen vom Anhänger geschoben werden. Zur Ladungssicherung waren an den Führungsbahnen vorne Absteckstifte und hinten Klauen vorhanden.

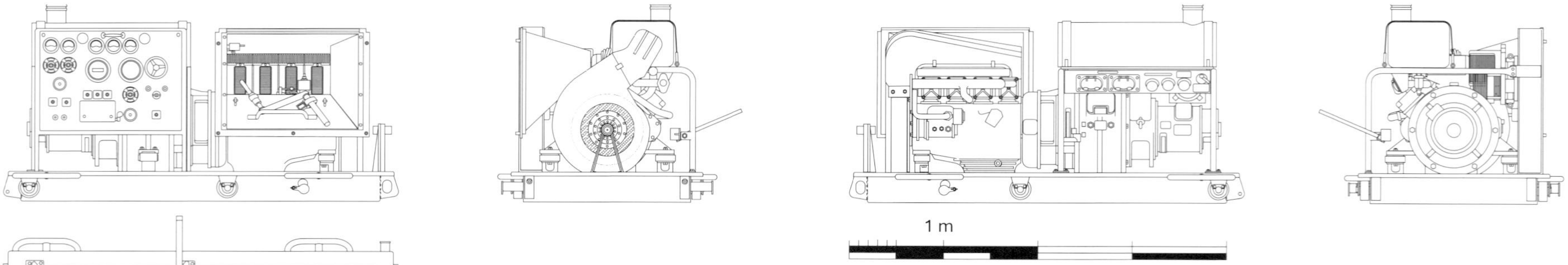

Scale / Maßstab 1:35 © Volker Ruff

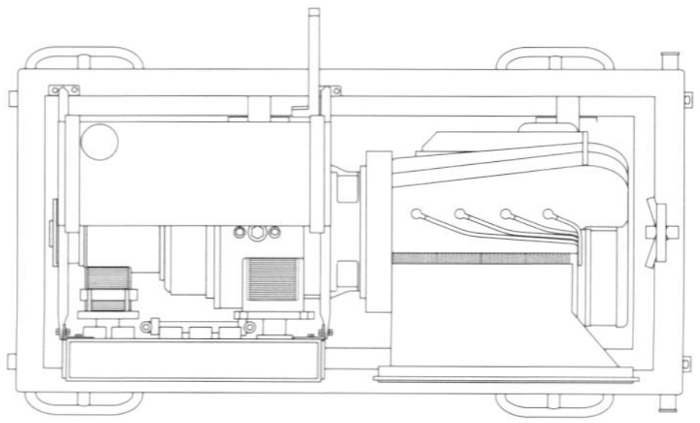

253 The machinery set shown is set up ready for operation. A grounding screw is visible in the foreground. The grounding screw was stored in a holder on the roof during the march.

253 Der abgebildete Maschinensatz ist betriebsbereit aufgestellt. Die im Vordergrund erkennbare Erdungsschraube ist zur Masseableitung in den Boden eingedreht. Diese war auf dem Marsch auf einer Halterung auf dem Dach gelagert.

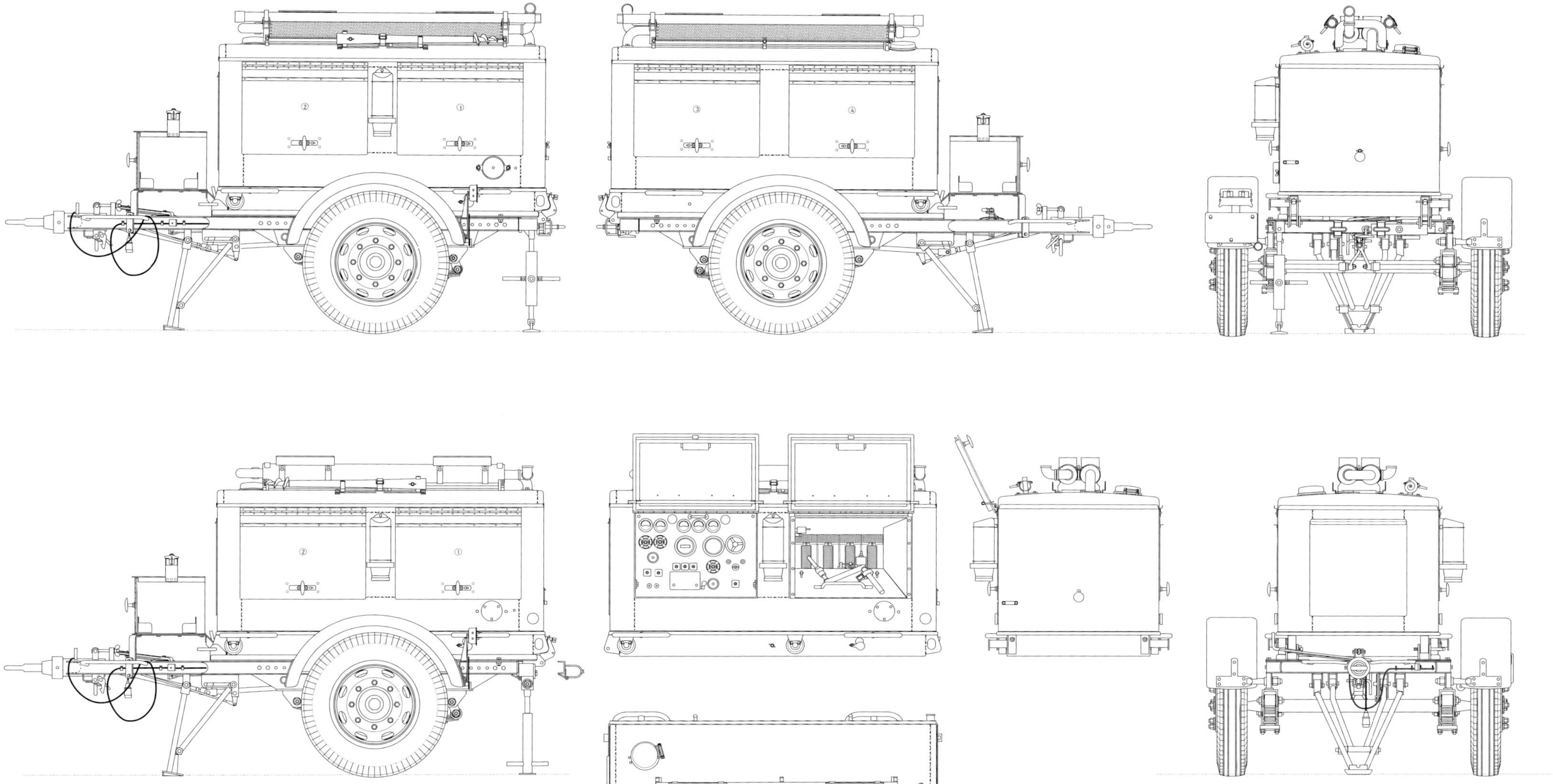

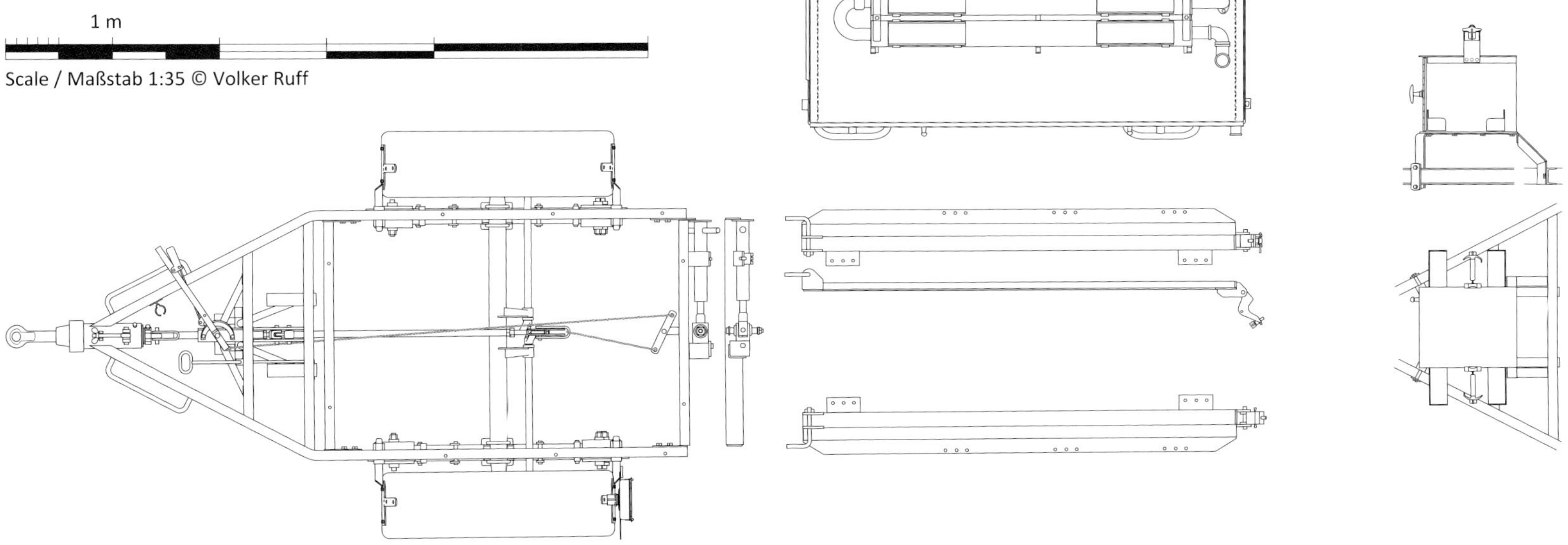

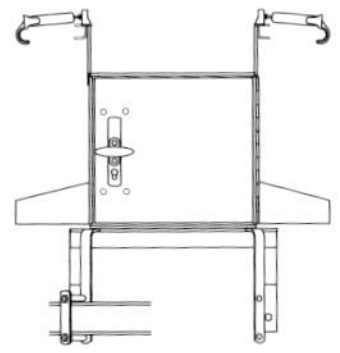

Fahrzeug 504

s. Maschinensatz I (nach Anlage N3670)

Partially used drawings acc. to D641/41 original vehicle

Teilweise verwendete Zeichnungen D641/41 Originalfahrzeug

Vehicle 505

Internal or former designation: Fz.-Pos.-Nr.505
Converter trailer 220/380 V to 27 V
Converter

DMA, GD639.5.18
HAP11/T 507
Karlshagen, 1. Juni 1944
Bd.Nr. E 3195/44 g

DMA, GD639.5.1
EW234
Fz.-Pos.-Nr.505

Intended use

For the operation of Test Vehicle II (Fz.501) a test voltage 27 V is required at the test console and the devices in the B-rack.

The drive motor on the converter trailer is powered by the s. Machine set (Fz.504) with a current at 220/380 V. The working current is regulated to a voltage of 27 V by a coupled direct current generator. The converter is housed in a carrying frame and can be loaded. A »Light vehicle trailer, single-axle, dismountable« A1 is used for transport. A base plate with slide-in frame is mounted on the trailer. The support frame with converter is pushed into the frame and secured with pins. A cover protects the converter from dust and moisture.

The converter trailer has the following dimensions and weights:

Total length:	2,900 mm without hitch
Total width:	1,650 mm
Total height:	1,230 mm
Distance:	2,300 mm from axle centre to eyelet centre
Track width:	1,450 mm
Tare weight:	140 kg
Payload:	525 kg
Maximum weight:	750 kg
Service weight:	530 kg
Wheels:	5.50-16

The main parts of the converter trailer are:

- Chassis LW trailer A1 c with cover
- Support frame – generator and direct current generator

The chassis

The trailer serves as a chassis
»Light vehicle trailer, single-axle, demountable« A1 (D. (Luft) T. 8204)

DMA, GD639.5.21
OKH/Wa Prüf10
Karlshagen, 1.4.1944
Bd.Nr. 789/44 gK
Auszug
Besprechung der
Fahrzeuggruppe 5
vom 1.4.44
in Karlshagen I

Production satus: Vehicle 505
Initial requirement 7
Available from Kdo. Basse 7 vehicles

Fahrzeug 505

Interne oder frühere Bezeichnung: Fz.-Pos.-Nr.505
Umformer-Anhänger 220/380 V auf 27 V
Umformer

Verwendungszweck

Für den Betrieb des Prüfwägens II (Fz.501) ist eine Prüfspannung von 27 V am Prüfpult und den Geräten im B-Gestell notwendig.

Der Antriebsmotor am Umformer-Anhänger wird vom s. Maschinensatz (Fz.504) mit Kraftstrom 220/380 V gespeist. Der Arbeitsstrom wird von einem angekoppelten Gleichstromerzeuger auf eine Spannung von 27 V geregelt. Der Umformer ist in einem Tragegerüst untergebracht und verlastbar. Zum Transport wird ein »Leichter Kfz.-Anhänger, einachsig, zerlegbar« A1 verwendet. Auf dem Kfz.-Anhänger ist eine Bodenplatte mit Einschieberahmen montiert. Das Tragegerüst mit Umformer wird in den Rahmen geschoben und mit Absteckstiften gesichert. Eine Abdeckhaube schützt den Umformer vor Staub und Nässe.

Der Umformer-Anhänger hat folgende Maße und Gewichte:

Gesamtlänge:	2.900 mm ohne Anhängerstück
Gesamtbreite:	1.650 mm
Gesamthöhe:	1.230 mm
Abstand:	2.300 mm von Achsmitte bis Ösenmitte
Spurweite:	1.450 mm
Leergewicht:	140 kg
Nutzlast:	525 kg
Zul. Höchstgew.	750 kg
Gefechtsgewicht:	530 kg gewogen
Räder:	5,50-16

Die Hauptteile des Umformer-Anhänger sind:

- Fahrgestell LW-Anhänger A1 c mit Abdeckhaube
- Traggestell – Generator und Gleichstromerzeuger

Das Fahrgestell

Als Fahrgestell dient der Anhänger
»Leichter Kfz.-Anhänger, einachsig, zerlegbar« A1 (D. (Luft) T. 8204)

Fertigungslage: Fahrzeug 505
Gesamtbedarf 7 Fahrzeuge
Verfügbar bei Kdo. Basse 7 Fahrzeuge

Chassis LW trailer A1 with cover

The trailer was developed as a »light dismountable motor vehicle trailer A1« for the Luftwaffe and served as a standard chassis for all independent light ground equipment which, due to their weight, could no longer be man-portable but had to be transported quickly. The LW Trailer A1 was chosen as the chassis for the converter. A base plate, with guide rails and four locking hooks, was bolted to the frame. A simple sheet metal cover was placed over the converter with four handles and secured to the base plate with locking hooks. Once set up the trailer was secured with simple floor supports, on the drawbar and on the rear frame section.

Fahrgestell LW-Anhänger A1 mit Abdeckhaube

Der Anhänger wurde als »leichter zerlegbarer Kfz. Anhänger A1« für die Luftwaffe entwickelt und diente als einheitliches Fahrgestell für alle selbständigen leichten Bodengeräte, die wegen ihres Gewichtes nicht mehr tragbar ausgeführt werden konnten und schnell befördert werden mussten. Der LW-Anhänger A1 wurde als Fahrgestell für den Umformer ausgewählt. Auf den Rahmen war eine Bodenplatte mit Führungsschienen und vier Verschlusshaken verschraubt. Eine einfache Abdeckhaube aus Blech wurde mit vier Haltegriffe über den Umformer gestülpt und mit den Verschlusshaken an der Bodenplatte gesichert. Einmal aufgestellt, wurde der Anhänger mit einfachen Bodenstützen an der Zugdeichsel und am hinteren Rahmenteil festgestellt.

Deutsches Museum, München, Archiv- FA014/B317/44TD5224 [CD74737]

254

Deutsches Museum, München, Archiv- FA014/B318/44 TD5224 [CD74738]

255

Deutsches Museum, München, Archiv- FA014/B319/44 TD5224 [CD74741]

256

Carrying frame, drive motor and direct current generator

The cover is removed. The converter unit is mounted on a base plate and skids and surrounded by a carrying frame. The carrying frame can be removed. The three-phase current supply line is mounted on the switch box in the direction of travel.

On the back of the switch box are a socket for the measuring line and a socket for direct current (27 V). Controls are provided on the side of the switch box for the ammeter, voltmeter, regulator, fuses, main switch and manual regulator.

Traggestell, Antriebsmotor und Gleichstrom-Erzeuger

Die Abdeckhaube ist entfernt. Die Umformereinheit ist auf einer Bodenplatte mit Kufen montiert und von einem Tragegestell umgeben. Der Tragerahmen kann entfernt werden. Am Schaltkasten ist in Fahrtrichtung die Drehstromzuleitung montiert.

Auf der Rückseite des Schaltkastens ist eine Steckdose für die Messleitung und eine Steckdose zur Gleichstromentnahme (27 V) vorhanden. Bedienelemente sind seitlich am Schaltkasten für Strom- und Spannungsmesser, Regler, Voltmeter, Sicherungen, Hauptschalter und Handregler vorhanden.

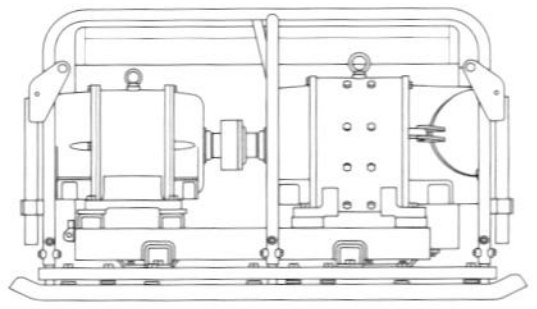

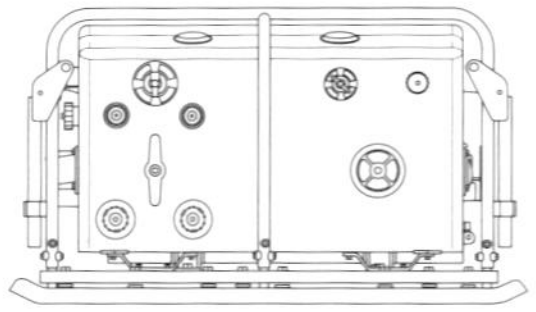

255, 253 The converter unit was supplied with 220/380 V by the Fz.504 machine set Fz.504, and the current was regulated to 27 V and delivered to the test vehicle II Fz.501. If necessary the converter unit could be removed from the trailer with the handles on its carrying frame and used as a stationary unit.

255, 256 Die Umformereinheit wurde vom s. Maschinensatz Fz.504 mit 220/380 V gespeist, der Strom auf 27 V geregelt und an den Prüfwagen II Fz.501 abgegeben. Im Bedarfsfall war es möglich, die Umformereinheit mit den Handgriffen am Tragegestell vom Anhänger zu nehmen und stationär zu verwenden.

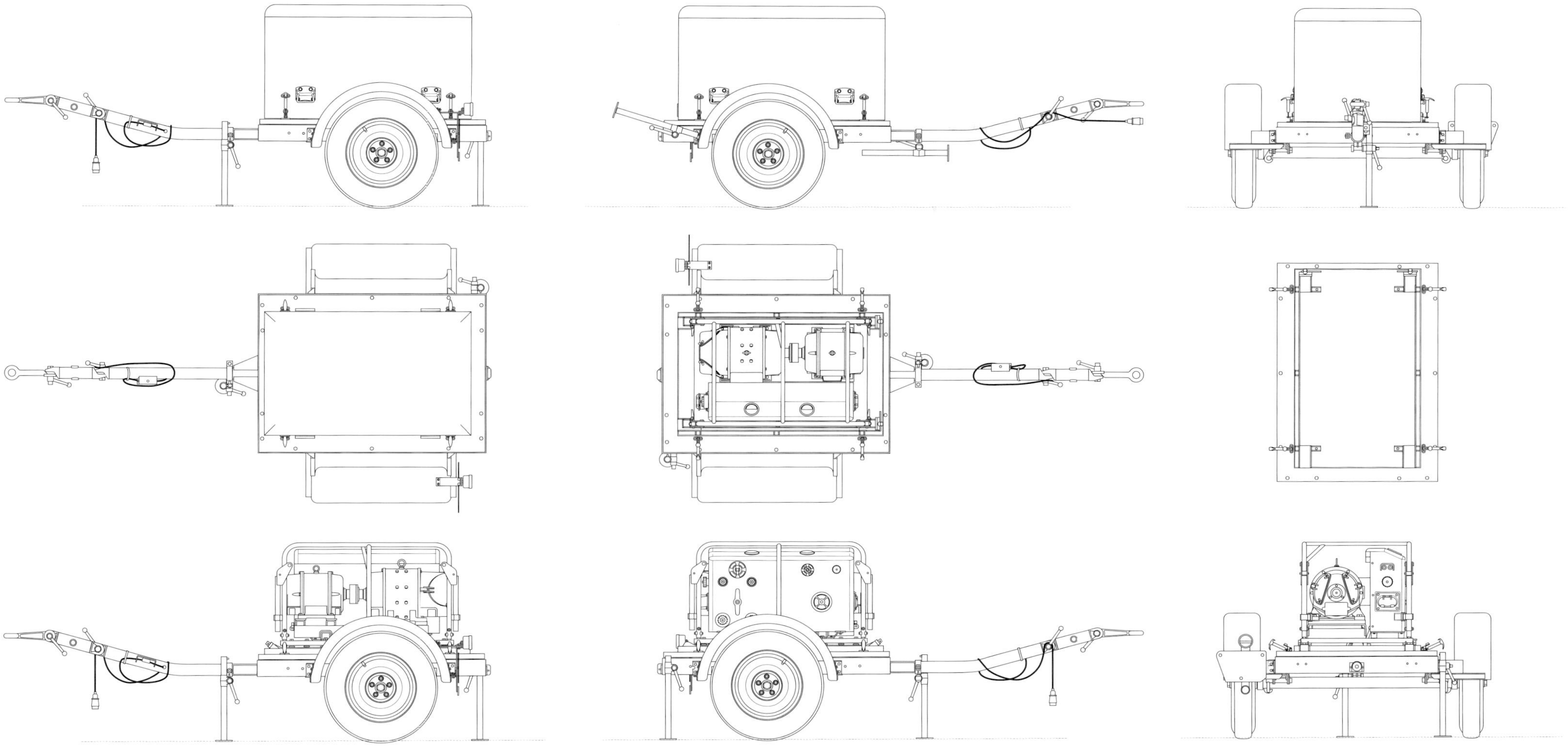

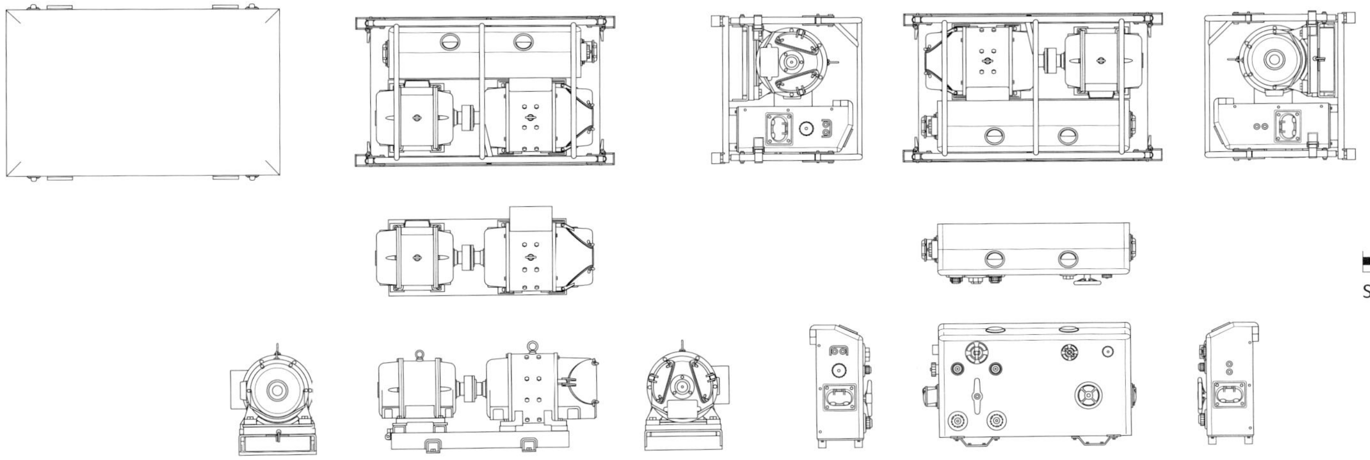

1 m

Scale / Maßstab 1:35 © Volker Ruff

Fahrzeug 505

Umformer-Anhanger 220/380 V
auf 27 V

Drawing set used n.a.
Verwendeter Zeichnungssatz k.A.

Vehicle 506

Internal or former designation: Fz.-Pos.-Nr.506
Compressed air generator trailer (2-axle)
Air compressor

DMA, GD639.5.18
HAP11/T 507
Karlshagen, 1.6.1944
Bd.Nr. E 3195/44 g

DMA, GD639.5.1
EW234
Fz.-Pos.-Nr.506

DMA, GD630.02
A4 Lehrbuch
der FR-Schule
Mai 1944
Auszug aus
technische Fahrzeuge

Intended use

A compressed air generator is carried on a two-axle trailer. The pressure cylinders required for its operation are also accommodated on the trailer. A multi-stage Junkers free piston compressor (diesel torpedo air pump) is used to generate the compressed air of 230 atü. The compressed air is alternately led through two gel cartridges for drying. While one cartridge is in operation, the other is regenerated by the exhaust gases. The compressed air generated by the air compressor is used in the technical battery to check the functionality of various valves on the horizontal Geraet, but also to fill the P-Stoff bottle battery on the Meillerwagen, which fills the compressed air supply in the *Gerät*.

The air compressor has the following dimensions and weights:

Total length:	5,910 mm
Total width:	2,210 mm
Total height:	2,950 mm
Wheel base :	2,650 mm (extended wheel base)
Track width:	1,573 mm
Tare weight:	2,000 kg
Payload:	2,580 kg
Maximum weight:	4,580 kg
Service weight:	4,265 kg
Wheels:	7,25-20
Unit:	High pressure air compressor unit 4FK115/230 Junkers Motoren Werke Munich-Allach / Deutsche Kühl und Kraftmaschinen Company Scharffenstein

The main parts of the air compressor are:

- Chassis
- Body
- Diesel torpedo air pump, air storage battery

Chassis

Essential components of the standard trailer 3 t, type E3 (type 02) (equipment description according to D 692/21) form the basis of the adapted chassis. Plans for a new production run foreseen the unchanged use of the Standard trailer 3 t, type E3. A draft drawing was made in 1945.

Fahrzeug 506

Interne oder frühere Bezeichnung: Fz.-Pos.-Nr.506
Luftverdichter-Anhänger (2-achs.)
Kompressor

Verwendungszweck

Auf einem zweiachsigen Anhänger wird ein Presslufterzeuger mitgeführt. Die zu seinem Betrieb erforderlichen Druckflaschen sind ebenfalls auf dem Anhänger untergebracht. Zur Erzeugung der Pressluft von 230 atü wird ein mehrstufiger Junkers-Freikolben-Verdichter (Diesel-Torpedoluftpumpe) verwendet. Die Pressluft wird zur Trocknung wechselweise durch zwei Gel-Patronen geleitet. Während die eine Patrone in Betrieb ist, wird die andere durch die Abgase regeneriert. Mit der erzeugten Pressluft des Luftverdichters werden in der technischen Batterie die Funktionsfähigkeit verschiedenster Ventile am liegenden Gerät überprüft, aber auch die P-Stoff Flaschenbatterie am Meillerwagen befüllt, die den Pressluftvorrat im Gerät befüllen.

Der Luftverdichter hat folgende Maße und Gewichte:

Gesamtlänge:	5.910 mm
Gesamtbreite:	2.210 mm
Gesamthöhe:	2.950 mm
Radstand :	2.650 mm (verlängerter Radstand)
Spurweite:	1.573 mm
Leergewicht:	2.000 kg
Nutzlast:	2.580 kg
Zul. Höchstgew.	4.580 kg
Gefechtsgewicht:	4.265 kg gewogen
Räder:	7,25-20
Gerät:	Hochdruck Luftverdichteranlage 4FK115/230 Junkers Motoren Werke München-Allach / Deutsche Kühl und Kraftmaschinen Gesellschaft Scharffenstein

Die Hauptteile des Luftverdichters sind:

- Fahrgestell
- Aufbau
- Diesel-Torpedoluftpumpe, Luftspeicherbatterie

Fahrgestell

Wesentliche Bauteile des Einheitsanhängers 3 t, Baumuster E3 (Typ 02) (Gerätebeschreibung nach D 692/21) dienen als Basis für das angepasste Fahrgestell. Planungen für eine Neufertigung sahen die unveränderte Verwendung des Einheitsanhängers 3 t, Baumuster E3, vor. Eine Entwurfszeichnung wurde noch in 1945 erstellt.

Deutsches Museum, München, Archiv- FA014/B976/44 T5224 [CD79114]

257

257 The chassis for the first air compressors was manufactured using components from the E3 standard trailer. In the front wall was a cut out, behind it was located the radiator for the diesel unit of the air compressor system. In the background, the firing position is being set up. On the left behind the trailer is an erected A4 with a Meiller trailer and on the right the rear of a Hanomag tractor is recognizable.

257 Das Fahrgestell für die ersten Luftverdichter wurde mit Bauteilen des Einheitsanhängers E3 gefertigt. In der vorderen Wand wurde ein Ausschnitt eingebracht, dahinter befand sich der Radiator für das Dieselaggregat der Luftverdichter-Anlage. Im Hintergrund wird gerade die Feuerstellung aufgebaut. Links hinter dem Anhänger ist eine aufgerichtete A4 mit Meillerwagen und rechts das Heck eines Hanomag-Schleppers erkennbar.

DMA, GD639.5.11
T66 (T404)
o.O., den 28.7.1944
Bd.Nr.00564/44 gK
Auszug
Teileliste Fz.506

Vehicles on order — Fahrzeugbestellungen

	Item Lfd. Nr.	Qty. ordered Best. Stückzahl	Qty. per vehicle Stück je Fahrzeug	Subject drawing number Gegenstand Zeichnungs-Nr.	Ordered by order number Besteller Bestell-Nr.	Supplier Lieferfirma
Fahrzeugbestellung	1	80	1	Sonder-Anhänger (2-achs.) mit geschlossenem Aufbau	WaPrüf11/III 361-0013/43 361-0047/43	Ackermann Wuppertal-Vohwinkel

Customizing/outfitting — Ausrüstung

	Item Lfd. Nr.	Qty. ordered Best. Stückzahl	Qty. per vehicle Stück je Fahrzeug	Subject drawing number Gegenstand Zeichnungs-Nr.	Ordered by order number Besteller Bestell-Nr.	Supplier Lieferfirma
Ausrüstungsfirma	1	26	1	Sonder-Anhänger als kompl. Luftverdichteranlage nach Zeichnung B 325	WaPrüf 10/H 0112-3330/43 WaPrüf 11/III SS 4949 0112-3842/43 WaPrüf 11/III SS 4949 0112-3843/43 WaPrüf 11/III SS 4949 0112-3844/44	Hugo Junkers GmbH Motorenbau München-Allach

DMA, GD639.5.21
OKH/Wa Prüf 10
Karlshagen, 1.4.1944
Bd.Nr. 789/44 gK
Auszug
Besprechung der
Fahrzeuggruppe 5
vom 1.4.44
in Karlshagen I

Production status: Vehicle 506
Total requirement 80 vehicles,
of which are delivered 13 vehicles,
of which available with Kdo. Basse 13.
Estimated demand 14, 2 each for 6 technical Batteries
and 2 for 1st Training Battery.
Still to be completed 1 at HAP 11.

Current defects and their remedy

The following changes and improvements compared to the previous compressor are included in the 14 new compressors, or will be until 15.4.44.

Supplements of all filters. Refueling-pumps are on the way for 10 sets. The heating devices for the filters can be supplied only after 15.4.44. The missing parts mentioned above do not affect the operation of the equipment. Operation is exactly the same as with the old system.

Fertigungslage: Fahrzeug 506
Gesamtbedarf: 80
Bisher ausgeliefert: 13 Stück
Davon: 13 Stück bei Kdo. Basse
Vorläufiger Bedarf 14, und zwar für 6 techn. Battr. je 2
und für 1. Ausb. Battr. 2 Stück
Noch zu ergänzen 1 Stück bei HAP 11

Derzeitige Mängel und ihre Behebung

Folgende Änderungen und Verbesserungen gegenüber dem früheren Kompressor sind bei den 14 neuen Kompressoren, bzw. werden bis 15.4.44 berücksichtigt.

Ergänzungen sämtlicher Filter. Für 10 Anlagen sind die Betankungspumpen unterwegs. Die Heizungsvorrichtungen für die Filter können erst nach dem 15.4.44 geliefert werden. Durch die vorgenannten fehlenden Teile wird der Betrieb der Anlagen nicht gestört. Die Aufrechterhaltung des Betriebes erfolgt dann genau wie bei der alten Anlage.

DM, 8001D
Tarnschlüssel für
Betriebsstoffe, mpe

P-Stoff = Stickstoff – nitrogen – blue – RAL 5009*
* Technical colour – Stealth term for operating materials

P-Stoff = Stickstoff – Stickstoff – blau – RAL 5009*
* Technische Farbe – Tarnbezeichnung für Betriebsstoffe

Body

The body itself consists of a framework made from profiles with corner reinforcements at the floor assembly and the arched roof. The front and rear walls are clad with profiled wood. A door on the back wall led to the interior. A diesel torpedo air pump 4 FK 11/230 atü was housed there.

Aufbau

Der Aufbau bestand aus einen Holzständerrahmen für die Wände und einem gewölbtem Dach. Die vordere und hintere Wand war mit Profilholz verkleidet die mit Blechwinkel und Eckverstärkungen an der Ladefläche verschraubt wurden. An der Rückwand führte eine Tür zum Innenbereich. Dort war eine Diesel-Torpedoluftpumpe 4 FK 11/230 atü untergebracht.

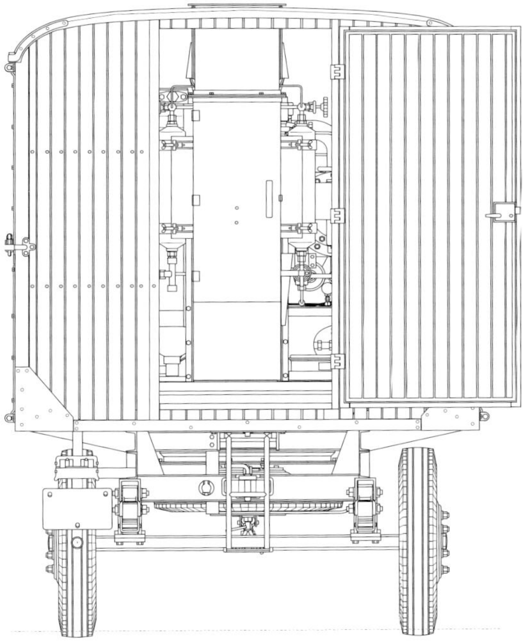

BArch, RH8II Bild-B1603-44, o.Ang

258

258 Through the rear access door you have a direct view of the ventilation shaft which leads to the roof and is closed with a flap. To the left and right of the ventilation shaft are the silica gel filters for drying the compressed air.

258 Durch die hintere Zugangstür blickt man direkt auf den Entlüftungsschacht, der zum Dach führt und mit einer Klappe verschlossen wird. Links und rechts vom Entlüftungsschacht befinden sich die Silica-Gel-Filter für die Trocknung der Pressluft.

Air storage battery

To operate the system, a permanently installed air storage battery with 6 compressed air cylinders of 55 l each is mounted.

Luftspeicherbatterie

Zum Betreiben der Anlage war eine festinstallierte Luftspeicherbatterie aus 6 Pressluftflaschen mit je 55 l montiert.

Deutsches Museum, München, Archiv- FA014/B975/44 T5224 [CD90455]

259

Deutsches Museum, München, Archiv- FA014/B974/44 TD5224 [CD90454]

260

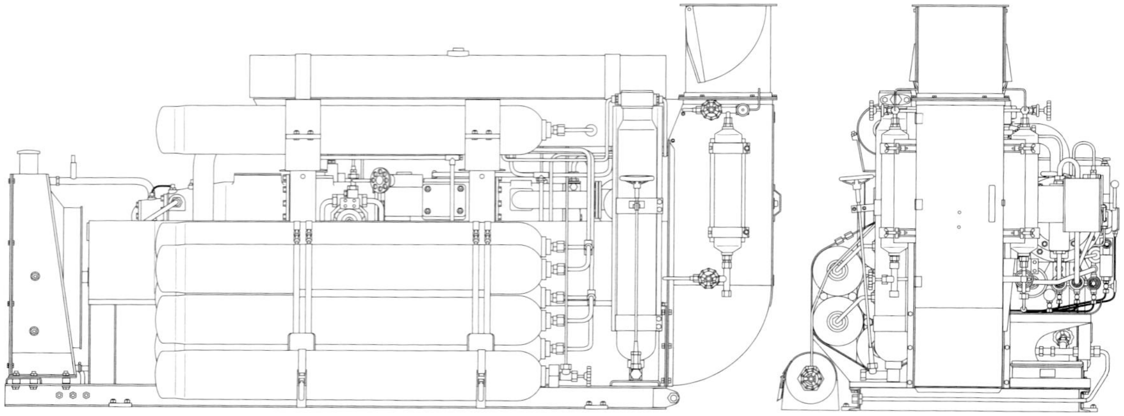

259, 260 The air storage battery with its 6 compressed air cylinders. A starting bottle is attached above. A standpipe and silica gel filter are mounted on the vent. Its exhaust hood is open on the roof. The side walls on the body are divided into an upper and lower side wall and can be opened. The upper side wall, folding upwards, is fixed when open with three locks on the wall brackets and the central support. The lower side wall folds down.

259, 260 Die Luftspeicherbatterie mit ihren 6 Pressluftflaschen. Darüber ist eine Anlassflasche angebracht. Ein Standrohr und Silica-Gel-Filter sind am Entlüftungsschacht montiert. Dessen Ablufthaube ist auf dem Dach geöffnet. Die Seitenwände am Aufbau waren in eine obere und untere Seiten wand geteilt und konnten geöffnet werden. Die obere Seitenwand wurde, wenn nach oben geöffnet, mit drei Feststellern an Haltern der Wände und zur Mittelstütze fixiert. Die untere Seitenwand wurde nach unten geklappt

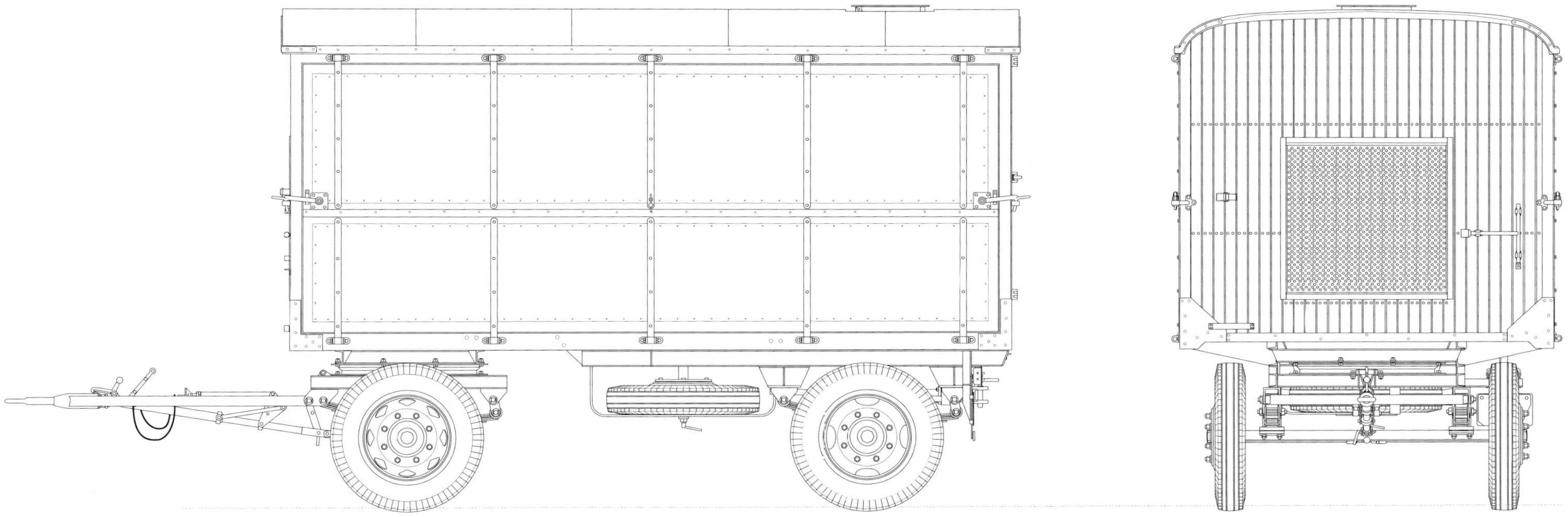

Scale / Maßstab 1:35 © Volker Ruff

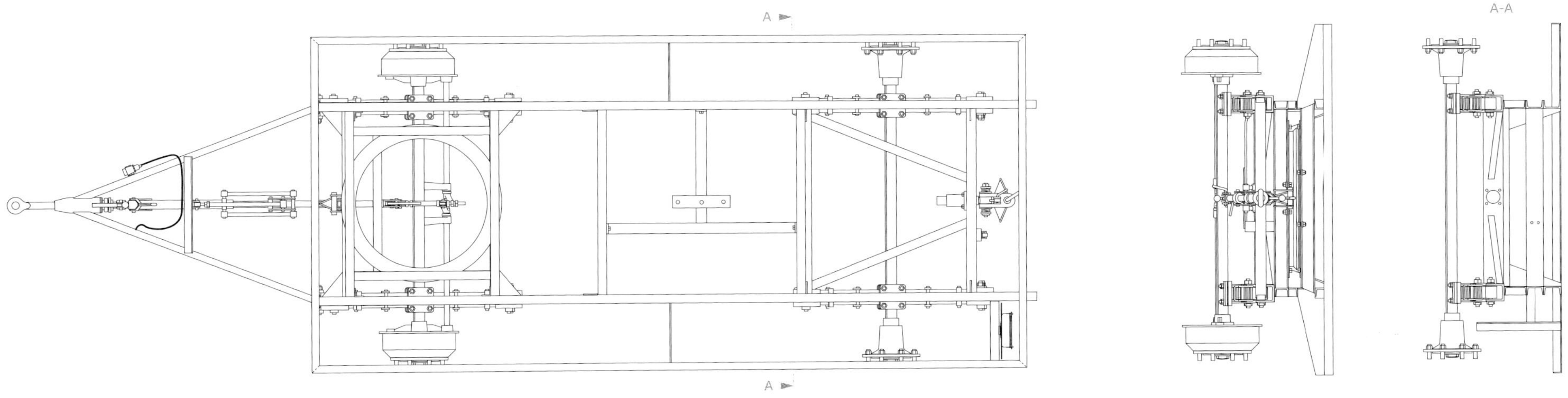

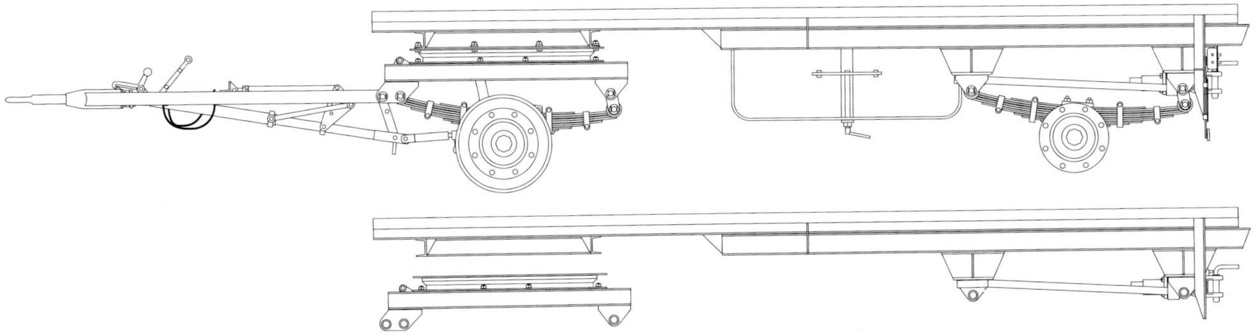

Fahrzeug 506

Luftverdichter-Anhänger (2-achs.)

Partially used drawings acc. to group list 21Gr161

Teilweise verwendete Zeichnungen nach Gruppenliste 21Gr161

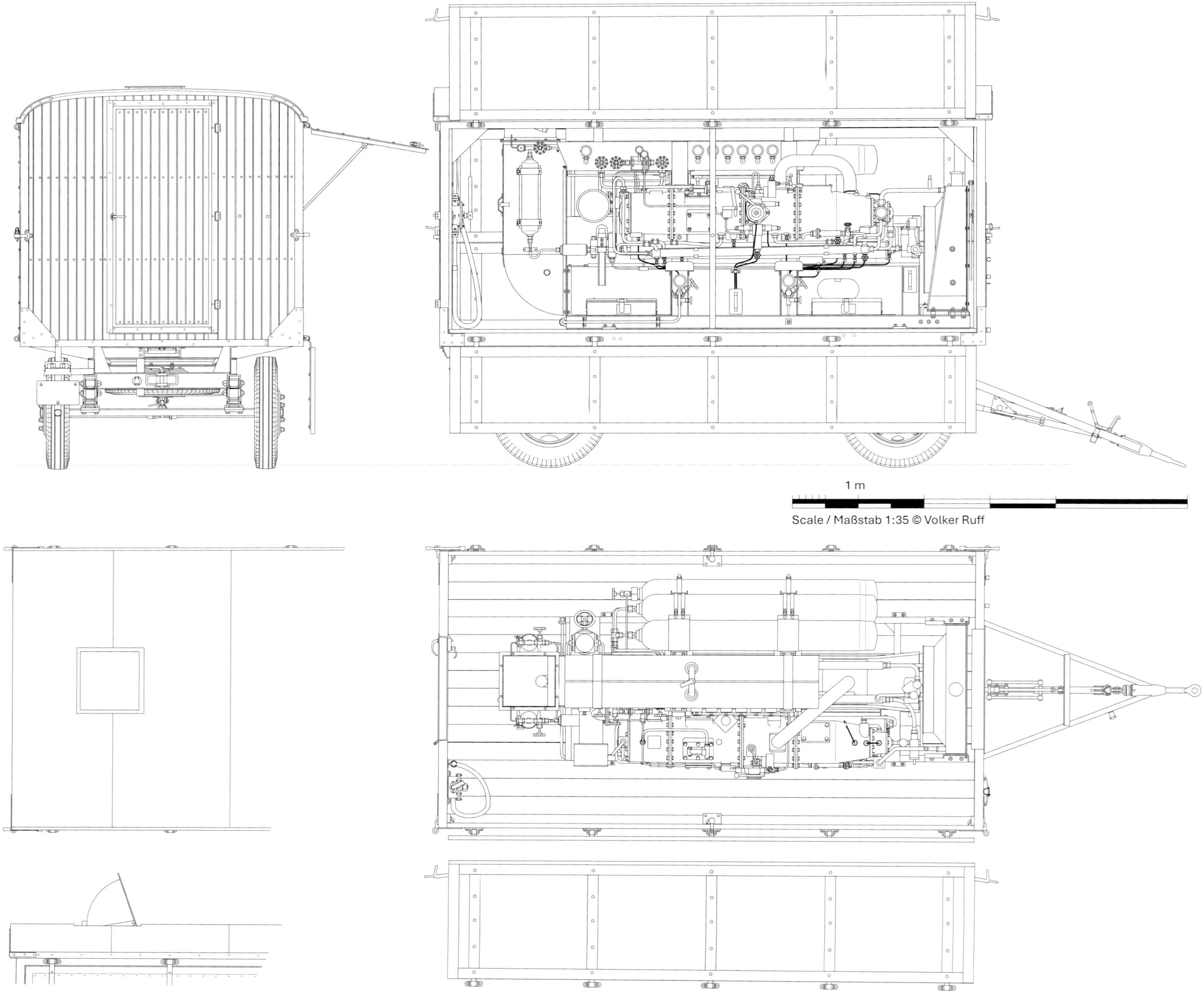

Scale / Maßstab 1:35 © Volker Ruff

Air compressor unit

The air compressor system with radiator housing, diesel torpedo air pump and ventilation shaft was mounted on a mounting frame and bolted to the floor assembly on the chassis frame. Above the torpedo air pump there are various pressure gauges, regulators and the fuel tank for the unit.

Luftverdichter-Anlage

Die Luftverdichter-Anlage war mit Radiatorgehäuse, Diesel-Torpedoluftpumpe und Entlüftungsschacht auf einem Montagerahmen befestigt und mit der Bodengruppe auf dem Fahrgestellrahmen verschraubt. Oberhalb der Torpedoluftpumpe befanden sich diverse Druckregler und der Kraftstofftank für das Aggregat.

Deutsches Museum, München, Archiv- FA014/B974/44T5224 [CD90454]

261

Deutsches Museum, München, Archiv- FA014/B975/44TD5224 [CD90455]

262

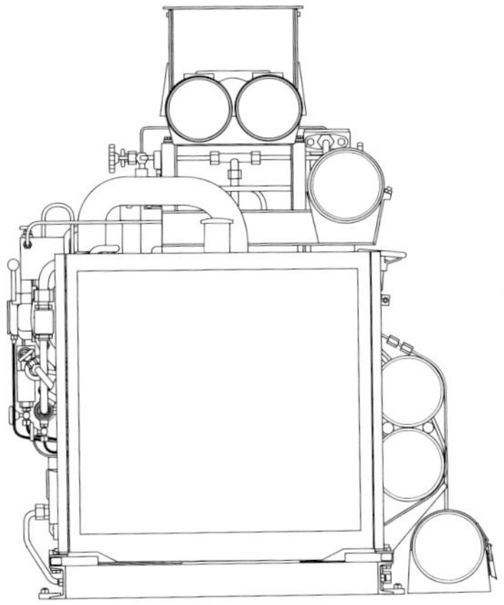

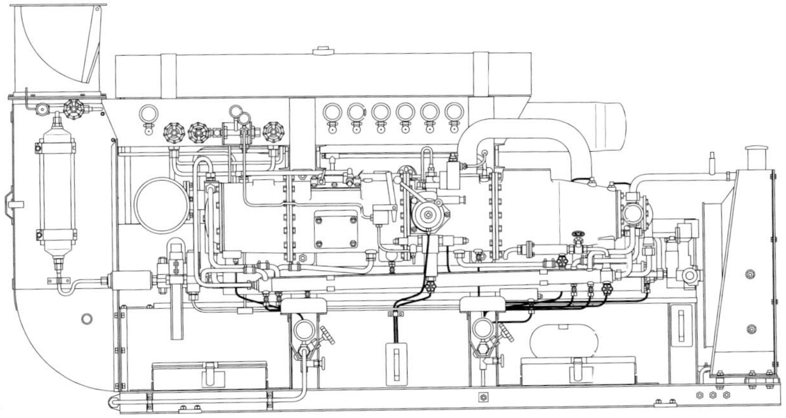

261, 262 The round air intake point of the air compressor can be seen in the middle left. The two compressed air extraction points above the mounting frame are attached in the middle to the left and right behind the support. A fuel hand pump for filling the diesel fuel tank is mounted on the inside of the rear wall.

261, 262 Von der Mitte links ist die runde Luftansaugstelle der Luftverdichteranlage erkennbar. Mittig sind links und rechts hinter der Stütze die beiden Pressluftent nahmestellen oberhalb dem Montagerahmen befestigt. Eine Handpumpe (Kraftstoff) zum Befüllen des Dieselkraftstofftanks ist an der Innenseite der Rückwand montiert.

DMA, GD639.5.18
HAP11/T 507
Karlshagen, 1.6.1944
Bd.Nr. E 3195/44 g

DMA, GD639.5.1
EW234
Fz.-Pos.-Nr. 508

DMA, GD630.02
A4 Lehrbuch
der FR-Schule
Mai 1944
Auszug aus
Technische Fahrzeuge
des A4-Gerätes

DMA, 639.5.21
OKH/Wa Prüf10
Karlshagen, 1.4.1944
Bd.Nr. 789/44 gK
Auszug
Besprechung der
Fahrzeuggruppe 5
vom 1.4.44
in Karlshagen I

Vehicle 508

Internal or former designation: Fz.-Pos.-Nr.508
Accumulator truck on 4,5 t truck closed (heated)
Accumulator truck

Intended use

The Accumulator-Kw is a heated vehicle. The 27-volt on-board battery and 50-volt nickel-cadmium battery are transported in it and kept at a temperature of 30 °C all the times. A charging system in the vehicle enables the batteries to be recharged continuously. The Z-Stoff canisters required for the shooting operation are also carried in the accumulator-KW.

The Accumulator-KW has the following dimensions and weights:

Total length:	7,570 mm
Width:	2,400 mm
Height:	3,005 mm
Wheelbase:	4,700 mm
Track width:	1,934/1,714 mm rear center distance between dual tires
Tare weight:	6,080 kg
Payload:	4,500 kg
Service weight:	9,250 kg
Wheels:	270-20

The Accumulator-KW (thermal truck) consists of the following main components:

- Chassis
- Closed standard superstructure with racks and oven with circulation heating

Chassis

The chassis is the chassis of the Büssing-NAG Typ 4500 A 4,5 t truck Type serie A 1 (see technical manual D 667/9)

Production status: Vehicle 508:

Estimated total requirement 14, as for 408. Available at Kdo. Basse 7 vehicles and a further 7 by 5.4.44 to be finished at LR (without Z-Stoff-cans).

Current defects and their remedy

TD531 maintains that installation of 6 extra boxes for 6 volt electric batteries in the accumulator vehicles is necessary. This action creates difficulties as it needs a fundamental change of all these vehicles. TD 5 to check the necessity by 5.4.44. Action by Prof. Wierer.

Fahrzeug 508

Interne oder frühere Bezeichnung: Fz.-Pos.-Nr.508
Sammler-Lkw auf Lkw 4,5 t geschl., (heizbar)
Sammler-Lkw

Verwendungszweck

Der Sammler-Kw ist ein beheizbares Fahrzeug. In ihm werden die 27 V Bordbatterie und 50 V Nickel-Kadmium-Batterie befördert und dauernd auf einer Temperatur von 30 °C gehalten. Eine im Fahrzeug befindliche Ladeanlage ermöglicht eine laufende Nachladung der Batterien. Gleichzeitig werden in den Sammler-KW die für den Schiessbetrieb erforderlichen Z-Stoff-Kanister mitgeführt.

Der Sammler-KW hat folgende Maße und Gewichte:

Gesamtlänge:	7.570 mm
Breite:	2.400 mm
Höhe:	3.005 mm
Radstand:	4.700 mm
Spurweite:	1.934/1.714 mm hinten Mittenabstand zwischen Doppelreifen
Leergewicht:	6.080 kg
Nutzlast:	4.500 kg
Gefechtsgewicht:	9.250 kg
Räder:	270-20

Der Sammler-Kw (Thermowagen) besteht aus folgenden Hauptbauteilen:

- Fahrgestell
- Geschlossener Einheitsaufbau mit Einbauten und Ofen mit Umlaufheizung

Fahrgestell

Als Fahrgestell dient das Fahrgestell des Büssing-NAG Typ 4500 A 4,5 t Lastkraftwagens Baureihe A 1 (Gerätebeschreibung siehe D 667/9).

Fertigungslage: Fahrzeug 508

Vorläufiger Gesamtbedarf 14 Stück, sinngemäß wie bei Fahrzeug 408. Verfügbar bei Kdo. Basse 7 Fahrzeuge. Weitere 7 Fahrzeuge bis 5.4.44 in LR fertig (ohne Z-Stoff-Kanister)

Derzeitige Mängel und ihre Behebung

TD531 hält den Einbau von weiteren 6 Kästen für 6 Volt Batterie im Sammler-Kw für erforderlich. Diese Maßnahme stößt insofern auf Schwierigkeiten als sie ein grundsätzlichen Umbau der Fahrzeuge nach sich zieht. TD 5 überprüft die Notwendigkeit bis zum 5.4.44. Durchführung Prof. Wierer.

BArch, RH8II Bild-B1439-44, o.Ang.

263

263 The accumulator truck (Fz.508) is identical in construction to the Fz.408. The 4.5 t Büssing pictured, already shows production simplifications such as the simple fender panels, shortened bumper, simple radiator grille panel, and side hood panels with two slots for ventilation. A large toolbox is mounted below the side door and a ventilation flap is fitted in the side wall. At the front corner of the box body is a stovepipe with deflector plate which belongs to a stove with circulation heating. The stove is feeded from outside and the heated air is led into the interior via a heating pipe.

263 Der Sammler-Lkw (Fz.508) ist baugleich mit dem Fz.408. Der abgebildete 4,5 t Büssing zeigt bereits Fertigungsvereinfachungen wie einfache Kotflügelbleche, eine gekürzte Stoßstange, das einfache Kühlergrillblech und seitliche Motorhauben-Bleche mit zwei Prägungen für die Entlüftung. Unterhalb der seitlichen Tür ist ein großer Werkzeugkasten montiert und in der Seitenwand eine Belüftungsklappe eingesetzt. An der vorderen Ecke des Kofferaufbaus befindet sich ein Ofenrohr mit Abweisblech. Der Ofen mit Umlaufheizung wird von außen beschickt und die erwärmte Luft über ein Heizrohr in den Innenraum geführt.

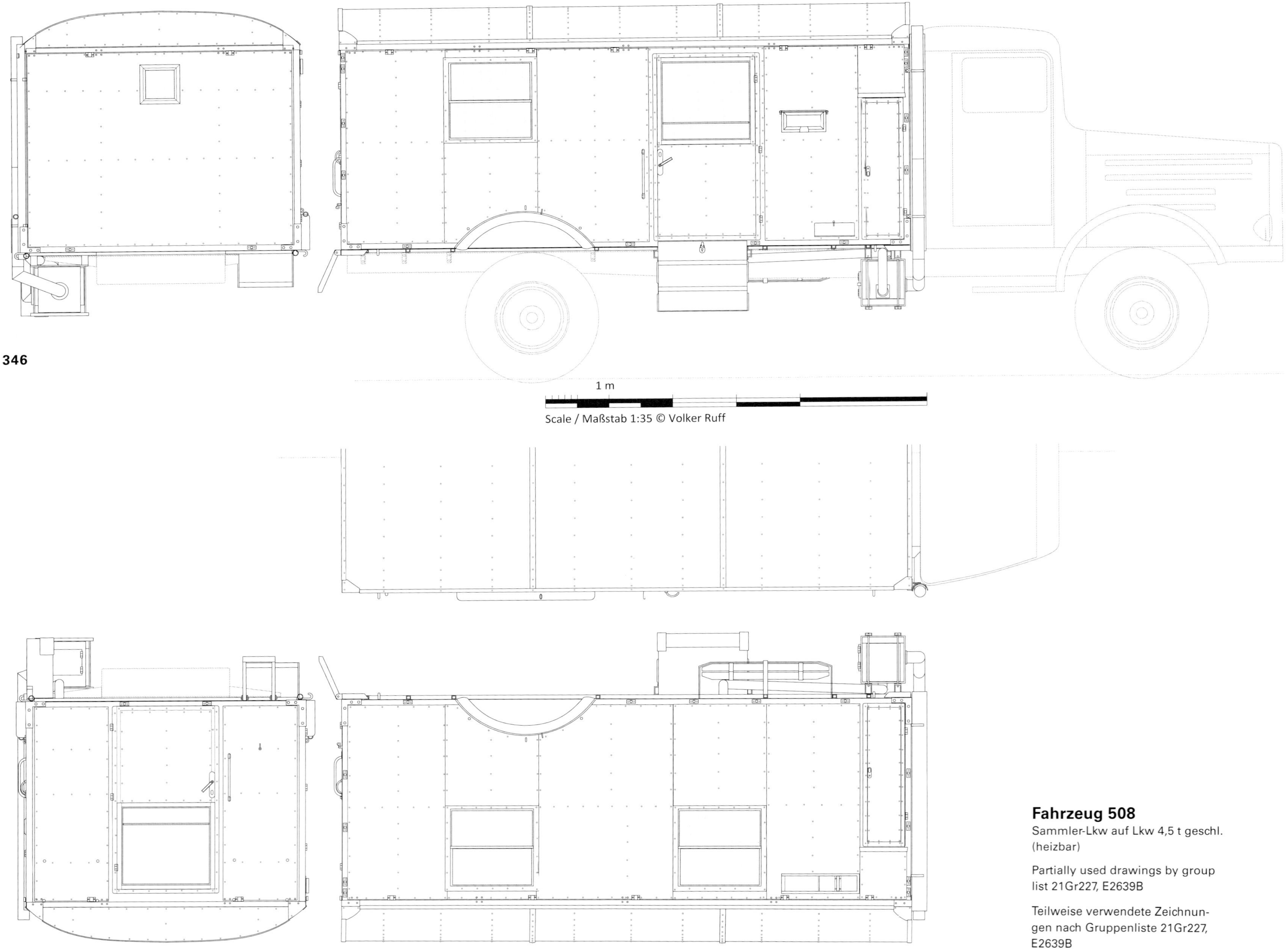

Scale / Maßstab 1:35 © Volker Ruff

Fahrzeug 508

Sammler-Lkw auf Lkw 4,5 t geschl. (heizbar)

Partially used drawings by group list 21Gr227, E2639B

Teilweise verwendete Zeichnungen nach Gruppenliste 21Gr227, E2639B

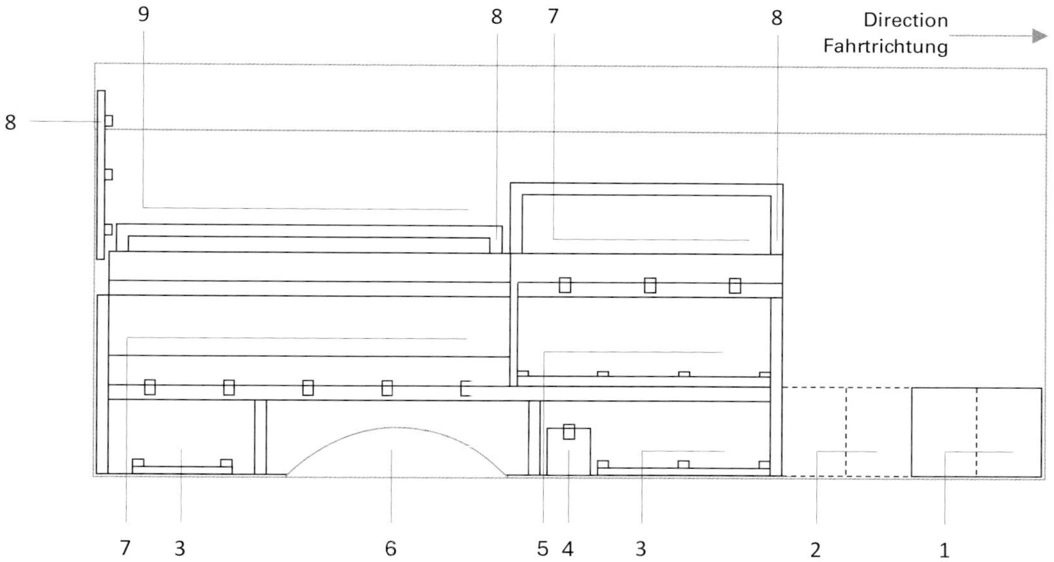

1 shelf for 22 Z-Stoff canisters
2 shelf for 22 Z-Stoff canisters (only on the right-hand side of carriage wall) shelves, mirror-inverted on both sides of the carriage for:
3 3 battery boxes I each contain 2x5 3 FS 2 (Stotz pin battery) 30V
4 3 EA collector 3 FS 1 (on-board measuring battery) 6 V
5 3 EA battery boxes II contain each 8 EA 3 FS 1 (control unit battery) 48 V 1 EA 3 FS 1 (ignition battery) 6 V
6 wheel arch
7 24 pieces of collector 3T-92-154 (on-board battery) 27 V
8 holder for collector
9 12 pieces of 50 Nc 1.3 (command transmitter battery) 50 V

1 Regal für 22 Z-Stoff Kanister
2 Regal für 22 Z-Stoff Kanister (nur an rechter Wagenwand) Regale, spiegelbildlich an beiden Wagenseiten gleich für:
3 3 St. Batteriekästen I enthalten je 2x5 St. 3 FS 2 (Stotzstecker-Batterie) 30 V
4 3 St. Sammler 3 FS 1 (Meßbatterie bordseitig) 6 V
5 3 St. Batteriekästen II enthalten je 8 St. 3 FS 1 (Kdo.-Geber-Batterie) 48 V 1 St. 3 FS 1 (Zünd-Batterie) 6 V
6 Radkasten
7 24 St. Sammler 3T-92-154 (Bordbatterie) 27 V
8 Halterung für Sammler
9 12 St. 50 Nc 1,3 (Kdo Geber-Batterie) 50 V

BArch, RH8II Bild-B1442-44, o. Ang.

264

264 In the interior there are at left and right built-in components which are mirror images of each other. In the shelves are stored 24 on-board batteries (27 V) and 12 command transmitter batteries (50 V). Other different batteries for Stotz plug, ignition battery and master battery are stowed under the shelves. On the rear wall, at left and right of the door, are the hold-downs for the accumulators hung on hooks. A thermometer for temperature monitoring is attached to the window frame on the left. Not shown in the picture are the shelves for 22+2 Z-Stoff containers on the rear wall of the driver's cab.

264 Im Innenraum sind links und rechts Einbauten eingesetzt, die spiegelbildlich angeordnet sind. In den Regalen werden 24 Bordbatterien (27 V) und 12 Kommando-Geber-Batterie (50 V) gelagert. Weitere verschiedene Batterien für Stotzstecker, Zünd-Batterie und Geber-Batterie werden unter den Regalböden verstaut. An der Rückwand sind links und rechts der Tür die Niederhalter für die Sammler an Haken aufgehängt. Am Fensterrahmen links ist ein Thermometer zur Temperaturüberwachung angebracht. Im Bild nicht gezeigt sind die Regale für 22+2 Z-Stoff-Behälter an der Rückwand zum Fahrerhaus.

Vehicle 509

Internal or former designation: Fz.-Pos.-Nr.509
Charging Station trailer Type B (2-axle)
Charging Station

DMA, GD639.5.18
HAP11/T 507
Karlshagen, 1.6.1944
Bd.Nr. E 3195/44 g

DMA, GD639.5.1
EW234
Fz.-Pos.-Nr.509

DMA, GD639.55.8
Auszug
Bedienungsanweisung
Fz.509

Intended use

The trailer was developed to carry special Luftwaffe bodies and for the economy, and designated »B2 chassis, two-axle, 3 t«. With the help of holes pre-drilled in the chassis – and only with those – closed bodies, platform bodies and special bodies of any kind could be placed on the chassis.

The mobile charging station was used for charging and maintenance of the command battery and vehicle electrical system accumulator, as well as any other accumulators. It was towed by a motor vehicle. The vehicle contained an electric machine set giving 6 kVA, consisting of an air-cooled 2-cylinder 4-stroke Otto engine of 17 HP at 3.000 rpm and a three-phase generator for 220/380 V.

The mobile charging station has the following dimensions and weights:

Total length:	7,070 mm
Total width:	2,250 mm
Total height:	3,090 mm
Wheelbase	2,860 mm
Track width:	1,700 mm
Tare weight:	2,720 kg
Payload:	2,280 kg
Maximum weight:	5,000 kg
Service weight:	3,710 kg weighted
Wheels:	190-20

The main parts of the mobile charging station are:

- Chassis
- Closed body
- Charging compartment for lead collectors and NC collectors

Chassis

The chassis was designated »B 2 chassis, two-axle, 3 t« (Equipment Manual D.(Luft) T.8207).

DMA, GD639.5.21
OKH/Wa Prüf10
Karlshagen, 1.4.1944
Bd.Nr. 789/44 gK
Auszug
Besprechung der
Fahrzeuggruppe 5
vom 1.4.44
in Karlshagen I

Production status: Vehicle 509
estimated total requirement 7 vehicles.
Available at Kdo. Basse 7 vehicles (completed).

Fahrzeug 509

Interne oder frühere Bezeichnung: Fz.-Pos.-Nr.509
Ladestation-Anhänger Type B (2-achs)
Ladestation

Verwendungszweck

Der Anhänger wurde für Sonderaufbauten der Luftwaffe und für die Wirtschaft entwickelt und trägt die Bezeichnung »B2 Fahrgestell, zweiachsig, 3 t«. Mit Hilfe der im Fahrgestell vorgebohrten Löcher – und zwar nur mit diesen – können geschlossene Aufbauten, Pritschenaufbauten und Sonderaufbauten jeder Art auf das Fahrgestell gesetzt werden.

Die fahrbare Ladestation dient zum Aufladen und zur Pflege der Kommandogeberbatterie, der Bordnetzsammler sowie auch beliebiger anderer Sammler. Die Beförderung erfolgt durch Anhängen an ein Kraftfahrzeug. Das Fahrzeug enthält einen Elektromaschinensatz von 6 kVA Leistung, bestehend aus einem luftgekühlten zweizylindrigen 4-Takt-Otto-Motor von 17 PS bei 3.000 Umdrehungen und einem Drehstromgenerator für 220/380 V.

Die fahrbare Ladestation hat folgende Maße und Gewichte:

Gesamtlänge:	7.070 mm
Gesamtbreite:	2.250 mm
Gesamthöhe:	3.090 mm
Radstand	2.860 mm
Spurweite:	1.700 mm
Leergewicht:	2.720 kg
Nutzlast:	2.280 kg
Zul. Höchstgew.	5.000 kg
Gefechtsgewicht:	3.710 kg gewogen
Räder:	190-20

Die Hauptteile der fahrbaren Ladestation sind:

- Fahrgestell
- Geschlossener Aufbau
- Aufladeraum für Bleisammler und NC-Sammler

Fahrgestell

Als Fahrgestell dient das »B 2 Fahrgestell, zweiachsig, 3 t« (Geräte-Handbuch D.(Luft) T.8207.

Fertigungslage: Fahrzeug 509
Voraussichtlicher Gesamtbedarf 7 Fahrzeuge.
Verfügbar bei Kdo. Basse 7 Fahrzeuge (fertig).

Current defects and their remedy

At present the accumulators are not sufficiently secured for transport in the charging station trailer. The molding are set too deep so vibration while moving can possibly tilt the accumulators. Change needed. For example by adding an extra molding, to be done immediately. Action by Kdo. Basse. Relevant requirements of Kdo. Basse for workshop assistance and material are to be completed urgently and immediately by the appropriate departments of HAP11.

DMA, GD639.55.8
Auszug
Bedienungsanweisung
Fz.509

Closed body

Similar to the LC box bodies, a closed body is mounted on the B2 chassis. The interior is divided into two charging compartments for lead accumulator (collector) and NC accumulator (collector). Access to the lead accumulator compartment is via a folding step and doors on the right-hand side wall.

Derzeitige Mängel und ihre Behebung

Z.Zt. sind im Ladestationsanhänger die Akkus für den Transport nicht genügend gesichert. Die Leisten sitzen zu tief, so dass bei groben Fahrerschütterungen ein Kippen der Sammler nicht ausgeschlossenen ist. Entsprechende Änderung. Beispielsweise durch Anbringung einer weiteren Leiste, ist sofort vorzunehmen. Veranlassung durch Kdo. Basse. Diesbezügliche Anforderungen des Kdo. Basse an Werkstatthilfe und Material sind seitens der entsprechenden Stellen des HAP11 vordringlich und umgehend zu erledigen.

Geschlossener Aufbau

Ähnlich der LC-Koffer ist ein geschlossener Aufbau auf das B2-Fahrgestell montiert. Der Innenraum ist aufgeteilt in zwei Abteilungen: den Laderäumen für Bleisammler und NC-Sammler. Der Zugang zum Bleisammler-Raum erfolgt über einen Klapptritt und eine Tür an der rechten Seitenwand.

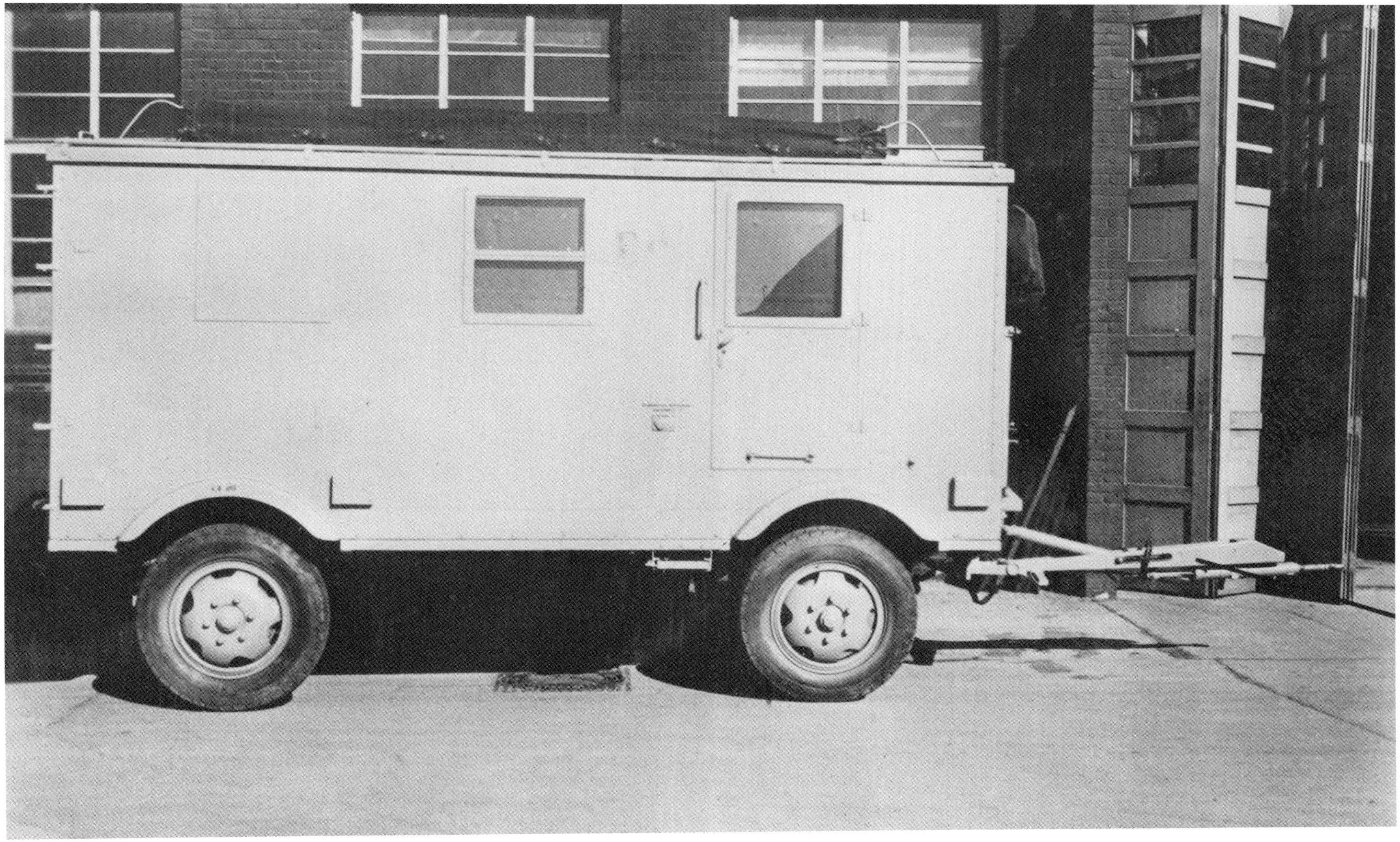

Deutsches Museum, München, Archiv- FA014/B242/44 TD5224 [CD74731]

265

Front and rear walls of closed body

Additional equipment was attached to the front wall. From left to right: cable reel with power cable and cover, spade and pick axe, two fuel canisters, spare wheel and an earth screw. The window in the front wall was closed by a filler.

Access to the NC-accumulator charging compartment was through a door in the rear wall, a folding step make it easier to get in. A fan window was inserted in the upper right corner. Another fan window for the lead accumulator compartment was located in the left side wall, at the top front. The roof with a roof rack and tarpaulin was accessible via steps at the two corners of the rear wall.

Vorder- und Rückwand am geschlossen Aufbau

An der Vorderwand wurden zusätzliche Ausrüstungsgegenstände angebracht. Von links nach rechts: Kabeltrommel mit Netzanschlusskabel mit Abdeckung, Spaten und Kreuzhacke, zwei Brennstoffkanister, Reserverad und eine Erdungsschraube. Das Fenster in der Vorderwand war mit einem Füllstück verschlossen.

Über die Tür in der Rückwand gelangte man in den NC-Sammler-Laderaum. Ein Klapptritt erleichterte den Einstieg. Im rechten oberen Eck war ein Ventilatorfenster eingesetzt. Ein weiteres Ventilatorfenster für den Bleisammlerraum befand sich oben vorne in der linken Seitenwand. Das Dach mit Dachträger und Plane war über Steigtritte an den beiden Ecken der Rückwand erreichbar.

Deutsches Museum, München, Archiv- FA014/B243/44 TD5224 [CD74728]

266

Deutsches Museum, München, Archiv- FA014/B241/44 TD5224 [CD74730]

267

The B2 chassis, two-axle, 3 t

The chassis frame consisted of a middle (central) tube with seven cross members and was supported on two tube axles connected to with swinging arms and axle stands. The axles were rigid, but moved like swing axles due to the spring suspension and wheel compensation. Suspension was provided by four torsion bar springs.

Das B2-Fahrgestell, zweiachsig, 3 t

Der Fahrgestellrahmen bestand aus einem Mittel- (Zentral)-Rohr mit sieben Querträgern und lagerte in Verbindung mit Schwingern und Achsständern auf zwei Rohrachsen. Die Achsen waren starr, bewegten sich aber infolge der Federaufhängung und des Radausgleiches wie Schwingachsen. Die Federung erfolgte durch vier Drehstabfedern.

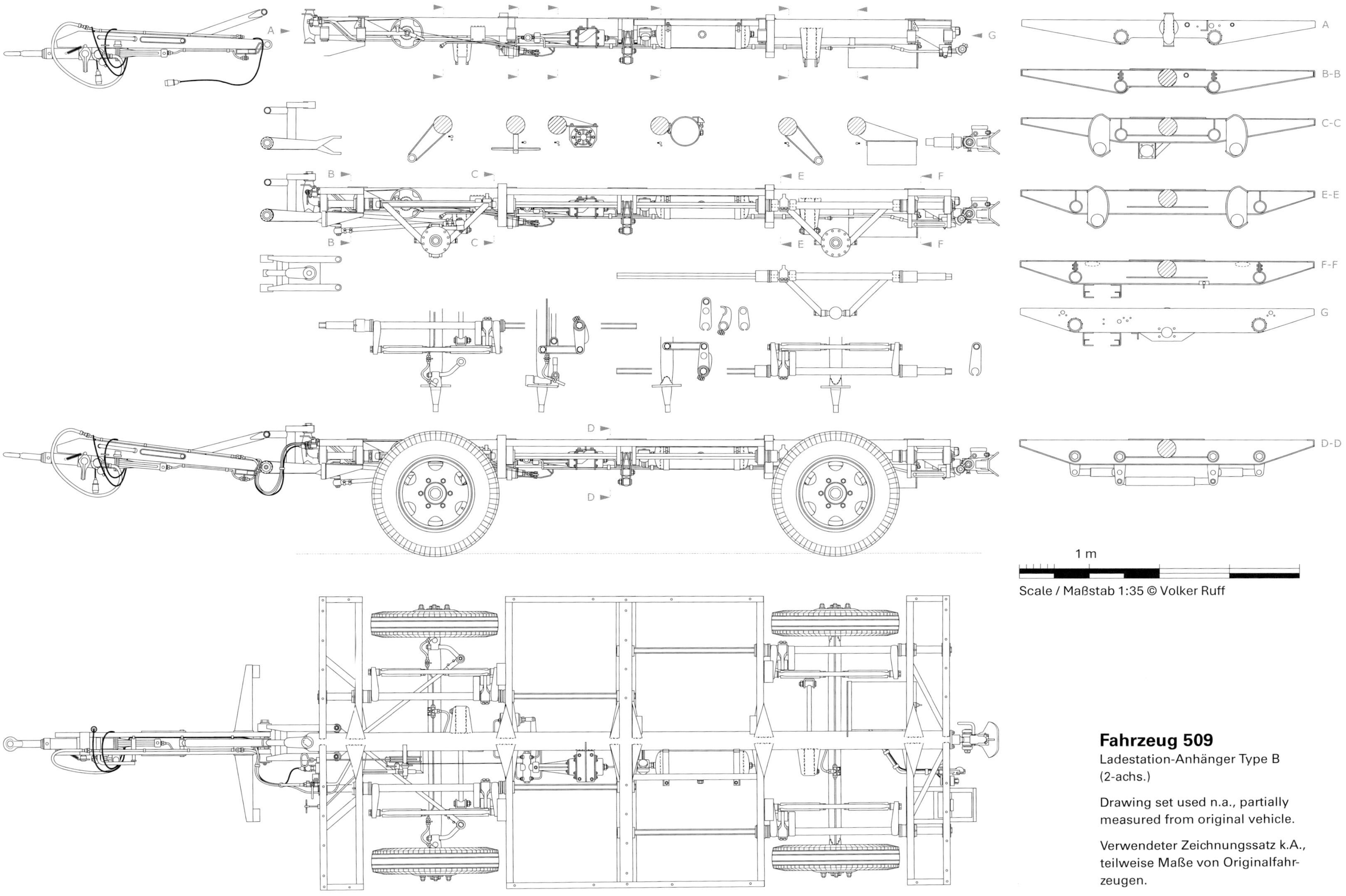

Scale / Maßstab 1:35 © Volker Ruff

Fahrzeug 509

Ladestation-Anhänger Type B (2-achs.)

Drawing set used n.a., partially measured from original vehicle.

Verwendeter Zeichnungssatz k.A., teilweise Maße von Originalfahrzeugen.

Lead accumulator and NC-accumulator charging compartment

The front compartment held for charging lead accumulators. It was equipped with two charging racks for a total of 50 lead accumulator (on-board power supply battery), a worktable with drawers in which the collector care device was located, as well as a high-performance fan for ventilation of the room. Access to the lead accumulator charging compartment was via a door on the right-hand side wall.

The rear compartment held for charging the NC-accumulators (command transmitter battery), a rectifier and work space. The equipment of this room consisted of 2 charging racks for 20 NC-accumulators each of type 50 NC, 2 multiple rectifiers, 1 main connection box, and 2 work tables with drawers.

Bleisammler- und NC-Sammler-Laderaum

Der vordere Raum diente als Bleisammler-Laderaum. Er war ausgerüstet mit zwei Ladegestellen zur Aufnahme von insgesamt 50 Bleisammlern (Bordnetzbatterie), einem Werktisch mit Schubfächern in dem das Sammler-Pflege-Gerät untergebracht war sowie einen Hochleistungsventilator zur Entlüftung des Raumes. Der Zutritt zum Bleisammler-Laderaum erfolgte durch eine Tür über die rechte Seitenwand.

Der hintere Raum diente als Laderaum für die NC-Sammler (Kommandogeberbatterie), einem Gleichrichter und Werkraum. Die Ausrüstung dieses Raumes bestand aus 2 Ladegestellen für je 20 NC-Sammler Type 50 NC, 2 Mehrfachrichtern, 1 Netzanschlusskasten, 2 Werktische mit Schubfächern.

Deutsches Museum, München, Archiv- FA014/B254/44 TD5224 [CD74740]

268

Deutsches Museum, München, Archiv- FA014/B255/44 TD5224 [CD74806]

269

Electric machine set

Through the open door in the rear wall you can look into the NC-accumulator recharging compartment. In front of the partition wall to the lead accumulator compartment were the switchboards (covered). For transport, the electric machine set* is placed in the aisle between the workbench and the washbasin and fixed with brackets to clamping rails rails embedded in the floor.

For use, the electric machine set was pushed out via a rampand set up outside. The electric machine set consists of a support frame with a petrol engine and a three-phase generator to generate the electricity required for the accumulator rooms.

* Electric maschine set – three phase generator – EVZ 6 d (3-phase current, 220/380 V, 6 kVA)

Elektromaschinensatz

Durch die geöffnete Tür in der Rückwand blickt man in den NC-Sammler-Laderaum. Vor der Trennwand zum Bleisammlerraum waren die Schalttafeln (abgedeckt) aufgestellt. Zum Transport ist der Elektromaschinensatz' im Gang zwischen dem Werk- und Waschtisch platziert und mit Haltern an im Boden eingelassenen Spannschienen fixiert.

Für die Inbetriebnahme wurde der Elektromaschinensatz über eine Rutschleiter herausgeschoben und außen aufgebaut. Der Elektromaschinensatz bestand aus einem Tragegerüst mit Otto-Motor – der Drehstromgenerator erzeugte die für die Sammlerräume notwendige elektrische Energie.

* Elektromaschinensatz – Drehstromgenerator EVZ 6 d (Drehstrom, 220/380 V, 6 kVA)

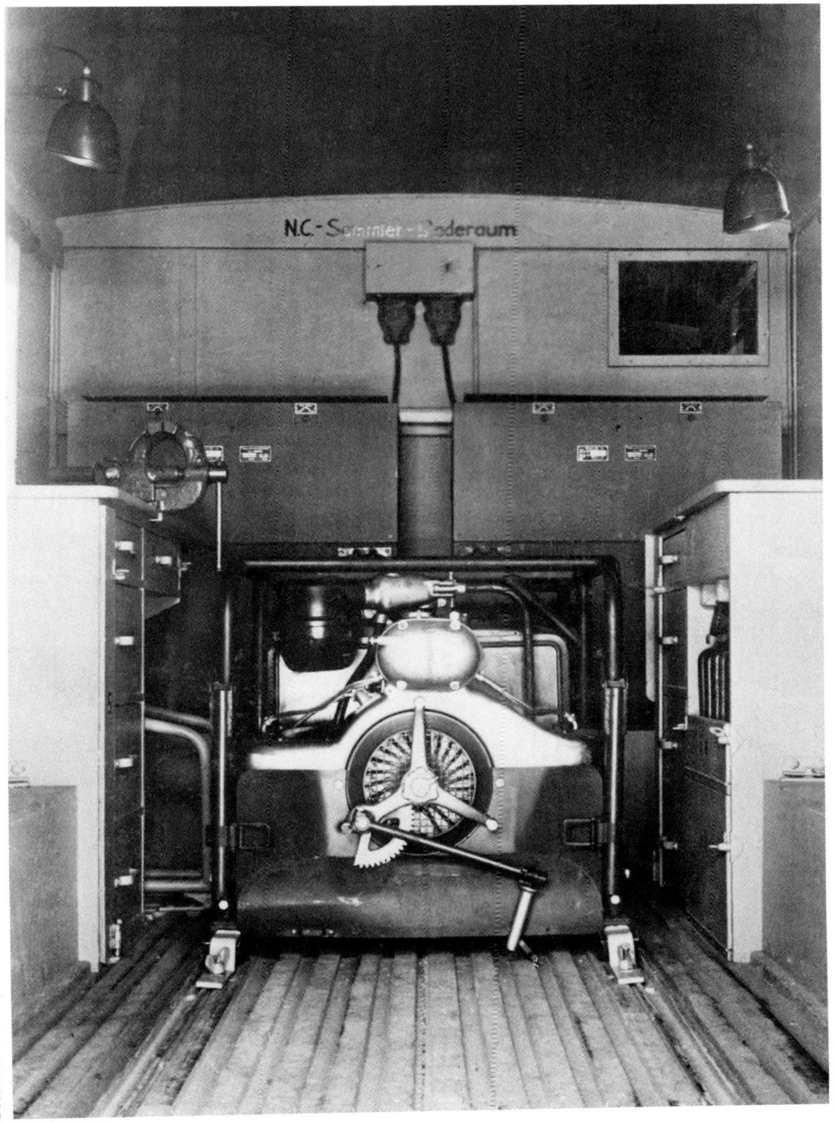

Deutsches Museum, München, Archiv- FA014/B244/44 TD5224 [CD74729]

270

Deutsches Museum, München, Archiv- FA014/B245/44 TD5224 [CD74734]

271

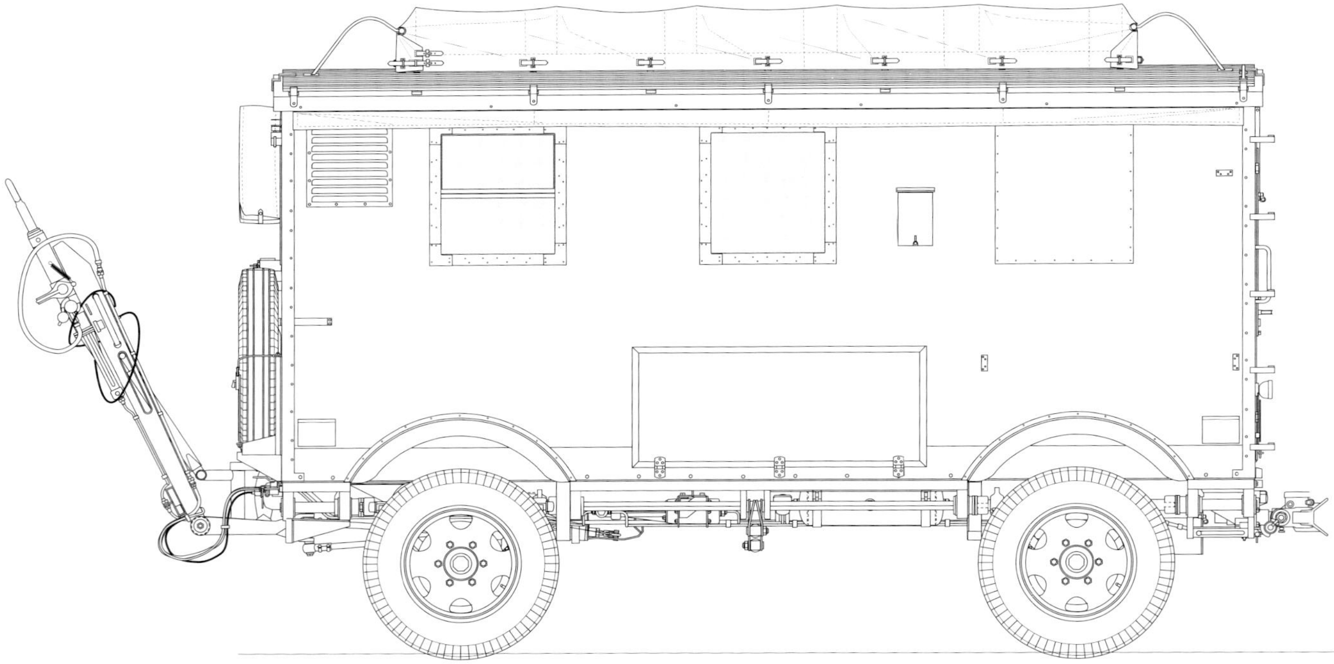

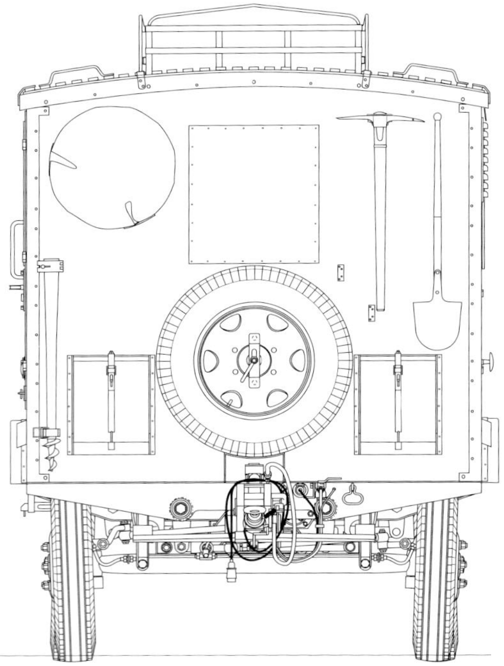

1 m

Scale / Maßstab 1:35 © Volker Ruff

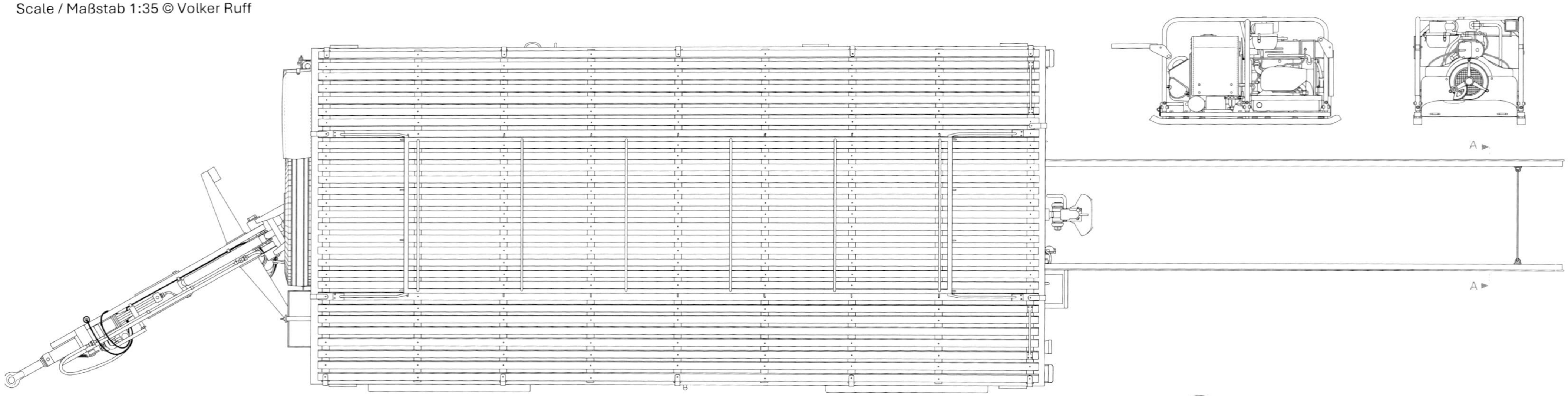

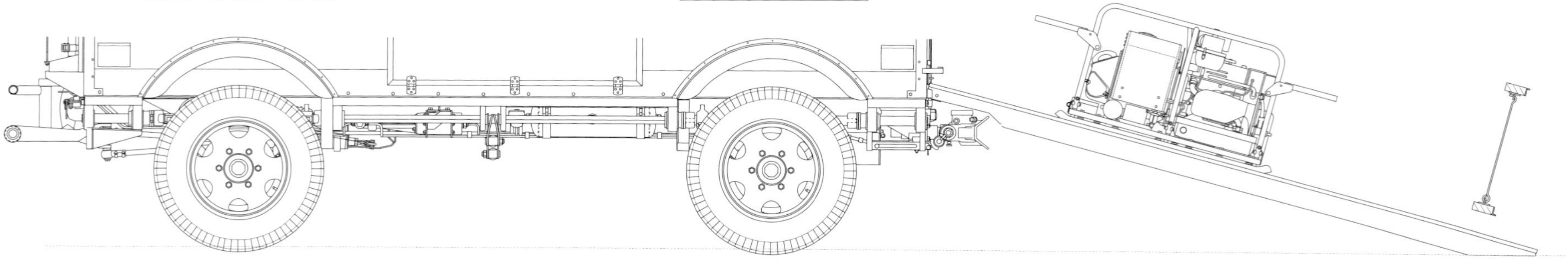

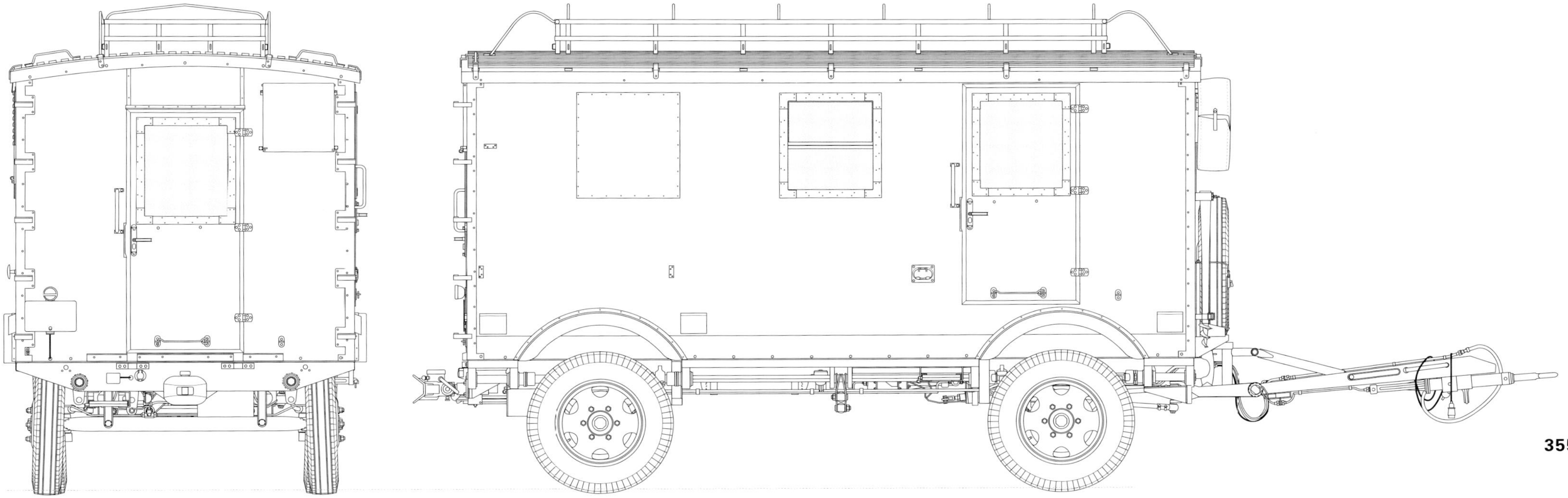

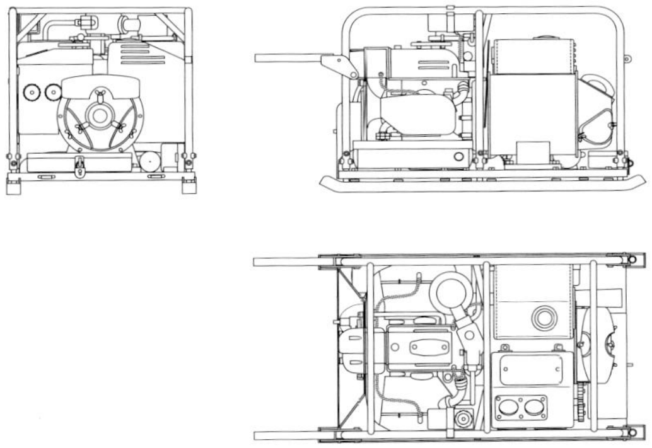

Electrical machine set EVZ 6 d
Elektromaschinensatz EVZ 6 d

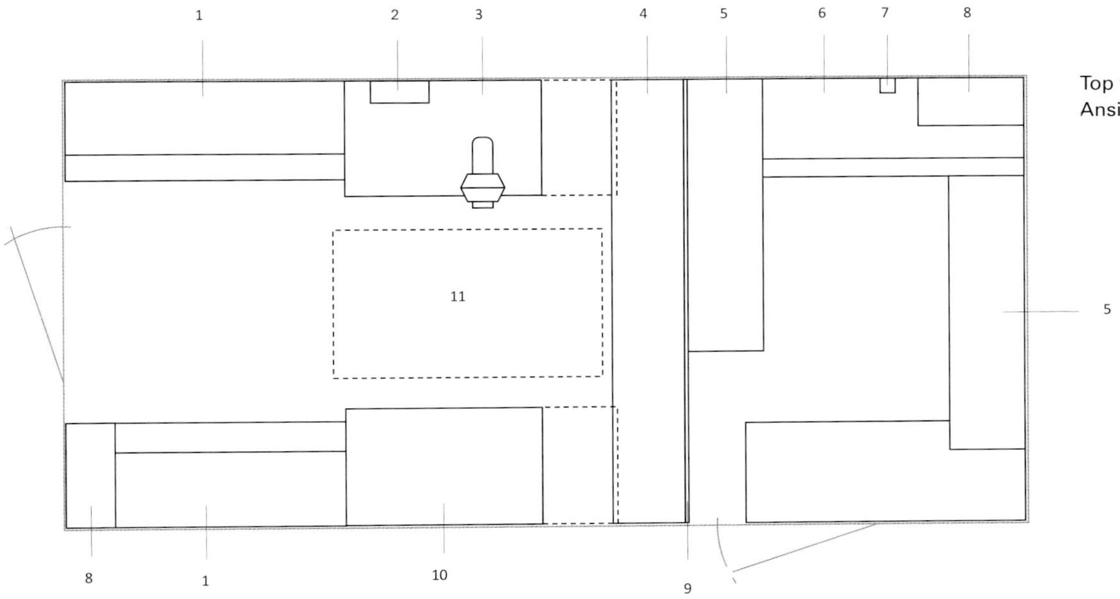

Top view
Ansicht von oben

1	Charging rack for NC collectors	1	Ladegestell NC Sammler
2	Power distribution box	2	Netzanschlusskasten
3	Workbench with vice	3	Werktisch mit Schraubstock
4	Switch and control panel	4	Schalt und Kontrolltafel
5	Charging rack for lead accumulator	5	Ladegestell für Bleisammler
6	Workbench with drawers	6	Werktisch mit Schubladen
7	Connection terminals for charging fields	7	Anschlussklemmen für Ladefelder
8	High-performance fan	8	Hochleistungsventilator
9	Partition wall charging areas	9	Trennwand Laderäume
10	Tool table	10	Werkzeugtisch
11	EMS electrical machine set	11	Elektro-Maschinen-Satz EMS

Fahrzeug 509
Ladestation-Anhänger Type B
(2-achs.)

Loading plan
Beladeplan

Vehicle 510

Internal or former designation: Fz.-Pos.-Nr.510
Spare parts truck III on 3 t truck closed (305)
FR spare parts

DMA, GD639.5.18
HAP11/T 507
Karlshagen, 1.6.1944
Bd.Nr. E 3195/44 g

DMA, GD639.5.1
EW234
Fz.-Pos.-Nr.510

Intended use
The spare parts truck III is intended for transporting the mechanical storage components for the A4 *Gerät*. For transportation the vehicle is equipped with 1 bench seat box and 3 standard racks.

These standardized racks are used to hold 18 transport boxes (4 boxes of engine parts and 14 boxes for electrical equipment). The standardizied racks are interchangeable and can be loaded onto other motor vehicles (from the same vehicle group) if required.

The stock items truck III has the following dimensions and weights:

Total length:	6,130 mm
Total width:	2,100 mm
Total height:	2,950 mm
Wheelbase	3,450 mm
Track width:	1,630/1,642 mm front/rear
Tare weight:	3,275 kg
Payload:	2,825 kg
Maximum weight :	6,100 kg
Service weight:	4,080 kg*
Wheels:	190-20

The main parts of the storage truck III are:
- Chassis
- Closed superstructure
- Installations

Chassis
The chassis used is that of the 3 t Opel 6700 Type A truck (equipment description according to D 666/5) but the normal chassis of the Opel 3 t Type 3.6/36 S truck (equipment description according to D 669/19B) was also used.

DMA, GD639.5.21
OKH/Wa Prüf10
Karlshagen, 1.4.1944
Bd.Nr. 789/44 gK
Auszug
Besprechung der
Fahrzeuggruppe 5
vom 1.4.44
in Karlshagen I

Production status: vehicle 510
Total demand 7 vehicles.
Vehicles held at Kdo. Basse.
Requested in the same way as for 501, 5 units until 15.4.1944.
Actual delivery itself assured, although development of the fittings has only just begun. Execution of this development is halted, waiting for decision by Ing. Horn.

* Service weight without 18 boxes with contents and hot air blower.

Fahrzeug 510

Interne oder frühere Bezeichnung: Fz.-Pos.-Nr.510
Vorratsteilewagen III auf Lkw 3 t geschl. (305)
FR-Ersatzteile

Verwendungszweck
Der Vorratsteilewagen III ist für den Transport der mechanischen Vorratsteile für das A4-Gerät vorgesehen. Als Transportmittel befinden sich im Kraftfahrzeug 1 Sitzbankkasten und 3 Einheitsgestelle.

Diese Einheitsgestelle dienen zur Aufnahme von 18 Transportkästen (davon 4 Kisten triebwerksseitig und 14 Kisten für elektrische Geräte). Die Einheitsgestelle sind austauschbar und können bei Bedarf auf andere Kraftfahrzeug (ähnlicher Gruppe) verlastet werden.

Der Vorratsteilewagen III hat folgende Maße und Gewichte:

Gesamtlänge:	6.130 mm
Gesamtbreite:	2.100 mm
Gesamthöhe:	2.950 mm
Radstand	3.450 mm
Spurweite:	1.630/1.642 mm vorne/hinten
Leergewicht:	3.275 kg
Nutzlast:	2.825 kg
Zul. Höchstgewicht	6.100 kg
Gefechtsgewicht:	4.080 kg*
Räder:	190-20

Die Hauptteile des Vorratsteilewagen III sind:
- Fahrgestell
- Geschlossener Aufbau (305) LC-Koffer
- Einbauten

Fahrgestell
Als Fahrgestell dient das Fahrgestell des Lastkraftwagen 3 t Opel 6700 Typ A (Gerätebeschreibung nach D 666/5), aber auch das normale Fahrgestell des Opel 3 t Lastkraftwagens Typ 3,6/36 S (Gerätebeschreibung nach D 669/19B) wurde verbaut.

Fertigungslage: Fahrzeug 510
Gesamtbedarf 7 Fahrzeuge.
Fahrzeuge selbst bei Kdo. Basse.
Gefordert sinngemäß wie bei 501, 5 Stück bis 15.4.1944.
Lieferung an sich sichergestellt, obwohl Entwicklung der Einbauten erst in Angriff genommen ist. Ausführung dieser Entwicklung wird gestoppt bis zur Aktion Horn.

* Gefechtsgewicht ohne 18 Kisten mit Inhalt und Heißluftbläser.

Closed superstructure

The spare parts truck III was externally indistinguishable from the Kfz.305. The closed superstructure of an LC box body was taken over unchanged from the Kfz.305. In the interior, the 3 unit shelves were fastened with screws and brackets to the reinforced side walls and the floor. The windows were covered from the inside with wooden panels to prevent damage by the cargo.

Geschlossener Aufbau

Der Vorratsteilewagen III war äußerlich vom Kfz.305 nicht zu unterscheiden. Der geschlossene Aufbau, ein LC-Koffer, wurde unverändert vom Kfz.305 übernommen. Im Innenraum waren die 3 Einheitsregale mit Schrauben und Haltern an Seitenwandverstärkungen und am Fußboden befestigt. Vor Beschädigungen durch Ladegut wurden die Fenster von innen durch Holztafeln verschlossen.

357

BArch, RH8II Bild-B1160-44, o.Angabe

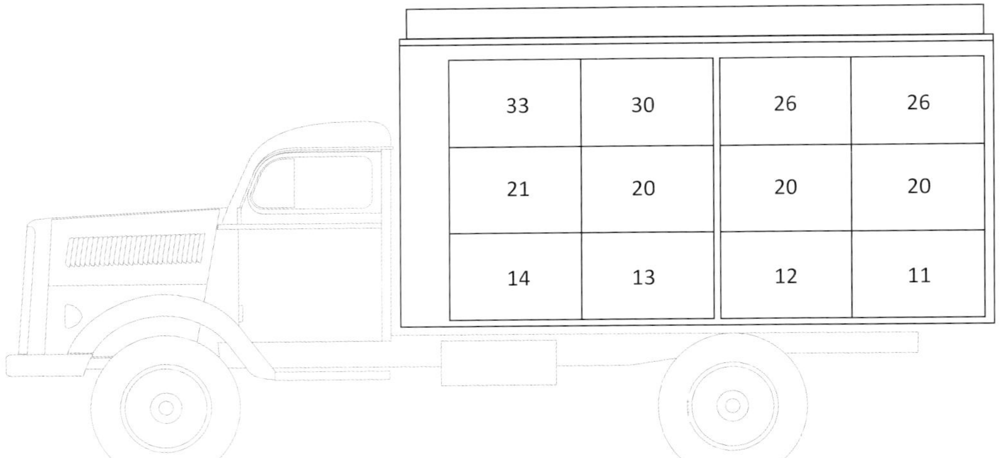

Rechte Seite

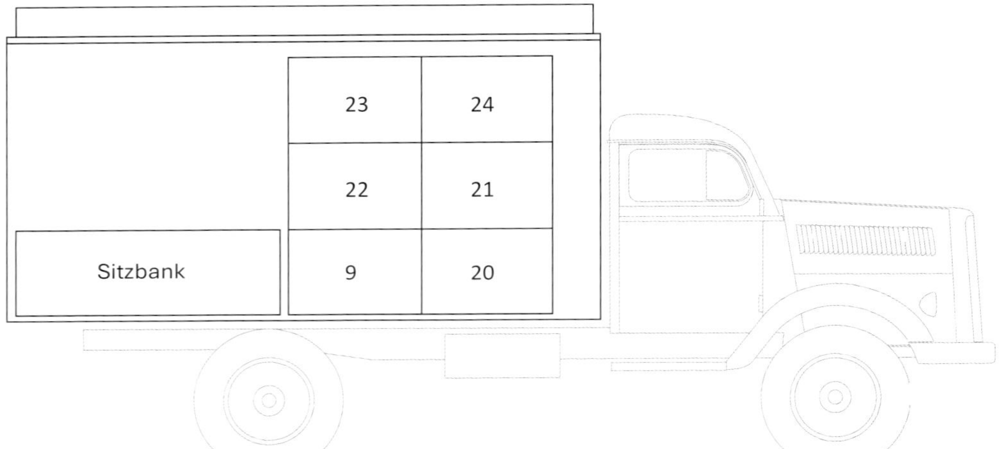

Linke Seite

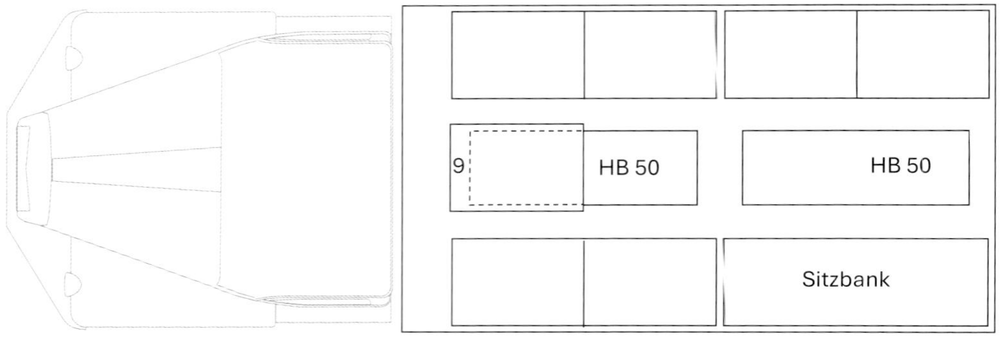

Schema für die Unterbringung der Kisten

Auszug GD639.5.1 – Abbildung 3

Loading plan

The means of transport in the vehicle contains a bench seat box (19) and three standard frames. These standard racks serve to accommodate 18 transport boxes (4 of which are for engine parts and 14 for electrical equipment). The unit racks are interchangeable and can be loaded onto other motor vehicles (similar vehicle) if required.

One transport box and two hot air blowers are located in the central aisle. More than 230 different spare parts are stored in the transport boxes.

Sondergerät

- Exterior parts
- equipment compartment (steering unit)
- Interior parts tail (exhaust) vanes, serevos, rudders etc.)
- Internal parts T-installation (reaction chamber)
- Standard parts and screw connections
- Electrical installations

Ground service

- Test panels
- Hose lines
- Five-way couplings
- HB50 Hot air blowers

Beladeplan

Als Transportmittel befindet sich im Kraftfahrzeug ein Sitzbankkasten (19) und drei Einheitsgestelle. Diese Einheitsgestelle dienen zur Aufnahme von 18 Transportkästen (davon 4 Kästen triebwerksseitig und 14 Kästen für elektrische Geräte). Die Einheitsgestelle sind austauschbar und können bei Bedarf auf andere Kraftfahrzeuge (ähnlicher Gruppe) verlastet werden.

Ein Transportkasten und zwei Heißluftbläser sind im Mittelgang untergebracht. Über 230 verschiedene Vorratsartikel sind in den Transportkästen verstaut.

Sondergerät

- Außenteile
- Innenteile Geräteraum (Steuermaschine)
- Innenteile Heck (Luftruder, Strahlruder etc.)
- Innenteile T-Anlage (Dampfanlage)
- Normteile und Verschraubungen
- Elektrische Einbauten

Bodenaufwand

- Prüftafeln
- Schlauchleitungen
- Fünffachkupplungen
- HB50-Heißluftbläser

Installations

The spare parts were packed in individual boxes in an ingenious system and stowed on shelves. The positions of the crates on the shelves took into account the total weight of the crates with spare parts, heavy crates at the bottom and lighter ones at the top. A unique rack with three shelves for six boxes was installed on the left hand side. On the right hand side there were two more racks for a total of 12 boxes. All transport boxes were numbered and provided with the vehicle number, and were secured with wooden bars The second No. 9 box was stowed in the middle aisle and space was provided for two HB50 hot air blowers.

Einbauten

Die Vorratsteile waren in einem ausgeklügelten System in einzelne Kisten gepackt und in Regalen verstaut. Die Position der Kisten im Regal berücksichtigte das Gesamtgewicht der Kisten mit Ersatzteilen. Schwere Kisten befanden sich unten, leichtere oben. Ein Einheitsgestell mit drei Fächern für sechs Kisten wurde an der linken Seite montiert. Auf der rechten Seite befanden sich zwei weitere Einheitsgestelle für insgesamt 12 Kisten. Alle Kisten waren durchnummeriert und mit der Fz. Nr. versehen, gesichert wurden sie mit Holzriegel. Im Mittelgang war die zweite Kiste Nr. 9 verstaut und Platz für zwei Heißluftbläser HB50 vorgesehen.

BArch, RH8II Bild-B1293-44, o. Ang.

273

BArch, RH8II Bild-B1294-44, o. Ang.

274

273, 274 If you entered the LC box body through the rear door, there was a bench seat box on the left. Its cover served as a bench seat for the crews, and underneath it were insert boxes for pipes and hoses.

273, 274 Wenn man den LC-Koffer durch die hintere Tür betrat, befand sich auf der linken Seite ein Sitzbankkasten. Seine Abdeckung diente den Mannschaften als Sitzmöglichkeit. Darunter befanden sich Einschubkästen für Rohre und Schläuche.

Addendum to the loading plan for a storage truck (KW 510)
Comparison of the ongoing numbering process the transport boxes according to the loading plan with now using transport box identification numbers.

Transport boxes	No. 1	get ident.-Nb	11 (SK Nb.7- T41 H)
"	No. 2	dto.	12 (SK.Nb.7 – T41 H)
"	No. 3	dto.	13 (SK.Nb.18 – T41 H)
"	No. 4	dto.	14 (SK.Nb.18 – T41 H) with cover sheet
"	No. 5	dto.	20 (Dwg.-Nb. 7557 B)
"	No. 6	dto.	20 (Dwg.-Nb. 7557 B)
"	No. 7	dto.	20 (Dwg.-Nb. 7557 B)
"	No. 8	dto.	21 (Dwg.-Nb. 7558 B)
"	No. 9	dto.	26 (Dwg.-Nb. 7614 B)
"	No. 10	dto.	26 (Dwg.-Nb. 7614 B)
"	No. 11	dto.	30 (Dwg.-Nb. 7613B)
"	No. 12	dto.	33 (Dwg.-Nb. 7612B)
"	No. 13	dto.	9 (Dwg.-Nb. Nil)
"	No. 14	dto.	20 (Dwg.-Nb. 7557 B)
"	No. 15	dto.	22 (Dwg.-Nb. 7559 B)
"	No. 16	dto.	21 (Dwg.-Nb. 7558 B)
"	No. 17	dto.	23 (Dwg.-Nb. 7604 B)
"	No. 18	dto.	24 (Dwg.-Nb. 7611 B)
"	No. 19	dto.	bench box
"	No. 20	dto.	9 (Dwg.-Nb. Nil) aisle

Box identification number	Content boxes	Qty.
9	Spare parts for hot air blowers	2
11	Main valve, breathers, valves, pressure reducer, filter, screws	1
12	Five-way couplings, valves, pressure reducers, switch batteries, fuel filler flap, flaps	1
13	6 Air rudders, screws, sealing rings, sealing cones, protective caps, pipes and tubes,	1
14	6 Air rudders, valves, vaporizer, lines	1
Bench box seat	Tubes and hoses	1
20	Command receiver	4
21	Doppler	2
22	2 Guide beam receiver	1
23	2 *Mischgerät* (mixing devices)	1
24	2 Vertikante with mounting bracket, 2 internal switchgear	1
26	2 Horizons with converter and controller	2
30	2 Converters with controller, timers, Sterg, key with plug	1
33	Standard parts and screw connections	1

View layout page 358

Nachtrag zum Beladeplan für einen Vorratsteilwagen (KW 510)
Gegenüberstellung der laufend durchgeführten Benummerung der Transportkästen laut Beladeplan mit nunmehr zu verwendenden Transportkisten-Kennnummern.

Transportkisten	Nr. 1	erhält Kenn-Nr.	11 (SK Nb.7- T41 H)
"	Nr. 2	dto.	12 (SK.Nr.7 – T41 H)
"	Nr. 3	dto.	13 (SK.Nr.18 – T41 H)
"	Nr. 4	dto.	14 (SK.Nr.18 – T41 H) mit Deckblatt
"	Nr. 5	dto.	20 (Zeichnung-Nr. 7557 B)
"	Nr. 6		20 (Zeichnung-Nr. 7557 B)
"	Nr. 7	dto.	20 (Zeichnung-Nr. 7557 B)
"	Nr. 8	dto.	21 (Zeichnung-Nr. 7558 B)
"	Nr. 9		26 (Zeichnung-Nr. 7614 B)
"	Nr. 10	dto.	26 (Zeichnung-Nr. 7614 B)
"	Nr. 11	dto.	30 (Zeichnung-Nr. 7613 B)
"	Nr. 12		33 (Zeichnung-Nr. 7612 B)
"	Nr. 13		9 (Zeichnung-Nr. ohne)
"	Nr. 14	dto.	20 (Zeichnung-Nr. 7557 B)
"	Nr. 15		22 (Zeichnung-Nr. 7559 B)
"	Nr. 16		21 (Zeichnung-Nr. 7558 B)
"	Nr. 17	dto.	23 (Zeichnung-Nr. 7604 B)
"	Nr. 18	dto.	24 (Zeichnung-Nr. 7611 B)
"	Nr. 19	dto.	Sitzbank
"	Nr. 20	dto.	9 (Zeichnung-Nr. ohne Gang)

Kenn-Nr. der Kisten	Inhalt der Kisten	Anzahl
9	Ersatzteile für Heißluftbläser	2
11	Hauptventil, Entlüfter, Ventile, Manometer, Filter, Schrauben	1
12	Fünffachkupplungen, Ventile, Druckmindrer, Schaltbatterien, Betankungstutzenklappe, Klappen	1
13	6 Segel, Schrauben,Dichtringe, Dichtkegel, Schutzkappen, Leitungen und Rohre,	1
14	6 Segel, Ventile, Zerstäuber, Leitungen	1
Sitzbankkiste	Rohre und Schläuche	1
20	Kommandoempfänger	4
21	Verdoppler	2
22	2 Leitstrahlempfänger	1
23	2 Mischgeräte	1
24	2 Vertikante mit Halterung, 2 Innenschaltgeräte	1
26	2 Horizonte mit Umformer und Regler	2
30	2 Umformer mit Regler, Zeitschaltwerke, Sterg, Schlüssel mit Stecker	1
33	Normteile und Verschraubungen	1

Siehe Layout Seite 358

Transport boxes

Example of the equipment of an transport box

Transport box 3, identification number of box 13 (3) ground service

The transport boxes were designed as simple wooden boxes with lids. Up to three inserted boxes could be stowed in the boxes. Special boxes had been developed for the sensitive electronic devices, and were suspended on springs.

The transport case 3 with two inserted boxes consisted of:
Insert box 3a was equipped with various pipes for A- (32,33), B- (34, 35, 36) and T-Stoff pipes (113) and other pipes, e.g. P-ND pipes (126, 127) for the control valves of the 25 t valve are also stored.

Insert box 3 contained e.g. valves (83, 85), shut-off valves (93), pressure gauges (106), and various small parts such as screws, cap nuts, sealing cones and support rings.
The bottom of the box was divided by an insert, where 6 »Segel« (air rudder) (37) for the *Gerät* were stowed.

Transportkästen

Beispiel für die Ausstattung eines Transportkastens

Transportkasten 3, Kennnummer der Kiste 13 (3) Bodenanlage

Die Transportkästen waren als einfache Holzkisten mit Deckel ausgeführt. Bis zu drei Einsatzkästen waren in den Kisten verstaubar. Für die empfindlichen elektronischen Geräte wurden spezielle Einsätze entwickelt. Die Einsatzkästen sind an Federn eingehängt.

Der Transportkasten 3 mit zwei Einsatzkästen bestand aus:
Der Einsatzkasten 3a war bestückt mit verschiedenen Rohrleitungen für A- (32, 33), B- (34, 35, 36) und T-Stoffleitungen (113). Weitere Leitungen, wie z.B. P-ND-Rohre (126, 127) für die Steuerventile des 25 t-Ventils waren ebenfalls bevorratet.

Im Einsatzkasten 3 waren Ventile (83, 85), Absperrventile (93), Manometer (106) und verschiedenste andere Kleinteile wie Schrauben, Überwurfmuttern, Dichtkegel und Stützringe enthalten.

Der Kistenboden war mit einem Einsatz unterteilt. Dort waren 6 »Segel« (Luftruder) (37) für das Gerät verstaut.

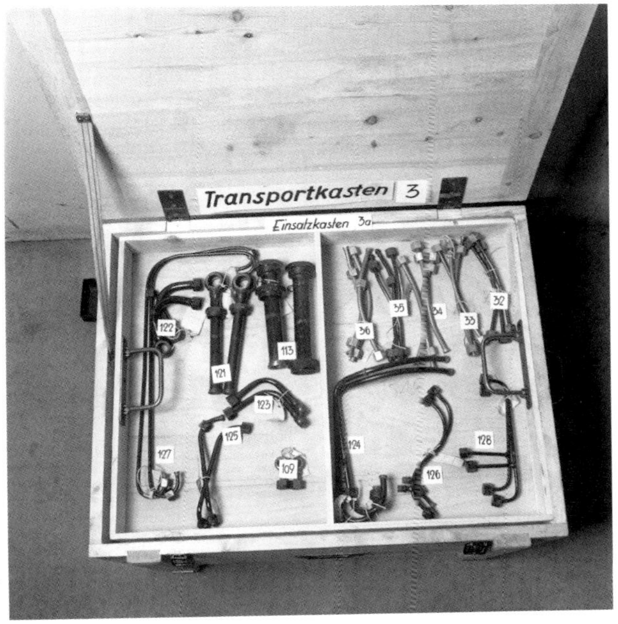

BArch, RH8II Bild-B1266-44, o. Ang.

275

BArch, RH8II Bild-B1265-44, o. Ang.

276

BArch, RH8II Bild-B1264-44, o. Ang.

277

Vehicle 511

Internal or former designation: Fz.-Pos.-Nr.511
Two-axle trailer for Kfz. workshop
Kfz. workshop trailer

DMA, GD639.5.18
HAP11/T 507
Karlshagen, 1. 6.1944
Bd.Nr. E 3195/44 g

DMA, GD639.5.1
EW234
Fz.-Pos.-Nr. 511

Intended use
A body is mounted on a two-axle trailer for work on motor vehicles. The interior accommodates workbenches, tools, a lathe, and a column drilling machine. The required power is generated by a mobile three-phase generator 220/380 V 3.2 kVA, which is stowed inside the workshop trailer during transport.

In addition to the motor vehicle workshop team of vehicle group 3, the technical troup is also able to carry out simple turning work, production of components and modifications to the special vehicles to a limited extent.

The workshop vehicle trailer has the following dimensions and weights:

Total length:	5,950 mm
Width:	2,200 mm
Height:	2,970 mm
Wheelbase:	4,700 mm
Track width:	1,573 mm
Tare weight:	2,000 kg
Payload:	2,500 kg
Combat weight:	3,400 kg
Wheels:	7.25-20

The main parts of the two-axle workshop trailer are:
- Chassis
- Body
- Installations, lathe and three-phase generator 220/380 V 3.2 kVA

Chassis
The chassis uses essential components of the trailer 3 t, type E3 (type 02) (equipment description according to D 692/21) and serves as a basis for the body.

DMA, GD639.5.21
OKH/Wa Prüf10
Karlshagen, 1.4.1944
Bd.Nr. 789/44 gK
Auszug
Besprechung der
Fahrzeuggruppe 5
vom 1.4.44
in Karlshagen I

Production status: vehicle 511
Total provisional requirements 7 vehicles.
Available at Kdo. Basse 4 cars (empty) on 20.4.44.
At firm 4 vehicles. These were intended for LR and must go directly to Kdo. Basse. Initiation by Wa Prüf 11/III.

Fahrzeug 511

Interne oder frühere Bezeichnung: Fz.-Pos.-Nr.511
Zwei-Achs-Anhänger Kfz.-Werkstattwagen
Kfz.-Werkstattanhänger

Verwendungszweck
Auf einem zweiachsigen Anhänger ist ein geschlossener Aufbau für den Betrieb als Kfz.-Werkstattwagen montiert. Im Innenraum sind Werkbänke, Werkzeuge, eine Drehbank und eine Ständerbohrmaschine untergebracht. Der benötigte Strom wird mit einem mobilen Drehstromaggregat 220/380 V 3,2 kVA erzeugt, das während des Transportes im Innenraum des Werkstattwagens verstaut wird.

In Ergänzung zum Kfz.-Werkstattrupp der Fahrzeuggruppe 3 ist es der Technischen Truppe im begrenzten Umfang möglich auch einfache Dreharbeiten, Anfertigung von Bauteilen und Änderungen an den Spezialfahrzeugen auszuführen.

Der Werkstattwagen-Anhänger hat folgende Maße und Gewichte:

Gesamtlänge:	5.950 mm
Breite:	2.200 mm
Höhe:	2.970 mm
Radstand:	4.700 mm
Spurweite:	1.573 mm
Leergewicht:	2.000 kg
Nutzlast:	2.500 kg
Gefechtsgewicht:	3.400 kg
Räder:	7,25-20

Die Hauptteile des zweiachsigen Werkstattwagens sind:
- Fahrgestell
- Aufbau
- Einbauten, Drehbank und Drehstromaggregat 220/380 V 3,2 kVA

Fahrgestell
Als Fahrgestell wurden wesentliche Bauteile des Einheitsanhängers 3 t, Baumuster E3 (Typ 02) (Gerätebeschreibung nach D 692/21) verwendet und dienen als Basis für das angepasste Fahrgestell.

Fertigungslage: Fahrzeug 511
Vorläufiger Gesamtbedarf 7 Fahrzeuge.
Verfügbar bei Kdo. Basse 4 Fahrzeuge (leer) am 20.4.44.
Bei Firma 4 Fahrzeuge. Diese waren vorgesehen für LR und müssen direkt zum Kdo. Basse. Veranlassung Wa Prüf 11/III.

BArch, RH8II Bild-B1478-44, o.Ang.

278

Chassis and body
Fahrgestell und Aufbau

278 The body, made of a wooden frame and curved roof assembly is attached to the floor assembly of the E3 chassis with corner reinforcements. A spare wheel basket is mounted under the floor. The side walls and the roof are covered with profiled wood and the roof surface is sealed with tar paper. Five windows are fitted in the walls and give enough light into the interior. On the front wall there are holder for entrenching equipment, wheel wrenches and a power connection.

278 Der Aufbau, gefertigt aus einem Holzrahmen und gewölbter Dachbaugruppe, ist mit Eckverstärkungen an der Bodengruppe des E3-Fahrgestells befestigt. Unter der Bodengruppe ist ein Ersatzradkorb angebracht. Die Seitenwände und das Dach sind mit Profilholz verkleidet. Die Dachfläche ist mit Teerpappe abgedichtet. Fünf Fenster sind in die Wände eingepasst und geben ausreichend Licht in den Innenraum. An der vorderen Wand sind Halter für Schanzzeug, Radschlüssel und ein Stromanschluss montiert.

BArch, RH8II Bild-B1477-44, o.Ang.

BArch, RH8II Bild-B1479-11, o.Ang.

279

280

Body – back side
Aufbau – Rückseite

279, 280 A wide access stair and a separate step are mounted at the rear. If the access stairway is lowered, access to the interior is possible via a large double-leaf door.

When the doors are open, the following fixtures are visible: On the left, various boxes for special tools, adjusting device and carpenter's tools. In the middle is the three-phase generator, and on the right a standard cabinet with lathe accessories and a compartment with five machinist toolboxes.

279, 280 An der Rückseite ist eine breite Zugangstreppe und ein Auftritt montiert. Bei herabgelassener Zugangstreppe ist der Zugang über eine große zweiflügelige Tür in den Innenraum möglich.

Bei geöffneten Türen sind folgende Einbauten sichtbar: Links: verschiedene Kisten für Spezialwerkzeug, Justiergerät und Tischlerwerkzeug. Mittig: der Drehstromgenerator. Rechts: ein Einheitsschrank mit Drehbankzubehör und ein Fach mit fünf Schlosser-Werkzeugkästen.

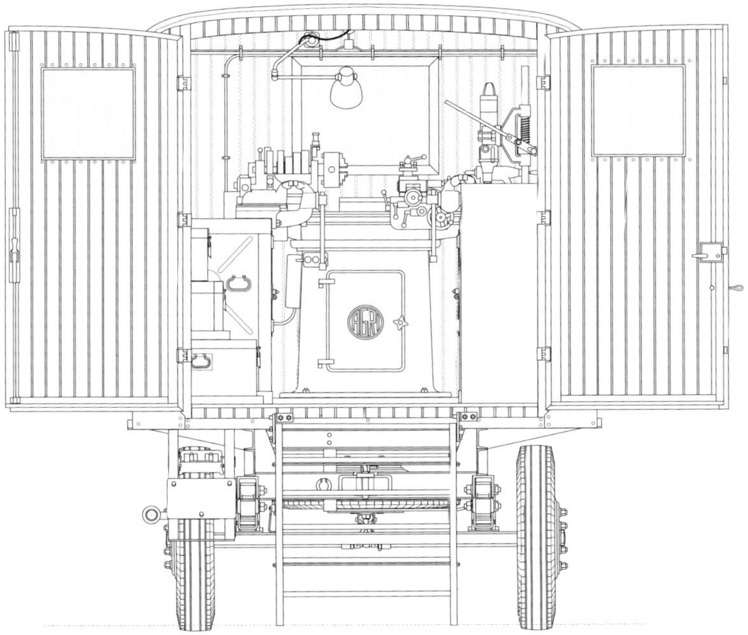

1 m

Scale / Maßstab 1:35 © Volker Ruff

BArch, RH8II Bild-B1170-44, o. Ang.

281

BArch, RH8II Bild-B1171-44, o. Ang.

282

Interior
Innenraum

281, 282 In addition to the above mentioned unit frame and tool boxes , there are at left and right »Matra« EWB2 unit workbenches with a machinist vice and a hand drill with drill stand.

On the front wall there is a lathe made by »Agri«, spindle height 135 mm, drilling length 600 mm. The current produced by the power generator is switched via a fuse panel located on the right wall above the lathe spindle. Lighting is provided by a ceiling light and three workshop lights above the workstations.

281, 282 Neben dem bereits erwähnten Einheitsgestell und den Werkzeugkisten sind links und rechts Einheitswerkbänke der Marke »Matra« EWB2 mit Parallelschraubstock und einer Handbohrmaschine mit Bohrständer vorhanden.

An der Stirnwand ist eine Drehbank Fabrikat »Agri« mit einer Spindelhöhe von 135 mm und einer Bohrlänge von 600 mm eingebaut. Der vom Drehstromgenerator erzeugte Strom wird über eine Sicherungstafel geschaltet, die sich an der rechten Wand oberhalb der Drehbankspindel befindet. Als Beleuchtung ist eine Deckenleuchte und drei Werkstattleuchten über den Arbeitsplätzen vorhanden.

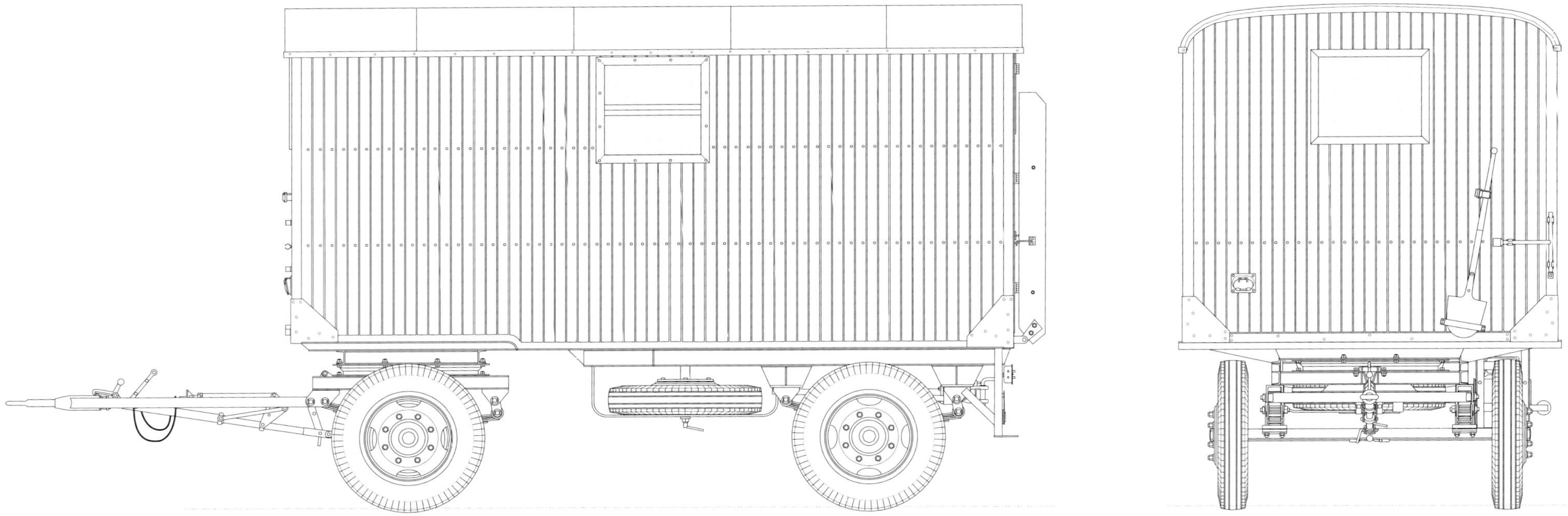

1 m

Scale / Maßstab 1:35 © Volker Ruff

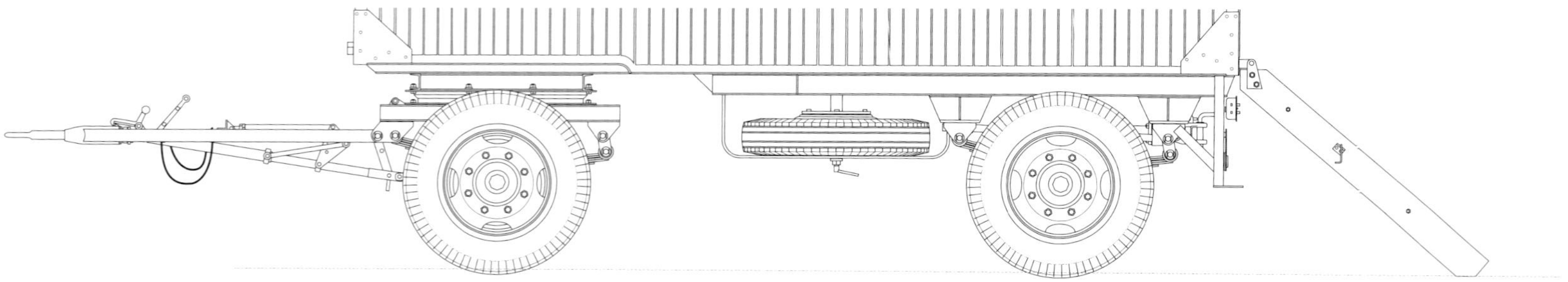

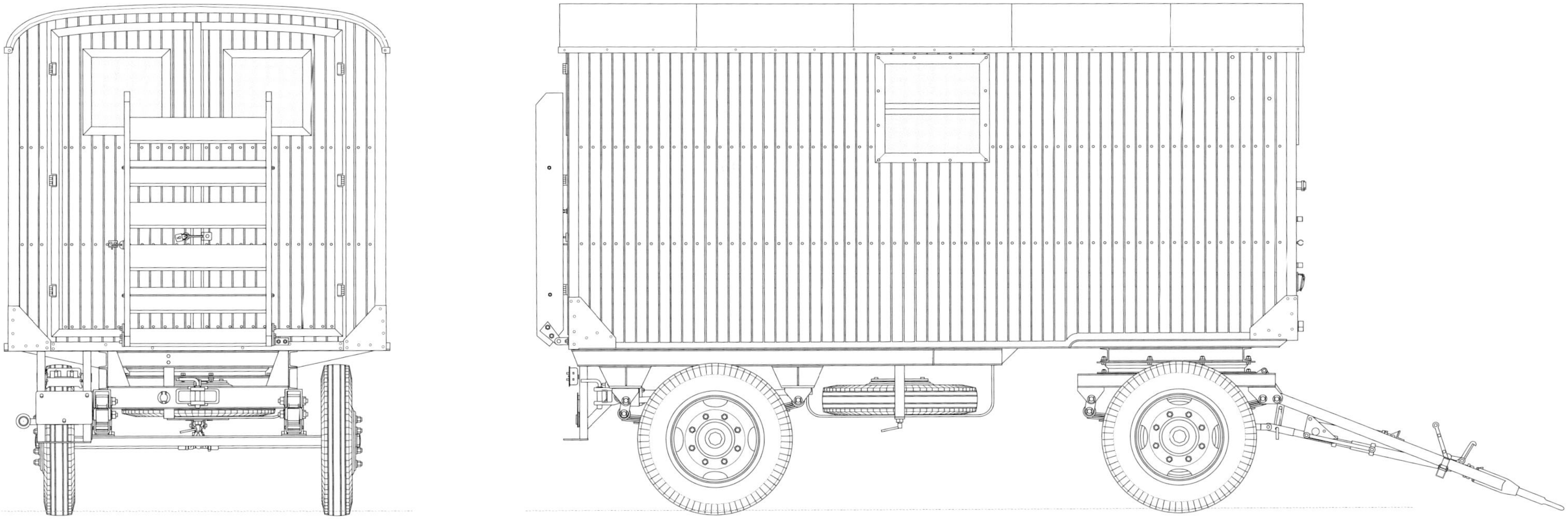

1 m

Scale / Maßstab 1:35 © Volker Ruff

EWB II

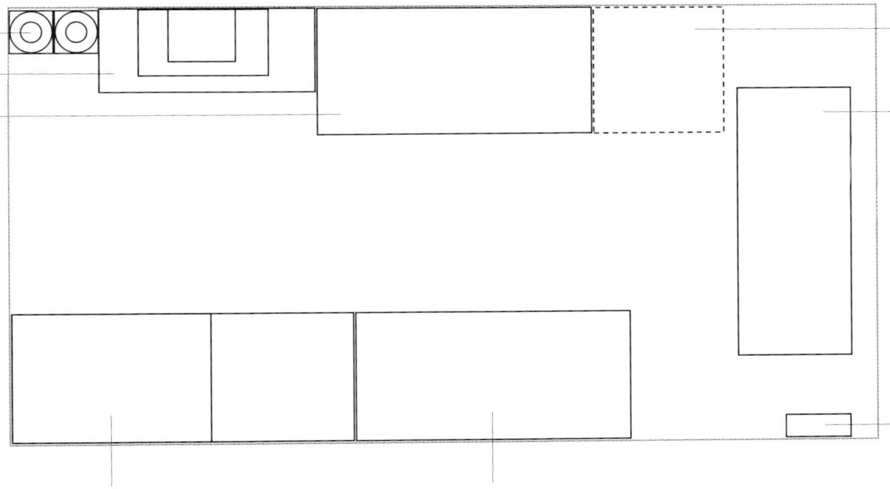

1 Welding gas cylinders
2 Testing- and measuring tools in boxes
3 EWB II standard work bench
4 Shelf
5 EWB II standard work bench
6 Switch board
7 Lathe type AGRI
8 Field forge (optional)

1 Gasflaschen Schweißen
2 Prüf – und Messwerkzeuge in Kisten
3 EWB II Einheitswerkbank
4 Regal
5 EWB II Einheitswerkbank
6 Sicherungskasten
7 Drehbank type AGRI
8 Feldschmiede (optional)

Fahrzeug 511

Zwei-Achs-Anhänger Kfz.-Werkstatt

Drawing set used D692/21, Baumuster trailer E3.

Verwendeter Zeichnungssatz D692/21, Einheitsanhänger E3.

VEHICLE GROUP VI
FAHRZEUGGRUPPE VI

DMA, GD630.02
A4 Lehrbuch
der FR-Schule
Mai 1944

Auszug aus
Kapitel
Nachschub Transport
Umschlag

Supply unit

Purpose and task

Special road transport trailers had been developed for road transport, of which two versions were available. The older version, called the Vidalwagen, had two axles and was lighter than the newer model with a third axle and bogie. The later version was similar to the FR trailer in the construction of its chassis and had a correspondingly greater overall height and thus a higher position of the A4 *Gerät*. The *Gerät* is placed with the tail section directly behind the prime mover, thus reducing the risk of damage to the sensitive outer parts of the tail section.* With the »Vidalwagen«, the A4 *Gerät* is positioned with the rear end to the rear. For the sake of completeness, the FR trailer should also be mentioned here as a special trailer for road transport over short distances. Its main task, however, was to lift the A4 *Gerät* and place on the launch platform.

BAMA RH8-1310
Spezialfahrzeuge
Der Vidalwagen war
dort nicht erfasst.

DMA, GD637.9.2
HAP11
Archiv N.72/36
Versuchsbericht
Versuche mit
Straßenfahrzeugen
und Feldbahnwagen

Special vehicles of the supply unit

Fz.-Pos.-Nr.601	FR transport trailer (2-axle) »Vidalwagen«
Fz.-Pos.-Nr.601*	FR transport trailer (3-axle) »Vidal«
Fz.-Pos.-Nr.601b	FR transport trailer (2-axle) »Vidalwagen«

* Damage to the tail section and fins of the A4 *Gerät* has been detected several times. These were obviously caused by vibrations of the *Gerät* during transport. By order of the H.A.P. the Reichsbahn-Versuchsamt carried out »Tests with road vehicles and light rail cars« in June 1944. Several A4 *Geräte* with and without an attached warhead »Elephant« were available for the tests. The prepared *Gerät* was loaded and secured on the special trailers Vidal 2-axle, new Vidal 3-axle and Meillerwagen. To establish the data the trailers were driven at varying speeds over roads with different characteristics. The result of the recorded data was sobering. Both the Meillerwagen and the new Vidal 3-axle wagon suffered damage to the tail section, which necessitated changes to the special road transport trailers. By reducing the driving speed of the Meillerwagen, changes to the erecting frame could be avoided. The changes to the Vidal 3-axle became so complex that a production stop was imposed and the »older« Vidal version was retained with a few modifications, e.g. a tubular frame for protection of the *Geräte* tail section.

Nachschubeinrichtung

Zweck und Aufgabe

Für den Transport auf der Straße sind spezielle Straßentransport-Sonderanhänger entwickelt worden, von denen zwei Ausführungen vorhanden sind. Die ältere Ausführung, genannt »Vidalwagen«, besitzt zwei Achsen und ist leichter ausgeführt als das neue Modell mit einer dritten Achse und einem Drehgestell. Die letztere Ausführung ähnelt im Aufbau des Fahrgestells dem FR-Wagen, hat eine entsprechend größere Bauhöhe und damit eine höhere Lage des A4-Gerätes. Auch wird hier das Gerät so aufgelegt, dass sich das Heck unmittelbar hinter der Zugmaschine befindet und dadurch die Gefahr einer Beschädigung der empfindlichen äußeren Heckteile vermindert wird.* Beim »Vidalwagen« liegt das A4-Gerät mit dem Heck nach hinten. Der Vollständigkeit halber sei hier auch der FR-Wagen als Sonderanhänger für Straßentransporte auf kurzen Wegstrecken erwähnt. Seine Hauptaufgabe ist jedoch das Aufrichten und Absetzen des A4-Gerätes auf die Abschussplattform.

Spezialfahrzeuge der Nachschubeinrichtung

Fz.-Pos.-Nr.601	FR-Transportanhänger (2-achs.) »Vidalwagen«
Fz.-Pos.-Nr.601*	FR-Transportanhänger (3-achs.) »Vidal«
Fz.-Pos.-Nr.601b	FR-Transportanhänger (2-achs.) »Vidalwagen«

* Es wurden mehrfach Beschädigungen am Heckteil und Flossen des Gerätes festgestellt. Diese wurden offensichtlich durch auftretende Schwingungen des Gerätes beim Transport verursacht. Im Auftrag des H.A.P fanden im Juni 1944 durch das Reichsbahn-Versuchsamt »Versuche mit Straßenfahrzeugen und Feldbahnwagen« statt. Für die Versuche standen mehrere A4-Geräte mit und ohne angebauter Nutzlastspitze »Elefant« zur Verfügung. Die vorbereiteten Geräte wurden auf die Sonderanhänger Vidalwagen 2-achs, neuer Vidal 3-achs und Meillerwagen verladen und gesichert. Zur Ermittlung der Daten wurden die Anhänger bei wechselnden Geschwindigkeiten über unterschiedlich beschaffene Straßen gefahren. Das Ergebnis der erfassten Daten war ernüchternd. Sowohl beim Meillerwagen als auch beim neuen Vidalwagen 3-achs traten Beschädigung am Heckteil auf, die Änderungen an den Straßentransport-Sonderanhängern erforderlich machten. Durch Herabsetzen der Fahrgeschwindigkeit des Meillerwagens konnten Änderungen am Kipperrahmen vermieden werden. Die Änderungen am Vidal 3-achs wurde derart aufwändig, dass hier ein Fertigungsstopp ausgesprochen wurde und man die »ältere« Vidal-Ausführung mit ein paar Anpassungen, z.B. einem Rohrrahmen zum Schutz für das Geräteheck beibehielt.

283

283 At the »Heidelager« military training area near Blizna in Southeast Poland, practiced the Art. Divisions until mid-June 1944 transportation and loading procedures. These exercises consisted of taking over the rocket, reloading it onto Vidalwagen and subsequent transfer from the Vidalwagen to the Meillerwagen. The A4 *Gerät* was brought up on the Vidalwagen in the »old« version Fz.601 and lifted up with the 16 t Strabo crane and hoist »Elephant«. One now waited to be reloaded onto the Meillerwagen. Meanwhile the Vidal wagen was being towed out of the crane area.

283 Auf dem Truppenübungsplatz »Heidelager« bei Blizna in Südostpolen übten die Art. Abteilungen bis Mitte Juni 1944 unter anderem auch die Transport- und Verladeabläufe. Diese Übungen bestanden aus der Übernahme der Rakete, dem Umladen auf Vidalwagen und anschließendem Umsetzen vom Vidalwagen auf den Meillerwagen. Das A4-Gerät wurde auf dem Vidalwagen in der »alten« Ausführung Fz.601 herangefahren und mit dem 16 t-Strabokran und Heißgeschirr »Elefant« entnommen. Man wartete nun auf das Umladen auf den Meillerwagen. In der Zwischenzeit wurde der Vidalwagen aus dem Kranbereich geschleppt.

DM, GD639.5.18
HAP11/T 507
Karlshagen, 1.6.1944
Bd.Nr. E 3195/44 g

DMA, GD630.02
A4 Lehrbuch
der FR-Schule
Stand Mai 1944
Auszug aus
Technische Fahrzeuge

Vehicle 601

Internal or former designation: Fz.-Pos.-Nr.601
FR Transport trailer (2-axle) »Vidalwagen«
Vidalwagen

Intended use
For the tasks – taking over the rocket from the delivery station, transport to the field storage area, mounting the warhead, delivery for reloading on the Meillerwagen and transport on the road – a special road transport trailer was developed.
This was called the Vidalwagen.

The Vidalwagen Fz.601 was a simple construction, designed for mass production. Tubes (commercial lengths of 6 m) were welded together to form an acute-angled triangular frame stiffened by cross tubes. At the apex of the triangle a plate with a ball head was welded on and a support saddle bolted on top of it. The whole thing was rotatably mounted on an axle with single tires, transverse spring package and drawbar. At the left and right ends of the frame spring assemblies and an axle with dual tires are attached. The rear axle had mechanical brakes and a brake spur to prevent rolling backwards. A mounting plate was placed over the axis and welded to the tubular frame. The large saddle for the A4 *Gerät* was bolted on this and braced.

The Vidalwagen had the following dimensions and weights:

Total length:	11,620 mm
Width:	2,140 mm
Height:	1,900 mm
Wheelbase:	6,500 mm
Track width:	1,860 mm
	1,715 mm distance between dual tires
Tare weight:	n.a.
Payload:	n.a.
Combat weight:	n.a.
Wheels:	190-20

The main parts of the Vidalwagen are:
- Chassis
- Support saddle front and rear
- Tooling, accessories and spare wheel

Fahrzeug 601

Interne oder frühere Bezeichnung: Fz.-Pos.-Nr.601
FR-Transportanhänger (2-achs.) »Vidalwagen«
Vidalwagen

Zweck und Aufgabe
Für die Aufgaben – Übernahme der Rakete vom Anlieferbahnhof, Transportieren zum Feldspeicher, Montage der Nutzlastspitze, Anliefern zum Umladen auf dem Meillerwagen und den Transport auf der Straße – wurde ein besonderer Straßentransport-Sonderanhänger entwickelt.
Dieser wurde Vidalwagen genannt.

Der Vidalwagen Fz.601 war eine einfache Konstruktion und für Massenfertigung ausgelegt. Mehrere Rohre (Handelslänge 6 m) sind zu einem spitzwinkligen Dreiecksrahmen zusammengeschweißt, Querrohre versteifen den Rahmen. An der Dreieckspitze ist eine Platte mit Kugelkopf angeschweißt und darauf ein Auflagesattel geschraubt. Das Ganze ist drehbar auf eine Achse mit Einfachbereifung, querliegendem Federpaket und Zugstange montiert. Am Rahmenende sind links und rechts Federpakete angebracht und eine Achse mit Doppelbereifung befestigt. Die Hinterachse besitzt eine mechanisch arbeitende Bremse, gegen ein Rückwärtsrollen ist ein Bremssporn montiert. Über der Achse ist eine Montageplatte gesetzt und mit dem Rohrrahmen verschweißt. Darauf ist der große Auflagesattel für das A4-Gerät montiert und abgestrebt.

Der Vidalwagen 601 hat folgende Maße und Gewichte:

Gesamtlänge:	11.620 mm
Breite:	2.140 mm
Höhe:	1.900 mm
Radstand:	6.500 mm
Spurweite:	1.860 mm
	1.715 mm hinten Mittenabstand zwischen Doppelreifen
Leergewicht:	k.A.
Nutzlast:	k.A.
Gewicht beladen:	k.A.
Räder:	190-20

Die Hauptteile des Vidalwagen:
- Rohrrahmen-Fahrgestell
- Aufnahmesattel vorne, Aufnahmesattel hinten
- Werkzeug, Zubehör und Reserverad

BArch, RH8II Bild-B1808-44, o.Ang.

284 The A4 *Gerät* without its »Elephant« warhead is protected with canvas tarpaulins. The multi-part tarpaulins, sprayed in an interesting camouflage pattern, are linked with straps. In order to further hide the contour of the *Gerät* a framework of tubes and angle sections was placed on the Vidalwagen Fz.601 and lashed down with tension wires, over which further tarpaulins were then drawn for camouflage.

284 Das A4-Gerät ohne Nutzlastspitze »Elefant« ist mit Segeltuchplanen geschützt. Die mehrteiligen Planen sind mit Riemen verknüpft und mit einem interessanten Tarnmuster versehen. Um die Kontur des Gerätes weiter zu verbergen, wurde ein Gerüst aus Rohren und Winkelprofilen auf den Vidalwagen Fz.601 gesetzt und mit Spanndrähten festgezurrt, über die dann weitere Planen zur Tarnung gezogen wurden.

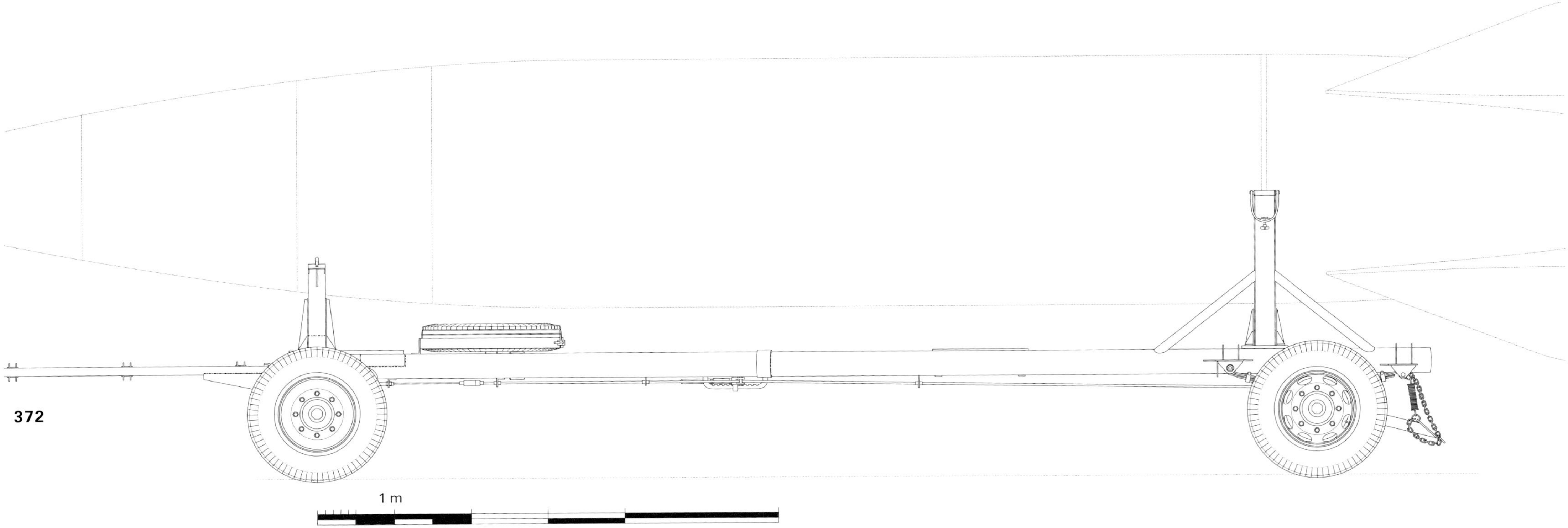

Scale / Maßstab 1:35 © Volker Ruff

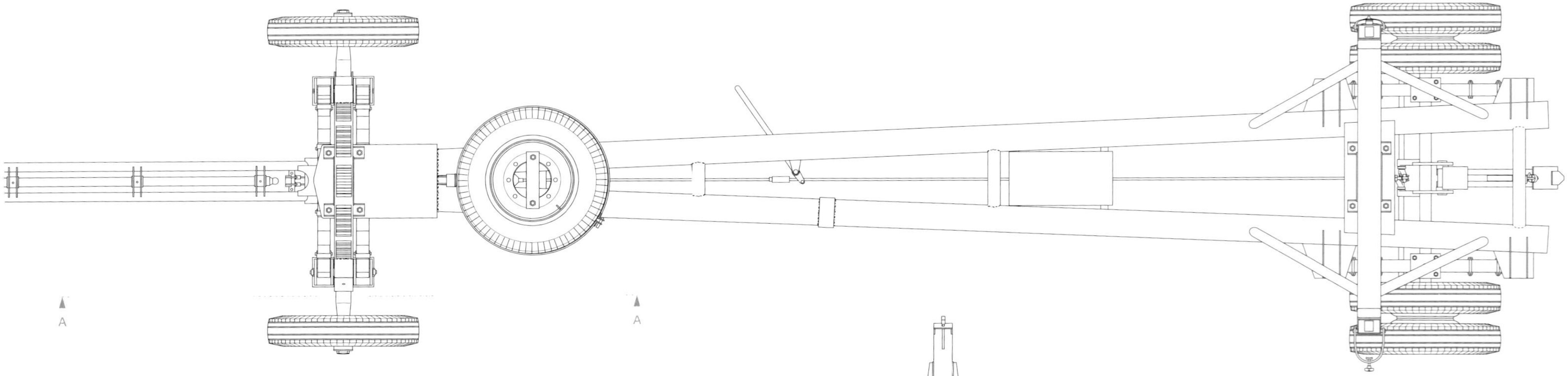

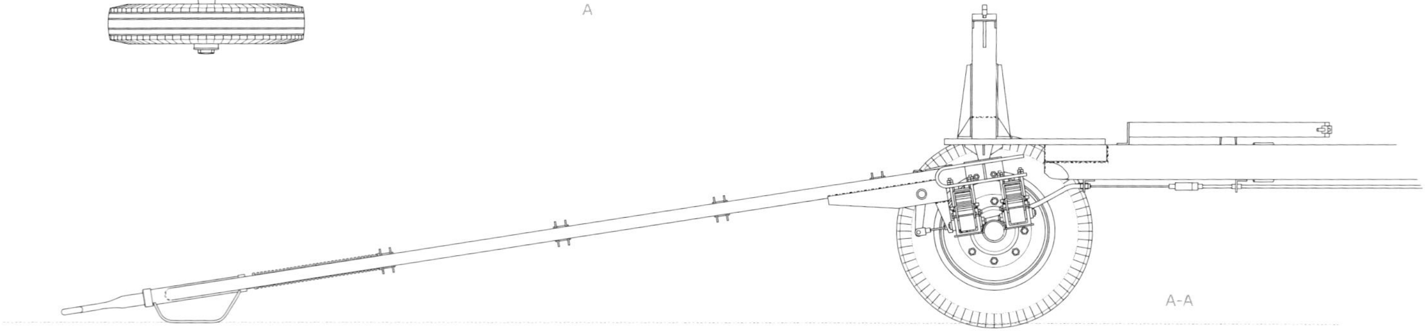

Fahrzeug 601

601 FR-Transportanhänger (2-achs.) »Vidal«

Partially used drawings E3156DP, measurements from original parts

Teilweise verwendete Zeichnungen E3156D, Maße von Originalteilen

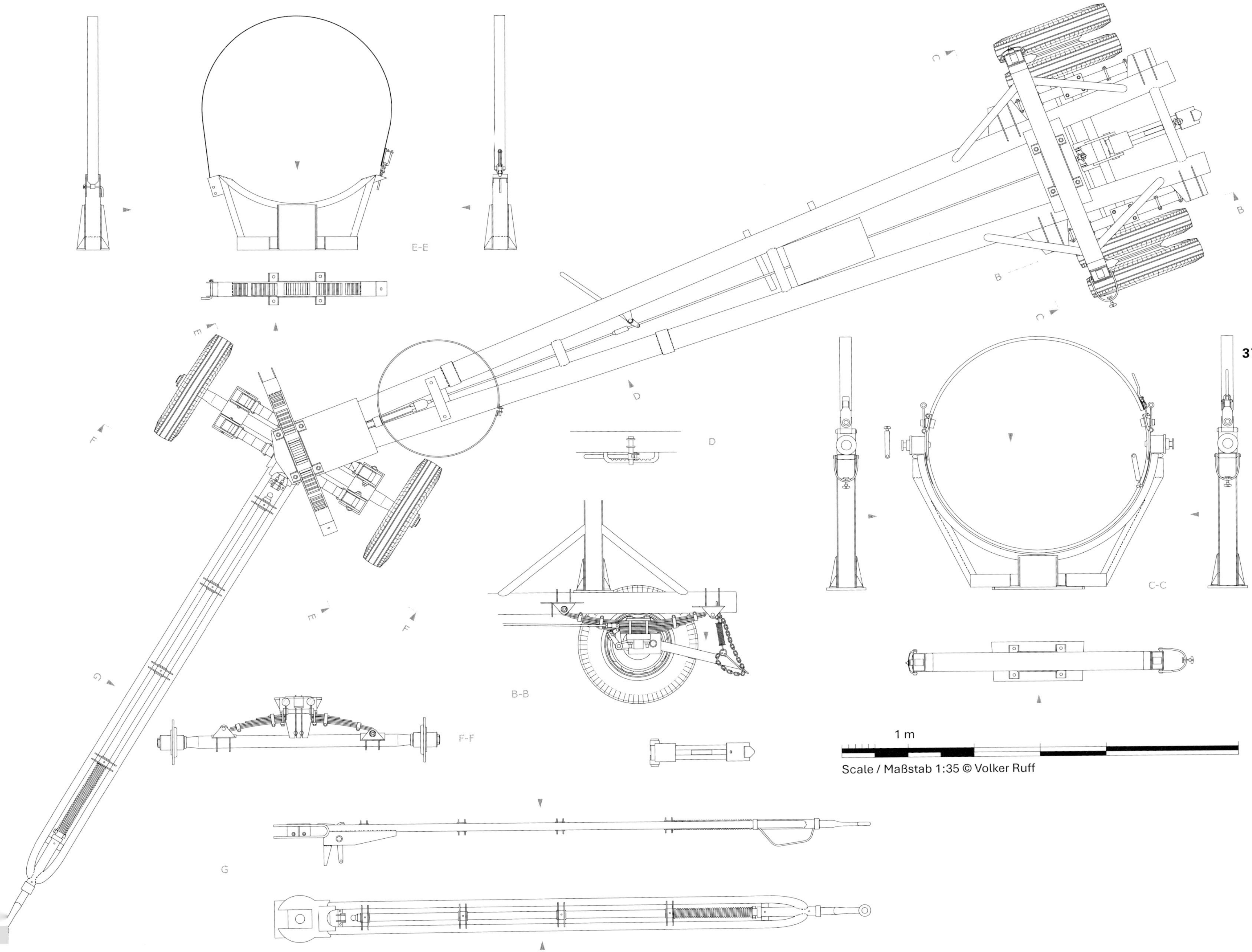
E-E
B-B
C-C
D
F-F
G
1 m
Scale / Maßstab 1:35 © Volker Ruff

BArch, RH8II Bild-B1582-44, o.Ang.

285

Vidalwagen – EW23 Modification
Vidalwagen – EW23 Modifikation

285 In an intermediate step, the Fz.601 was tested with a modified framework. For this purpose the acute-angled triangular frame was separated in the middle and extended to the front with a slightly upwards bend H-frame. On the new front part a lower support saddle with height adjustment was mounted above the pivot axis and another support was attached to the frame head. To the rear the frame tubes were extended and a protective frame was welded on.

285 In einem Zwischenschritt wurde der Fz.601 mit einem geänderten Rahmen erprobt. Dazu wurde der spitzwinklige Dreiecksrahmen in der Mitte getrennt und nach vorne mit einen leicht ansteigenden H-Rahmen verlängert. Am neuen Vorderteil wurde über der Drehachse ein niedrigerer Auflagesattel mit Höhen-Verstellung montiert und am Rahmenkopf eine weitere Auflage angebracht. Nach hinten wurden die Rahmenrohre verlängert und ein Schutzrahmen angeschweißt.

286 The *Gerät* is protected with multi-part tarpaulins. The belt straps for the »Elephant« hoist harness remain on the *Gerät* ; the belt strap at the separation point has locating pins that fit into their counterparts on the support saddles of Vidal- and Meillerwagens. In addition, a belt strap holds the *Gerät* on the support above the front axle; the head support supports the warhead. The picture clearly shows the slight inclination of the *Gerät* structure with the tip pointing downwards, above the support point of the rear support saddle.

286 Das Gerät ist mit mehrteiligen Abdeckplanen geschützt. Die Gurtbänder für das Heißgeschirr »Elefant« sind am Gerät verblieben. Das Gurtband an der Trennstelle hat Aufnahmezapfen, die in die Gegenstücke am Auflagesattel von Vidal- und Meillerwagen passen. Zusätzlich hält ein Gurtband das Gerät am Auflager oberhalb der Vorderachse. Die Kopflagerung unterstützt den Gefechtskopf. Das Bild zeigt deutlich die leichte Neigung der Gerätestruktur mit der Spitze nach unten über den Aufnahmepunkt des hinteren Auflagesattels.

BArch, RH8II Bild-B1808-44, o.Ang.

286

DMA, GD630.02
A4 Lehrbuch
der FR-Schule
Stand Mai 1944
Auszug aus
Technische Fahrzeuge

DMA, GD637.9.2
HAP11
Archiv N.72/36
Versuchsbericht
Versuche mit
Straßenfahrzeugen
und Feldbahnwagen

Vehicle 601 *

Internal or former designation: Fz.-Pos.-Nr.601
FR transport trailer (3-axle) »Vidal«

The new version of the FR transport trailer Vidal with 3 axles was designed similarly to the Meillerwagen with one front axle and one bogie suspension. Both axles were rotatable, and attached to a chassis frame made of a bundle of three tubes. A pluggable frame work for tarpaulins was mounted around the chassis. The complex construction was much more difficult to drive and steer than the Vidal 2-axle. In some accidents the double axle broke in half, also the frame was twisted and similar damages and problems which were found during the driving tests of the Reichsbahn-Versuchsamt, finally led to the 3-axle version being abandoned.

Fahrzeug 601 *

Interne oder frühere Bezeichnung: Fz.-Pos.-Nr.601
FR-Transportanhänger (3-achs.) »Vidal«

Die neue Ausführung, der FR-Transportanhänger »Vidal« mit 3 Achsen, wurde ähnlich dem Meillerwagen mit einer Vorder- und einer Doppelachse ausgelegt. Beide Achsen waren drehbar an einen Fahrgestellrahmen aus einem Bündel von drei Rohren befestigt. Um das Fahrgestell war ein steckbares Gerüst für Abdeckplanen montiert. Die komplexe Konstruktion war wesentlich schwieriger zu fahren und zu lenken als der Vidal 2-Achser. Bei einigen Unfällen brach die Doppelachse in zwei Hälften, auch der Rahmen wurde verdreht. Ähnliche Schäden und Probleme, die bei den Fahrversuchen des Reichsbahn-Versuchsamtes festgestellt wurden, führten schließlich dazu, dass die 3-achsige Version aufgegeben wurde.

BArch, RH8II Bild-D06-44

287

* If the designation logic is followed, FR transport trailer (3-axle) would have to be listed as Fz.601a, but there is no proof of this.

* Folgt man der Bezeichnungslogik, müsste der FR-Transportanhänger (3-Achser) als Fz.601a geführt werden. Hierfür fehlt der Nachweis.

Vehicle 601b

Internal or former designation: Fz.-Pos.-Nr.601b
FR transport trailer (2-axle) »Vidalwagen«

The following changes were made to the Vidalwagen (2-axle): By lowering the front support saddle and slightly adjusting the inclination the rear support saddle, the axis of the rocket tilts over the rear support point. A tubular frame was added to protect the rear of the *Gerät*, and the drawbar was reinforced. This revised version was named Fz.601b. The previously delivered Fz.601 was exchanged for newly manufactured Fz.601b. The »old« Fz.601 were modified and returned to the supply units Kf.Abt 899 and 900.

Fahrzeug 601b

Interne oder frühere Bezeichnung: Fz.-Pos.-Nr.601b
FR-Transportanhänger (2-achs.) »Vidalwagen«

Am Vidalwagen (2-Achser) wurden folgende Änderungen ausgeführt: Durch Absenken des vorderen Auflagesattels und geringer Anpassung der Neigung des hinteren Auflagesattels neigt sich die Achse der Rakete über den Aufnahmepunkt. Ein Rohrrahmen wurde zum Schutz des Gerätehecks angebracht und die Zugstange verstärkt. Diese überarbeitete Variante wurde als Fz.601b bezeichnet. Die bisher ausgelieferten Fz.601 wurden gegen neu gefertigte Fz.601b getauscht. Die »alten« Fz.601 wurden modifiziert und den Nachschubeinheiten Kf.Abt 899 und 900 wieder zugeführt.

288

288 Fifty years of flying show – german V2 Rocket – Hendon Aerodrome – UK 1951. Together with other aircraft, an V2 rocket from the collection of Operation Backfire was presented to the public

288 Fifty years of flying show – deutsche V2-Rakete – Hendon Aerodrome, UK 1951. Neben anderen Flugzeugen wurde auch eine V2-Rakete aus dem Fundus der Operation Backfire der Öffentlichkeit präsentiert.

ANNEX
ANHANG

Abbreviations / Abkürzungen

Abbreviations		Abkürzungen
(E)	E = railroad	E = Eisenbahn
(F)	F = fortress, bunker	F = Festung , Bunker
(S)	S = road	S = Straße
A4	Aggregate 4, see also *Gerät*, *Sondergerät*	Aggregat 4, siehe auch Gerät, Sondergerät
Abnahme	Army Acceptance (Service) H.Abn.	Heeresabnahme (Dienststelle) H.Abn.
Aggregat 4	aggregate 4 see also A4, *Gerät*, *Sondergerät*, V2, FR	Gerät weitere Bezeichnung für A4
AHA	General Army Office	Allgemeines Heeresamt
AI	Aero-Dynamic Institute at EW	Aero-Dynamisches Institut im EW
Art.Abt.	Artillery Abteilung, Artillery Division	Artillerie Abteilung
ASL, AS Lubmin	Branch office of Wa Prüf 10 in Lubmin	Aussenstelle des Wa Prüf 10 in Lubmin
A-Stoff	A-Stoff (substance) = LOX liquid Oxygen – 183 Grad Celcius	A-Stoff = Sauerstoff flüssig - 183 Grad Celcius
atü	Atmosphere overpressure	Atmosphäre Überdruck
B.d.E., BdE	Commander of the Replacement Training Army	Befehlshaber des Ersatzheeres
B.f.E.	Authorised representative for development in Wa Prüf 10	Bevollmächtigter für Entwicklung im Wa Prüf 10
B.f.F.	Authorised representative for research in Wa Prüf 10	Bevollmächtigter für Forschung im Wa Prüf 10
B.z.b.V.	Authorized representative for special use (army) (Generalmajor Dr. Ing. Dornberger)	Bevollmächtigter zur besonderen Verwendung (Heer) (Generalmajor Dr. Ing. Dornberger)
BA, B.A.	Ground facility	Boden-Anlage
Batt.	Battery	Batterie
B-Gestell	Control unit rack	Bediengeräte Gestell
Blinza	SS-Military training ground (Heidelager) near Blizna Poland	SS-Truppenübungsplatz (Heidelager) nähe Blizna/Polen
BN, B.N.	Ground network (power distributor, junction)	Bodennetz (Stromverteiler)
BN	Büssing-NAG	Büssing-NAG
BS	Combustion cut off	Brennschluß
BSM	On-board equipment, control system, Measurement technology and on-board system	Bordausrüstung, Steuerung, Meßtechnik und Bordnetz
B-Stoff	B-Stoff (substance) = Spiritus – Aethyl alcohol aqueous solution consists of 40 % ethyl alcohol + 35 % methyl alcohol + 25 % water	B-Stoff = Spiritus – Aethylalkohol wässrige Lösung bestehend aus 40 % Aethylalokohol + 35 % Methylalkohol + 25 % Wasser
BW	Aluminum treatment (pickling) workshop at EW	Beizwerkstatt im EW
cot	Manufacture code of Franz Xaver Meiller machine factory and wagon manufacturing plant Munich	Fertigungskennzeichen für Franz Xaver Meiller Maschinenfabrik und Wagenbauanstalt München
DIN	German Industry Standard	Deutsche Industrie Norm
dxl	Manufacture code Gottfried Lindner AG, Ammendorf b. Halle, a.d. Saale	Fertigungskennzeichen Gottfried Lindner AG, Ammendorf b. Halle, a.d. Saale

E3-Anhänger	Standard trailer 3 t type E 3	Einheitsanhänger 3 t Baumuster E 3
E5-Anhänger	Standard trailer 5 t type E 5	Einheitsanhänger 5 t Baumuster E 5
EW	Development Works at HAP, see also layout	Entwicklungswerk im HAP, HAP11 bis Ende Juni 44
EW	Electromechanical Works Karlshagen as of July 44	Elektromechanische Werke Karlshagen ab Juli 44
EW/T	Development Works/technology	Entwicklungswerk/Technik
EW1	Management	Leitung
EW2	Development	Entwicklung
EW21	Design department	Konstruktionsabteilung
EW22	Dept. of electrotechnical devices	Abt. Elektrotechnische Geräte
EW23	Dept. of ground installations	Abt. Bodeneinrichtungen
EW24	Production supervision	Fertigungsaufsicht
EW3	Manufacturing	Fertigung
EW4	Testing	Prüfung
EW5	Administration	Verwaltung
EWB	Unit workbench type I or II	Einheitswerkbank Typ I oder II
F.d.R.	For the correctness	Für die Richtigkeit
F.X.Meiller	Franz Xaver Meiller machine factory and wagon construction company Munich	Franz Xaver Meiller Maschinenfabrik und Wagenbauanstalt München
FF 33	Field telephone 33	Feldfernsprecher 33
FF, FFK, FF-Kabel	Distance cable, long-distance cable	Fernkabel, Fernfeldkabel
FL, FLK, FL-Kabel	108 wire AA cable (transmission line)	108-adriges Flakkabel (Übertragungsleitung)
FL-Panzer	Armored fire control vehicle (*Feuerleitpanzer*)	Sonder-Kfz.7 mit gepanzertem Aufbau (Feuerleitpanzer)
FL-Stecker, Flakstecker	AA plug 104 pins	FL = Flak, Flakstecker 104-polig
FL-Steckdose, Flaksteckdose	AA socket 104 pins	FL = Flak, Flaksteckdose 104-polig
FR	Long-range rocket	Fernrakete
FS	Teletype	Fernschreiben
FW	Production Works at EW	Fertigungswerk im EW
Fz In	Vehicle repair and maintenance	Fahrzeug Instandsetzung
Fz.	Vehicle	Fahrzeug
g, geh.	secret	geheim
Gerät	*Gerät* see also A4, Aggregat 4, *Sondergerät*, V2, FR	Gerät weitere Bezeichnung für A4
g.K, g.Ko, g.Kos	secret command matter	geheime Kommandosache
G+S	Company Gollnow and Son	Firma Gollnow und Sohn
GD	German Documents	German Documents
gep.	armored	gepanzert
gez	drawn, signed	gezeichnet, signiert
gl.	cross-country	geländegängig
glv	manufacture code Robert Schenk wagon factory, Stuttgart-Feuerbach	Fertigungskennzeichen Robert Schenk, Wagenfabrik Stuttgart-Feuerbach
Görkau	(Jirkov, Czechia – Bohemia) Location of TELEFUNKEN GmbH for highfrequency technology	(Jirkov, Tschechien – Böhmen) Standort der TELEFUNKEN GmbH für Hochfrequenztechnik
Gross-Born	Military training ground Groß Born, Eastern Pomerania, today in Poland (Borne Sulinowo)	Truppenübungsplatz Groß Born, Hinterpommern, heute in Polen (Borne Sulinowo)
H.Abn.	Army Acceptance (Service)	Heeresabnahme (Dienststelle)
HAP	Army Artillery Park Peenemünde	Heeres Artillerie Park Peenemünde
HAP11	Home Artillery Park 11	Heimat Artillerie Park 11
HF	High Frequency	Hochfrequenz
HKP	Army motor transport park	Heeres Kraftfahr Park
HKp	Armoured half-track vehicle	Halbkettenfahrzeug gepanzert
hnj	Manufacture code MAKO Maschinenbau und Co GmbH, Erfurt	Fertigungskennzeichen MAKO Maschinenbau & Co GmbH, Erfurt
Hptm.	Captain	Hauptmann
HVP	Army Research Station Peenemünde	Heeres-Versuchsanstalt Peenemünde
HWA	Army ordnance branch (Reichswehr)	Heereswaffenamt (Reichswehr)
Ia	First General Staff Officer	Erster Generalstabsoffizier
IG2 (I-Gerät)	Integration device, combustion cut-off by timer	Integration-Gerät, Brennschlussgabe mit Zeitschaltuhr
IW	Maintenance Works at EW	Instandsetzungswerk im EW
Ing.	Engineer	Ingenieur
k.A.	no details	keine Angaben
K.Kw	tank truck	Kessel-Kraftwagen
Kdo. Basse	Command Basse – Captain Basse of tech. Abt.91	Kommando Basse – Hauptmann Basse der tech. Abt.91

Kfz.	Motor vehicle	Kraftfahrzeug
Kitz	code name for Pre production factory (South)	Deckname für Versuch- und Vorserienwerk Süd
kVA	Kilovolt Ampere	Kilovolt Ampere
Kw	Motor vehicle, truck	Kraftwagen
kW	kilowatt	Kilowatt
L	Management	L = Leitung
Lager Rebstock	code name »Lager Rebstock« Camp Rebstock facility of G+S to manufacture the final outfitting and installations into the ground support vehicles located at the Ahr valley	Deckname der Betriebsstätte von G+S, Fertigung von Teilen der Endausrüstung und Einbau in die Bodenfahrzeuge, Standort im Ahrtal
LC-Koffer	Luftwaffe design of an closed superstructure	Luftwaffe design für einen geschlossenen Aufbau (Kfz.305)
Leitfirma	Leading firm coordination of suppliers for design, production, sourcing includes subcontracting.	Leitfirma – koordiert die Entwicklung, Fertigung, Beschaffung, und Unterbeauftragung.
Lindner	Gottfried Lindner AG, Ammendorf b. Halle, a.d. Saale	Gottfried Lindner AG, Ammendorf b. Halle, a.d. Saale
LR	Lager Rebstock – former railroad tunnels in the Ahr valley	Lager Rebstock – ehemalige Eisenbahntunnels im Ahrtal
LS	Guide beam	Leit-Strahl
Lt.	Lieutenant	Leutnant
LW-Anhänger	Air Force Trailer	Luftwaffen-Anhänger
m Zgkw	Medium tractor trailer	mittlerer Zugkraftwagen
MA	Military trailer, internal use of F.X.M	Militäranhänger, interne Verwendung F.X.M
Mako	Mako Maschinenbau und Co Gmbh, Erfurt	Mako Maschinenbau und Co Gmbh, Erfurt
MAN	Machine factory Augsburg Nuremberg	Maschinenfabrik Augsburg Nürnberg
MAN-Wagen	four-axle rail car SSla type Köln with Meiller tipping frame	vierachsiger Schienenwagen SSla Gattung Köln mit Meiller Kipprahmen
Marinesteckdose	Socket water and dust resistant with cap-protection	Wasser- und staubdichte Steckdose mit Schutzkappe
Mittelwerk GmbH (MW)	Founded on 21st of September 1943, located at Ilfeld Harz, responsible for the A4 assembly line	Gegründet am 21. Septmber 1943 mit Sitz in Ilfeld Harz, verantwortlich für die A4-Montagelinie
mot.	motorised	motorisiert
mpe	manufacture code Heimat Artillerie Park 11 Werk Nord (EW)	Fertigungskennzeichen Heimat Artillerie Park 11 Werk Nord (EW)
mpf	manufacture code Heimat Artillerie Park 11 Werk Süd (FW)	Fertigungskennzeichen Heimat Artillerie Park 11 Werk Süd (FW)
MU	Materials testing house at EW	Materialuntersuchungshaus im EW
MW	Adit system in the Kohnstein – facility for A4 production	Stollenanlage im Kohnstein – Serienfertigung der A4
N.A.	not applicable	nicht zutreffend
NC-collector	Nickel-cadmium (NiCd) collector, rechargeable battery	Nickel-Cadmium (NiCd) Sammler, wiederaufladbare Batterie, Akku
ND-Rohre	Low pressure pipes	Niederdruckrohre
NF	Low frequency	Niederfrequenz
NMH, NMH Kabel	Rubber hose line for medium stress	Gummischlauchleitung für mittlere Beanspruchung
NSH, NSH Kabel	Strong rubber hose line for particularly high mechanical stress	starke Gummischlauchleitung für besonders hohe mechanische Beanspruchung
Nutzlastspitze »Elefant«	Elephant = warhead (payload) with ~ 700 kg explosive »Amatol«	Elefant = Nutzlastspitze mit ~ 700 kg Sprengstoff »Amatol«
NW	Nominal size e.g. for inner diameter of hose	Nennweite z.B. für Schlauchinnendurchmesser
o.O.	without location	ohne Ortsangabe
o.U.	without accommodation (place, location)	ohne Unterkunft (Ort, Ortsangabe)
O_2-Anlage	LOX (liquid oxygen) facility at EW	Sauerstoff (flüssig) Produktionsanlage im EW
OB	Local area participants	Ortsbereich Teilnehmer
OB/ZG/W-Betrieb	Local-, central battery operation, dial-up operation for several (similar) interconnected, interconnected devices	Orts-, Zentralbatteriebetrieb, Wählbetrieb für mehrere (gleichartige) zusammengeschaltete, zusammengekoppelte Geräte
OKH	High Command of the Army	Oberkommando des Heeres
PL	Command office for location affairs	Kommandostelle für Platzangelegenheit
Produktenmesser	electrodynamic measuring device for AC voltage and DC voltage measurment, moving coil meter	elektrodynamische Meßwerk zur Spannungsmessung von Wechsel- und Gleichstrom, Drehspulmessgerät
Prozentenmesser	Voltmeter	Spannungsmesser (Voltmeter)
PS	Horsepower	Pferdestärke
P-Stoff	P-Stoff (substance) = Nitrogen – Nitrogen	P-Stoff = Stickstoff – Stickstoff
PW	Test workshop at EW	Prüfwerkstatt im EW
RAL	Reich Committee for Delivery Conditions (founded 1925)	Reichsausschuss für Lieferbedingungen (gegründet 1925)
RLM	Reich Aviation Ministry	Reichsluftfahrtministerium
S	HAP Commanding Officer	HAP- Kommandeur
SAERG	Special committee for electrotechnical accessories for Rockets	Sonderausschuß elektrotechnisches Zubehör für R-Gerät
Schenk	Robert Schenk wagon factory, Stuttgart-Feuerbach	Robert Schenk, Wagenfabrik Stuttgart-Feuerbach

Schlosser	Mechanic	Schlosser
SF, SF-Kabel	heavy field cable	schweres Feldkabel
Sd.Ah.	Special trailer	Sonderanhänger
SK	Special design	Sonderkonstruktion
Spi	Spirit = B-Stoff (substance)	Spiritus = B-Stoff
Sondergerät	*Sondergerät* see also A4, Aggregat, *Gerät*, V2, FR	Sondergerät weitere Bezeichnung für A4
SS	SS Special class = urgency classification of the procurement/order	SS Sonderstufe = Dringlichkeitseinstufung der Beschaffung/Bestellung
SS-Wagen	four-axle rail car SS type Köln	vierachsiger Schienenwagen SS Gattung Köln
St	Steel e.g. St 34 unalloyed structural steel	Stahl z.B. St 34 unlegierter Baustahl
Stotz I	Multi-pin connector / - connection of the A4 for relay of the control signals up to for tests respectively, up to launch	Vielpolstecker / - Anschluß der A4 zur Übertragung der Steuerungssignale für Tests bzw, bis zum Schuß
Stotz II	Multi-pin connector / - connection of the A4 for relay of the engine control signals and power supply for tests or until launch. The Stotz I + II connection at the A4 is locked by a magnet with guide pin, disconnection is made by current pulse at firing.	Vielpolstecker / - Anschluß der A4 zur Übertragung der Triebwerkssteuerungssignale und Stromversorgung für Tests bzw. bis zum Schuß. Die Stotz I + II Verbindung an der A4 ist durch ein Magnet mit Führungsstift gehalten, Trennung erfolgt durch Stromimpuls beim Abschuß.
T/Ass	Technology/Assistance	Technik/Assistenz
T1	Directorate of Operations	Betriebsdirektion
T2	Directorate Technic	Direktion Technik
T3	Directorate Development and Construction	Direktion Entwicklung und Konstruktion
T4	Directorate Testing	Direktion Prüfung
T5	Directorate electrical equipment	Direktion elektrische Geräte
TD	Technical Directorate (Management)	Technische Direktion
tech.	technical	technische
T-Kolonne	Fuel column, vehicles and pumps for A-, B- and T-Stoff	Tank-Kolonne, Fahrzeuge und Pumpen für A-, B- und T-Stoff
Tombak	Brass with copper content > 67 %	Messing mit Kupferanteil > 67 %
T-Stoff	T-Stoff (substance) = Thymiol – Hydrogen peroxide. Watery solution of 79 % hydrogen peroxide + 21 % water	T-Stoff = Thymiol – Wasserstoffperoxid. Wässrige Lösung aus 79 % Wasserstoffsuperoxid + 21 % Wasser
TW	Parts workshop at EW	Teilewerkstatt im EW
V	Volt = electrical voltage	Volt = elektrische Spannung
V K	Administation motor vehicles	Verwaltung Kraftfahrzeuge
V2	Vengeance Weapon 2 (V2), propaganda name of the A4 in service	Vergeltungswaffe 2 (V2), Propagandaname der A4 im Einsatz
Vers D	Test Dept.D, D = Implementation und testing	Versuch Abtlg.D D = Durchführung und Erprobung
Vers.	Test (Test Division)	Versuch (Abteilung Versuch)
VfR	Society for Space Travel founded in Breslau, 1927	Verein für Raumschifffahrt gegründet in Breslau 1927
Vowe Zange	movable tongs to hold the *Gerät*, designed by Ing. Vowe	bewegliche Zange zum Halten des Gerätes, Entwurf Ing. Vowe
VTL	provisional technical delivery specification	vorläufige technische Lieferbedingung
VW	Pre-series plant (south) in the HAP	Vorserienwerk (Süd) im HAP
Wa	ordnance, weapon	Waffen
Wa J Rü	Industrial mobilization Section in the Army Ordnance Office	Amtsgruppe für Industrielle Rüstung im WaA
Wa Prüf	Development and Testing Section in the Army Ordnance Office	Amtsgruppe für Entwicklung und Prüfung im WaA
Wa Prüf 6	Motor transport and Motorization Section of Army Ordnance Office	Abteilung Kraftfahr- und Motorisierungsabteilung im WaA
Wa Prüf 10	Flak Development Section in the Army Ordnance Office	Abteilung für Flakentwicklung im WaA
Wa Prüf 11	Special Equipment Section of Army Ordnance Office	Abteilung für Sondergerät im WaA
WaA	Army Ordnance Office	Waffenamt (Heereswaffenamt)
WKW	Wehrmacht Motor Vehicles Department	Wehrmachtkraftfahrwesen
WuG	Weapons and Equipment	Waffen und Gerät
WuG 5	Office group for Engineer and Fortress Construction	Amtsgruppe Pionier und Festungsbau
WuG 6	Office group for Tank and Motorization	Amtsgruppe Panzer und Motorisierung
WuG 7	Office group for radio equipment	Amtsgruppe Nachrichtengeräte
WuG 10	Office group for Flak Development	Amtsgruppe für Flakentwicklung
WuG 11	Office group for Special equipment	Amtsgruppe Sondergeräte
WuG 12	Office group for Test and Experimental grounds	Amtsgruppe Versuchsplätze
Z	Central Administration	Zentralverwaltung
Z-Stoff	Z-Stoff (substance) = Rubide – Sodium Permanganate aqueous solution of 27 % sodium permanganate + 73 % water	Z-Stoff = Rubid – Natriumpermanganat wässrige Lösung bestehend aus 27 % Natriumpermanganat + 73 % Wasser
ZW	Assembly workshop at EW	Zusammenbauwerkstatt im EW

Annotations

Page/Seite 10

[1] »The Rocket to Planetary Spaces 1923, Paths to Space Flight 1929,« by Herman Oberth, rocket vehicles (Opel, Valier) and smaller rocket models were built and successfully tested.

[2] In 1927 the »Verein für Raumschifffahrt« (VfR) was founded in Breslau with the aim to explore space and other planets. Members were among others Max Valier, Johannes Winkler, Hermann Oberth, Rudolf Nebel, Wernher von Braun.

Page/Seite 12

[1] Heeresversuchsanstalt Kummersdorf. Development and Testing Centre for New Weapon Systems and Equipment.

[2] Paris cannon WWI – manufacturer the Friedrich Krupp AG, Essen. Range of projectiles approx. 130 km.

Page/Seite 14

[1] In the »Mission planning A4 according to the state of development at the end of 1942«, the vehicle requirement for a FR-Abteilung (mot) was largely confirmed. Bd.Nr3/43 g.k. Chefs Wa Prüf

Page/Seite 36

[1] Construction stopped in 1923, after which the tunnel was used for civilian purposes. During WWII it was used as »Lager Rebstock« and as an air-raid shelter. From 1960 onwards, it was converted into a government bunker, and between the beginning of 2000 and 2005, it was deconstructed. Today it offers in Ahrweiler a »documentation site government bunker« and in Marienthal the »Rebstock Camp memorial site«.

[2] A total of 1,500 forced laborers and concentration camp inmates were used in the »Rebstock Camp«: 500 Italian military internees, 120 SS front-line workers, 367 Dutch forced laborers, 213 concentration camp inmates from Buchenwald and 300 concentration camp inmates from Auschwitz.

Page/Seite 134

[1] Please note that all test vehicles I (106) can be modified to test vehicle III (107) and that test vehicle III can be used as test vehicle I at any time, but not vice versa. B.z.b.V. Heer's demand that this modification is possible for the unit at short notice and easily has not yet been met present.

Page/Seite 160

[1] The trajectory of the A4 *Gerät* consists of two parts:
1 The propulsion trajectory, which extends from the launch point to the burn termination point, and
2 The propulsionless trajectory, which covers the path from the firing end point to the impact at the target.
On the propulsion trajectory, the thrust accelerates the *Gerät* from the rest position = 0 to its maximum speed V_o, at the same time the angle of attack of 47 degrees is reached by program in several time steps with small angle corrections after approx. 52 seconds. When V_o is reached after approx. 57 to 62 seconds, the remote steering position transmits »combustion cut-off« by radio command* with the BS ground installation; when using a guide beam, the lateral deviation is corrected by radio commands if necessary.
The trajectory without propulsion - after passing the combustion cut-off point, the A4 *Gerät* behaves like an artillery projectile fired at the combustion cut-off point with a certain Vo according to the program with the elevation. As a result of the arrow stabilization, the *Gerät* is turned with the tip even further in the firing direction on the free flight path until the atmosphere ceases. When the *Gerät* enters the atmosphere, the drag turn it back into the trajectory with the tip. For the scattering of the A4 *Gerät*, the unavoidable deviation of the so-called "combustion cut-off conditions" on the other hand, but also external ballistic reasons are responsible. By flight control an accuracy of +/- 5 km in the firing range could be achieved at a flight distance of 275 km.
* When shooting with the integration unit, the »combustion cut-off« by radio command is ommitted.

Page/Seite 312

[1] Repair work and minor repairs are also carried out on ground vehicles. Extensive spare parts are kept in stock and stored in the spare parts vehicle for this purpose.

Anmerkungen

[1] »Die Rakete zu den Planetenräumen 1923, Wege zur Raumschifffahrt 1929«, von Herman Oberth, Raketenfahrzeuge (Opel, Valier) und kleinere Raketenmodelle wurden gebaut und erfolgreich getestet.

[2] In Breslau gründete sich 1927 der »Verein für Raumschifffahrt« mit dem Ziel den Weltraum und andere Planeten zu erforschen. Mitglieder waren u.a. Max Valier, Johannes Winkler, Hermann Oberth, Rudolf Nebel, Wernher von Braun.

[1] Heeresversuchsanstalt Kummersdorf. Entwicklungs- und Erprobungszentrum für neue Waffensysteme und Ausrüstung.

[2] Pariser Kanone WW1. Hersteller: Friedrich Krupp AG, Essen. Reichweite der Geschosse: ca.130 km.

[1] In der »Einsatzplanung A4 nach dem Entwicklungsstand Ende 1942« wurde der Fahrzeugbedarf für eine FR-Abteilung (mot) weitestgehend bestätigt. Bd.Nr3/43 g.k. Chefs Wa Prüf 11

[1] Der Baustopp erfolgte im Jahre 1923. Danach wurden die Tunnel zivil, im Zweiten Weltkrieg als »Lager Rebstock« und Luftschutzraum genutzt. Ab 1960 erfolgte der Umbau zum Regierungsbunker, Anfang 2000 bis 2005 dessen Rückbau. Heute existiert in Ahrweiler die »Dokumentationsstätte Regierungsbunker« und in Marienthal die »Erinnerungsstätte Lager Rebstock«.

[2] Im »Lager Rebstock« kamen insgesamt 1500 Zwangsarbeiter und KZ-Häftlinge zum Einsatz: 500 Italienische Militärinternierte, 120 SS-Frontarbeiter, 367 niederländische Zwangsarbeiter, 213 KZ-Häftlinge aus Buchenwald und 300 KZ-Häftlinge aus Auschwitz.

[1] Es wird darauf hingewiesen, dass sämtliche Prüfwagen I (106) in Prüfwagen III (107) abgeändert werden können, und das Prüfwagen III jederzeit als Prüfwagen I verwendbar ist, aber nicht umgekehrt. Die Forderung von B.z.b.V. Heer, dass diese Umänderung durch einfaches Einschieben entsprechend der Baugruppen, also kurzfristig und einfach für die Truppe möglich ist, ist zur Zeit noch nicht erfüllt.

[1] Die Flugbahn des A4-Gerätes besteht aus zwei Teilen:
1 Der Antriebsbahn, die sich erstreckt von der Abschussstelle bis Brennschlusspunkt und
2 der antriebslosen Bahn, die den Weg vom Brennschlusspunkt bis zum Einschlag im Ziel umfasst.
Auf der Antriebsbahn beschleunigt der Schub das Gerät aus der Ruhelage = 0 auf seine Höchstgeschwindigkeit V_o, gleichzeitig wird per Programm der Anstellwinkel von 47 Grad in mehreren zeitlichen Schritten mit kleinen Winkelkorrekturen nach ca. 52 Sekunden erreicht. Die Fernlenkstellung überträgt bei Erreichen der V_o nach circa 57 bis 62 Sekunden mit der BS-Bodenanlage per Funk* »Brennschluss«, bei Einsatz von Leitstrahl wird ggf die Seitenabweichung durch Funkkommandos korrigiert.
Die antriebslose Bahn - Nach Passieren des Brennschlusspunktes benimmt sich das A4-Gerät wie ein Artilleriegeschoss, das im Brennschlusspunkt mit einer bestimmte Vo nach Maßgabe des Programms mit der Erhöhung abgefeuert wurde. Infolge der Pfeilstabilisierung wird das Gerät auf der Freiflugbahn bis zum Aufhören der Atmosphäre mit der Spitze noch weiter in die Schussrichtung eingedreht. Beim Eintauchen in die Atmosphäre wird das Gerät von den Luftkräften wieder mit der Spitze in die Flugbahn eingelenkt. Für die Streuung das A4-Gerätes sind also in erster Linie die unvermeidliche Abweichung der sogenannten »Brennschlussbedingungen« anderseits aber auch außenballistische Gründe verantwortlich. Durch Flugsteuerung konnte eine Genauigkeit von +/- 5 km in der Schussweite bei einer Flugstrecke von 275 km erreicht werden.
* Beim Schießen mit dem Integrationsgerät entfällt die Brennschlussgabe per Funk.

[1] Ferner werden Instandsetzungsarbeiten und kleinere Reparaturen an den Bodenfahrzeugen durchgeführt. Hierfür sind umfangreiche Ersatzteile bevorratet und in den Ersatzteilwagen eingelagert.

Leading firms

Leading firm coordination of suppliers for design, production, sourcing includes subcontracting.
Company Mittelwerk GmbH (MW) was accting as prime contractor for the ground support vehicles, by placing orders to the *Leitfirmen* (leading firms).

Leitfirmen

Leitfirma – koordinierte die Entwicklung, Fertigung, Beschaffung, und Unterbeauftragung.
Die Mittelwerk GmbH (MW) fungierte als Hauptauftragnehmer für die Bodenfahrzeuge, indem es Aufträge an die Leitfirmen vergab.

Vehicle no. **Fahrzeug-Nr.**	**Leading firm** **Leitfirma**	**Production agent with registered place of business** **Sachbearbeiter der Fertigung mit Dienstsitz**	**Remark** **Bemerkungen**
101	Gollnow	Staat, Rebstock	
102	Lindner	Heinisch,Wa Prüf 4, Berlin	
103	Lindner	Heinisch,Wa Prüf 4, Berlin	
105	Gollnow	Paula, Rebstock	
106/107	Gollnow	Kempini, Rebstock	For electrical parts Ellner, Kitz. Für elektrische Teile Ellner, Kitz.
108	Gollnow	Staat, Rebstock	
110	Gollnow	Paula, Rebstock	
201/202	Gollnow	Vogt, Rebstock	
204	Telefunken, Goerkau	Vogt/Hoberg, Rebstock	
205/206	Telefunken, Goerkau		The control systems* developed by Telefunken will be withdrawn for the time being from the MW order or special committee. Die von Telefunken entwickelten Leitanlagen* scheiden aus der Bestellung MW bzw. Sonderausschuss vorläufig aus.
208	Gollnow	Ellner, Kitz	
301/303	Gollnow	n/a	Is not part of the support contract for ground installations. Gehört nicht zum Betreuungsauftrag Bodenanlagen.
304	Gollnow	Bürger, Kitz	
402	Mako	Production clerk: Herr Fiege, responsible: Wa Prüf 4. Liaison already cleared by Dr. Simon.	
404	Schwelmer Eisenwerk	Ellner, Kitz	
406	Volkmar-Heinig, Dresden	Ellner, Kitz	
406	Schwelmer Eisenwerk	Ellner, Kitz	
407	Pintch	Ellner, Kitz	
501	Gollnow	Schwedes, Rebstock	
502	Gollnow	Schwedes, Rebstock	
503	Gollnow	Bürger, Kitz	
504	Gollnow		Is not supported by MW. Wird nicht durch MW betreut.
505	Still Hamburg	Staat, Rebstock	
506	Junkers-Allach	Papenhausen, Kitz	
508	Gollnow	Schwedes, Rebstock	
509	Fahrgeb	Schwedes, Rebstock	
510	Gollnow	Bürger, Kitz	
511	Gollnow	-	The question of manufacturing is still being reviewed by HAP. No support from MW until clarification. Frage der Fertigung wird noch von HAP geprüft. Bis zur Klärung keine Betreuung durch MW.

In preparation

In Spezialfahrzeuge 2 the railroad vehicles, vehicle new developments to mot.III, test firing V29 and »rail«, alternative use bunker and earthworm position are presented. List of the Art. Divisions, equipment with the special vehicles and experiences of the Art.Abt.836 with reports documented. Furthermore, an overview of the mission preparation and last modifications of the vehicles before the mission is shown. A short summary of the operations »Group North and South« of the Division z.V. conclude this documentation.

Ca. 384 pages, hardcover in landscape format, text both in English and German with more than 200 mostly unpublished photographs, transcripts in excerpts from original documents, tables and detailed 1/35 scale drawings, will give historical technical researchers and scale modelers a great new source with valuable information.

In Vorbereitung

In Spezialfahrzeuge 2 werden die Eisenbahnfahrzeuge, Fahrzeug Neuentwicklungen zu mot.III, Versuchsschießen V29 und »Schiene«, alternative Einsatzmöglichkeiten Bunker und Regenwurm Stellung vorgestellt. Aufstellung der Art. Abteilungen, Ausstattung mit den Sonderfahrzeugen und Erfahrungen der Art.Abt.836 mit Berichten dokumentiert. Darüber hinaus wird ein Überblick zu den Einsatzvorbereitungen und letzten Änderungen an den Fahrzeugen vor dem Einsatz gezeigt. Eine kurze Zusammenfassung der Einsätze »Gruppe Nord und Süd« der Division z.V. schließen diese Dokumentation ab.

Ca. 384 Seiten, Hardcover im Querformat, Texte in Deutsch und Englisch, mit mehr als 200 meist unveröffentlichten Bildern, Abschriften in Auszügen von Originaldokumenten, K.St.N, Tabellen und Detailzeichnungen im Maßstab 1/35 bieten dem an historischer Militärtechnik Interessierten und Modellbauer eine großartige neue Quelle mit wertvollen Informationen.